Discrete Mathematical Structures

DISCRETE MATHEMATICAL STRUCTURES

Fourth Edition

Bernard Kolman

Drexel University

Robert C. Busby

Drexel University

Sharon Cutler Ross

Georgia Perimeter College

PRENTICE HALL, Upper Saddle River, NJ 07458

Library of Congress Cataloging-in-Publication Data

Kolman, Bernard
    Discrete mathematical structures. -- 4th ed. / Bernard Kolman,
  Robert C. Busby, Sharon Cutler Ross.
      p.   cm.
    Includes index.
    ISBN 0-13-083143-3
    1. Computer Science--Mathematics.   I. Busby, Robert C.   II. Ross,
    Sharon Cutler
  III. Title.
  QA76.9.M35 K64 2000
  511'.6--dc21   99-051581

Acquisitions Editor: George Lobell
Assistant Vice President of Production and Manufacturing: David W. Riccardi
Executive Managing Editor: Kathleen Schiaparelli
Senior Managing Editor: Linda Mihatov Behrens
Production Editor: Betsy A. Williams
Manufacturing Buyer: Alan Fischer
Manufacturing Manager: Trudy Pisciotti
Marketing Manager: Melody Marcus
Marketing Assistant: Vince Jansen
Director of Marketing: John Tweeddale
Associate Editor, Mathematics/Statistics Media: Audra J. Walsh
Editorial Assistant/Supplements Editor: Gale Epps
Art Director/Cover Designer: Ann France
Assistant to Art Director: John Christiana
Interior Designer: Donna Wickes
Art Editor: Grace Hazeldine
Art Manager: Gus Vibal
Director of Creative Services: Paul Belfanti
Cover Image: Tony Smith "Untitled," 1953–54, Oil on Board.
  19-1/2 × 23-3/8″ (49.5 × 59.4 cm). New York/Photo credit Thomas Powell,
  courtesy Agnes Gund Collection
Compositor/Illustrator: Dennis Kletzing

ISBN 0-13-083143-3

Prentice-Hall International (UK) Limited, *London*
Prentice-Hall of Australia Pty. Limited, *Sydney*
Prentice-Hall Canada Inc., *Toronto*
Prentice-Hall Hispanoamericana, S.A., *Mexico*
Prentice-Hall of India Private Limited, *New Delhi*
Prentice-Hall of Japan, Inc., *Tokyo*
Pearson Education Asia Pte. Ltd.
Editora Prentice-Hall do Brasil, Ltda., *Rio de Janeiro*

To the memory of Lillie
B.K.

To my wife, Patricia, and our sons, Robert and Scott
R.C.B.

To Bill and bill
S.C.R.

CONTENTS

PREFACE

Discrete mathematics is a difficult course to teach and to study at the freshman and sophomore level for several reasons. It is a hybrid course. Its content is mathematics, but many of its applications and more than half its students are from computer science. Thus careful motivation of topics and previews of applications are important and necessary strategies. Moreover, the number of substantive and diverse topics covered in the course is high, so that student must absorb them rather quickly. At the same time, the student may also be expected to develop proof-writing skills.

APPROACH

First, we have limited both the areas covered and the depth of coverage to what we deemed prudent in a *first course* taught at the freshman and sophomore level. We have identified a set of topics that we feel are of genuine use in computer science and elsewhere and that can be presented in a logically coherent fashion. We have presented an introduction to these topics along with an indication of how they can be pursued in greater depth.

For example, we cover the simpler finite-state machines, not Turing machines. We have limited the coverage of abstract algebra to a discussion of semigroups and groups and have given application of these to the important topics of finite-state machines and error-detecting and error-correcting codes. Error-correcting codes, in turn, have been primarily restricted to simple linear codes.

Second, the material has been organized and interrelated to minimize the mass of definitions and the abstraction of some of the theory. Relations and digraphs are treated as two aspects of the same fundamental mathematical idea, with a directed graph being a pictorial representation of a relation. This fundamental idea is then used as the basis of virtually all the concepts introduced in the book, including functions, partial orders, graphs, and algebraic structures. Whenever possible, each new idea introduced in the text uses previously encountered material and, in turn, is developed in such a way that it simplifies the more complex ideas that follow. Thus partial orders, lattices, and Boolean algebras develop from general relations. This material in turn leads naturally to other algebraic structures.

WHAT IS NEW IN THE FOURTH EDITION

We continue to be pleased by the reception given to earlier editions of this book. We still believe that the book works well in the classroom because of the unifying role played by two key concepts: relations and digraphs. For this edition we have modified the order of topics slightly and made extensive revisions of the exercise sets. The discourse on proof has been expanded in several ways. One of these is the insertion of comments on nearly every proof in the book. Whatever changes we have made, our goal continues to be that of maximizing the clarity of presentation. As the audience for an introductory discrete mathematics course changes and as the course is increasingly used as a bridge course, we have added the following features.

- A new section, Transport Networks, introduces this topic using ideas from Chapter 4.
- A new section, Matching Problems, applies the techniques of transport networks to a broad class of problems.
- The section on mathematical induction now includes the strong form of induction as well.
- The discussion of proofs and proof techniques is now woven throughout the book with comments on most proofs, more exercises related to the mechanics of proving statements, and Tips for Proofs sections. Tips for Proofs highlight the types of proofs commonly seen for that chapter's material and methods for selecting fruitful proof strategies.
- A Self-Test is provided for each chapter with answers for all problems given at the back of the book.
- Exercise Sets have a broader range of problems: more routine problems and more challenging problems. More exercises focus on the mechanics of proof and proof techniques. As with writing in general, students learn to write proofs not only by reading, analyzing, and recognizing the structure of proofs, but especially by writing, re-writing, and writing more proofs themselves.

EXERCISES

The exercises form an integral part of the book. Many are computational in nature, whereas others are of a theoretical type. Many of the latter and the experiments, to be further described below, require verbal solutions. Exercises to help develop proof-writing skills ask the student to analyze proofs, amplify arguments, or complete partial proofs. Answers to all odd-numbered exercises appear in the back of the book. Solutions to all exercises appear in the **Instructor's Manual**, which is available (to instructors only) gratis from the publisher. The Instructor's Manual also includes notes on the pedagogical ideas underlying each chapter, goals and grading guidelines for the experiments further described below, and a test bank.

EXPERIMENTS

Appendix B contains a number of assignments that we call experiments. These provide an opportunity for discovery and exploration, or a more-in-depth look at various topics discussed in the text. These are suitable for group work. Content prerequisites for each experiment are given in the Instructor's Manual.

END OF CHAPTER MATERIAL

Each chapter contains Tips for Proofs, a summary of Key Ideas, a set of Coding Exercises, and a Self-Test covering the chapter's material.

CONTENT

Chapter 1 contains a miscellany of basic material required in the course. This includes sets, subsets, and their operations; sequences; division in the integers; matrices; and mathematical structures. A goal of this chapter is to help students develop skills in identifying patterns on many levels. Chapter 2 covers logic and related material, including methods of proof and mathematical induction. Although the discussion of proof is based on this chapter, the commentary continues throughout the book. Chapter 3, on counting, deals with permutations, combinations, the pigeonhole principle, elements of probability, and recurrence relations. Chapter 4 presents basic types and properties of relations, along with their representation as directed graphs. Connections with matrices and other data structures are also explored in this chapter. The power of multiple representations for the concept of relation is fully exploited. Chapter 5 deals with the notion of a function and gives important examples of functions, including functions of special interest in computer science. An introduction to the growth of functions is developed.

Chapter 6 covers partially ordered sets, including lattices and Boolean algebras. Chapter 7 introduces directed and undirected trees along with applications of these ideas. Elementary graph theory is the focus of Chapter 8. New to this edition are sections on Transport Networks and Matching Problems; these build on the foundation of Chapter 4.

In Chapter 9 we give the basic theory of semigroups and groups. These ideas are applied in Chapters 10 and 11. Chapter 10 is devoted to finite-state machines. It complements and makes effective use of ideas developed in previous chapters. Chapter 11 treats the subject of binary coding.

Appendix A discusses algorithms and pseudocode. The simplified pseudocode presented here is used in some text examples and exercises; these may be omitted without loss of continuity. Appendix B gives a collection of experiments dealing with extensions or previews of topics in various parts of the course.

USE OF THIS TEXT

This text can be used by students in mathematics as an introduction to the fundamental ideas of discrete mathematics, and as a foundation for the development of more advanced mathematical concepts. If used in this way, the topics dealing with specific computer science applications can be ignored or selected independently as important examples. The text can also be used in a computer science or computer engineering curriculum to present the foundations of many basic computer-related concepts, and provide a coherent development and common theme for these ideas.

The instructor can easily develop a suitable course by referring to the chapter prerequisites, which identify material needed by that chapter.

ACKNOWLEDGMENTS

We are pleased to express our thanks to the following reviewers of the first three editions: Harold Fredrickson, Naval Postgraduate School; Thomas E. Gerasch, George Mason University ; Samuel J. Wiley, La Salle College; Kenneth B. Reid, Louisiana Sate University; Ron Sandstrom, Fort Hays State University; Richard H. Austing, University of Maryland; Nina Edelman, Temple University; Paul Gormley, Villanova University; Herman Gollwitzer and Loren N. Argabright, both at Drexel University; Bill Sands, University of Calgary, who brought to our attention a number of errors in the second edition; Moshe Dror, University of Arizona, Tucson; Lloyd Gavin, California State University at Sacramento; Robert H. Gilman, Stevens Institute of Technology; Earl E. Kymala, California State University at Sacramento; and Art Lew, University of Hawaii, Honolulu; and of the fourth edition: Ashok T. Amin, University of Alabama at Huntsville; Donald S. Hart, Rochester Institute of Technology; Minhua Liu, William Rainey Harper College; Charles Parry, Virginia Polytechnic Institute & University; Arthur T. Poe, Temple University; Suk Jai Seo, University of Alabama at Huntsville; Paul Weiner, St. Mary's University of Minnesota. The suggestions, comments, and criticisms of these people greatly improved the manuscript.

We thank Dennis R. Kletzing, Stetson University, who carefully typeset the entire manuscript; Nina Edelman, Temple University, for critically reading page proofs; Blaise DeSesa, Allentown College of St. Frances de Sales, who checked the answers and solutions to all the exercises in the book; and instructors and students from many institutions in the United States and other countries, for sharing with us their experiences with the book and for offering helpful suggestions.

Finally, a sincere expression of thanks goes to Betsy Williams, George Lobell, Gale Epps, and the entire staff at Prentice Hall for their enthusiasm, interest, and unfailing cooperation during the conception, design, production, and marketing phases of this edition.

B.K.
R.C.B.
S.C.R.

Discrete Mathematical Structures

1

FUNDAMENTALS

Prerequisites: There are no formal prerequisites for this chapter; the reader is encouraged to read carefully and work through all examples.

In this chapter we introduce some of the basic tools of discrete mathematics. We begin with sets, subsets, and their operations, notions with which you may already be familiar. Next we deal with sequences, using both explicit and recursive patterns. Then we review some of the basic divisibility properties of the integers. Finally we introduce matrices and matrix operations. This gives us the background needed to begin our exploration of mathematical structures.

1.1 SETS AND SUBSETS

Sets

A **set** is any well-defined collection of objects called the **elements** or **members of the set**. For example, the collection of all wooden chairs, the collection of all one-legged black birds, or the collection of real numbers between zero and one is each a set. Well-defined just means that it is possible to decide if a given object belongs to the collection or not. Almost all mathematical objects are first of all sets, regardless of any additional properties they may possess. Thus set theory is, in a sense, the foundation on which virtually all of mathematics is constructed. In spite of this, set theory (at least the informal brand we need) is quite easy to learn and use.

One way of describing a set that has a finite number of elements is by listing the elements of the set between braces. Thus the set of all positive integers that are less than 4 can be written as

$$\{1, 2, 3\}. \tag{1}$$

The order in which the elements of a set are listed is not important. Thus $\{1, 3, 2\}$, $\{3, 2, 1\}$, $\{3, 1, 2\}$, $\{2, 1, 3\}$, and $\{2, 3, 1\}$ are all representations of the set given in (1). Moreover, repeated elements in the listing of the elements of a set can be ignored. Thus, $\{1, 3, 2, 3, 1\}$ is another representation of the set given in (1).

We use uppercase letters such as A, B, C to denote sets, and lowercase letters such as a, b, c, x, y, z, t to denote the members (or elements) of sets.

1

We indicate the fact that x is an element of the set A by writing $x \in A$, and we indicate the fact that x is not an element of A by writing $x \notin A$.

EXAMPLE 1   Let $A = \{1, 3, 5, 7\}$. Then $1 \in A$, $3 \in A$, but $2 \notin A$.   ■

Sometimes it is inconvenient or impossible to describe a set by listing all its elements. Another useful way to define a set is by specifying a property that the elements of the set have in common. We use the notation $P(x)$ to denote a sentence or statement P concerning the variable object x. The set defined by $P(x)$, written $\{x \mid P(x)\}$, is just the collection of all objects for which P is sensible and true. For example, $\{x \mid x$ is a positive integer less than 4$\}$ is the set $\{1, 2, 3\}$ described in (1) by listing its elements.

EXAMPLE 2   The set consisting of all the letters in the word "byte" can be denoted by $\{b, y, t, e\}$ or by $\{x \mid x$ is a letter in the word "byte"$\}$.   ■

EXAMPLE 3   We introduce here several sets and their notations that will be used throughout this book.

(a) $Z^+ = \{x \mid x$ is a positive integer$\}$.
   Thus Z^+ consists of the numbers used for counting: $1, 2, 3 \ldots$.
(b) $N = \{x \mid x$ is a positive integer or zero$\}$.
   Thus N consists of the positive integers and zero: $0, 1, 2, \ldots$.
(c) $Z = \{x \mid x$ is an integer$\}$.
   Thus Z consists of all the integers: $\ldots, -3, -2, -1, 0, 1, 2, 3, \ldots$.
(d) $\mathbb{Q} = \{x \mid x$ is a rational number$\}$.
   Thus $\mathbb{Q}$ consists of numbers that can be written as $\dfrac{a}{b}$, where a and b are integers and b is not 0.
(e) $\mathbb{R} = \{x \mid x$ is a real number$\}$.
(f) The set that has no elements in it is denoted either by $\{\ \}$ or the symbol $\varnothing$ and is called the **empty set**.   ■

EXAMPLE 4   Since the square of a real number is always nonnegative,

$$\{x \mid x \text{ is a real number and } x^2 = -1\} = \varnothing.$$ ■

Sets are completely known when their members are all known. Thus we say two sets A and B are **equal** if they have the same elements, and we write $A = B$.

EXAMPLE 5   If $A = \{1, 2, 3\}$ and $B = \{x \mid x$ is a positive integer and $x^2 < 12\}$, then $A = B$.   ■

EXAMPLE 6   If $A = \{\text{BASIC, PASCAL, ADA}\}$ and $B = \{\text{ADA, BASIC, PASCAL}\}$, then $A = B$.   ■

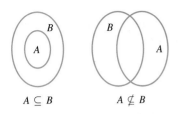

$A \subseteq B$ $A \nsubseteq B$

Figure 1.1

Subsets

If every element of A is also an element of B, that is, if whenever $x \in A$ then $x \in B$, we say that A is a **subset** of B or that A is **contained in** B, and we write $A \subseteq B$. If A is not a subset of B, we write $A \nsubseteq B$. (See Figure 1.1.)

Diagrams, such as those in Figure 1.1, which are used to show relationships between sets, are called **Venn diagrams** after the British logician John Venn. Venn diagrams will be used extensively in Section 1.2.

EXAMPLE 7 We have $Z^+ \subseteq Z$. Moreover, if $\mathbb{Q}$ denotes the set of rational numbers, then $Z \subseteq \mathbb{Q}$. ∎

EXAMPLE 8 Let $A = \{1, 2, 3, 4, 5, 6\}$, $B = \{2, 4, 5\}$, and $C = \{1, 2, 3, 4, 5\}$. Then $B \subseteq A$, $B \subseteq C$, and $C \subseteq A$. However, $A \nsubseteq B$, $A \nsubseteq C$, and $C \nsubseteq B$. ∎

EXAMPLE 9 If A is any set, then $A \subseteq A$. That is, every set is a subset of itself. ∎

EXAMPLE 10 Let A be a set and let $B = \{A, \{A\}\}$. Then, since A and $\{A\}$ are elements of B, we have $A \in B$ and $\{A\} \in B$. It follows that $\{A\} \subseteq B$ and $\{\{A\}\} \subseteq B$. However, it is not true that $A \subseteq B$. ∎

For any set A, since there are no elements of $\varnothing$ that are not in A, we have $\varnothing \subseteq A$. (We will look at this again in Section 2.1.)

It is easy to see that $A = B$ if and only if $A \subseteq B$ and $B \subseteq A$.

The collection of everything, it turns out, cannot be considered a set without presenting serious logical difficulties. To avoid this and other problems, which need not concern us here, we will assume that for each discussion there is a "universal set" U (which will vary with the discussion) containing all objects for which the discussion is meaningful. Any other set mentioned in the discussion will automatically be assumed to be a subset of U. Thus, if we are discussing real numbers and we mention sets A and B, then A and B must (we assume) be sets of real numbers, not matrices, electronic circuits, or rhesus monkeys. In most problems, a universal set will be apparent from the setting of the problem. In Venn diagrams, the universal set U will be denoted by a rectangle, while sets within U will be denoted by circles as shown in Figure 1.2.

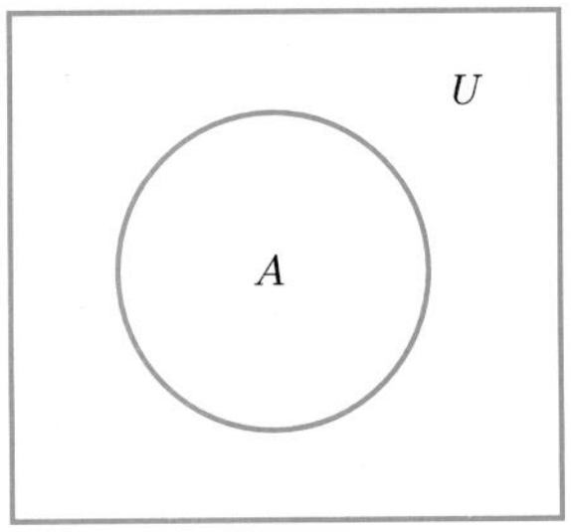

A set A is called **finite** if it has n distinct elements, where $n \in N$. In this case, n is called the **cardinality** of A and is denoted by $|A|$. Thus, the sets of Examples 1, 2, 4, 5, and 6 are finite. A set that is not finite is called **infinite**. The sets introduced in Example 3 (except $\varnothing$) are infinite sets.

If A is a set, then the set of all subsets of A is called the **power set** of A and is denoted by $P(A)$.

Figure 1.2

EXAMPLE 11 Let $A = \{1, 2, 3\}$. Then $P(A)$ consists of the following subsets of A: $\{\ \}$, $\{1\}$, $\{2\}$, $\{3\}$, $\{1, 2\}$, $\{1, 3\}$, $\{2, 3\}$, and $\{1, 2, 3\}$ (or A). In a later section, we will count the number of subsets that a set can have. ∎

1.1 Exercises

1. Let $A = \{1, 2, 4, a, b, c\}$. Identify each of the following as true or false.

    (a) $2 \in A$ (b) $3 \in A$ (c) $c \notin A$

    (d) $\varnothing \in A$ (e) $\{\ \} \notin A$ (f) $A \in A$

2. Let $A = \{x \mid x \text{ is a real number and } x < 6.\}$. Identify each of the following as true or false.

    (a) $3 \in A$ (b) $6 \in A$ (c) $5 \notin A$

    (d) $8 \notin A$ (e) $-8 \in A$ (f) $3.4 \notin A$

3. In each part, give the set of letters in each word by listing the elements of the set.

    (a) AARDVARK   (b) BOOK

    (c) MISSISSIPPI

4. Give the set by listing its elements.

    (a) The set of all positive integers that are less than ten.

    (b) $\{x \mid x \in Z \text{ and } x^2 < 12\}$

5. Let $A = \{1, \{2, 3\}, 4\}$. Identify each of the following as true or false.

    (a) $3 \in A$ (b) $\{1, 4\} \subseteq A$ (c) $\{2, 3\} \subseteq A$

    (d) $\{2, 3\} \in A$ (e) $\{4\} \in A$ (f) $\{1, 2, 3\} \subseteq A$

In Exercises 6 through 9, write the set in the form $\{x \mid P(x)\}$, where $P(x)$ is a property that describes the elements of the set.

6. $\{2, 4, 6, 8, 10\}$ 7. $\{a, e, i, o, u\}$

8. $\{1, 8, 27, 64, 125\}$ 9. $\{-2, -1, 0, 1, 2\}$

10. Let $A = \{1, 2, 3, 4, 5\}$. Which of the following sets are equal to A?

    (a) $\{4, 1, 2, 3, 5\}$ (b) $\{2, 3, 4\}$ (c) $\{1, 2, 3, 4, 5, 6\}$

    (d) $\{x \mid x \text{ is an integer and } x^2 \leq 25\}$

    (e) $\{x \mid x \text{ is a positive integer and } x \leq 5\}$

    (f) $\{x \mid x \text{ is a positive rational number and } x \leq 5\}$

11. Which of the following sets are the empty set?

    (a) $\{x \mid x \text{ is a real number and } x^2 - 1 = 0\}$

    (b) $\{x \mid x \text{ is a real number and } x^2 + 1 = 0\}$

    (c) $\{x \mid x \text{ is a real number and } x^2 = -9\}$

    (d) $\{x \mid x \text{ is a real number and } x = 2x + 1\}$

    (e) $\{x \mid x \text{ is a real number and } x = x + 1\}$

12. List all the subsets of $\{a, b\}$.

13. List all the subsets of $\{$BASIC, PASCAL, ADA$\}$.

14. List all the subsets of $\{\ \}$.

15. Let $A = \{1, 2, 5, 8, 11\}$. Identify each of the following as true or false.

    (a) $\{5, 1\} \subseteq A$ (b) $\{8, 1\} \in A$

    (c) $\{1, 8, 2, 11, 5\} \not\subseteq A$ (d) $\varnothing \subseteq A$

    (e) $\{1, 6\} \not\subseteq A$ (f) $\{2\} \subseteq A$

    (g) $\{3\} \notin A$ (h) $A \subseteq \{11, 2, 5, 1, 8, 4\}$

16. Let $A = \{x \mid x \text{ is an integer and } x^2 < 16\}$. Identify each of the following as true or false.

    (a) $\{0, 1, 2, 3\} \subseteq A$ (b) $\{-3, -2, -1\} \subseteq A$

    (c) $\{\ \} \subseteq A$

    (d) $\{x \mid x \text{ is an integer and } |x| < 4\} \subseteq A$

    (e) $A \subseteq \{-3, -2, -1, 0, 1, 2, 3\}$

17. Let $A = \{1\}$, $B = \{1, a, 2, b, c\}$, $C = \{b, c\}$, $D = \{a, b\}$, and $E = \{1, a, 2, b, c, d\}$. For each part, replace the symbol $\square$ with either $\subseteq$ or $\not\subseteq$ to give a true statement.

    (a) $A \square B$ (b) $\varnothing \square A$ (c) $B \square C$

    (d) $C \square E$ (e) $D \square C$ (f) $B \square E$

In Exercises 18 through 20, find the set of smallest cardinality that contains the given sets as subsets.

18. $\{a, b, c\}$, $\{a, d, e, f\}$, $\{b, c, e, g\}$

19. $\{1, 2\}$, $\{1, 3\}$, $\varnothing$

20. $\{2, 4, 6, \ldots, 20\}$, $\{3, 6, 9, \ldots, 21\}$

21. Is it possible to have two different (appropriate) universal sets for a collection of sets? Would having different universal sets create any problems? Explain.

22. Use the Venn diagram in Figure 1.3 to identify each of the following as true or false.

    (a) $A \subseteq B$ (b) $B \subseteq A$ (c) $C \subseteq B$

    (d) $x \in B$ (e) $x \in A$ (f) $y \in B$

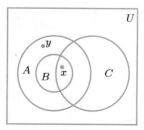

Figure 1.3

23. Use the Venn diagram in Figure 1.4 to identify each of the following as true or false.

   (a) $B \subseteq A$ (b) $A \subseteq C$ (c) $C \subseteq B$

   (d) $w \in A$ (e) $t \in A$ (f) $w \in B$

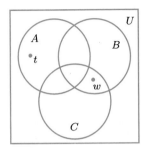

Figure 1.4

24. (a) If $A = \{3, 7\}$, find $P(A)$.

   (b) What is $|A|$?   (c) What is $|P(A)|$?

25. (a) If $A = \{3, 7, 2\}$, find $P(A)$.

   (b) What is $|A|$?   (c) What is $|P(A)|$?

In Exercises 26 *through* 28, *draw a Venn diagram that represents these relationships.*

26. $A \subseteq B$, $A \subseteq C$, $B \nsubseteq C$, and $C \nsubseteq B$

27. $x \in A$, $x \in B$, $x \notin C$, $y \in B$, $y \in C$, and $y \notin A$

28. $A \subseteq B$, $x \notin A$, $x \in B$, $A \nsubseteq C$, $y \in B$, $y \in C$

29. Describe all the subset relationships that hold for the sets given in Example 3.

30. Show that if $A \subseteq B$ and $B \subseteq C$, then $A \subseteq C$.

31. Suppose we know that set A has n subsets, $S_1, S_2, \ldots, S_n$. If set B consists of the elements of A and one more element so $|B| = |A| + 1$, show that B must have $2n$ subsets.

32. Compare the results of Exercises 12, 13, 24, and 25 and complete the following: Any set with two elements has _____ subsets. Any set with three elements has _____ subsets.

1.2 OPERATIONS ON SETS

In this section we will discuss several operations that will combine given sets to yield new sets. These operations, which are analogous to the familiar operations on the real numbers, will play a key role in the many applications and ideas that follow.

    If A and B are sets, we define their **union** as the set consisting of all elements that belong to A or B and denote it by $A \cup B$. Thus

$$A \cup B = \{x \mid x \in A \text{ or } x \in B\}.$$

Observe that $x \in A \cup B$ if $x \in A$ or $x \in B$ or x belongs to both A and B.

EXAMPLE 1   Let $A = \{a, b, c, e, f\}$ and $B = \{b, d, r, s\}$. Find $A \cup B$.

    Solution   Since $A \cup B$ consists of all the elements that belong to either A or B, $A \cup B = \{a, b, c, d, e, f, r, s\}$.   ■

    We can illustrate the union of two sets with a Venn diagram as follows. If A and B are the sets in Figure 1.5(a), then $A \cup B$ is the set represented by the shaded region in Figure 1.5(b).

    If A and B are sets, we define their **intersection** as the set consisting of all elements that belong to both A and B and denote it by $A \cap B$. Thus

$$A \cap B = \{x \mid x \in A \text{ and } x \in B\}.$$

EXAMPLE 2   Let $A = \{a, b, c, e, f\}$, $B = \{b, e, f, r, s\}$, and $C = \{a, t, u, v\}$. Find $A \cap B$, $A \cap C$, and $B \cap C$.

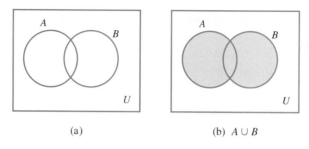

(a)   (b) $A \cup B$

Figure 1.5

Solution   The elements b, e, and f are the only ones that belong to both A and B, so $A \cap B = \{b, e, f\}$. Similarly, $A \cap C = \{a\}$. There are no elements that belong to both B and C, so $B \cap C = \{\ \}$.   ∎

Two sets that have no common elements, such as B and C in Example 2, are called **disjoint sets**.

We can illustrate the intersection of two sets by a Venn diagram as follows. If A and B are the sets given in Figure 1.6(a), then $A \cap B$ is the set represented by the shaded region in Figure 1.6(b). Figure 1.7 illustrates a Venn diagram for two disjoint sets.

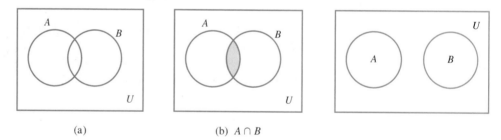

(a)   (b) $A \cap B$

Figure 1.6   Figure 1.7

The operations of union and intersection can be defined for three or more sets in an obvious manner:

$$A \cup B \cup C = \{x \mid x \in A \text{ or } x \in B \text{ or } x \in C\}$$

and

$$A \cap B \cap C = \{x \mid x \in A \text{ and } x \in B \text{ and } x \in C\}.$$

The shaded region in Figure 1.8(b) is the union of the sets A, B, and C shown in Figure 1.8(a), and the shaded region in Figure 1.8(c) is the intersection of the sets A, B, and C. Note that Figure 1.8(a) says nothing about possible relationships between the sets, but allows for all possible relationships. In general, if $A_1, A_2, \ldots, A_n$ are subsets of U, then $A_1 \cup A_2 \cup \cdots A_n$ will be denoted by $\bigcup\limits_{k=1}^{n} A_k$ and $A_1 \cap A_2 \cap \cdots \cap A_n$ will be denoted by $\bigcap\limits_{k=1}^{n} A_k$.

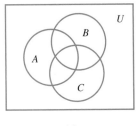
(a)

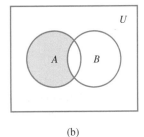
(b) $A \cup B \cup C$

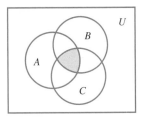
(c) $A \cap B \cap C$

Figure 1.8

EXAMPLE 3   Let $A = \{1, 2, 3, 4, 5, 7\}$, $B = \{1, 3, 8, 9\}$, and $C = \{1, 3, 6, 8\}$. Then $A \cap B \cap C$ is the set of elements that belong to A, B, and C. Thus $A \cap B \cap C = \{1, 3\}$.   ∎

If A and B are two sets, we define the **complement of B with respect to A** as the set of all elements that belong to A but not to B, and we denote it by $A - B$. Thus

$$A - B = \{x \mid x \in A \text{ and } x \notin B\}.$$

EXAMPLE 4   Let $A = \{a, b, c\}$ and $B = \{b, c, d, e\}$. Then $A - B = \{a\}$ and $B - A = \{d, e\}$.   ∎

If A and B are the sets in Figure 1.9(a), then $A - B$ and $B - A$ are represented by the shaded regions in Figures 1.9(b) and 1.9(c), respectively.

If U is a universal set containing A, then $U - A$ is called the **complement** of A and is denoted by $\overline{A}$. Thus $\overline{A} = \{x \mid x \notin A\}$.

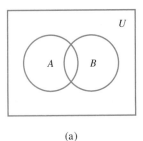
(a)

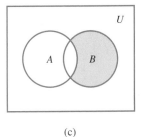
(b)

(c)

Figure 1.9

EXAMPLE 5   Let $A = \{x \mid x \text{ is an integer and } x \leq 4\}$ and $U = Z$. Then $\overline{A} = \{x \mid x \text{ is an integer and } x > 4\}$.   ∎

If A is the set in Figure 1.10, its complement is the shaded region in that figure.

If A and B are two sets, we define their **symmetric difference** as the set of all elements that belong to A or to B, but not to both A and B, and we denote it by $A \oplus B$. Thus

$$A \oplus B = \{x \mid (x \in A \text{ and } x \notin B) \text{ or } (x \in B \text{ and } x \notin A)\}.$$

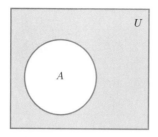

Figure 1.10

EXAMPLE 6   Let $A = \{a, b, c, d\}$ and $B = \{a, c, e, f, g\}$. Then $A \oplus B = \{b, d, e, f, g\}$.   ■

If A and B are as indicated in Figure 1.11(a), their symmetric difference is the shaded region shown in Figure 1.11(b). It is easy to see that

$$A \oplus B = (A - B) \cup (B - A).$$

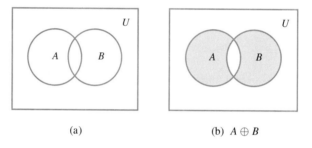

(a)   (b) $A \oplus B$

Figure 1.11

Algebraic Properties of Set Operations

The operations on sets that we have just defined satisfy many algebraic properties, some of which resemble the algebraic properties satisfied by the real numbers and their operations. All the principal properties listed here can be proved using the definitions given and the rules of logic. We shall prove only one of the properties and leave proofs of the remaining ones as exercises for the reader. Proofs are fundamental to mathematics. We discuss proof techniques in Chapter 2, but in this chapter some proofs are given as examples for later work. Some simple proofs are required in the exercises. Venn diagrams are often useful to suggest or justify the method of proof.

Theorem 1   The operations defined on sets satisfy the following properties:
Commutative Properties

1. $A \cup B = B \cup A$
2. $A \cap B = B \cap A$

Associative Properties

   3. $A \cup (B \cup C) = (A \cup B) \cup C$
   4. $A \cap (B \cap C) = (A \cap B) \cap C$

Distributive Properties

   5. $A \cap (B \cup C) = (A \cap B) \cup (A \cap C)$
   6. $A \cup (B \cap C) = (A \cup B) \cap (A \cup C)$

Idempotent Properties

   7. $A \cup A = A$
   8. $A \cap A = A$

Properties of the Complement

   9. $(\overline{\overline{A}}) = A$
  10. $A \cup \overline{A} = U$
  11. $A \cap \overline{A} = \varnothing$
  12. $\overline{\varnothing} = U$
  13. $\overline{U} = \{\ \}$
  14. $\overline{A \cup B} = \overline{A} \cap \overline{B}$
  15. $\overline{A \cap B} = \overline{A} \cup \overline{B}$ Properties 14 and 15 are known as
                                                De Morgan's laws

Properties of a Universal Set

  16. $A \cup U = U$
  17. $A \cap U = A$

Properties of the Empty Set

  18. $A \cup \varnothing = A$ or $A \cup \{\ \} = A$
  19. $A \cap \varnothing = \varnothing$ or $A \cap \{\ \} = \{\ \}$

Proof We prove Property 14 here and leave proofs of the remaining properties as exercises for the reader. A common style of proof for statements about sets is to choose an element in one of the sets and see what we know about it. Suppose that $x \in \overline{A \cup B}$. Then we know that $x \notin A \cup B$, so $x \notin A$ and $x \notin B$. (Why?) This means $x \in \overline{A} \cap \overline{B}$ (why?), so each element of $\overline{A \cup B}$ belongs to $\overline{A} \cap \overline{B}$. Thus $\overline{A \cup B} \subseteq \overline{A} \cap \overline{B}$. Conversely, suppose that $x \in \overline{A} \cap \overline{B}$. Then $x \notin A$ and $x \notin B$ (why?), so $x \notin A \cup B$, which means that $x \in \overline{A \cup B}$. Thus each element of $\overline{A} \cap \overline{B}$ also belongs to $\overline{A \cup B}$, and $\overline{A} \cap \overline{B} \subseteq \overline{A \cup B}$. Now we see that $\overline{A \cup B} = \overline{A} \cap \overline{B}$. ▼

The Addition Principle

Suppose now that A and B are finite subsets of a universal set U. It is frequently useful to have a formula for $|A \cup B|$, the cardinality of the union. If A and B are disjoint sets, that is, if $A \cap B = \varnothing$, then each element of $A \cup B$ appears in either A or B, but not in both; therefore, $|A \cup B| = |A| + |B|$. If A and B overlap, as shown in Figure 1.12, then elements in $A \cap B$ belong to both sets, and the sum $|A| + |B|$ counts

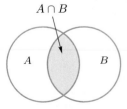

$A \cap B$

Figure 1.12

these elements twice. To correct for this double counting, we subtract $|A \cap B|$. Thus we have the following theorem, sometimes called the **addition principle**. Because of Figure 1.12, this is also called the **inclusion-exclusion principle**.

Theorem 2   If A and B are finite sets, then $|A \cup B| = |A| + |B| - |A \cap B|$.   ●

EXAMPLE 7   Let $A = \{a, b, c, d, e\}$ and $B = \{c, e, f, h, k, m\}$. Verify Theorem 2.

Solution   We have $A \cup B = \{a, b, c, d, e, f, h, k, m\}$ and $A \cap B = \{c, e\}$. Also, $|A| = 5$, $|B| = 6$, $|A \cup B| = 9$, and $|A \cap B| = 2$. Then $|A| + |B| - |A \cap B| = 5 + 6 - 2$ or 9 and Theorem 2 is verified.   ■

If A and B are disjoint sets, $A \cap B = \varnothing$ and $|A \cap B| = 0$, so the formula in Theorem 2 now becomes $|A \cup B| = |A| + |B|$. This special case can be stated in a way that is useful in a variety of counting situations.

The Addition Principle for Disjoint Sets

If a task T_1 can be performed in exactly n ways, and a different task T_2 can be performed in exactly m ways, then the number of ways of performing task T_1 or task T_2 is $n + m$.

The situation for three sets is shown in Figure 1.13. We state the three-set addition principle without discussion.

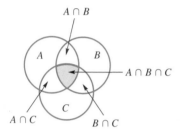

Figure 1.13

Theorem 3   Let A, B, and C be finite sets. Then $|A \cup B \cup C| = |A| + |B| + |C| - |A \cap B| - |B \cap C| - |A \cap C| + |A \cap B \cap C|$.   ●

Theorem 3 can be generalized for more than three sets. This is done in Exercises 37 and 38.

EXAMPLE 8   Let $A = \{a, b, c, d, e\}$, $B = \{a, b, e, g, h\}$, and $C = \{b, d, e, g, h, k, m, n\}$. Verify Theorem 3.

Solution   We have $A \cup B \cup C = \{a, b, c, d, e, g, h, k, m, n\}$, $A \cap B = \{a, b, e\}$, $A \cap C = \{b, d, e\}$, $B \cap C = \{b, e, g, h\}$, and $A \cap B \cap C = \{b, e\}$, so $|A| = 5$, $|B| = 5$, $|C| = 8$, $|A \cup B \cup C| = 10$, $|A \cap B| = 3$, $|A \cap C| = 3$, $|B \cap C| = 4$, and $|A \cap B \cap C| = 2$. Thus $|A| + |B| + |C| - |A \cap B| - |B \cap C| - |A \cap C| + |A \cap B \cap C| = 5 + 5 + 8 - 3 - 3 - 4 + 2$ or 10, and Theorem 3 is verified.   ■

EXAMPLE 9   A computer company wants to hire 25 programmers to handle systems programming jobs and 40 programmers for applications programming. Of those hired, ten will be expected to perform jobs of both types. How many programmers must be hired?

Solution   Let A be the set of systems programmers hired and B be the set of applications programmers hired. The company must have $|A| = 25$ and $|B| = 40$, and $|A \cap B| = 10$. The number of programmers that must be hired is $|A \cup B|$, but $|A \cup B| = |A| + |B| - |A \cap B|$. So the company must hire $25 + 40 - 10$ or 55 programmers.   ■

EXAMPLE 10 A survey has been taken on methods of commuter travel. Each respondent was asked to check BUS, TRAIN, or AUTOMOBILE as a major method of traveling to work. More than one answer was permitted. The results reported were as follows: BUS, 30 people; TRAIN, 35 people; AUTOMOBILE, 100 people; BUS and TRAIN, 15 people; BUS and AUTOMOBILE, 15 people; TRAIN and AUTOMOBILE, 20 people; and all three methods, 5 people. How many people completed a survey form?

Solution Let B, T, and A be the sets of people who checked BUS, TRAIN, and AUTOMOBILE, respectively. We know $|B| = 30$, $|T| = 35$, $|A| = 100$, $|B \cap T| = 15$, $|B \cap A| = 15$, $|T \cap A| = 20$, and $|B \cap T \cap A| = 5$. So $|B| + |T| + |A| - |B \cap T| - |B \cap A| - |T \cap A| + |B \cap T \cap A| = 30 + 35 + 100 - 15 - 15 - 20 + 5$ or 120 is $|A \cup B \cup C|$, the number of people who responded. ∎

1.2 Exercises

In Exercises 1 through 4, let $U = \{a, b, c, d, e, f, g, h, k\}$, $A = \{a, b, c, g\}$, $B = \{d, e, f, g\}$, $C = \{a, c, f\}$, *and* $D = \{f, h, k\}$.

1. Compute
- (a) $A \cup B$
- (b) $B \cup C$
- (c) $A \cap C$
- (d) $B \cap D$
- (e) $(A \cup B) - C$
- (f) $A - B$
- (g) $\overline{A}$
- (h) $A \oplus B$
- (i) $A \oplus C$
- (j) $(A \cap B) - C$

2. Compute
- (a) $A \cup D$
- (b) $B \cup D$
- (c) $C \cap D$
- (d) $A \cap D$
- (e) $(A \cup B) - (C \cup B)$
- (f) $B - C$
- (g) $\overline{B}$
- (h) $C - B$
- (i) $C \oplus D$
- (j) $(A \cap B) - (B \cap D)$

3. Compute
- (a) $A \cup B \cup C$
- (b) $A \cap B \cap C$
- (c) $A \cap (B \cup C)$
- (d) $(A \cup B) \cap C$
- (e) $\overline{A \cup B}$
- (f) $\overline{A \cap B}$

4. Compute
- (a) $A \cup \emptyset$
- (b) $A \cup U$
- (c) $B \cup B$
- (d) $C \cap \{\ \}$
- (e) $\overline{C \cup D}$
- (f) $\overline{C \cap D}$

In Exercises 5 through 8, let $U = \{1, 2, 3, 4, 5, 6, 7, 8, 9\}$, $A = \{1, 2, 4, 6, 8\}$, $B = \{2, 4, 5, 9\}$, $C = \{x \mid x$ *is a positive integer and* $x^2 \le 16\}$, *and* $D = \{7, 8\}$.

5. Compute
- (a) $A \cup B$
- (b) $A \cup C$
- (c) $A \cup D$
- (d) $B \cup C$
- (e) $A \cap C$
- (f) $A \cap D$
- (g) $B \cap C$
- (h) $C \cap D$

6. Compute
- (a) $A - B$
- (b) $B - A$
- (c) $C - D$
- (d) $\overline{C}$
- (e) $\overline{A}$
- (f) $A \oplus B$
- (g) $C \oplus D$
- (h) $B \oplus C$

7. Compute
- (a) $A \cup B \cup C$
- (b) $A \cap B \cap C$
- (c) $A \cap (B \cup C)$
- (d) $(A \cup B) \cap D$
- (e) $\overline{A \cup B}$
- (f) $\overline{A \cap B}$

8. Compute
- (a) $B \cup C \cup D$
- (b) $B \cap C \cap D$
- (c) $A \cup A$
- (d) $A \cap \overline{A}$
- (e) $A \cup \overline{A}$
- (f) $A \cap (\overline{C} \cup D)$

In Exercises 9 and 10, let $U = \{a, b, c, d, e, f, g, h\}$, $A = \{a, c, f, g\}$, $B = \{a, e\}$, *and* $C = \{b, h\}$.

9. Compute
- (a) $\overline{A}$
- (b) $\overline{B}$
- (c) $\overline{A \cup B}$
- (d) $\overline{A \cap B}$
- (e) $\overline{U}$
- (f) $A - B$

10. Compute
- (a) $\overline{A} \cap \overline{B}$
- (b) $\overline{B} \cup \overline{C}$
- (c) $\overline{A \cup A}$
- (d) $\overline{C \cap C}$
- (e) $A \oplus B$
- (f) $B \oplus C$

11. Let U be the set of real numbers, $A = \{x \mid x$ is a solution of $x^2 - 1 = 0\}$, and $B = \{-1, 4\}$. Compute
- (a) $\overline{A}$
- (b) $\overline{B}$
- (c) $\overline{A \cup B}$
- (d) $\overline{A \cap B}$

In Exercises 12 and 13, refer to Figure 1.14.

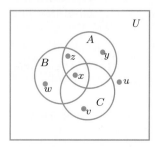

Figure 1.14

12. Identify the following as true or false.

(a) $y \in A \cap B$ (b) $x \in B \cup C$

(c) $w \in B \cap C$ (d) $u \notin C$

13. Identify the following as true or false.

(a) $x \in A \cap B \cap C$ (b) $y \in A \cup B \cup C$

(c) $z \in A \cap C$ (d) $v \in B \cap C$

14. Describe the shaded region shown in Figure 1.15 using unions and intersections of the sets A, B, and C. (Several descriptions are possible.)

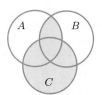

Figure 1.15

15. Let A, B, and C be finite sets with $|A| = 6$, $|B| = 8$, $|C| = 6$, $|A \cup B \cup C| = 11$, $|A \cap B| = 3$, $|A \cap C| = 2$, and $|B \cap C| = 5$. Find $|A \cap B \cap C|$.

In Exercises 16 through 18, verify Theorem 2 for the given sets.

16. (a) $A = \{1, 2, 3, 4\}$, $B = \{2, 3, 5, 6, 8\}$

(b) $A = \{1, 2, 3, 4\}$, $B = \{5, 6, 7, 8, 9\}$

17. (a) $A = \{a, b, c, d, e, f\}$, $B = \{a, c, f, g, h, i, r\}$

(b) $A = \{a, b, c, d, e\}$, $B = \{f, g, r, s, t, u\}$

18. (a) $A = \{x \mid x \text{ is a positive integer} < 8\}$, $B = \{x \mid x \text{ is an integer such that } 2 \le x \le 5\}$

(b) $A = \{x \mid x \text{ is a positive integer and } x^2 \le 16\}$, $B = \{x \mid x \text{ is a negative integer and } x^2 \le 25\}$

19. If A and B are disjoint sets such that $|A \cup B| = |A|$, what must be true about B?

In Exercises 20 through 22, verify Theorem 3 for the given sets.

20. $A = \{a, b, c, d, e\}$, $B = \{d, e, f, g, h, i, k\}$, $C = \{a, c, d, e, k, r, s, t\}$

21. $A = \{1, 2, 3, 4, 5, 6\}$, $B = \{2, 4, 7, 8, 9\}$, $C = \{1, 2, 4, 7, 10, 12\}$

22. $A = \{x \mid x \text{ is a positive integer} < 8\}$, $B = \{x \mid x \text{ is an integer such that } 2 \le x \le 4\}$, $C = \{x \mid x \text{ is an integer such that } x^2 < 16\}$

23. In a survey of 260 college students, the following data were obtained:

64 had taken a mathematics course,

94 had taken a computer science course,

58 had taken a business course,

28 had taken both a mathematics and a business course,

26 had taken both a mathematics and a computer science course,

22 had taken both a computer science and a business course, and

14 had taken all three types of courses.

(a) How many students were surveyed who had taken none of the three types of courses?

(b) Of the students surveyed, how many had taken only a computer science course?

24. A survey of 500 television watchers produced the following information: 285 watch football games, 195 watch hockey games, 115 watch basketball games, 45 watch football and basketball games, 70 watch football and hockey games, 50 watch hockey and basketball games, and 50 do not watch any of the three kinds of games.

(a) How many people in the survey watch all three kinds of games?

(b) How many people watch exactly one of the sports?

25. In a psychology experiment, the subjects under study were classified according to body type and gender as follows:

	ENDO-MORPH	ECTO-MORPH	MESO-MORPH
Male	72	54	36
Female	62	64	38

(a) How many male subjects were there?

(b) How many subjects were ectomorphs?

(c) How many subjects were either female or endo-morphs?

(d) How many subjects were not male mesomorphs?

(e) How many subjects were either male, ectomorph, or mesomorph?

26. Complete the following proof that $A \subseteq A \cup B$. Suppose $x \in A$. Then $x \in A \cup B$, because _____. Thus by the definition of subset $A \subseteq A \cup B$.

27. Complete the following proof that $A \cap B \subseteq A$. Suppose $x \in A \cap B$. Then x belongs to _____. Thus $A \cap B \subseteq A$.

28. (a) Draw a Venn diagram to represent the situation $C \subseteq A$ and $C \subseteq B$.

(b) To prove $C \subseteq A \cup B$, we should choose an element from which set?

(c) Prove that if that if $C \subseteq A$ and $C \subseteq B$, then $C \subseteq A \cup B$.

29. (a) Draw a Venn diagram to represent the situation $A \subseteq C$ and $B \subseteq C$.

(b) To prove $A \cup B \subseteq C$, we should choose an element from which set?

(c) Prove that if $A \subseteq C$ and $B \subseteq C$, then $A \cup B \subseteq C$.

30. Prove that $A - (A - B) \subseteq B$.

31. Suppose that $A \oplus B = A \oplus C$. Does this guarantee that $B = C$? Justify your conclusion.

32. Prove that $A - B = A \cap \overline{B}$.

33. If $A \cup B = A \cup C$, must $B = C$? Explain.

34. If $A \cap B = A \cap C$, must $B = C$? Explain.

35. Prove that if $A \subseteq B$ and $C \subseteq D$, then $A \cup C \subseteq B \cup D$ and $A \cap C \subseteq B \cap D$.

36. When is $A - B = B - A$? Explain.

37. Explain the last term in the sum in Theorem 3. Why is $|A \cap B \cap C|$ added and $|B \cap C|$ subtracted?

38. (a) Write the four-set version of Theorem 3; that is, $|A \cup B \cup C \cup D| = \cdots$.

(b) Describe in words the n-set version of Theorem 3.

1.3 SEQUENCES

Some of the most important sets arise in connection with sequences. A **sequence** is simply a list of objects arranged in a definite order; a first element, second element, third element, and so on. The list may stop after n steps, $n \in N$, or it may go on forever. In the first case we say that the sequence is **finite**, and in the second case we say that it is **infinite**. The elements may all be different, or some may be repeated.

EXAMPLE 1 The sequence 1, 0, 0, 1, 0, 1, 0, 0, 1, 1, 1 is a finite sequence with repeated items. The digit zero, for example, occurs as the second, third, fifth, seventh, and eighth elements of the sequence. ∎

EXAMPLE 2 The list 3, 8, 13, 18, 23, ... is an infinite sequence. The three dots in the expression mean "and so on," that is, continue the pattern established by the first few elements. ∎

EXAMPLE 3 Another infinite sequence is 1, 4, 9, 16, 25, ..., the list of the squares of all positive integers. ∎

It may happen that how a sequence is to continue is not clear from the first few terms. Also, it may be useful to have a compact notation to describe a sequence. Two kinds of formulas are commonly used to describe sequences. In Example 2, a natural description of the sequence is that successive terms are produced by adding 5 to the previous term. If we use a subscript to indicate a term's position in the sequence, we can describe the sequence in Example 2 as $a_1 = 3$, $a_n = a_{n-1} + 5$,

$2 \leq n < \infty$. A formula, like this one, that refers to previous terms to define the next term is called **recursive**. Every recursive formula must include a starting place.

On the other hand, in Example 3 it is easy to describe a term using only its position number. In the nth position is the square of n; $b_n = n^2$, $1 \leq n < \infty$. This type of formula is called **explicit**, because it tells us exactly what value any particular term has.

EXAMPLE 4   The recursive formula $c_1 = 5$, $c_n = 2c_{n-1}$, $2 \leq n \leq 6$, defines the finite sequence 5, 10, 20, 40, 80, 160. ∎

EXAMPLE 5   The infinite sequence 3, 7, 11, 15, 19, 23, ... can be defined by the recursive formula $d_1 = 3$, $d_n = d_{n-1} + 4$. ∎

EXAMPLE 6   The explicit formula $s_n = (-4)^n$, $1 \leq n < \infty$, describes the infinite sequence -4, 16, -64, 256, ... ∎

EXAMPLE 7   The finite sequence 87, 82, 77, 72, 67 can be defined by the explicit formula $t_n = 92 - 5n$, $1 \leq n \leq 5$. ∎

EXAMPLE 8   An ordinary English word such as "sturdy" can be viewed as the finite sequence

$$s, t, u, r, d, y$$

composed of letters from the ordinary English alphabet. ∎

In examples such as Example 8, it is common to omit the commas and write the word in the usual way, if no confusion results. Similarly, even a meaningless word such as "abacabcd" may be regarded as a finite sequence of length 8. Sequences of letters or other symbols, written without the commas, are also referred to as **strings**.

EXAMPLE 9   An infinite string such as *abababab*... may be regarded as the infinite sequence a, $b, a, b, a, b, \ldots$. ∎

EXAMPLE 10   The sentence "now is the time for the test" can be regarded as a finite sequence of English words: now, is, the, time, for, the, test. Here the elements of the sequence are themselves words of varying length, so we would not be able simply to omit the commas. The custom is to use spaces instead of commas in this case. ∎

The **set corresponding to a sequence** is simply the set of all distinct elements in the sequence. Note that an essential feature of a sequence is the order in which the elements are listed. However, the order in which the elements of a set are listed is of no significance at all.

EXAMPLE 11   (a) The set corresponding to the sequence in Example 3 is $\{1, 4, 9, 16, 25, \ldots\}$.
(b) The set corresponding to the sequence in Example 9 is simply $\{a, b\}$. ∎

The idea of a sequence is important in computer science, where a sequence is sometimes called a **linear array** or **list**. We will make a slight but useful distinction between a sequence and an array, and use a slightly different notation. If we have a sequence S: $s_1, s_2, s_3, \ldots$, we think of all the elements of S as completely determined. The element s_4, for example, is some fixed element of S, located in position four. Moreover, if we change any of the elements, we have a new sequence and will probably name it something other than S. Thus if we begin with the finite sequence S: 0, 1, 2, 3, 2, 1, 1 and we change the 3 to a 4, getting 0, 1, 2, 4, 2, 1, 1, we would think of this as a different sequence, say S'.

An array, on the other hand, may be viewed as a sequence of positions, which we represent in Figure 1.16 as boxes.

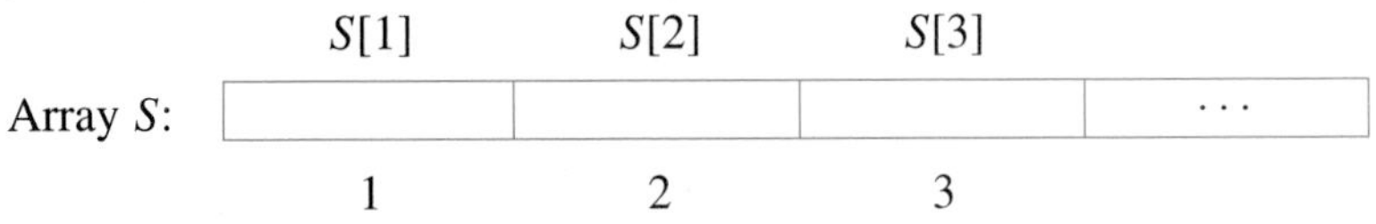

Figure 1.16

The positions form a finite or infinite list, depending on the desired size of the array. Elements from some set may be assigned to the positions of the array S. The element assigned to position n will be denoted by $S[n]$, and the sequence $S[1], S[2], S[3], \ldots$ will be called the **sequence of values** of the array S. The point is that S is considered to be a well-defined object, even if some of the positions have not been assigned values, or if some values are changed during the discussion. The following shows one use of arrays.

Characteristic Functions

A very useful concept for sets is the characteristic function. We discuss functions in Section 5.1, but for now we can proceed intuitively, and think of a function on a set as a rule that assigns some "value" to each element of the set. If A is a subset of a universal set U, the **characteristic function** f_A of A is defined for each $x \in U$ as follows:

$$f_A(x) = \begin{cases} 1 & \text{if } x \in A \\ 0 & \text{if } x \notin A. \end{cases}$$

We may add and multiply characteristic functions, since their values are numbers, and these operations sometimes help us prove theorems about properties of subsets.

Theorem 1   Characteristic functions of subsets satisfy the following properties:

(a) $f_{A \cap B} = f_A f_B$; that is, $f_{A \cap B}(x) = f_A(x) f_B(x)$ for all x.

(b) $f_{A \cup B} = f_A + f_B - f_A f_B$; that is, $f_{A \cup B}(x) = f_A(x) + f_B(x) - f_A(x) f_B(x)$ for all x.

(c) $f_{A \oplus B} = f_A + f_B - 2 f_A f_B$; that is, $f_{A \oplus B}(x) = f_A(x) + f_B(x) - 2 f_A(x) f_B(x)$ for all x.

Proof

(a) $f_A(x)f_B(x)$ equals 1 if and only if both $f_A(x)$ and $f_B(x)$ are equal to 1, and this happens if and only if x is in A and x is in B, that is, x is in $A \cap B$. Since $f_A f_B$ is 1 on $A \cap B$ and 0 otherwise, it must be $f_{A \cap B}$.

(b) If $x \in A$, then $f_A(x) = 1$, so $f_A(x) + f_B(x) - f_A(x)f_B(x) = 1 + f_B(x) - f_B(x) = 1$. Similarly, when $x \in B$, $f_A(x) + f_B(x) - f_A(x)f_B(x) = 1$. If x is not in A or B, then $f_A(x)$ and $f_B(x)$ are 0, so $f_A(x) + f_B(x) - f_A(x)f_B(x) = 0$. Thus $f_A + f_B - f_A f_B$ is 1 on $A \cup B$ and 0 otherwise, so it must be $f_{A \cup B}$.

(c) We leave the proof of (c) as an exercise.   ▼

Note that the proof of Theorem 4 proceeds by direct application of the definition of the characteristic function.

Computer Representation of Sets and Subsets

Another use of characteristic functions is in representing sets in a computer. To represent a set in a computer, the elements of the set must be arranged in a sequence. The particular sequence selected is of no importance. When we list the set $A = \{a, b, c, \ldots, r\}$ we normally assume no particular ordering of the elements in A. Let us identify for now the set A with the sequence $a, b, c, \ldots, r$.

When a universal set U is finite, say $U = \{x_1, x_2, \ldots, x_n\}$, and A is a subset of U, then the characteristic function assigns 1 to an element that belongs to A and 0 to an element that does not belong to A. Thus f_A can be represented by a sequence of 0's and 1's of length n.

EXAMPLE 12   Let $U = \{1, 2, 3, 4, 5, 6\}$, $A = \{1, 2\}$, $B = \{2, 4, 6\}$, and $C = \{4, 5, 6\}$. Then $f_A(x)$ has value 1 when x is 1 or 2, and otherwise is 0. Hence f_A corresponds to the sequence 1, 1, 0, 0, 0, 0. In a similar way, the finite sequence 0, 1, 0, 1, 0, 1 represents f_B and 0, 0, 0, 1, 1, 1 represents f_C.   ■

Any set with n elements can be arranged in a sequence of length n, so each of its subsets corresponds to a sequence of zeros and ones of length n, representing the characteristic function of that subset. This fact allows us to represent a universal set in a computer as an array A of length n. Assignment of a zero or one to each location $A[k]$ of the array specifies a unique subset of U.

EXAMPLE 13   Let $U = \{a, b, e, g, h, r, s, w\}$. The array of length 8 shown in Figure 1.17 represents U, since $A[k] = 1$ for $1 \le k \le 8$.

If $S = \{a, e, r, w\}$, then

$$f_S(x) = \begin{cases} 1 & \text{if } x = a, e, r, w \\ 0 & \text{if } x = b, g, h, s. \end{cases}$$

Figure 1.17

Hence the array in Figure 1.18 represents the subset S.

1	0	1	0	0	1	0	1

Figure 1.18   ∎

A set is called **countable** if it is the set corresponding to some sequence. In-formally, this means that the members of the set can be arranged in a list, with a first, second, third, ..., element, and the set can therefore be "counted." We shall show in Section 2.4 that all finite sets are countable. However, not all infinite sets are countable. A set that is not countable is called **uncountable**.

The most accessible example of an uncountable set is the set of all real numbers that can be represented by an infinite decimal of the form $0.a_1a_2a_3\ldots$, where a_i is an integer and $0 \le a_i \le 9$. We shall now show that this set is uncountable. We will prove this result by contradiction; that is, we will show the countability of this set implies an impossible situation. (We will look more closely at proof by contradiction in Chapter 2.)

Assume that the set of all decimals $0.a_1a_2a_3\ldots$ is countable. Then we could form the following list (sequence), containing all such decimals:

$$d_1 = 0.a_1a_2a_3\ldots$$
$$d_2 = 0.b_1b_2b_3\ldots$$
$$d_3 = 0.c_1c_2c_3\ldots$$

$$\vdots$$

Each of our infinite decimals must appear somewhere on this list. We shall establish a contradiction by constructing an infinite decimal of this type that is not on the list. Now construct a number x as follows: $x = 0.x_1x_2x_3\ldots$, where x_1 is 1 if $a_1 = 2$, otherwise x_1 is 2; $x_2 = 1$ if $b_2 = 2$, otherwise $x_2 = 2$; $x_3 = 1$ if $c_3 = 2$, otherwise $x_3 = 2$. This process can clearly be continued indefinitely. The resulting number is an infinite decimal consisting of 1's and 2's, but by its construction x differs from each number in the list at some position. Thus x is not on the list, a contradiction to our assumption. Hence no matter how the list is constructed, there is some real number of the form $0.x_1x_2x_3\ldots$ that is not in the list. On the other hand, it can be shown that the set of rational numbers is countable.

Strings and Regular Expressions

Given a set A, we can construct the set A^* consisting of all finite sequences of elements of A. Often, the set A is not a set of numbers, but some set of symbols. In this case, A is called an **alphabet**, and the finite sequences in A^* are called **words** from A, or sometimes strings from A. For this case in particular, the sequences in A^* are *not* written with commas. We assume that A^* contains the **empty sequence** or **empty string**, containing no symbols, and we denote this string by Λ. This string will be useful in Chapters 9 and 10.

EXAMPLE 14   Let $A = \{a, b, c, \ldots, z\}$, the usual English alphabet. Then A^* consists of all ordinary words, such as ape, sequence, antidisestablishmentarianism, and so on, as well as "words" such as yxaloble, zigadongdong, cay, and pqrst. All finite sequences from A are in A^*, whether they have meaning or not. ∎

If $w_1 = s_1 s_2 s_3 \ldots s_n$ and $w_2 = t_1 t_2 t_3 \ldots t_k$ are elements of A^* for some set A, we define the **catenation** of w_1 and w_2 as the sequence $s_1 s_2 s_3 \ldots s_n t_1 t_2 t_3 \ldots t_k$. The catenation of w_1 with w_2 is written as $w_1 \cdot w_2$ or $w_1 w_2$, and is another element of A^*. Note that if w belongs to A^*, then $w \cdot \Lambda = w$ and $\Lambda \cdot w = w$. This property is convenient and is one of the main reasons for defining the empty string Λ.

EXAMPLE 15   Let $A = \{$John, Sam, Jane, swims, runs, well, quickly, slowly$\}$. Then A^* contains real sentences such as "Jane swims quickly" and "Sam runs well," as well as nonsense sentences such as "Well swims Jane slowly John." Here we separate the elements in each sequence with spaces. This is often done when the elements of A are words. ∎

The idea of a recursive formula for a sequence is useful in more general settings as well. In the formal languages and the finite state machines we discuss in Chapter 10, the concept of regular expression plays an important role, and regular expressions are defined recursively. A **regular expression over A** is a string constructed from the elements of A and the symbols (,), ∨,*, Λ, according to the following definition.

RE1. The symbol Λ is a regular expression.

RE2. If $x \in A$, the symbol x is a regular expression.

RE3. If α and β are regular expressions, then the expression $\alpha\beta$ is regular.

RE4. If α and β are regular expressions, then the expression $(\alpha \vee \beta)$ is regular.

RE5. If α is a regular expression, then the expression $(\alpha)^*$ is regular.

Note here that RE1 and RE2 provide initial regular expressions. The other parts of the definition are used repetitively to define successively larger sets of regular expressions from those already defined. Thus the definition is recursive.

By convention, if the regular expression α consists of a single symbol x, where $x \in A$, or if α begins and ends with parentheses, then we write $(\alpha)^*$ simply as α^*. When no confusion results, we will refer to a regular expression over A simply as a **regular expression** (omitting reference to A).

EXAMPLE 16   Let $A = \{0, 1\}$. Show that the following expressions are all regular expressions over A.

(a) $0^*(0 \vee 1)^*$ (b) $00^*(0 \vee 1)^*1$ (c) $(01)^*(01 \vee 1^*)$

Solution

(a) By RE2, 0 and 1 are regular expressions. Thus $(0 \vee 1)$ is regular by RE4, and so 0^* and $(0 \vee 1)^*$ are regular by RE5 (and the convention mentioned previously). Finally, we see that $0^*(0 \vee 1)^*$ is regular by RE3.

(b) We know that 0, 1, and $0^*(0 \vee 1)^*$ are all regular. Thus, using RE3 twice, $00^*(0 \vee 1)^*1$ must be regular.

(c) By RE3, 01 is a regular expression. Since 1^* is regular, $(01 \vee 1^*)$ is regular by RE4, and $(01)^*$ is regular by RE5. Then the regularity of $(01)^*(01 \vee 1^*)$ follows from RE3. ∎

Associated with each regular expression over A, there is a corresponding subset of A^*. Such sets are called **regular subsets** of A^* or just **regular sets** if no reference to A is needed. To compute the regular set corresponding to a regular expression, we use the following correspondence rules.

1. The expression Λ corresponds to the set $\{\Lambda\}$, where Λ is the empty string in A^*.

2. If $x \in A$, then the regular expression x corresponds to the set $\{x\}$.

3. If α and β are regular expressions corresponding to the subsets M and N of A^*, then $\alpha\beta$ corresponds to $M \cdot N = \{s \cdot t \mid s \in M \text{ and } t \in N\}$. Thus $M \cdot N$ is the set of all catenations of strings in M with strings in N.

4. If the regular expressions α and β correspond to the subsets M and N of A^*, then $(\alpha \vee \beta)$ corresponds to $M \cup N$.

5. If the regular expression α corresponds to the subset M of A^*, then $(\alpha)^*$ corresponds to the set M^*. Note that M is a set of strings from A. Elements from M^* are finite sequences of such strings, and thus may themselves be interpreted as strings from A. Note also that we always have $\Lambda \in M^*$.

EXAMPLE 17 Let $A = \{a, b, c\}$. Then the regular expression a^* corresponds to the set of all finite sequences of a's, such as aaa, $aaaaaaa$, and so on. The regular expression $a(b \vee c)$ corresponds to the set $\{ab, ac\} \subseteq A^*$. Finally, the regular expression $ab(bc)^*$ corresponds to the set of all strings that begin with ab, and then repeat the symbols bc n times, where $n \geq 0$. This set includes the strings ab, $abbcbc$, $abbcbcbcbc$, and so on. ∎

EXAMPLE 18 Let $A = \{0, 1\}$. Find the regular sets corresponding to the three regular expressions in Example 16.

Solution

(a) The set corresponding to $0^*(0 \vee 1)^*$ consists of all sequences of 0's and 1's. Thus, the set is A^*.

(b) The expression $00^*(0 \vee 1)^*1$ corresponds to the set of all sequences of 0's and 1's that begin with at least one 0 and end with at least one 1.

(c) The expression $(01)^*(01 \vee 1^*)$ corresponds to the set of all sequences of 0's and 1's that either repeat the string 01 a total of $n \geq 1$ times, or begin with a total of $n \geq 0$ repetitions of 01 and end with some number $k \geq 0$ of 1's. This set includes, for example, the strings $1111, 01, 010101$, 0101010111111, and 011. ∎

1.3 Exercises

In Exercises 1 through 4, give the set corresponding to the sequence.

1. 1, 2, 1, 2, 1, 2, 1, 2, 1

2. 0, 2, 4, 6, 8, 10, . . .

3. *aabbccddee. . . zz*

4. *abbcccdddd*

5. Give three different sequences that have $\{x, y, z\}$ as a corresponding set.

6. Give three different sequences that have $\{1, 2, 3, \dots, \}$ as a corresponding set.

In Exercises 7 through 10, write out the first four terms (begin with $n = 1$) of the sequence whose general term is given.

7. $a_n = 5^n$

8. $b_n = 3n^2 + 2n - 6$

9. $c_1 = 2.5, c_n = c_{n-1} + 1.5$

10. $d_1 = -3, d_n = -2d_{n-1} + 1$

In Exercises 11 through 16, write a formula for the nth term of the sequence. Identify your formula as recursive or explicit.

11. 1, 3, 5, 7, . . .

12. 0, 3, 8, 15, 24, 35, . . .

13. 1, −1, 1, −1, 1, −1, . . .

14. 0, 2, 0, 2, 0, 2, . . .

15. 1, 4, 7, 10, 13, 16

16. $1, \frac{1}{2}, \frac{1}{4}, \frac{1}{8}, \frac{1}{16}, \dots$

17. Write an explicit formula for the sequence 2, 5, 8, 11, 14, 17, ...

18. Write a recursive formula for the sequence 2, 5, 7, 12, 19, 31, ...

19. Let $A = \{x \mid x \text{ is a real number and } 0 < x < 1\}$, $B = \{x \mid x \text{ is a real number and } x^2 + 1 = 0\}$, $C = \{x \mid x = 4m, m \in Z\}$, $D = \{(x, 3) \mid x \text{ is an English word whose length is } 3\}$, and $E = \{x \mid x \in Z \text{ and } x^2 \le 100\}$. Identify each set as finite, countable, or uncountable.

20. Let $A = \{ab, bc, ba\}$. In each part, tell whether the string belongs to A^*.

    (a) *ababab*   (b) *abc*   (c) *abba*

(d) *abbcbaba*   (e) *bcabbab*   (f) *abbbcba*

21. Let $U = \{$FORTRAN, PASCAL, ADA, COBOL, LISP, BASIC, C^{++}, FORTH$\}$, $B = \{$C^{++} BASIC, ADA$\}$, $C = \{$PASCAL, ADA, LISP, C$^{++}\}$, $D = \{$FORTRAN, PASCAL, ADA, BASIC, FORTH$\}$, $E = \{$PASCAL, ADA, COBOL, LISP, C$^{++}\}$. In each of the following, represent the given set by an array of zeros and ones.

    (a) $B \cup C$ (b) $C \cap D$

    (c) $B \cap (D \cap E)$ (d) $\overline{B} \cup E$

    (e) $\overline{C} \cap (B \cup E)$

22. Let $U = \{b, d, e, g, h, k, m, n\}$, $B = \{b\}$, $C = \{d, g, m, n\}$, and $D = \{d, k, n\}$.

    (a) What is $f_B(b)$?   (b) What is $f_C(e)$?

    (c) Find the sequences of length 8 that correspond to f_B, f_C, and f_D.

    (d) Represent $B \cup C$, $C \cup D$, and $C \cap D$ by arrays of zeros and ones.

23. Complete the proof that $f_{A \oplus B} = f_A + f_B - 2f_A f_B$ [Theorem 4(c)]. Suppose $x \in A$ and $x \notin B$. Then $f_A(x) = \text{\_\_\_\_}, f_B(x) = \text{\_\_\_\_}$, and $f_A(x)f_B(x) = \text{\_\_\_\_}$, so $f_A(x) + f_B(x) - 2f_A(x)f_B(x) = \text{\_\_\_\_}$. Now suppose $x \notin A$ and $x \in B$. Then $f_A(x) = \text{\_\_\_\_}, f_B(x) = \text{\_\_\_\_}$, and $f_A(x)f_B(x) = \text{\_\_\_\_}$, so $f_A(x) + f_B(x) - 2f_A(x)f_B(x) = \text{\_\_\_\_}$. The remaining case to check is $x \notin A \oplus B$. If $x \notin A \oplus B$, then $x \in \text{\_\_\_\_}$ and $f_A(x) + f_B(x) - 2f_A(x)f_B(x) = \text{\_\_\_\_}$. Explain how these steps prove Theorem 4(c).

24. Using characteristic functions, prove that $(A \oplus B) \oplus C = A \oplus (B \oplus C)$.

25. Let $A = \{+, \times, a, b\}$. Show that the following expressions are regular over A.

    (a) $a + b(ab)^*(a \times b \vee a)$

    (b) $a + b \times (a^* \vee b)$

    (c) $(a^*b \vee +)^* \vee \times b^*$

In Exercises 26 and 27, let $A = \{a, b, c\}$. In each exercise a string in A^* is listed, and a regular expression over A. In each case, tell whether or not the string on the left belongs to the regular set corresponding to the regular expression on the right.

26. (a) *ac* a^*b^*c (b) *abcc* $(abc \vee c)^*$

    (c) *aaabc* $((a \vee b) \vee c)^*$

27. (a) *ac* $(a^*b \vee c)$ (b) *abab* $(ab)^*c$

    (c) *aaccc* $(a^* \vee b)c^*$

28. Give three expressions that are not regular over the A given for Exercise 26.

29. Let $A = \{p, q, r\}$. Give the regular set corresponding to the regular expression given.

  (a) $(p \vee q)rq^*$ (b) $p(qq)^*r$

30. Let $S = \{0, 1\}$. Give the regular expression corresponding to the regular set given.

  (a) $\{00, 010, 0110, 011110, \ldots\}$

  (b) $\{0, 001, 000, 00001, 00000, 0000001, \ldots\}$

31. We define T-numbers recursively as follows:

  1. 0 is a T-number.

  2. If X is a T-number, $X + 3$ is a T-number.

Write a description of the set of T-numbers.

32. Define an S-number by

  1. 8 is an S-number.

  2. If X is an S-number and Y is a multiple of X, then Y is an S-number.

3. If X is an S-number and X is a multiple of Y, then Y is an S-number.

Describe the set of S-numbers.

33. Let F be a function defined for all nonnegative integers by the following recursive definition.

$$F(0) = 0, \qquad F(1) = 1$$
$$F(N + 2) = 2F(N) + F(N + 1), \quad N \geq 0$$

Compute the first six values of F; that is, write the values of $F(N)$ for $N = 0, 1, 2, 3, 4, 5$.

34. Let G be a function defined for all nonnegative integers by the following recursive definition.

$$G(0) = 1, \qquad G(1) = 2$$
$$G(N + 2) = G(N)^2 + G(N + 1), \quad N \geq 0$$

Compute the first five values of G.

1.4 DIVISION IN THE INTEGERS

We shall now discuss some results needed later about division and factoring in the integers. If m is an integer and n is a positive integer, we can plot the nonnegative integer multiples of n on a line, and locate m as in Figure 1.19. If m is a multiple of n, say $m = qn$, then we can write $m = qn + r$, where r is 0. On the other hand (as shown in Figure 1), if m is not a multiple of n, we let qn be the first multiple of n lying to the left of m and let r be $m - qn$. Then r is the distance from qn to m, so clearly $0 < r < n$, and again we have $m = qn + r$. We state these observations as a theorem.

Figure 1.19

Theorem 1   If n and m are integers and $n > 0$, we can write $m = qn + r$ for integers q and r with $0 \leq r < n$. Moreover, there is just one way to do this.   ●

EXAMPLE 1   (a) If n is 3 and m is 16, then $16 = 5(3) + 1$ so q is 5 and r is 1.

  (b) If n is 10 and m is 3, then $3 = 0(10) + 3$ so q is 0 and r is 3.

  (c) If n is 5 and m is -11, then $-11 = -3(5) + 4$ so q is -3 and r is 4.   ■

If the r in Theorem 1 is zero, so that m is a multiple of n, we write $n \mid m$, which is read "n divides m." If $n \mid m$, then $m = qn$ and $n \leq m$. If m is not a multiple of n, we write $n \nmid m$, which is read "n does not divide m." We now prove some simple properties of divisibility.

Theorem 2   Let a, b, and c be integers.

(a) If $a \mid b$ and $a \mid c$, then $a \mid (b + c)$.

(b) If $a \mid b$ and $a \mid c$, where $b > c$, then $a \mid (b - c)$.

(c) If $a \mid b$ or $a \mid c$, then $a \mid bc$.

(d) If $a \mid b$ and $b \mid c$, then $a \mid c$.   ●

Proof

(a) If $a \mid b$ and $a \mid c$, then $b = k_1 a$ and $c = k_2 a$ for integers k_1 and k_2. So $b + c = (k_1 + k_2)a$ and $a \mid (b + c)$.

(b) This can be proved in exactly the same way as (a).

(c) As in (a), we have $b = k_1 a$ or $c = k_2 a$. Then either $bc = k_1 ac$ or $bc = k_2 ab$, so in either case bc is a multiple of a and $a \mid bc$.

(d) If $a \mid b$ and $b \mid c$, we have $b = k_1 a$ and $c = k_2 b$, so $c = k_2 b = k_2(k_1 a) = (k_2 k_1)a$ and hence $a \mid c$.   ▼

Note that again we have a proof that proceeds directly from a definition by restating the original conditions. As a consequence of Theorem 2, we have that if $a \mid b$ and $a \mid c$, then $a \mid (mb + nc)$, for any integers m and n.

A number $p > 1$ in Z^+ is called **prime** if the only positive integers that divide p are p and 1.

EXAMPLE 2   The numbers 2, 3, 5, 7, 11, and 13 are prime, while 4, 10, 16, and 21 are not prime.   ■

It is easy to write a set of steps, or an **algorithm***, to determine if a positive integer $n > 1$ is a prime number. First we check to see if n is 2. If $n > 2$, we could divide by every integer from 2 to $n - 1$, and if none of these is a divisor of n, then n is prime. To make the process more efficient, we note that if $mk = n$, then either m or k is less than or equal to $\sqrt{n}$. This means that if n is not prime, it has a divisor k satisfying the inequality $1 < k \leq \sqrt{n}$, so we need only test for divisors in this range. Also, if n has any even number as a divisor, it must have 2 as a divisor. Thus after checking for divisibility by 2, we may skip all even integers.

◆ **ALGORITHM**   to test whether an integer $N > 1$ is prime:

Step 1   Check whether N is 2. If so, N is prime. If not, proceed to

Step 2   Check whether $2 \mid N$. If so, N is not prime; otherwise, proceed to

Step 3   Compute the largest integer $K \leq \sqrt{N}$. Then

Step 4   Check whether $D \mid N$, where D is any odd number such that $1 < D \leq K$. If $D \mid N$, then N is not prime; otherwise, N is prime.

---

*Algorithms are discussed in Appendix A.

Testing whether an integer is prime is a common task for computers. The algorithm given here is too inefficient for testing very large numbers, but there are many other algorithms for testing whether an integer is prime.

Theorem 3 Every positive integer $n > 1$ can be written uniquely as $p_1^{k_1} p_2^{k_2} \cdots p_s^{k_s}$, where $p_1 < p_2 < \cdots < p_s$ are distinct primes that divide n and the k's are positive integers giving the number of times each prime occurs as a factor of n. ●

We leave the proof of Theorem 3 to Section 2.4, but we give several illustrations.

EXAMPLE 3
(a) $9 = 3 \cdot 3 = 3^2$
(b) $24 = 8 \cdot 3 = 2 \cdot 2 \cdot 2 \cdot 3 = 2^3 \cdot 3$
(c) $30 = 2 \cdot 3 \cdot 5$ ■

Greatest Common Divisor

If a, b, and k are in Z^+, and $k \mid a$ and $k \mid b$, we say that k is a **common divisor** of a and b. If d is the largest such k, d is called the **greatest common divisor**, or GCD, of a and b, and we write $d = \text{GCD}(a, b)$. This number has some interesting properties. It can be written as a combination of a and b, and it is not only larger than all the other common divisors, it is a multiple of each of them.

Theorem 4 If d is $\text{GCD}(a, b)$, then

(a) $d = sa + tb$ for some integers s and t. (These are not necessarily positive.)
(b) If c is any other common divisor of a and b, then $c \mid d$. ●

Proof Let x be the smallest positive integer that can be written as $sa + tb$ for some integers s and t, and let c be a common divisor of a and b. Since $c \mid a$ and $c \mid b$, it follows from Theorem 2 that $c \mid x$, so $c \leq x$. If we can show that x is a common divisor of a and b, it will then be the greatest common divisor of a and b and both parts of the theorem will have been proved. By Theorem 1, $a = qx + r$ with $0 \leq r < x$. Solving for r, we have

$$r = a - qx = a - q(sa + tb) = a - qsa - qtb = (1 - qs)a + (-qt)b.$$

If r is not zero, then since $r < x$ and r is the sum of a multiple of a and a multiple of b, we will have a contradiction to the fact that x is the smallest positive number that is a sum of multiples of a and b. Thus r must be 0 and $x \mid a$. In the same way we can show that $x \mid b$, and this completes the proof. ▼

This proof is more complex than the earlier ones. At this stage you should focus on understanding the details of each step. We will discuss the structure of this proof later.

From the definition of greatest common divisor and Theorem 4(b), we have the following result: Let a, b, and d be in Z^+. The integer d is the greatest common divisor of a and b if and only if

(a) $d \mid a$ and $d \mid b$.

(b) Whenever $c \mid a$ and $c \mid b$, then $c \mid d$.

EXAMPLE 4

(a) The common divisors of 12 and 30 are 1, 2, 3, and 6, so that

$$\text{GCD}(12, 30) = 6 \quad \text{and} \quad 6 = 1 \cdot 30 + (-2) \cdot 12.$$

(b) It is clear that $\text{GCD}(17, 95) = 1$ since 17 is prime and $17 \nmid 95$, and the reader may verify that $1 = 28 \cdot 17 + (-5) \cdot 95.$ ∎

If $\text{GCD}(a, b) = 1$, as in Example 4(b), we say a and b are **relatively prime**.

One remaining question is that of how to compute the GCD conveniently in general. Repeated application of Theorem 1 provides the key to doing this.

We now present a procedure, called the **Euclidean algorithm**, for finding $\text{GCD}(a, b)$. Suppose that $a > b > 0$ (otherwise interchange a and b). Then by Theorem 1, we may write

$$a = k_1 b + r_1, \qquad \text{where } k_1 \text{ is in } Z^+ \text{ and } 0 \le r_1 < b. \qquad (1)$$

Now Theorem 2 tells us that if n divides a and b, then it must divide r_1, since $r_1 = a - k_1 b$. Similarly, if n divides b and r_1, then it must divide a. We see that the common divisors of a and b are the same as the common divisors of b and r_1, so $\text{GCD}(a, b) = \text{GCD}(b, r_1)$.

We now continue using Theorem 1 as follows:

$$
\begin{array}{llll}
\text{divide } b \text{ by } r_1: & b = k_2 r_1 + r_2 & 0 \le r_2 < r_1 \\
\text{divide } r_1 \text{ by } r_2: & r_1 = k_3 r_2 + r_3 & 0 \le r_3 < r_2 \\
\text{divide } r_2 \text{ by } r_3: & r_2 = k_4 r_3 + r_4 & 0 \le r_4 < r_3 \\
\qquad \vdots & \qquad \vdots & \qquad \vdots \\
\text{divide } r_{n-2} \text{ by } r_{n-1}: & r_{n-2} = k_n r_{n-1} + r_n & 0 \le r_n < r_{n-1} \\
\text{divide } r_{n-1} \text{ by } r_n: & r_{n-1} = k_{n+1} r_n + r_{n+1} & 0 \le r_{n+1} < r_n.
\end{array}
\qquad (2)
$$

Since $a > b > r_1 > r_2 > r_3 > r_4 > \cdots$, the remainder will eventually become zero, so at some point we have $r_{n+1} = 0$.

We now show that $r_n = \text{GCD}(a, b)$. We saw previously that

$$\text{GCD}(a, b) = \text{GCD}(b, r_1).$$

Repeating this argument with b and r_1, we see that

$$\text{GCD}(b, r_1) = \text{GCD}(r_1, r_2).$$

Upon continuing, we have

$$\text{GCD}(a, b) = \text{GCD}(b, r_1) = \text{GCD}(r_1, r_2) = \cdots = \text{GCD}(r_{n-1}, r_n).$$

Since $r_{n-1} = k_{n+1} r_n$, we see that $\text{GCD}(r_{n-1}, r_n) = r_n$. Hence $r_n = \text{GCD}(a, b)$.

EXAMPLE 5 Let a be 190 and b be 34. Then, using the Euclidean algorithm, we

divide 190 by 34: $190 = 5 \cdot 34 + 20$
divide 34 by 20: $34 = 1 \cdot 20 + 14$
divide 20 by 14: $20 = 1 \cdot 14 + 6$
divide 14 by 6: $14 = 2 \cdot 6 + 2$
divide 6 by 2: $6 = 3 \cdot 2 + 0$

so GCD(190, 34) = 2, the last of the nonzero divisors.   ∎

In Theorem 4(a), we observed that if $d = \text{GCD}(a, b)$, we can find integers s and t such that $d = sa + tb$. The integers s and t can be found as follows. Solve the next-to-last equation in (2) for r_n:

$$r_n = r_{n-2} - k_n r_{n-1}. \tag{3}$$

Now solve the second-to-last equation in (2), $r_{n-3} = k_{n-1}r_{n-2} + r_{n-1}$ for r_{n-1}:

$$r_{n-1} = r_{n-3} - k_{n-1}r_{n-2}$$

and substitute this expression in (3):

$$r_n = r_{n-2} - k_n[r_{n-3} - k_{n-1}r_{n-2}].$$

Continue to work up through the equations in (2) and (1), replacing r_i by an expression involving r_{i-1} and r_{i-2}, and finally arriving at an expression involving only a and b.

EXAMPLE 6 (a) Let $a = 190$ and $b = 34$ as in Example 5. Then

$$
\begin{aligned}
\text{GCD}(190, 34) = 2 &= 14 - 2(6) \\
&= 14 - 2[20 - 1(14)] &&6 = 20 - 1 \cdot 14 \\
&= 3(14) - 2(20) \\
&= 3[34 - 1(20)] - 2(20) &&14 = 34 - 1 \cdot 20 \\
&= 3(34) - 5(190 - 5 \cdot 34) &&20 = 190 - 5 \cdot 34 \\
&= 28(34) - 5(190)
\end{aligned}
$$

Hence $s = -5$ and $t = 28$. Note that the key is to carry out the arithmetic only partially.

(b) Let $a = 108$ and $b = 60$. Then

$$
\begin{aligned}
\text{GCD}(108, 60) = 12 &= 60 - 1(48) \\
&= 60 - 1[108 - 1(60)] &&48 = 108 - 1 \cdot 60 \\
&= 2(60) - 108.
\end{aligned}
$$

Hence $s = -1$ and $t = 2$.   ∎

Theorem 5   If a and b are in Z^+, then $\text{GCD}(a, b) = \text{GCD}(b, b \pm a)$.   ●

Proof   If c divides a and b, it divides $b \pm a$, by Theorem 2. Since $a = b - (b - a) = -b + (b + a)$, we see, also by Theorem 2, that a common divisor of b and $b \pm a$ also divides a and b. Since a and b have the same common divisors as b and $b \pm a$, they must have the same greatest common divisor.   ▼

This is another direct proof, but one that uses a previous theorem as well as definitions.

Least Common Multiple

If a, b, and k are in Z^+, and $a \mid k$, $b \mid k$, we say k is a **common multiple** of a and b. The smallest such k, call it c, is called the **least common multiple**, or LCM, of a and b, and we write $c = \text{LCM}(a, b)$. The following result shows that we can obtain the least common multiple from the greatest common divisor, so we do not need a separate procedure for finding the least common multiple.

Theorem 6   If a and b are two positive integers, then $\text{GCD}(a, b) \cdot \text{LCM}(a, b) = ab$.   ●

Proof   Let $p_1, p_2, \ldots, p_k$ be all the prime factors of either a or b. Then we can write

$$a = p_1^{a_1} p_2^{a_2} \cdots p_k^{a_k} \quad \text{and} \quad b = p_1^{b_1} p_2^{b_2} \cdots p_k^{b_k}$$

where some of the a_i and b_i may be zero. It then follows that

$$\text{GCD}(a, b) = p_1^{\min(a_1, b_1)} p_2^{\min(a_2, b_2)} \cdots p_k^{\min(a_k, b_k)}$$

and

$$\text{LCM}(a, b) = p_1^{\max(a_1, b_1)} p_2^{\max(a_2, b_2)} \cdots p_k^{\max(a_k, b_k)}.$$

Hence

$$\text{GCD}(a, b) \cdot \text{LCM}(a, b) = p_1^{a_1 + b_1} p_2^{a_2 + b_2} \cdots p_k^{a_k + b_k}$$
$$= (p_1^{a_1} p_2^{a_2} \cdots p_k^{a_k}) \cdot (p_1^{b_1} p_2^{b_2} \cdots p_k^{b_k})$$
$$= ab.$$ ▼

EXAMPLE 7   Let $a = 540$ and $b = 504$. Factoring a and b into primes, we obtain

$$a = 540 = 2^2 \cdot 3^3 \cdot 5 \quad \text{and} \quad b = 504 = 2^3 \cdot 3^2 \cdot 7.$$

Thus all the prime numbers that are factors of either a or b are $p_1 = 2$, $p_2 = 3$, $p_3 = 5$, and $p_4 = 7$. Then $a = 2^2 \cdot 3^3 \cdot 5^1 \cdot 7^0$ and $b = 2^3 \cdot 3^2 \cdot 5^0 \cdot 7^1$. We then have

$$\text{GCD}(540, 504) = 2^{\min(2,3)} \cdot 3^{\min(3,2)} \cdot 5^{\min(1,0)} \cdot 7^{\min(0,1)}$$
$$= 2^2 \cdot 3^2 \cdot 5^0 \cdot 7^0$$
$$= 2^2 \cdot 3^2 \text{ or } 36.$$

Also,

$$LCM(540, 504) = 2^{\max(2,3)} \cdot 3^{\max(3,2)} \cdot 5^{\max(1,0)} \cdot 7^{\max(0,1)}$$
$$= 2^3 \cdot 3^3 \cdot 5^1 \cdot 7^1 \text{ or } 7560.$$

Then

$$GCD(540, 504) \cdot LCM(540, 504) = 36 \cdot 7560 = 272,160 = 540 \cdot 504.$$

As a verification, we can also compute GCD(540, 504) by the Euclidean algorithm and obtain the same result.   ∎

If n and m are integers and $n > 1$, Theorem 1 tells us we can write $m = qn + r$, $0 \leq r < n$. Sometimes the remainder r is more important than the quotient q.

EXAMPLE 8   If the time is now 4 o'clock, what time will it be 101 hours from now?

Solution   Let $n = 12$ and $m = 4 + 101$ or 105. Then we have $105 = 8 \cdot 12 + 9$. The remainder 9 answers the question. In 101 hours it will be 9 o'clock.   ∎

For each $n \in Z^+$, we define a function f_n, the mod-n function, as follows: If z is a nonnegative integer, $f_n(z) = r$, the remainder when z is divided by n. (Again, functions are formally defined in Section 5.1, but as in Section 1.3, we need only think of a function as a rule that assigns some "value" to each member of a set.) The naming of these functions is made clear in Section 4.5.

EXAMPLE 9   (a) $f_3(14) = 2$, because $14 = 4 \cdot 3 + 2$ and $14 \equiv 2 \pmod 3$.
(b) $f_7(153) = 6$ ∎

Pseudocode Versions

An alternative to expressing an algorithm in ordinary English as we did in this section is to express it in something like a computer language. Throughout the book we use a **pseudocode** language, which is described fully in Appendix A. Here we give pseudocode versions for an algorithm that determines if an integer is prime and for an algorithm that calculates the greatest common divisor of two integers.

In the pseudocode for the algorithm to determine if an integer is prime, we assume the existence of functions SQR and INT, where SQR(N) returns the greatest integer not exceeding $\sqrt{N}$, and INT(X) returns the greatest integer not exceeding X. For example, SQR(10) = 3, SQR(25) = 5, INT(7.124) = 7, and INT(8) = 8.

SUBROUTINE PRIME(N)
1. **IF** ($N = 2$) **THEN**
    a. **PRINT** ('PRIME')
    b. **RETURN**
2. **ELSE**
    a. **IF** ($N/2 = $ INT($N/2$)) **THEN**
        1. **PRINT** ('NOT PRIME')
        2. **RETURN**

b. **ELSE**
    1. **FOR** $D = 3$ **THRU** SQR(N) **BY** 2
        a. **IF** $(N/D = \text{INT}(N/D))$ **THEN**
            1. **PRINT** ('NOT PRIME')
            2. **RETURN**
    2. **PRINT** ('PRIME')
    3. **RETURN**
END OF SUBROUTINE PRIME

The following gives a pseudocode program for finding the greatest common divisor of two positive integers. This procedure is different from the Euclidean algorithm, but in Chapter 2, we will see how to prove that this algorithm does indeed find the greatest common divisor.

FUNCTION GCD(X, Y)
1. **WHILE** $(X \neq Y)$
    a. **IF** $(X > Y)$ **THEN**
        1. $X \leftarrow X - Y$
    b. **ELSE**
        1. $Y \leftarrow Y - X$
2. **RETURN** (X)
END OF FUNCTION GCD

EXAMPLE 10   Use the pseudocode for GCD to calculate the greatest common divisor of 190 and 34 (Example 5).

Solution   The following table gives the values of X, Y, $X - Y$, or $Y - X$ as we go through the program.

X	Y	$X - Y$	$Y - X$
190	34	156	
156	34	122	
122	34	88	
88	34	54	
54	34	20	
20	34		14
20	14	6	
6	14		8
6	8		2
6	2	4	
4	2	2	
2	2		

Since the last value of X is 2, the greatest common divisor of 190 and 34 is 2. ■

1.4 Exercises

In Exercises 1 through 4, for the given integers m and n, write m as qn + r, with 0 ≤ r < n.

1. $m = 20, n = 3$

2. $m = 64, n = 37$

3. $m = 3, n = 22$

4. $m = 48, n = 12$

5. Write each integer as a product of powers of primes (as in Theorem 3).

   (a) 828   (b) 1666   (c) 1781

   (d) 1125   (e) 107

In Exercises 6 through 9, find the greatest common divisor d of the integers a and b, and write d as sa + tb.

6. $a = 60, b = 100$

7. $a = 45, b = 33$

8. $a = 34, b = 58$

9. $a = 77, b = 128$

In Exercises 10 through 13, find the least common multiple of the integers.

10. 72, 108

11. 150, 70

12. 175, 245

13. 32, 27

14. If f is the mod-7 function, compute each of the following.

   (a) $f(17)$ (b) $f(48)$ (c) $f(1207)$

   (d) $f(130)$ (e) $f(93)$ (f) $f(169)$

15. If f is the mod-11 function, compute each of the following.

   (a) $f(39)$ (b) $f(386)$ (c) $f(1232)$

   (d) $f(573)$ (e) $2f(87)$ (f) $f(175) + 4$

16. If g is the mod-5 function, solve each of the following.

   (a) $g(n) = 2$ (b) $g(n) = 4$

17. If g is the mod-6 function, solve each of the following.

   (a) $g(n) = 3$ (b) $g(n) = 1$

18. Complete the following proof. Let a and b be integers. If p is a prime and $p \mid ab$, then $p \mid a$ or $p \mid b$. We need to show that if $p \nmid a$, then p must divide b. If $p \nmid a$, then $\text{GCD}(a, p) = 1$, because _____. By Theorem 4, we can write $1 = sa + tp$ for some integers s and t. Then $b = sab + tpb$. (Why?) Then p must divide $sab + tpb$, because _____. So $p \mid b$. (Why?)

19. Show that if $\text{GCD}(a, c) = 1$ and $c \mid ab$, then $c \mid b$. (*Hint:* Model the proof on the one in Exercise 18.)

20. Show that if $\text{GCD}(a, c) = 1$, $a \mid m$, and $c \mid m$, then $ac \mid m$. (*Hint:* Use Exercise 19.)

21. Show that if $d = \text{GCD}(a, b)$, $a \mid b$, and $c \mid b$, then $ac \mid bd$.

22. Show that $\text{GCD}(ca, cb) = c\,\text{GCD}(a, b)$.

23. Show that $\text{LCM}(a, ab) = ab$.

24. Show that if $\text{GCD}(a, b) = 1$, then $\text{LCM}(a, b) = ab$.

25. Let $c = \text{LCM}(a, b)$. Show that if $a \mid k$ and $b \mid k$, then $c \mid k$.

26. Prove that if a and b are positive integers such that $a \mid b$ and $b \mid a$, then $a = b$.

27. Let a be an integer and let p be a positive integer. Prove that if $p \mid a$, then $p = \text{GCD}(a, p)$.

28. Theorem 2(c) says that if $a \mid b$ or $a \mid c$, then $a \mid bc$. Is the converse true; that is, if $a \mid bc$, then $a \mid b$ or $a \mid c$? Justify your conclusion.

29. Prove that if m and n are relatively prime and mn is a perfect square, then m and n are each perfect squares.

30. Is the statement in Exercise 29 true for cubes? For any fixed power? Justify your conclusion.

In Exercises 31 through 33, let $U = \{1, 2, 3 \ldots, 1689\}$, $A = \{x \mid x \in U \text{ and } 3 \mid x\}$, $B = \{y \mid y \in U \text{ and } 5 \mid y\}$, and $C = \{z \mid z \in U \text{ and } 11 \mid z\}$. Compute each of the following.

31. (a) $|A|$ (b) $|B|$ (c) $|C|$

32. (a) The number of elements in U that are divisible by 15

   (b) The number of elements of U that are divisible by 165

   (c) The number of elements of U that are divisible by 55

33. Use the results of Exercises 31 and 32 to compute each of the following.

   (a) $|A \cup B|$ (b) $|A \cup B \cup C|$

1.5   MATRICES

A **matrix** is a rectangular array of numbers arranged in m horizontal **rows** and n vertical **columns**:

$$\mathbf{A} = \begin{bmatrix} a_{11} & a_{12} & \cdots & a_{1n} \\ a_{21} & a_{22} & \cdots & a_{2n} \\ \vdots & \vdots & & \vdots \\ a_{m1} & a_{m2} & \cdots & a_{mn} \end{bmatrix} \tag{1}$$

The **ith row** of $\mathbf{A}$ is $\begin{bmatrix} a_{i1} & a_{i2} & \cdots & a_{in} \end{bmatrix}$, $1 \le i \le m$, and the **jth column** of $\mathbf{A}$ is

$$\begin{bmatrix} a_{1j} \\ a_{2j} \\ \vdots \\ a_{mj} \end{bmatrix}, 1 \le j \le n.$$

We say that $\mathbf{A}$ is **m by n**, written $m \times n$. If $m = n$, we say $\mathbf{A}$ is a **square matrix** of order n and that the numbers $a_{11}, a_{22}, \ldots, a_{nn}$ form the **main diagonal** of $\mathbf{A}$. We refer to the number a_{ij}, which is in the ith row and jth column of $\mathbf{A}$ as the **i, jth element of $\mathbf{A}$** or as the **(i, j) entry of $\mathbf{A}$**, and we often write (1) as $\mathbf{A} = \begin{bmatrix} a_{ij} \end{bmatrix}$.

EXAMPLE 1   Let

$$\mathbf{A} = \begin{bmatrix} 2 & 3 & 5 \\ 0 & -1 & 2 \end{bmatrix}, \quad \mathbf{B} = \begin{bmatrix} 2 & 3 \\ 4 & 6 \end{bmatrix}, \quad \mathbf{C} = \begin{bmatrix} 1 & -1 & 3 & 4 \end{bmatrix}$$

$$\mathbf{D} = \begin{bmatrix} -1 \\ 2 \\ 0 \end{bmatrix}, \quad \text{and} \quad \mathbf{E} = \begin{bmatrix} 1 & 0 & -1 \\ -1 & 2 & 3 \\ 2 & 4 & 5 \end{bmatrix}.$$

Then $\mathbf{A}$ is 2×3 with $a_{12} = 3$ and $a_{23} = 2$, $\mathbf{B}$ is 2×2 with $b_{21} = 4$, $\mathbf{C}$ is 1×4, $\mathbf{D}$ is 3×1, and $\mathbf{E}$ is 3×3. ■

A square matrix $\mathbf{A} = \begin{bmatrix} a_{ij} \end{bmatrix}$ for which every entry off the main diagonal is zero, that is, $a_{ij} = 0$ for $i \ne j$, is called a **diagonal matrix**.

EXAMPLE 2   Each of the following is a diagonal matrix.

$$\mathbf{F} = \begin{bmatrix} 4 & 0 \\ 0 & 3 \end{bmatrix}, \quad \mathbf{G} = \begin{bmatrix} 2 & 0 & 0 \\ 0 & -3 & 0 \\ 0 & 0 & 5 \end{bmatrix}, \quad \text{and} \quad \mathbf{H} = \begin{bmatrix} 0 & 0 & 0 \\ 0 & 7 & 0 \\ 0 & 0 & 6 \end{bmatrix} \quad ■$$

Matrices are used in many applications in computer science, and we shall see them in our study of relations and graphs. At this point we present the following simple application showing how matrices can be used to display data in a tabular form.

EXAMPLE 3   The following matrix gives the airline distances between the cities indicated.

$$
\begin{array}{c}
\\
\text{London} \\
\text{Madrid} \\
\text{New York} \\
\text{Tokyo}
\end{array}
\begin{array}{cccc}
\text{London} & \text{Madrid} & \text{New York} & \text{Tokyo} \\
\left[\begin{array}{cccc}
0 & 785 & 3469 & 5959 \\
785 & 0 & 3593 & 6706 \\
3469 & 3593 & 0 & 6757 \\
5959 & 6706 & 6757 & 0
\end{array}\right]
\end{array}
$$
∎

Two $m \times n$ matrices $\mathbf{A} = \left[\, a_{ij} \,\right]$ and $\mathbf{B} = \left[\, b_{ij} \,\right]$ are said to be **equal** if $a_{ij} = b_{ij}$, $1 \le i \le m$, $1 \le j \le n$; that is, if corresponding elements are the same. Notice how easy it is to state the definition using generic elements a_{ij}, b_{ij}.

EXAMPLE 4   If

$$
\mathbf{A} = \begin{bmatrix} 2 & -3 & -1 \\ 0 & 5 & 2 \\ 4 & -4 & 6 \end{bmatrix} \quad \text{and} \quad \mathbf{B} = \begin{bmatrix} 2 & x & -1 \\ y & 5 & 2 \\ 4 & -4 & z \end{bmatrix},
$$

then $\mathbf{A} = \mathbf{B}$ if and only if $x = -3$, $y = 0$, and $z = 6$. ∎

If $\mathbf{A} = \left[\, a_{ij} \,\right]$ and $\mathbf{B} = \left[\, b_{ij} \,\right]$ are $m \times n$ matrices, then the **sum** of $\mathbf{A}$ and $\mathbf{B}$ is the matrix $\mathbf{C} = \left[\, c_{ij} \,\right]$ defined by $c_{ij} = a_{ij} + b_{ij}$, $1 \le i \le m$, $1 \le j \le n$. That is, $\mathbf{C}$ is obtained by adding the corresponding elements of $\mathbf{A}$ and $\mathbf{B}$. Once again the use of generic elements makes it easy to state the definition.

EXAMPLE 5   Let $\mathbf{A} = \begin{bmatrix} 3 & 4 & -1 \\ 5 & 0 & -2 \end{bmatrix}$ and $\mathbf{B} = \begin{bmatrix} 4 & 5 & 3 \\ 0 & -3 & 2 \end{bmatrix}$. Then

$$
\mathbf{A} + \mathbf{B} = \begin{bmatrix} 3+4 & 4+5 & -1+3 \\ 5+0 & 0+(-3) & -2+2 \end{bmatrix} = \begin{bmatrix} 7 & 9 & 2 \\ 5 & -3 & 0 \end{bmatrix}.
$$
∎

Observe that the sum of the matrices $\mathbf{A}$ and $\mathbf{B}$ is defined only when $\mathbf{A}$ and $\mathbf{B}$ have the same number of rows and the same number of columns. We agree to write $\mathbf{A} + \mathbf{B}$ only when the sum is defined.

A matrix all of whose entries are zero is called a **zero matrix** and is denoted by $\mathbf{0}$.

EXAMPLE 6   Each of the following is a zero matrix.

$$
\begin{bmatrix} 0 & 0 \\ 0 & 0 \end{bmatrix} \qquad \begin{bmatrix} 0 & 0 & 0 \\ 0 & 0 & 0 \end{bmatrix} \qquad \begin{bmatrix} 0 & 0 & 0 \\ 0 & 0 & 0 \\ 0 & 0 & 0 \end{bmatrix}
$$
∎

The following theorem gives some basic properties of matrix addition; the proofs are omitted.

Theorem 1   (a) $\mathbf{A} + \mathbf{B} = \mathbf{B} + \mathbf{A}$.

(b) $(\mathbf{A} + \mathbf{B}) + \mathbf{C} = \mathbf{A} + (\mathbf{B} + \mathbf{C})$.

(c) $\mathbf{A} + \mathbf{0} = \mathbf{0} + \mathbf{A} = \mathbf{A}$.   ●

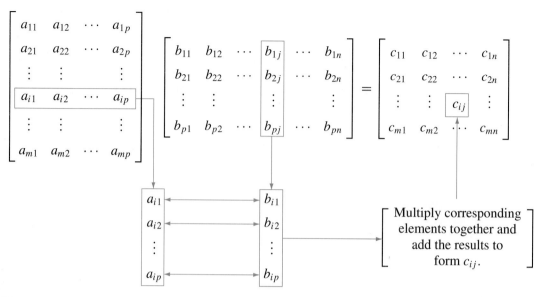

Figure 1.20

If $\mathbf{A} = \begin{bmatrix} a_{ij} \end{bmatrix}$ is an $m \times p$ matrix and $\mathbf{B} = \begin{bmatrix} b_{ij} \end{bmatrix}$ is a $p \times n$ matrix, then the **product** of $\mathbf{A}$ and $\mathbf{B}$, denoted $\mathbf{AB}$, is the $m \times n$ matrix $\mathbf{C} = \begin{bmatrix} c_{ij} \end{bmatrix}$ defined by

$$c_{ij} = a_{i1}b_{1j} + a_{i2}b_{2j} + \cdots + a_{ip}b_{pj} \qquad 1 \le i \le m, 1 \le j \le n. \qquad (2)$$

Let us explain (2) in more detail. The elements $a_{i1}, a_{i2}, \ldots, a_{ip}$ form the ith row of $\mathbf{A}$, and the elements $b_{1j}, b_{2j}, \ldots, b_{pj}$ form the jth column of $\mathbf{B}$. Then (2) states that for any i and j, the element c_{ij} of $\mathbf{C} = \mathbf{AB}$ can be computed in the following way, illustrated in Figure 1.20.

1. Select row i of $\mathbf{A}$ and column j of $\mathbf{B}$, and place them side by side.
2. Multiply corresponding entries and add all the products.

EXAMPLE 7   Let $\mathbf{A} = \begin{bmatrix} 2 & 3 & -4 \\ 1 & 2 & 3 \end{bmatrix}$ and $\mathbf{B} = \begin{bmatrix} 3 & 1 \\ -2 & 2 \\ 5 & -3 \end{bmatrix}$. Then

$$\mathbf{AB} = \begin{bmatrix} (2)(3) + (3)(-2) + (-4)(5) & (2)(1) + (3)(2) + (-4)(-3) \\ (1)(3) + (2)(-2) + (3)(5) & (1)(1) + (2)(2) + (3)(-3) \end{bmatrix}$$
$$= \begin{bmatrix} -20 & 20 \\ 14 & -4 \end{bmatrix}.$$

An **array of dimension two** is a modification of the idea of a matrix, in the same way that a linear array is a modification of the idea of a sequence. By an **$m \times n$ array A** we will mean an $m \times n$ matrix $\mathbf{A}$ of mn positions. We may assign numbers to these positions later, make further changes in these assignments, and still refer to the array as $\mathbf{A}$. This is a model for two-dimensional storage of information in a computer. The number assigned to row i and column j of an array $\mathbf{A}$ will be denoted $\mathbf{A}[i, j]$.

As we have seen, the properties of matrix addition resemble the familiar properties for the addition of real numbers. However, some of the properties of matrix multiplication do not resemble those of real number multiplication. First, observe that if $\mathbf{A}$ is an $m \times p$ matrix and $\mathbf{B}$ is a $p \times n$ matrix, then $\mathbf{AB}$ can be computed and is an $m \times n$ matrix. As for $\mathbf{BA}$, we have the following four possibilities:

1. $\mathbf{BA}$ may not be defined; we may have $n \neq m$.
2. $\mathbf{BA}$ may be defined and then $\mathbf{BA}$ is $p \times p$, while $\mathbf{AB}$ is $m \times m$ and $p \neq m$. Thus $\mathbf{AB}$ and $\mathbf{BA}$ are not equal.
3. $\mathbf{AB}$ and $\mathbf{BA}$ may both be the same size, but not be equal as matrices.
4. $\mathbf{AB} = \mathbf{BA}$.

We agree as before to write $\mathbf{AB}$ only when the product is defined.

EXAMPLE 8   Let $\mathbf{A} = \begin{bmatrix} 2 & 1 \\ 3 & -2 \end{bmatrix}$ and $\mathbf{B} = \begin{bmatrix} 1 & -1 \\ 2 & -3 \end{bmatrix}$. Then $\mathbf{AB} = \begin{bmatrix} 4 & -5 \\ -1 & 3 \end{bmatrix}$ and $\mathbf{BA} = \begin{bmatrix} -1 & 3 \\ -5 & 8 \end{bmatrix}$. ∎

The basic properties of matrix multiplication are given by the following theorem.

Theorem 2   (a) $\mathbf{A}(\mathbf{BC}) = (\mathbf{AB})\mathbf{C}$.
(b) $\mathbf{A}(\mathbf{B} + \mathbf{C}) = \mathbf{AB} + \mathbf{AC}$.
(c) $(\mathbf{A} + \mathbf{B})\mathbf{C} = \mathbf{AC} + \mathbf{BC}$. ●

The $n \times n$ diagonal matrix

$$\mathbf{I} = \begin{bmatrix} 1 & 0 & \cdots & 0 \\ 0 & 1 & \cdots & 0 \\ \vdots & \vdots & & \vdots \\ 0 & 0 & \cdots & 1 \end{bmatrix},$$

all of whose diagonal elements are 1, is called the **identity matrix** of order n. If $\mathbf{A}$ is an $m \times n$ matrix, it is easy to verify that $\mathbf{I}_m\mathbf{A} = \mathbf{AI}_n = \mathbf{A}$. If $\mathbf{A}$ is an $n \times n$ matrix and p is a positive integer, we define

$$\mathbf{A}^p = \underbrace{\mathbf{A} \cdot \mathbf{A} \cdots \mathbf{A}}_{p \text{ factors}} \quad \text{and} \quad \mathbf{A}^0 = \mathbf{I}_n.$$

If p and q are nonnegative integers, we can prove the following laws of exponents for matrices:

$$\mathbf{A}^p\mathbf{A}^q = \mathbf{A}^{p+q} \quad \text{and} \quad (\mathbf{A}^p)^q = \mathbf{A}^{pq}.$$

Observe that the rule $(\mathbf{AB})^p = \mathbf{A}^p\mathbf{B}^p$ does not hold for square matrices. However, if $\mathbf{AB} = \mathbf{BA}$, then $(\mathbf{AB})^p = \mathbf{A}^p\mathbf{B}^p$.

If $\mathbf{A} = \begin{bmatrix} a_{ij} \end{bmatrix}$ is an $m \times n$ matrix, then the $n \times m$ matrix $\mathbf{A}^T = \begin{bmatrix} a_{ij}^T \end{bmatrix}$, where $a_{ij}^T = a_{ji}$, $1 \le i \le m$, $1 \le j \le n$, is called the **transpose of A**. Thus the transpose of $\mathbf{A}$ is obtained by interchanging the rows and columns of $\mathbf{A}$.

EXAMPLE 9　Let $\mathbf{A} = \begin{bmatrix} 2 & -3 & 5 \\ 6 & 1 & 3 \end{bmatrix}$ and $\mathbf{B} = \begin{bmatrix} 3 & 4 & 5 \\ 2 & -1 & 0 \\ 1 & 6 & -2 \end{bmatrix}$. Then

$$\mathbf{A}^T = \begin{bmatrix} 2 & 6 \\ -3 & 1 \\ 5 & 3 \end{bmatrix} \quad \text{and} \quad \mathbf{B}^T = \begin{bmatrix} 3 & 2 & 1 \\ 4 & -1 & 6 \\ 5 & 0 & -2 \end{bmatrix}.$$

The following theorem summarizes the basic properties of the transpose operation.

Theorem 3　If $\mathbf{A}$ and $\mathbf{B}$ are matrices, then

(a) $(\mathbf{A}^T)^T = \mathbf{A}$.

(b) $(\mathbf{A} + \mathbf{B})^T = \mathbf{A}^T + \mathbf{B}^T$.

(c) $(\mathbf{AB})^T = \mathbf{B}^T \mathbf{A}^T$.

A matrix $\mathbf{A} = \begin{bmatrix} a_{ij} \end{bmatrix}$ is called **symmetric** if $\mathbf{A}^T = \mathbf{A}$. Thus, if $\mathbf{A}$ is symmetric, it must be a square matrix. It is easy to show that $\mathbf{A}$ is symmetric if and only if $a_{ij} = a_{ji}$. That is, $\mathbf{A}$ is symmetric if and only if the entries of $\mathbf{A}$ are symmetric with respect to the main diagonal of $\mathbf{A}$.

EXAMPLE 10　If $\mathbf{A} = \begin{bmatrix} 1 & 2 & -3 \\ 2 & 4 & 5 \\ -3 & 5 & 6 \end{bmatrix}$ and $\mathbf{B} = \begin{bmatrix} 1 & 2 & -3 \\ 2 & 4 & 0 \\ 3 & 2 & 1 \end{bmatrix}$, then $\mathbf{A}$ is symmetric and $\mathbf{B}$ is not symmetric.

Boolean Matrix Operations

A **Boolean matrix** is an $m \times n$ matrix whose entries are either zero or one. We shall now define three operations on Boolean matrices that have useful applications in Chapter 4.

Let $\mathbf{A} = \begin{bmatrix} a_{ij} \end{bmatrix}$ and $\mathbf{B} = \begin{bmatrix} b_{ij} \end{bmatrix}$ be $m \times n$ Boolean matrices. We define $\mathbf{A} \vee \mathbf{B} = \mathbf{C} = \begin{bmatrix} c_{ij} \end{bmatrix}$, the **join** of $\mathbf{A}$ and $\mathbf{B}$, by

$$c_{ij} = \begin{cases} 1 & \text{if } a_{ij} = 1 \text{ or } b_{ij} = 1 \\ 0 & \text{if } a_{ij} \text{ and } b_{ij} \text{ are both } 0 \end{cases}$$

and $\mathbf{A} \wedge \mathbf{B} = \mathbf{D} = \begin{bmatrix} d_{ij} \end{bmatrix}$, the **meet** of $\mathbf{A}$ and $\mathbf{B}$, by

$$d_{ij} = \begin{cases} 1 & \text{if } a_{ij} \text{ and } b_{ij} \text{ are both } 1 \\ 0 & \text{if } a_{ij} = 0 \text{ or } b_{ij} = 0. \end{cases}$$

Note that these operations are only possible when **A** and **B** have the same size, just as in the case of matrix addition. Instead of adding corresponding elements in **A** and **B**, to compute the entries of the result, we simply examine the corresponding elements for particular patterns.

EXAMPLE 11   Let $\mathbf{A} = \begin{bmatrix} 1 & 0 & 1 \\ 0 & 1 & 1 \\ 1 & 1 & 0 \\ 0 & 0 & 0 \end{bmatrix}$ and $\mathbf{B} = \begin{bmatrix} 1 & 1 & 0 \\ 1 & 0 & 1 \\ 0 & 0 & 1 \\ 1 & 1 & 0 \end{bmatrix}$.

(a) Compute $\mathbf{A} \vee \mathbf{B}$.   (b) Compute $\mathbf{A} \wedge \mathbf{B}$.

Solution

(a) Let $\mathbf{A} \vee \mathbf{B} = \begin{bmatrix} c_{ij} \end{bmatrix}$. Then, since a_{43} and b_{43} are both 0, we see that $c_{43} = 0$. In all other cases, either a_{ij} or b_{ij} is 1, so c_{ij} is also 1. Thus

$$\mathbf{A} \vee \mathbf{B} = \begin{bmatrix} 1 & 1 & 1 \\ 1 & 1 & 1 \\ 1 & 1 & 1 \\ 1 & 1 & 0 \end{bmatrix}.$$

(b) Let $\mathbf{A} \wedge \mathbf{B} = \begin{bmatrix} d_{ij} \end{bmatrix}$. Then, since a_{11} and b_{11} are both 1, $d_{11} = 1$, and since a_{23} and b_{23} are both 1, $d_{23} = 1$. In all other cases, either a_{ij} or b_{ij} is 0, so $d_{ij} = 0$. Thus

$$A \wedge B = \begin{bmatrix} 1 & 0 & 0 \\ 0 & 0 & 1 \\ 0 & 0 & 0 \\ 0 & 0 & 0 \end{bmatrix}. \qquad \blacksquare$$

Finally, suppose that $\mathbf{A} = \begin{bmatrix} a_{ij} \end{bmatrix}$ is an $m \times p$ Boolean matrix and $\mathbf{B} = \begin{bmatrix} b_{ij} \end{bmatrix}$ is a $p \times n$ Boolean matrix. Notice that the condition on the sizes of **A** and **B** is exactly the condition needed to form the matrix product **AB**. We now define another kind of product.

The **Boolean product** of **A** and **B**, denoted $\mathbf{A} \odot \mathbf{B}$, is the $m \times n$ Boolean matrix $\mathbf{C} = \begin{bmatrix} c_{ij} \end{bmatrix}$ defined by

$$c_{ij} = \begin{cases} 1 & \text{if } a_{ik} = 1 \text{ and } b_{kj} = 1 \text{ for some } k, 1 \le k \le p \\ 0 & \text{otherwise.} \end{cases}$$

This multiplication is similar to ordinary matrix multiplication. The preceding formula states that for any i and j the element c_{ij} of $\mathbf{C} = \mathbf{A} \odot \mathbf{B}$ can be computed in the following way, as illustrated in Figure 1.21. (Compare this with Figure 1.20.)

1. Select row i of **A** and column j of **B**, and arrange them side by side.
2. Compare corresponding entries. If even a single pair of corresponding entries consists of two 1's, then $c_{ij} = 1$. If this is not the case, then $c_{ij} = 0$.

We can easily perform the indicated comparisons and checks for each position of the Boolean product. Thus, at least for human beings, the computation of elements in $\mathbf{A} \odot \mathbf{B}$ is considerably easier than the computation of elements in **AB**.

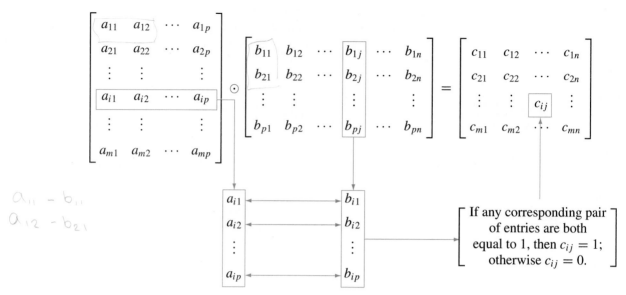

Figure 1.21

$a_{11} - b_{11}$
$a_{12} - b_{21}$

EXAMPLE 12　Let $\mathbf{A} = \begin{bmatrix} 1 & 1 & 0 \\ 0 & 1 & 0 \\ 1 & 1 & 0 \\ 0 & 0 & 1 \end{bmatrix}$ and $\mathbf{B} = \begin{bmatrix} 1 & 0 & 0 & 0 \\ 0 & 1 & 1 & 0 \\ 1 & 0 & 1 & 1 \end{bmatrix}$. Compute $\mathbf{A} \odot \mathbf{B}$.

down to over

Solution　Let $\mathbf{A} \odot \mathbf{B} = \begin{bmatrix} e_{ij} \end{bmatrix}$. Then $e_{11} = 1$, since row 1 of $\mathbf{A}$ and column 1 of $\mathbf{B}$ each have a 1 as the first entry. Similarly, $e_{12} = 1$, since $a_{12} = 1$ and $b_{22} = 1$; that is, the first row of $\mathbf{A}$ and the second column of $\mathbf{B}$ have a 1 in the second position. In a similar way we see that $e_{13} = 1$. On the other hand, $e_{14} = 0$, since row 1 of $\mathbf{A}$ and column 4 of $\mathbf{B}$ do not have common 1's in any position. Proceeding in this way, we obtain

$$\mathbf{A} \odot \mathbf{B} = \begin{bmatrix} 1 & 1 & 1 & 0 \\ 0 & 1 & 1 & 0 \\ 1 & 1 & 1 & 0 \\ 1 & 0 & 1 & 1 \end{bmatrix}.$$

The following theorem, whose proof is left as an exercise, summarizes the basic properties of the Boolean matrix operations just defined.

Theorem 4　If $\mathbf{A}$, $\mathbf{B}$, and $\mathbf{C}$ are Boolean matrices of compatible sizes, then

1. (a) $\mathbf{A} \vee \mathbf{B} = \mathbf{B} \vee \mathbf{A}$.

   (b) $\mathbf{A} \wedge \mathbf{B} = \mathbf{B} \wedge \mathbf{A}$.

2. (a) $(\mathbf{A} \vee \mathbf{B}) \vee \mathbf{C} = \mathbf{A} \vee (\mathbf{B} \vee \mathbf{C})$.

   (b) $(\mathbf{A} \wedge \mathbf{B}) \wedge \mathbf{C} = \mathbf{A} \wedge (\mathbf{B} \wedge \mathbf{C})$.

3. (a) $\mathbf{A} \wedge (\mathbf{B} \vee \mathbf{C}) = (\mathbf{A} \wedge \mathbf{B}) \vee (\mathbf{A} \wedge \mathbf{C})$.

   (b) $\mathbf{A} \vee (\mathbf{B} \wedge \mathbf{C}) = (\mathbf{A} \vee \mathbf{B}) \wedge (\mathbf{A} \vee \mathbf{C})$.

4. $(\mathbf{A} \odot \mathbf{B}) \odot \mathbf{C} = \mathbf{A} \odot (\mathbf{B} \odot \mathbf{C})$.

1.5 Exercises

1. Let $\mathbf{A} = \begin{bmatrix} 3 & -2 & 5 \\ 4 & 1 & 2 \end{bmatrix}$, $\mathbf{B} = \begin{bmatrix} 3 \\ -2 \\ 4 \end{bmatrix}$, and

   $\mathbf{C} = \begin{bmatrix} 2 & 3 & 4 \\ 5 & 6 & -1 \\ 2 & 0 & 8 \end{bmatrix}$.

   (a) What is a_{12}, a_{22}, a_{23}?

   (b) What is b_{11}, b_{31}?

   (c) What is c_{13}, c_{23}, c_{33}?

   (d) List the elements on the main diagonal of C.

2. Which of the following are diagonal matrices?

   (a) $\mathbf{A} = \begin{bmatrix} 2 & 3 \\ 0 & 0 \end{bmatrix}$ (b) $\mathbf{B} = \begin{bmatrix} 3 & 0 & 0 \\ 0 & -2 & 0 \\ 0 & 0 & 5 \end{bmatrix}$

   (c) $\mathbf{C} = \begin{bmatrix} 0 & 0 & 0 \\ 0 & 0 & 0 \\ 0 & 0 & 0 \end{bmatrix}$

   (d) $\mathbf{D} = \begin{bmatrix} 2 & 6 & -2 \\ 0 & -1 & 0 \\ 0 & 0 & 3 \end{bmatrix}$

   (e) $\mathbf{E} = \begin{bmatrix} 4 & 0 & 0 \\ 0 & 4 & 0 \\ 0 & 0 & 4 \end{bmatrix}$

3. If $\begin{bmatrix} a+b & c+d \\ c-d & a-b \end{bmatrix} = \begin{bmatrix} 4 & 6 \\ 10 & 2 \end{bmatrix}$, find $a, b, c,$ and d.

4. If $\begin{bmatrix} a+2b & 2a-b \\ 2c+d & c-2d \end{bmatrix} = \begin{bmatrix} 4 & -2 \\ 4 & -3 \end{bmatrix}$, find $a, b, c,$ and d.

In Exercises 5 through 10, let

$$\mathbf{A} = \begin{bmatrix} 2 & 1 & 3 \\ 4 & 1 & -2 \end{bmatrix}, \quad \mathbf{B} = \begin{bmatrix} 0 & 1 \\ 1 & 2 \\ 2 & 3 \end{bmatrix},$$

$$\mathbf{C} = \begin{bmatrix} 1 & -2 & 3 \\ 4 & 2 & 5 \\ 3 & 1 & 2 \end{bmatrix}, \quad \mathbf{D} = \begin{bmatrix} -3 & 2 \\ 4 & 1 \end{bmatrix},$$

$$\mathbf{E} = \begin{bmatrix} 3 & 2 & -1 \\ 5 & 4 & -3 \\ 0 & 1 & 2 \end{bmatrix}, \quad \mathbf{F} = \begin{bmatrix} -2 & 3 \\ 4 & 5 \end{bmatrix}.$$

5. If possible, compute each of the following.

   (a) $\mathbf{C} + \mathbf{E}$ (b) $\mathbf{AB}$

   (c) $\mathbf{CB} + \mathbf{F}$ (d) $\mathbf{AB} + \mathbf{DF}$

6. If possible, compute each of the following.

   (a) $\mathbf{A(BD)}$ and $\mathbf{(AB)D}$

   (b) $\mathbf{A(C + E)}$ and $\mathbf{AC} + \mathbf{AE}$

   (c) $\mathbf{FD} + \mathbf{AB}$

7. If possible, compute each of the following.

   (a) $\mathbf{EB} + \mathbf{FA}$

   (b) $\mathbf{A(B + D)}$ and $\mathbf{AB} + \mathbf{AD}$

   (c) $\mathbf{(F + D)A}$ (d) $\mathbf{AC} + \mathbf{DE}$

8. If possible, compute each of the following.

   (a) $\mathbf{A}^T$ and $(\mathbf{A}^T)^T$

   (b) $(\mathbf{C} + \mathbf{E})^T$ and $\mathbf{C}^T + \mathbf{E}^T$

   (c) $(\mathbf{AB})^T$ and $\mathbf{B}^T \mathbf{A}^T$ (d) $(\mathbf{B}^T \mathbf{C}) + \mathbf{A}$

9. If possible, compute each of the following.

   (a) $\mathbf{A}^T (\mathbf{D} + \mathbf{F})$ (b) $(\mathbf{BC})^T$ and $\mathbf{C}^T \mathbf{B}^T$

   (c) $(\mathbf{B}^T + \mathbf{A})\mathbf{C}$ (d) $(\mathbf{D}^T + \mathbf{E})\mathbf{F}$

10. Compute $\mathbf{D}^3$.

11. Let $\mathbf{A}$ be an $m \times n$ matrix. Show that $\mathbf{I}_m \mathbf{A} = \mathbf{A} \mathbf{I}_n = \mathbf{A}$. (*Hint:* Choose a generic element of $\mathbf{I}_m \mathbf{A}$.)

12. Let $\mathbf{A} = \begin{bmatrix} 2 & 1 \\ 3 & -2 \end{bmatrix}$ and $\mathbf{B} = \begin{bmatrix} -1 & 2 \\ 3 & 4 \end{bmatrix}$. Show that $\mathbf{AB} \neq \mathbf{BA}$.

13. Let $\mathbf{A} = \begin{bmatrix} 3 & 0 & 0 \\ 0 & -2 & 0 \\ 0 & 0 & 4 \end{bmatrix}$.

    (a) Compute $\mathbf{A}^3$.   (b) What is $\mathbf{A}^k$?

14. Show that $\mathbf{A0} = \mathbf{0}$ for any matrix $\mathbf{A}$.

15. Show that $\mathbf{I}_n^T = \mathbf{I}_n$.

16. (a) Show that if $\mathbf{A}$ has a row of zeros, then $\mathbf{AB}$ has a corresponding row of zeros. (*Hint:* Use the generic element definition of $\mathbf{AB}$ given in this section.)

(b) Show that if **B** has a column of zeros, then **AB** has a corresponding column of zeros.

17. Show that the jth column of the matrix product **AB** is equal to the matrix product $\mathbf{AB}_j$, where $\mathbf{B}_j$ is the jth column of **B**.

18. If **0** is the 2×2 zero matrix, find two 2×2 matrices **A** and **B**, with $\mathbf{A} \neq \mathbf{0}$ and $\mathbf{B} \neq \mathbf{0}$, such that $\mathbf{AB} = \mathbf{0}$.

19. If $\mathbf{A} = \begin{bmatrix} 0 & 1 \\ 1 & 0 \end{bmatrix}$, show that $\mathbf{A}^2 = \mathbf{I}_2$.

20. Determine all 2×2 matrices $\mathbf{A} = \begin{bmatrix} 0 & a \\ b & c \end{bmatrix}$ such that $\mathbf{A}^2 = \mathbf{I}_2$.

21. Let **A** and **B** be symmetric matrices.
(a) Show that $\mathbf{A} + \mathbf{B}$ is also symmetric.
(b) Is **AB** also symmetric?

22. Let **A** be an $n \times n$ matrix.
(a) Show that $\mathbf{AA}^T$ and $\mathbf{A}^T\mathbf{A}$ are symmetric.
(b) Show that $\mathbf{A} + \mathbf{A}^T$ is symmetric.

23. Prove Theorem 3. (*Hint*: For part (c), show that the i, jth element of $(\mathbf{AB})^T$ equals the i, jth element of $\mathbf{B}^T\mathbf{A}^T$.)

In Exercises 24 and 25, compute $\mathbf{A} \vee \mathbf{B}$, $\mathbf{A} \wedge \mathbf{B}$, *and* $\mathbf{A} \odot \mathbf{B}$ *for the given matrices* **A** *and* **B**.

24. (a) $\mathbf{A} = \begin{bmatrix} 1 & 0 \\ 0 & 1 \end{bmatrix}$, $\mathbf{B} = \begin{bmatrix} 1 & 1 \\ 0 & 1 \end{bmatrix}$

(b) $\mathbf{A} = \begin{bmatrix} 1 & 1 \\ 0 & 1 \end{bmatrix}$, $\mathbf{B} = \begin{bmatrix} 0 & 0 \\ 1 & 1 \end{bmatrix}$

(c) $\mathbf{A} = \begin{bmatrix} 1 & 1 \\ 1 & 1 \end{bmatrix}$, $\mathbf{B} = \begin{bmatrix} 0 & 0 \\ 1 & 0 \end{bmatrix}$

25. (a) $\mathbf{A} = \begin{bmatrix} 1 & 0 & 0 \\ 0 & 1 & 1 \\ 1 & 0 & 0 \end{bmatrix}$, $\mathbf{B} = \begin{bmatrix} 1 & 1 & 1 \\ 0 & 0 & 1 \\ 1 & 0 & 1 \end{bmatrix}$

(b) $\mathbf{A} = \begin{bmatrix} 0 & 0 & 1 \\ 1 & 1 & 0 \\ 1 & 0 & 0 \end{bmatrix}$, $\mathbf{B} = \begin{bmatrix} 0 & 1 & 1 \\ 1 & 1 & 0 \\ 1 & 0 & 1 \end{bmatrix}$

(c) $\mathbf{A} = \begin{bmatrix} 1 & 0 & 0 \\ 0 & 0 & 1 \\ 1 & 0 & 1 \end{bmatrix}$, $\mathbf{B} = \begin{bmatrix} 1 & 1 & 1 \\ 1 & 1 & 1 \\ 1 & 0 & 0 \end{bmatrix}$

26. Complete the following proofs.
(a) $\mathbf{A} \vee \mathbf{A} = \mathbf{A}$. Proof: Let b_{ij} be an element of $\mathbf{A} \vee \mathbf{A}$. If $b_{ij} = 0$, then $a_{ij} = $ _____, because _____. If $b_{ij} = 1$, then $a_{ij} = $ _____ because _____. Hence $b_{ij} = a_{ij}$ for each i, j pair.

(b) $\mathbf{A} \wedge \mathbf{A} = \mathbf{A}$. Proof: Let b_{ij} be an element of $\mathbf{A} \wedge \mathbf{A}$. If $b_{ij} = 0$, then _____. If $b_{ij} = 1$, then _____. (Explain.) Hence $b_{ij} = a_{ij}$ for each i, j pair.

27. Show that $\mathbf{A} \vee \mathbf{B} = \mathbf{B} \vee \mathbf{A}$.

28. Show that $\mathbf{A} \wedge \mathbf{B} = \mathbf{B} \wedge \mathbf{A}$.

29. Show that $\mathbf{A} \vee (\mathbf{B} \vee \mathbf{C}) = (\mathbf{A} \vee \mathbf{B}) \vee \mathbf{C}$.

30. Show that $\mathbf{A} \wedge (\mathbf{B} \wedge \mathbf{C}) = (\mathbf{A} \wedge \mathbf{B}) \wedge \mathbf{C}$.

31. Show that $A \odot (\mathbf{B} \odot \mathbf{C}) = (\mathbf{A} \odot \mathbf{B}) \odot \mathbf{C}$.

32. Show that $A \wedge (\mathbf{B} \vee \mathbf{C}) = (\mathbf{A} \wedge \mathbf{B}) \vee (\mathbf{A} \wedge \mathbf{C})$.

33. Show that $\mathbf{A} \vee (\mathbf{B} \wedge \mathbf{C}) = (\mathbf{A} \vee \mathbf{B}) \wedge (\mathbf{A} \vee \mathbf{C})$.

34. What fact does Example 8 illustrate?

35. Let $\mathbf{A} = \begin{bmatrix} a_{ij} \end{bmatrix}$ and $\mathbf{B} = \begin{bmatrix} b_{ij} \end{bmatrix}$ be two $n \times n$ matrices and let $\mathbf{C} = \begin{bmatrix} c_{ij} \end{bmatrix}$ represent **AB**. Prove that if k is an integer and $k \mid a_{ij}$ for all i, j, then $k \mid c_{ij}$ for all i, j.

36. Let p be a prime number with $p > 2$, and let **A** and **B** be matrices all of whose entries are integers. Suppose that p divides all the entries of $\mathbf{A} + \mathbf{B}$ and all the entries of $\mathbf{A} - \mathbf{B}$. Prove that p divides all the entries of **A** and all the entries of **B**.

1.6 MATHEMATICAL STRUCTURES

A situation we have seen several times in this chapter, and will see many more times in later chapters, is the following. A new kind of mathematical object is defined; for example, a set or a matrix. Then notation is introduced for representing the new type of object and a way to determine whether two objects are the same is described. Usually the next topic is ways to classify objects of the new type; for example, finite or infinite for sets, and Boolean or symmetric for matrices. Then operations are defined for the objects and the properties of these operations are examined.

A collection of objects with operations defined on them and the accompanying properties form a **mathematical structure** or **system**. In this book we deal only with discrete mathematical structures.

EXAMPLE 1   The collection of sets with the operations of union, intersection, and complement and their accompanying properties is a (discrete) mathematical structure. We denote this structure by [sets, $\cup$, $\cap$, $^-$]. ∎

EXAMPLE 2   The collection of 3×3 matrices with the operations of addition, multiplication, and transpose is a mathematical structure denoted by [3×3 matrices, $+$, $*$, ^{T}]. ∎

An important property we have not identified before is closure. A structure is **closed with respect to** an operation if that operation always produces another member of the collection of objects.

EXAMPLE 3   The structure [5×5 matrices, $+$, $*$, ^{T}] is closed with respect to addition because the sum of two 5×5 matrices is another 5×5 matrix. ∎

EXAMPLE 4   The structure [odd integers, $+$, $*$] is not closed with respect to addition. The sum of two odd integers is an even integer. This structure does have the closure property for multiplication, since the product of two odd numbers is an odd number. ∎

An operation that combines two objects is a **binary operation**. An operation that requires only one object is a **unary operation**. Binary operations often have similar properties, as we have seen earlier.

EXAMPLE 5   (a) Set intersection is a binary operation since it combines two sets to produce a new set.

(b) Producing the transpose of matrix is a unary operation. ∎

Common properties have been given names. For example, if the order of the objects does not affect the outcome of a binary operation, we say that the operation is **commutative**. That is, if $x \square y = y \square x$, where $\square$ is some binary operation, $\square$ is commutative.

EXAMPLE 6   (a) Join and meet for Boolean matrices are commutative operations.

$$\mathbf{A} \vee \mathbf{B} = \mathbf{B} \vee \mathbf{A} \quad \text{and} \quad \mathbf{A} \wedge \mathbf{B} = \mathbf{B} \wedge \mathbf{A}.$$

(b) Ordinary matrix multiplication is not a commutative operation. $\mathbf{AB} \neq \mathbf{BA}$. ∎

Note that when we say an operation has a property, this means that the statement of the property is true when the operation is used with any objects in the structure. If there is even one case when the statement is not true, the operation does not have that property. If $\square$ is a binary operation, then $\square$ is **associative** or **has the associative property** if

$$(x \square y) \square z = x \square (y \square z).$$

EXAMPLE 7   Set union is an associative operation, since $(A \cup B) \cup C = A \cup (B \cup C)$ is always true. ∎

If a mathematical structure has two binary operations, say $\square$ and $\triangledown$, a **distributive property** has the following pattern:

$$x \square (y \triangledown z) = (x \square y) \triangledown (x \square z).$$

EXAMPLE 8

(a) We are familiar with the distributive property for real numbers; if a, b, and c are real numbers, then $a \cdot (b + c) = a \cdot b + a \cdot c$. Note that because we have an agreement about real number arithmetic to multiply before adding, parentheses are not needed on the right-hand side.

(b) The structure [sets, $\cup$, $\cap$, $^-$] has two distributive properties:

$$A \cup (B \cap C) = (A \cup B) \cap (A \cup C)$$

and

$$A \cap (B \cup C) = (A \cap B) \cup (A \cap C). \qquad \blacksquare$$

Several of the structures we have seen have a unary operation and two binary operations. For such structures we can ask whether De Morgan's laws are properties of the system. If the unary operation is $*$ and the binary operations are $\square$ and $\triangledown$, then **De Morgan's laws** are

$$(x \square y)^* = x^* \triangledown y^* \quad \text{and} \quad (x \triangledown y)^* = x^* \square y^*.$$

EXAMPLE 9

(a) As we saw in Section 1.2, sets satisfy De Morgan's laws for union, intersection, and complement: $\overline{(A \cup B)} = \overline{A} \cap \overline{B}$ and $\overline{(A \cap B)} = \overline{A} \cup \overline{B}$.

(b) The structure [real numbers, $+$, $*$, $\sqrt{\ }$] does not satisfy De Morgan's laws, since $\sqrt{x + y} \neq \sqrt{x} * \sqrt{y}$. $\qquad \blacksquare$

A structure with a binary operation $\square$ may contain a distinguished object e, with the property $x \square e = e \square x = x$ for all x in the collection. We call e an **identity for** $\square$. In fact, an identity for an operation must be unique.

Theorem 1   If e is an identity for a binary operation $\square$, then e is unique. $\qquad \bullet$

Proof   Assume another object i also has the identity property, so $x \square i = i \square x = x$. Then $e \square i = e$, but since e is an identity for $\square$, $i \square e = e \square i = i$. Thus, $i = e$. There is at most one object with the identity property for $\square$. $\qquad \blacktriangledown$

This is one of our first examples of a proof that does not proceed directly. We assumed that there were two identity elements and showed that they were in fact the same element.

EXAMPLE 10   For [$n \times n$ matrices, $+$, $*$, ^{T}], $\mathbf{I}_n$ is the identity for matrix multiplication and the $n \times n$ zero matrix is the identity for matrix addition. $\qquad \blacksquare$

If a binary operation $\square$ has an identity e, we say y is a $\square$-**inverse** of x if $x \square y = y \square x = e$.

Theorem 2   If □ is an associative operation and x has a □-inverse y, then y is unique.   ●

Proof   Assume there is another □-inverse for x, say z. Then $(z \square x) \square y = e \square y = y$ and $z \square (x \square y) = z \square e = z$. Since □ is associative, $(z \square x) \square y = z \square (x \square y)$ and so $y = z$.   ▼

EXAMPLE 11

(a) In the structure [3×3 matrices, $+, *,^T$], each matrix $\mathbf{A} = \left[\, a_{ij} \,\right]$ has a $+$-inverse, or additive inverse, $-\mathbf{A} = \left[\, -a_{ij} \,\right]$.

(b) In the structure [integers, $+, *$], only the integers 1 and -1 have multiplicative inverses.   ■

EXAMPLE 12

Let □, ▽, and $*$ be defined for the set $\{0, 1\}$ by the following tables.

□	0	1
0	0	1
1	1	0

▽	0	1
0	0	0
1	0	1

x^*	x
0	1
1	0

Thus $1 \square 0 = 1$, $0 \bigtriangledown 1 = 0$, and $1^* = 0$. Determine if each of the following is true for [$\{0, 1\}$, □, ▽, $*$].

(a) □ is commutative.   (b) ▽ is associative.

(c) De Morgan's laws hold.

(d) Two distributive properties hold for the structure.

Solution

(a) The statement $x \square y = y \square x$ must be true for all choices of x and y. Here there is only one case to check: Is $0 \square 1 = 1 \square 0$ true? Since both $0 \square 1$ and $1 \square 0$ are 1, □ is commutative.

(b) The eight possible cases to be checked are left as an exercise. See Exercise 6(b).

(c) $(0 \square 0)^* = 0^* = 1$ $0^* \bigtriangledown 0^* = 1 \bigtriangledown 1 = 1.$
$(0 \square 1)^* = 1^* = 0$ $0^* \bigtriangledown 1^* = 1 \bigtriangledown 0 = 0.$
$(1 \square 1)^* = 0^* = 1$ $1^* \bigtriangledown 1^* = 0 \bigtriangledown 0 = 0.$

The last pair shows that De Morgan's laws do not hold in this structure.

(d) One possible distributive property is $x \square (y \bigtriangledown z) = (x \square y) \bigtriangledown (x \square z)$. We must check all possible cases. One way to organize this is shown in a table.

x	y	z	$y \bigtriangledown z$	$x \square (y \bigtriangledown z)$	$x \square y$	$x \square z$	$(x \square y) \bigtriangledown (x \square z)$
0	0	0	0	0	0	0	0
0	0	1	0	0	0	1	0
0	1	0	0	0	1	0	0
0	1	1	1	1	1	1	1
1	0	0	0	1	1	1	1
1	0	1	0	1	1	0	0
1	1	0	0	1	0	1	0
1	1	1	1	0	0	0	0
				(A)			(B)

Since columns (A) and (B) are not identical, this possible distributive property does not hold in this structure. The check for the other distributive property is Exercise 7. ∎

In later sections, we will find it useful to consider mathematical structures themselves as objects and to classify them according to the properties associated with their operations.

1.6 Exercises

In Exercises 1 and 2, tell whether the structure has the closure property with respect to the operation.

1. (a) [sets, ∪, ∩, ⁻]   union

    (b) [sets, ∪, ∩, ⁻]   complement

2. (a) [4 × 4 matrices, +, *, ^{T}]   multiplication

    (b) [3 × 5 matrices, +, *, ^{T}]   transpose

In Exercises 3 and 4, tell whether the structure has the closure property with respect to the operation.

3. (a) [integers, +, −, *, ÷]   division

    (b) [A^*, catenation]   catenation

4. (a) [$n × n$ Boolean matrices, ∨, ∧, ^{T}]   meet

    (b) [prime numbers, +, *]   addition

5. Show that ⊕ is a commutative operation for sets.

6. Using the definitions in Example 12, (a) show that □ is associative. (b) Show that ▽ is associative.

7. Using the definitions in Example 12, determine if the other possible distributive property holds.

8. Give the identity element, if one exists, for each binary operation in the given structure.

    (a) [real numbers, +, *, $\sqrt{}$]

    (b) [sets, ∪, ∩, ⁻]

    (c) [{0, 1}, □, ▽, *] as defined in Example 12

    (d) [subsets of a finite set A, ⊕, ⁻]

9. Give the identity element, if one exists, for each binary operation in the structure [5 × 5 Boolean matrices, ∨, ∧, ⊙].

*In Exercises 10 through 16, use the structure $S = [n × n$ diagonal matrices, +, *, ^{T}].*

10. Show that S is closed with respect to addition.

11. Show that S is closed with respect to multiplication.

12. Show that S is closed with respect to the transpose operation.

13. Does S have an identity for addition? If so, what is it?

14. Does S have an identity for multiplication? If so, what is it?

15. Let **A** be an $n × n$ diagonal matrix. Describe the additive inverse of **A**.

16. Let **A** be an $n × n$ diagonal matrix. Describe the multiplicative inverse of **A**.

In Exercises 17 through 23, use the structure $R = [M, +, *, ^T]$, where M is the set of matrices of the form $\begin{bmatrix} a & 0 \\ 0 & 0 \end{bmatrix}$, where a is a real number.

17. Show that R is closed with respect to addition.

18. Show that R is closed with respect to multiplication.

19. Show that R is closed with respect to the transpose operation.

20. Does R have an identity for addition? If so, what is it?

21. Does R have an identity for multiplication? If so, what is it?

22. Let **A** be an element of M. Describe the additive inverse for **A**.

23. Let **A** be an element of M. Describe the multiplicative inverse for **A**.

In Exercises 24 through 28, let $R = [Q, □]$, where $x □ y = \dfrac{x + y}{2}$. Determine which of the following properties hold for this structure:

24. Closure

25. Commutative

26. Associative

27. An identity element

28. An inverse for every element

29. Let $R = [2 \times 1 \text{ matrices}, \triangledown]$, where

$$\begin{bmatrix} x \\ y \end{bmatrix} \triangledown \begin{bmatrix} w \\ z \end{bmatrix} = \begin{bmatrix} x + w \\ y + z + 1 \end{bmatrix}.$$

Determine which of the following properties hold for this structure:

(a) Closure   (b) Commutative

(c) Associative

(d) An identity element

(e) An inverse for every element

30. For a Boolean matrix **B**, we define comp **B** to be the matrix formed by changing each 0 entry of **B** to 1 and each 1 entry of **B** to 0. Let $R = [5 \times 5 \text{ Boolean matrices}, \wedge, \vee, \text{comp}]$. Do De Morgan's laws hold for R? Justify your answer.

The properties of a mathematical structure can be used to rewrite expressions just as is done in ordinary algebra. In Exercises 31 through 34, rewrite the given expression to produce the requested result.

31. $(A \cup B) \cap (A \cup \overline{B})$ one set, no operations

32. $\overline{(A \cap B)} \cap A$ two sets, two operations

33. $\overline{(A \cup B)} \cup (\overline{A} \cap \overline{B})$ two sets, two operations

34. $\overline{(A \cup \overline{B})} \cap \overline{(\overline{A} \cup B)}$ one set, no operations

TIPS FOR PROOFS

Many exercises in this chapter ask that you show, prove, or verify a statement. To show or prove a statement means to give a written explanation demonstrating that the statement is always true. To verify a statement, in this book, means to check its truth for a particular case; see, for example, Section 1.2, Exercises 14 and 16.

Most proofs required in this chapter proceed directly from the given conditions using definitions and previously proven facts; an example is Section 1.4, Theorem 2. A powerful tool for constructing a proof is to choose a generic object of the type in the statement and to see what you know about this object. Remember that you must explain why the statement is always true, so choosing a specific object will only verify the statement for that object.

The most common way to show that two sets are equal is to show each is a subset of the other (Section 1.2, Theorem 1).

In proving statements about sets or matrix operations, try to work at the level of object names rather than at the element or entry-level. For example, Section 1.5, Exercise 22 is more easily proved by using the facts that if **A** is symmetric, then $\mathbf{A}^T = \mathbf{A}$ and Theorem 3 rather than by using the fact that if $\mathbf{A} = [a_{ij}]$ is symmetric, then $a_{ij} = a_{ji}$ for each i and j.

One other style of direct proof is seen in Section 1.6, Example 12. Sometimes we show the statement is always true by examining all possible cases.

KEY IDEAS FOR REVIEW

- Set: a well-defined collection of objects
- $\varnothing$ (empty set): the set with no elements
- Equal sets: sets with the same elements
- $A \subseteq B$ (A is a subset of B): Every element of A is an element of B.
- $|A|$ (cardinality of A): the number of elements of A
- Infinite set: see page 3
- $P(A)$ (power set of A): the set of all subsets of A
- $A \cup B$ (union of A and B): $\{x \mid x \in A \text{ or } x \in B\}$
- $A \cap B$ (intersection of A and B): $\{x \mid x \in A \text{ and } x \in B\}$
- Disjoint sets: two sets with no elements in common

- $A - B$ (complement of B with respect to A): $\{x \mid x \in A \text{ and } x \notin B\}$
- $\overline{A}$ (complement of A): $\{x \mid x \notin A\}$
- Algebraic properties of set operations: see pages 8–9
- Theorem (the addition principle): If A and B are finite sets, then $|A \cup B| = |A| + |B| - |A \cap B|$.
- Theorem (the three-set addition principle): If A, B, and C are finite sets, then $|A \cup B \cup C| = |A| + |B| + |C| - |A \cap B| - |A \cap C| - |B \cap C| + |A \cap B \cap C|$.
- Inclusion-Exclusion Principle: see page 10
- Sequence: list of objects arranged in a definite order

- Recursive formula: formula that uses previously defined terms
- Explicit formula: formula that does not use previously defined terms
- Linear array: see page 15
- Characteristic function of a set A: $f_A(x) = \begin{cases} 1 & \text{if } x \in A \\ 0 & \text{if } x \notin A \end{cases}$
- Countable set: a set that corresponds to a sequence
- Word: finite sequence of elements of A
- Regular expression: see page 18
- Theorem: If n and m are integers and $n > 0$, we can write $m = qn + r$ for integers q and r with $0 \leq r < n$. Moreover, there is just one way to do this.
- GCD(a, b): $d = $ GCD(a, b) if $d \mid a$, $d \mid b$, and d is the largest common divisor of a and b.
- Theorem: If d is GCD(a, b), then
  (a) $d = sa + tb$ for some integers s and t.
  (b) If $c \mid a$ and $c \mid b$, then $c \mid d$.
- Relatively prime: two integers a and b with GCD$(a, b) = 1$
- Euclidean algorithm: method used to find GCD(a, b); see page 24
- LCM(a, b): $c = $ LCM(a, b) if $a \mid c$, $b \mid c$, and c is the smallest common multiple of a and b
- GCD$(a, b) \cdot$ LCM$(a, b) = ab$
- mod-n function: $f_n(z) = r$, where r is the remainder when z is divided by n
- Matrix: rectangular array of numbers
- Size of a matrix: A is $m \times n$ if it has m rows and n columns
- Diagonal matrix: a square matrix with zero entries off the main diagonal
- Equal matrices: matrices of the same size whose corresponding entries are equal

- $A + B$: the matrix obtained by adding corresponding entries of A and B
- Zero matrix: a matrix all of whose entries are zero
- AB: see page 32
- I_n (identity matrix): a square matrix with ones on the diagonal and zeros elsewhere
- A^T: the matrix obtained from A by interchanging the rows and columns of A
- Symmetric matrix: $A^T = A$
- Array of dimension 2: see page 32
- Boolean matrix: a matrix whose entries are either one or zero
- $A \vee B$: see page 34
- $A \wedge B$: see page 34
- $A \odot B$: see page 35
- Properties of Boolean matrix operations: see page 36
- Mathematical structure: a collection of objects with operations defined on them and the accompanying properties
- Binary operation: an operation that combines two objects
- Unary operation: an operation that requires only one object
- Closure property: each application of the operation produces another object in the collection
- Associative property: $(x \square y) \square z = x \square (y \square z)$
- Distributive property: $x \square (y \triangledown z) = (x \square y) \triangledown (x \square z)$
- De Morgan's laws: $(x \square y)^* = x^* \triangledown y^*$ and $(x \triangledown y)^* = x^* \square y^*$
- Identity for $\square$: an element e such that $x \square e = e \square x = x$ for all x in the structure
- $\square$-inverse for x: an element y such that $x \square y = y \square x = e$, where e is the identity for $\square$

CODING EXERCISES

For each of the following, write the requested program or subroutine in pseudocode (as described in Appendix A) or in a programming language that you know. Test your code either with a paper-and-pencil trace or with a computer run.

In Exercises 1 through 3, assume that A and B are finite sets of integers. Write a subroutine to compute the specified set.

1. $A \cup B$

2. $A \cap B$

3. $A - B$

4. Consider the sequence recursively defined by $g(0) = 1$, $g(1) = -1$, $g(n) = 3g(n-1) - 2g(n-2)$.
   (a) Write a subroutine that will print the first 20 terms of the sequence.
   (b) Write a subroutine that will print the first n terms of the sequence. The user should be able to supply the value of n at runtime.

5. Write a subroutine to find the least common multiple of two positive integers.

CHAPTER 1 SELF-TEST

1. Let $A = \{x \mid x \text{ is a real number and } 0 < x < 1\}$,
   $B = \{x \mid x \text{ is a real number and } x^2 + 1 = 0\}$,
   $C = \{x \mid x = 4m, m \in Z\}$, $D = \{0, 2, 4, 6, \dots\}$, and
   $E = \{x \mid x \in Z \text{ and } x^2 \leq 100\}$.

   (a) Tell if each of the following is true or false.

       (i) $C \subseteq D$ (ii) $\{4, 16\} \subseteq C$
       (iii) $\{4, 16\} \subseteq E$ (iv) $D \subseteq D$
       (v) $B \subseteq \varnothing$

   (b) Tell if each of the following is true or false.

       (i) $C \cap E \nsubseteq (C \cup E)$
       (ii) $\varnothing \subseteq (A \cap B)$ (iii) $C \cap D = D$
       (iv) $C \cup E \subseteq D$
       (v) $A \cap D \subseteq A \cap C$

2. Let $A = \{x \mid x = 2n, n \in Z^+\}$,
   $B = \{x \mid x = 2n+1, n \in Z^+\}$, $C = \{x \mid x = 4n, n \in Z^+\}$,
   and $D = \{x \mid x(x^2 - 6x + 8) = 0, x \in Z\}$. Use Z as the
   universal set and find

   (a) $A \cup B$ (b) $\overline{A}$

   (c) $(A \cap D) \oplus (A \cap B)$ (d) $A \cup C$

   (e) $A - C$

3. Draw a Venn diagram to represent (a) $A \cap \overline{B}$ and
   (b) $\overline{A} \cap \overline{B}$.

4. Under what conditions will $A \cap B = A \cup B$?

5. Suppose that 109 of the 150 mathematics students at Verysmall College take at least one of the following computer languages: PASCAL, BASIC, C++. Suppose 45 study BASIC, 61 study PASCAL, 53 study C++, 18 study BASIC and PASCAL, 15 study BASIC and C++, and 23 study PASCAL and C++.

   (a) How many students study all three languages?

   (b) How many students study only BASIC?

   (c) How many students do not study any of the languages?

6. Define a sequence as follows: $a_0 = 0$, $a_1 = 0$,
   $a_n = 1 - 3a_{n-1} + 2a_{n-2}$. Compute the first six terms of
   this sequence.

7. Let $U = \{a, b, c, d, e, f, g, h, i, j\}$, $A = \{a, b, d, f\}$,
   $B = \{a, b, c, h, j\}$, $C = \{b, c, f, h, i\}$, and $D = \{g, h\}$.
   Represent each of the following sets by an array of zeros
   and ones.

   (a) $A \cup B$ (b) $A \cap B$

   (c) $A \cap (B \cup C)$ (d) $(\overline{A} \cap B) \cup D$

8. Let $I = \{a, b, c\}$. In each part that follows is listed a
   string in I^* and a regular expression over I. For each, state
   whether the string belongs to the regular set corresponding
   to the expression.

   (a) ab a^*bc^* (b) $acbb$ $((acb) \vee b)^*$

   (c) bc $((ab^*) \vee c)$ (d) $abaca$ $(ab)^*ac$

9. Use the Euclidean algorithm to compute $\text{GCD}(4389, 7293)$
   and write it as $s(7293) + t(4389)$.

10. Let $\mathbf{A} = \begin{bmatrix} 2 & 6 & 4 \\ -1 & 3 & 2 \end{bmatrix}$ and $\mathbf{B} = \begin{bmatrix} 2 & 0 \\ -3 & 1 \end{bmatrix}$. Compute, if
    possible, each of the following.

    (a) $\mathbf{AB}$ (b) $\mathbf{BA}$

    (c) $\mathbf{B}^T$ (d) $\mathbf{A} + \mathbf{B}$

    (e) $\mathbf{A}^T\mathbf{B}$

11. Let $\mathbf{C} = \begin{bmatrix} 1 & 0 & 1 \\ 1 & 1 & 0 \\ 0 & 1 & 1 \end{bmatrix}$ and $\mathbf{D} = \begin{bmatrix} 1 & 1 & 0 \\ 0 & 1 & 0 \\ 1 & 1 & 0 \end{bmatrix}$.
    Compute each of the following.

    (a) $\mathbf{C} \odot \mathbf{D}$ (b) $\mathbf{C} \vee \mathbf{D}$ (c) $\mathbf{C} \wedge \mathbf{D}$

12. Let $S = [2 \times 2 \text{ Boolean matrices}, \wedge, \vee, \odot]$ and $\mathbf{A}$ be a
    2×2 Boolean matrix. Describe the $\wedge$-inverse of $\mathbf{A}$ in S.

LOGIC

Prerequisites: *Chapter 1*

Logic is the discipline that deals with the methods of reasoning. On an elementary level, logic provides rules and techniques for determining whether a given argument is valid. Logical reasoning is used in mathematics to prove theorems, in computer science to verify the correctness of programs and to prove theorems, in the natural and physical sciences to draw conclusions from experiments, in the social sciences, and in our everyday lives to solve a multitude of problems. Indeed, we are constantly using logical reasoning. In this chapter we discuss a few of the basic ideas.

2.1 PROPOSITIONS AND LOGICAL OPERATIONS

A **statement** or **proposition** is a declarative sentence that is either true or false, but not both.

EXAMPLE 1

Which of the following are statements?

(a) The earth is round.
(b) $2 + 3 = 5$
(c) Do you speak English?
(d) $3 - x = 5$
(e) Take two aspirins.
(f) The temperature on the surface of the planet Venus is $800°$F.
(g) The sun will come out tomorrow.

Solution

(a) and (b) are statements that happen to be true.

(c) is a question, so it is not a statement.

(d) is a declarative sentence, but not a statement, since it is true or false depending on the value of x.

(e) is not a statement; it is a command.

(f) is a declarative sentence whose truth or falsity we do not know at this time; however, we can in principle determine if it is true or false, so it is a statement.

(g) is a statement since it is either true or false, but not both, although we would have to wait until tomorrow to find out if it is true or false. ∎

Logical Connectives and Compound Statements

In mathematics, the letters x, y, z, ... often denote variables that can be replaced by real numbers, and these variables can be combined with the familiar operations $+$, $\times$, $-$, and $\div$. In logic, the letters p, q, r, ... denote **propositional variables**; that is, variables that can be replaced by statements. Thus we can write p: The sun is shining today. q: It is cold. Statements or propositional variables can be combined by logical connectives to obtain **compound statements**. For example, we may combine the preceding statements by the connective *and* to form the compound statement *p and q*: The sun is shining *and* it is cold. The truth value of a compound statement depends only on the truth values of the statements being combined and on the types of connectives being used. We shall look at the most important connectives.

If p is a statement, the **negation** of p is the statement *not p*, denoted by $\sim p$. Thus $\sim p$ is the statement "it is not the case that p." From this definition, it follows that if p is true, then $\sim p$ is false, and if p is false, then $\sim p$ is true. The truth value of $\sim p$ relative to p is given in Table 2.1. Such a table, giving the truth values of a compound statement in terms of its component parts, is called a **truth table**. Strictly speaking, *not* is not a connective, since it does not join two statements, and $\sim p$ is not really a compound statement. However, *not* is a unary operation for the collection of statements and $\sim p$ is a statement if p is.

Table 2.1

p	$\sim p$
T	F
F	T

EXAMPLE 2 Give the negation of the following statements:

(a) p: $2 + 3 > 1$ (b) q: It is cold.

Solution

(a) $\sim p$: $2 + 3$ is not greater than 1. That is, $\sim p$: $2 + 3 \leq 1$. Since p is true in this case, $\sim p$ is false.

(b) $\sim q$: It is not the case that it is cold. More simply, $\sim q$: It is not cold. ∎

Table 2.2

p	q	$p \wedge q$
T	T	T
T	F	F
F	T	F
F	F	F

If p and q are statements, the **conjunction** of p and q is the compound statement "p and q," denoted by $p \wedge q$. The connective *and* is denoted by the symbol $\wedge$. In the language of Section 1.6, *and* is a binary operation on the set of statements. The compound statement $p \wedge q$ is true when both p and q are true; otherwise, it is false. The truth values of $p \wedge q$ in terms of the truth values of p and q are given in the truth table shown in Table 2.2. Observe that in giving the truth table of $p \wedge q$ we need to look at four possible cases. This follows from the fact that each of p and q can be true or false.

EXAMPLE 3 Form the conjunction of p and q for each of the following.

(a) p: It is snowing. q: I am cold.

(b) p: $2 < 3$ \qquad\qquad q: $-5 > -8$

(c) p: It is snowing. \qquad q: $3 < 5$

Solution

(a) $p \wedge q$: It is snowing and I am cold.

(b) $p \wedge q$: $2 < 3$ and $-5 > -8$

(c) $p \wedge q$: It is snowing and $3 < 5$. ■

Example 3(c) shows that in logic, unlike in everyday English, we may join two totally unrelated statements by the connective *and*.

If p and q are statements, the **disjunction** of p and q is the compound statement "p or q," denoted by $p \vee q$. The connective *or* is denoted by the symbol $\vee$. The compound statement $p \vee q$ is true if at least one of p or q is true; it is false when both p and q are false. The truth values of $p \vee q$ are given in the truth table shown in Table 2.3.

Table 2.3

p	q	$p \vee q$
T	T	T
T	F	T
F	T	T
F	F	F

EXAMPLE 4

Form the disjunction of p and q for each of the following.

(a) p: 2 is a positive integer \qquad q: $\sqrt{2}$ is a rational number.

(b) p: $2 + 3 \neq 5$ \qquad q: London is the capital of France.

Solution

(a) $p \vee q$: 2 is a positive integer or $\sqrt{2}$ is a rational number. Since p is true, the disjunction $p \vee q$ is true, even though q is false.

(b) $p \vee q$: $2 + 3 \neq 5$ or London is the capital of France. Since both p and q are false, $p \vee q$ is false. ■

Example 4(b) shows that in logic, unlike in ordinary English, we may join two totally unrelated statements by the connective *or*.

The connective *or* is more complicated than the connective *and* because it is used in two different ways in English. Suppose that we say "I left for Spain on Monday or I left for Spain on Friday." In this compound statement we have the disjunction of the statements p: I left for Spain on Monday and q: I left for Spain on Friday. Of course, exactly one of the two possibilities occurred. Both could not have occurred, so the connective *or* is being used in an *exclusive* sense. On the other hand, consider the disjunction "I passed mathematics or I failed French." In this case, at least one of the two possibilities occurred. However, both could have occurred, so the connective *or* is being used in an *inclusive* sense. In mathematics and computer science we agree to use the connective *or* always in the inclusive manner.

In general, a compound statement may have many component parts, each of which is itself a statement, represented by some propositional variable. The statement s: $p \vee (q \wedge (p \vee r))$ involves three propositions, p, q, and r, each of which may independently be true or false. There are altogether 2^3 or 8 possible combinations of truth values for p, q, and r, and a truth table for s must give the truth or falsity of s in all these cases. If a compound statement s contains n component statements, there will need to be 2^n rows in the truth table for s. (In Section 3.1 we look at how to count the possibilities in such cases.) Such a truth table may be systematically constructed in the following way.

STEP 1: The first n columns of the table are labeled by the component propositional variables. Further columns are included for all intermediate combinations of
the variables, culminating in a column for the full statement.

STEP 2: Under each of the first n headings, we list the 2^n possible n-tuples of
truth values for the n component statements.

STEP 3: For each of the remaining columns, we compute, in sequence, the remaining truth values.

EXAMPLE 5   Make a truth table for the statement $(p \wedge q) \vee (\sim p)$.

Solution   Because two propositions are involved, the truth table will have 2^2
or 4 rows. In the first two columns we list all possible pairs of truth values for
p and q. The numbers below the remaining columns show the order in which
the columns were filled.

p	q	$p \wedge q$	$\vee$	$\sim p$
T	T	T	T	F
T	F	F	F	F
F	T	F	T	T
F	F	F	T	T
		(1)	(3)	(2)

∎

Quantifiers

In Section 1.1, we defined sets by specifying a property $P(x)$ that elements of the
set have in common. Thus, an element of $\{x \mid P(x)\}$ is an object t for which the
statement $P(t)$ is true. Such a sentence $P(x)$ is called a **predicate**, because in English the property is grammatically a predicate. $P(x)$ is also called a **propositional
function**, because each choice of x produces a proposition $P(x)$ that is either true
or false. Another use of predicates is in programming. Two common constructions
are "if $P(x)$, then execute certain steps" and "while $Q(x)$, do specified actions." The
predicates $P(x)$ and $Q(x)$ are called the **guards** for the block of programming code.
Often the guard for a block is a conjunction or disjunction.

EXAMPLE 6   Let $A = \{x \mid x$ is an integer less than $8\}$. Here $P(x)$ is the sentence "x is an integer
less than 8." The common property is "is an integer less than 8." Since $P(1)$ is true,
$1 \in A$.

∎

The **universal quantification** of a predicate $P(x)$ is the statement "For all
values of x, $P(x)$ is true." We assume here that only values of x that make sense in
$P(x)$ are considered. The universal quantification of $P(x)$ is denoted $\forall x \ P(x)$. The
symbol $\forall$ is called the universal quantifier.

EXAMPLE 7   (a) The sentence $P(x)$: $-(-x) = x$ is a predicate that makes sense for real numbers
x. The universal quantification of $P(x)$, $\forall x \ P(x)$, is a true statement, because
for all real numbers, $-(-x) = x$.

(b) Let $Q(x)$: $x + 1 < 4$. Then $\forall x \ Q(x)$ is a false statement, because $Q(5)$ is not
true.

∎

Universal quantification can also be stated in English as "for every x," "every x," or "for any x."

A predicate may contain several variables. Universal quantification may be applied to each of the variables. For example, a commutative property can be expressed as $\forall x \; \forall y \; x \; \Box \; y = y \; \Box \; x$. The order in which the universal quantifiers are considered does not change the truth value. Often mathematical statements contain implied universal quantifications (for example in Theorem 1, Section 1.2).

In some situations we only require that there be at least one value for which the predicate is true. The **existential quantification** of a predicate P(x) is the statement "There exists a value of x for which P(x) is true." The existential quantification of P(x) is denoted $\exists x$ P(x). The symbol $\exists$ is called the existential quantifier.

EXAMPLE 8

(a) Let Q(x): $x + 1 < 4$. The existential quantification of Q(x), $\exists x$ Q(x), is a true statement, because Q(2) is a true statement.

(b) The statement $\exists y \; y + 2 = y$ is false. There is no value of y for which the propositional function $y + 2 = y$ produces a true statement. ∎

In English $\exists x$ can also be read "there is an x," "there is some x," "there exists an x," or "there is at least one x."

Existential quantification may be applied to several variables in a predicate and the order in which the quantifications are considered does not affect the truth value. For a predicate with several variables we may apply both universal and existential quantification. In this case the order does matter.

EXAMPLE 9

Let **A** and **B** be $n \times n$ matrices.

(a) The statement $\forall \mathbf{A} \; \exists \mathbf{B} \; \mathbf{A} + \mathbf{B} = \mathbf{I}_n$ is read "for every **A** there is a **B** such that $\mathbf{A} + \mathbf{B} = \mathbf{I}_n$." For a given $\mathbf{A} = \begin{bmatrix} a_{ij} \end{bmatrix}$, define $\mathbf{B} = \begin{bmatrix} b_{ij} \end{bmatrix}$ as follows: $b_{ii} = 1 - a_{ii}$, $1 \le i \le n$ and $b_{ij} = -a_{ij}$, $i \ne j$, $1 \le i \le n$, $1 \le j \le n$. Then $\mathbf{A} + \mathbf{B} = \mathbf{I}_n$ and we have shown that $\forall \mathbf{A} \; \exists \mathbf{B} \; \mathbf{A} + \mathbf{B} = \mathbf{I}_n$ is a true statement.

(b) $\exists \mathbf{B} \; \forall \mathbf{A} \; \mathbf{A} + \mathbf{B} = \mathbf{I}_n$ is the statement "there is a **B** such that for all **A** $\mathbf{A} + \mathbf{B} = \mathbf{I}_n$." This statement is false; no single **B** has this property for all **A**'s.

(c) $\exists \mathbf{B} \; \forall \mathbf{A} \; \mathbf{A} + \mathbf{B} = \mathbf{A}$ is true. What is the value for **B** that makes the statement true? ∎

Let p: $\forall x$ P(x). The negation of p is false when p is true, and true when p is false. For p to be false there must be at least one value of x for which P(x) is false. Thus, p is false if $\exists x \sim$P(x) is true. On the other hand, if $\exists x \sim$P(x) is false, then for every x, $\sim$P(x) is false; that is, $\forall x$ P(x) is true.

EXAMPLE 10

(a) Let p: For all positive integers n, $n^2 + 41n + 41$ is a prime number. Then $\sim p$ is There is at least one positive integer n for which $n^2 + 41n + 41$ is not prime.

(b) Let q: There is some integer k for which $12 = 3k$. Then $\sim q$: For all integers k, $12 \ne 3k$. ∎

2.1 Exercises

1. Which of the following are statements?
   (a) Is 2 a positive number?
   (b) $x^2 + x + 1 = 0$
   (c) Study logic.
   (d) There will be snow in January.
   (e) If stock prices fall, then I will lose money.

2. Give the negation of each of the following statements.
   (a) $2 + 7 \le 11$
   (b) 2 is an even integer and 8 is an odd integer.

3. Give the negation of each of the following statements.
   (a) It will rain tomorrow or it will snow tomorrow.
   (b) If you drive, then I will walk.

4. In each of the following, form the conjunction and the disjunction of p and q.
   (a) p: $3 + 1 < 5$ q: $7 = 3 \times 6$
   (b) p: I am rich. q: I am happy.

5. In each of the following, form the conjunction and the disjunction of p and q.
   (a) p: I will drive my car. q: I will be late.
   (b) p: NUM > 10 q: NUM ≤ 15

6. Determine the truth or falsity of each of the following statements.
   (a) $2 < 3$ and 3 is a positive integer.
   (b) $2 \ge 3$ and 3 is a positive integer.
   (c) $2 < 3$ and 3 is not a positive integer.
   (d) $2 \ge 3$ and 3 is not a positive integer.

7. Determine the truth or falsity of each of the following statements.
   (a) $2 < 3$ or 3 is a positive integer.
   (b) $2 \ge 3$ or 3 is a positive integer.
   (c) $2 < 3$ or 3 is not a positive integer.
   (d) $2 \ge 3$ or 3 is not a positive integer.

In Exercises 8 and 9, find the truth value of each proposition if p and r are true and q is false.

8. (a) $\sim p \wedge \sim q$ (b) $(\sim p \vee q) \wedge r$
   (c) $p \vee q \vee r$ (d) $\sim (p \vee q) \wedge r$

9. (a) $\sim p \wedge (q \vee r)$ (b) $p \wedge (\sim(q \vee \sim r))$
   (c) $(r \wedge \sim q) \vee (p \vee r)$ (d) $(q \wedge r) \wedge (p \vee \sim r)$

10. Which of the following statements is the negation of the statement "2 is even and -3 is negative"?
    (a) 2 is even and -3 is not negative.
    (b) 2 is odd and -3 is not negative.
    (c) 2 is even or -3 is not negative.
    (d) 2 is odd or -3 is not negative.

11. Which of the following statements is the negation of the statement "2 is even or -3 is negative"?
    (a) 2 is even or -3 is not negative.
    (b) 2 is odd or -3 is not negative.
    (c) 2 is even and -3 is not negative.
    (d) 2 is odd and -3 is not negative.

In Exercises 12 and 13 use p: Today is Monday; q: The grass is wet; and r: The dish ran away with the spoon.

12. Write each of the following in terms of p, q, r, and logical connectives.
    (a) Today is Monday and the dish did not run away with the spoon.
    (b) Either the grass is wet or today is Monday.
    (c) Today is not Monday and the grass is dry.
    (d) The dish ran away with the spoon, but the grass is wet.

13. Write an English sentence that corresponds to each of the following.
    (a) $\sim r \wedge q$ (b) $\sim q \vee r$
    (c) $\sim (p \vee q)$ (d) $p \vee \sim r$

In Exercises 14 through 19, use P(x): x is even; Q(x): x is a prime number; R(x, y): x + y is even. The variables x and y represent integers.

14. Write an English sentence corresponding to each of the following.
    (a) $\forall x\ P(x)$ (b) $\exists x\ Q(x)$

15. Write an English sentence corresponding to each of the following.
    (a) $\forall x\ \exists y\ R(x, y)$ (b) $\exists x\ \forall y\ R(x, y)$

16. Write an English sentence corresponding to each of the following.
    (a) $\forall x\ (\sim Q(x))$ (b) $\exists y\ (\sim P(y))$

17. Write an English sentence corresponding to each of the following.

(a) $\sim(\exists x \, P(x))$ (b) $\sim(\forall x \, Q(x))$

18. Write each of the following in terms of $P(x)$, $Q(x)$, $R(x, y)$, logical connectives, and quantifiers.

(a) Every integer is an odd integer.

(b) The sum of any two integers is an even number.

(c) There are no even prime numbers.

(d) Every integer is even or a prime.

19. Determine the truth value of each statement given in Exercises 14 through 18.

In Exercises 20 through 23, make a truth table for the statement.

20. $(\sim p \wedge q) \vee p$

21. $(p \vee q) \vee \sim q$

22. $(p \vee q) \wedge r$

23. $(\sim p \vee q) \wedge \sim r$

For Exercises 24 through 26, define $p \downarrow q$ to be a true statement if neither p nor q is true.

p	q	$p \downarrow q$
T	T	F
T	F	F
F	T	F
F	F	T

24. Make a truth table for $(p \downarrow q) \downarrow r$.

25. Make a truth table for $(p \downarrow q) \wedge (p \downarrow r)$.

26. Make a truth table for $(p \downarrow q) \downarrow (p \downarrow r)$.

For Exercises 27 through 29, define $p \triangle q$ to be true if either p or q, but not both, is true. Make a truth table for the statement.

27. (a) $p \triangle q$ (b) $p \triangle \sim p$

28. $(p \wedge q) \triangle p$

29. $(p \triangle q) \triangle (q \triangle r)$

In Exercises 30 and 31, revision of the given programming block is needed. Replace the guard $P(x)$ with $\sim P(x)$.

30. IF $(x \neq max$ and $y > 4)$ THEN take action

31. WHILE $(key = $ "open" or $t < $ limit$)$ take action

2.2 CONDITIONAL STATEMENTS

If p and q are statements, the compound statement 'if p then q,' denoted $p \Rightarrow q$, is called a **conditional statement**, or **implication**. The statement p is called the **antecedent** or **hypothesis**, and the statement q is called the **consequent** or **conclusion**. The connective *if ... then* is denoted by the symbol $\Rightarrow$.

EXAMPLE 1   Write the implication $p \Rightarrow q$ for each of the following.

(a) p: I am hungry. q: I will eat.
(b) p: It is snowing. q: $3 + 5 = 8$

Solution

(a) If I am hungry, then I will eat.
(b) If it is snowing, then $3 + 5 = 8$.

Example 1(b) shows that in logic we use conditional statements in a more general sense than is customary. Thus in English, when we say "if p then q," we are tacitly assuming there is a cause-and-effect relationship between p and q. That is, we would never use the statement in Example 1(b) in ordinary English, since there is no way statement p can have any effect on statement q.

In logic, implication is used in a much weaker sense. To say the compound statement $p \Rightarrow q$ is true simply asserts that if p is true, then q will also be found to be true. In other words, $p \Rightarrow q$ says only that we will not have p true and q false

Table 2.4

p	q	$p \Rightarrow q$
T	T	T
T	F	F
F	T	T
F	F	T

at the same time. It does not say that p "caused" q in the usual sense. Table 2.4 describes the truth values of $p \Rightarrow q$ in terms of the truth of p and q. Notice that $p \Rightarrow q$ is considered false only if p is true and q is false. In particular, if p is false, then $p \Rightarrow q$ is true for any q. This fact is sometimes described by the statement: "A false hypothesis implies any conclusion." This statement is misleading, since it seems to say that if the hypothesis is false, the conclusion must be true, an obviously silly statement. Similarly, if q is true, then $p \Rightarrow q$ will be true for any statement p. The implication "If $2 + 2 = 5$, then I am the king of England" is true, simply because p: $2 + 2 = 5$ is false, so it is not the case that p is true and q is false simultaneously.

In the English language, and in mathematics, each of the following expressions is an equivalent form of the conditional statement $p \Rightarrow q$: p implies q; q, if p; p only if q; p is a sufficient condition for q; q is a necessary condition for p.

If $p \Rightarrow q$ is an implication, then the **converse** of $p \Rightarrow q$ is the implication $q \Rightarrow p$, and the **contrapositive** of $p \Rightarrow q$ is the implication $\sim q \Rightarrow \sim p$.

EXAMPLE 2 Give the converse and the contrapositive of the implication "If it is raining, then I get wet."

Solution We have p: It is raining; and q: I get wet. The converse is $q \Rightarrow p$: If I get wet, then it is raining. The contrapositive is $\sim q \Rightarrow \sim p$: If I do not get wet, then it is not raining. ∎

Table 2.5

p	q	$p \Leftrightarrow q$
T	T	T
T	F	F
F	T	F
F	F	T

If p and q are statements, the compound statement p if and only if q, denoted by $p \Leftrightarrow q$, is called an **equivalence** or **biconditional**. The connective *if and only if* is denoted by the symbol $\Leftrightarrow$. The truth values of $p \Leftrightarrow q$ are given in Table 2.5. Observe that $p \Leftrightarrow q$ is true only when both p and q are true or when both p and q are false. The equivalence $p \Leftrightarrow q$ can also be stated as p is a necessary and sufficient condition for q.

EXAMPLE 3 Is the following equivalence a true statement? $3 > 2$ if and only if $0 < 3 - 2$.

Solution Let p be the statement $3 > 2$ and let q be the statement $0 < 3 - 2$. Since both p and q are true, we conclude that $p \Leftrightarrow q$ is true. ∎

EXAMPLE 4 Compute the truth table of the statement $(p \Rightarrow q) \Leftrightarrow (\sim q \Rightarrow \sim p)$.

Solution The following table is constructed using steps 1, 2, and 3 as given in Section 2.1. The numbers below the columns show the order in which they were constructed.

p	q	$p \Rightarrow q$	$\sim q$	$\sim p$	$\sim q \Rightarrow \sim p$	$(p \Rightarrow q) \Leftrightarrow (\sim q \Rightarrow \sim p)$
T	T	T	F	F	T	T
T	F	F	T	F	F	T
F	T	T	F	T	T	T
F	F	T	T	T	T	T
		(1)	(2)	(3)	(4)	(5)

∎

A statement that is true for all possible values of its propositional variables is called a **tautology**. A statement that is always false is called a **contradiction** or an **absurdity**, and a statement that can be either true or false, depending on the truth values of its propositional variables, is called a **contingency**.

EXAMPLE 5

(a) The statement in Example 4 is a tautology.

(b) The statement $p \wedge \sim p$ is an absurdity. (Verify this.)

(c) The statement $(p \Rightarrow q) \wedge (p \vee q)$ is a contingency. ∎

We have now defined a new mathematical structure with two binary operations and one unary operation, [propositions, $\wedge$, $\vee$, $\sim$]. It makes no sense to say two propositions are equal; instead we say p and q are **logically equivalent**, or simply **equivalent**, if $p \Leftrightarrow q$ is a tautology. When an equivalence is shown to be a tautology, this means its two component parts are always either both true or both false, for any values of the propositional variables. Thus the two sides are simply different ways of making the same statement and can be regarded as "equal." We denote that p is equivalent to q by $p \equiv q$. Now we can adapt our properties for operations to say this structure has a property if using equivalent in place of equal gives a true statement.

EXAMPLE 6

The binary operation $\vee$ has the commutative property; that is, $p \vee q \equiv q \vee p$. The truth table for $(p \vee q) \Leftrightarrow (q \vee p)$ shows the statement is a tautology.

p	q	$p \vee q$	$q \vee p$	$(p \vee q) \Leftrightarrow (q \vee p)$
T	T	T	T	T
T	F	T	T	T
F	T	T	T	T
F	F	F	F	T

∎

Another way to use a truth table to determine if two statements are equivalent is to construct a column for each statement and compare these to see if they are identical. In Example 6 the third and fourth columns are identical, and this will guarantee that the statements they represent are equivalent.

Forming $p \Rightarrow q$ from p and q is another binary operation for statements, but we can express it in terms of the operations in Section 2.1.

EXAMPLE 7

The conditional statement $p \Rightarrow q$ is equivalent to $(\sim p) \vee q$. Columns 1 and 3 in the following table show that for any truth values of p and q, $p \Rightarrow q$ and $(\sim p) \vee q$ have the same truth values.

p	q	$p \Rightarrow q$	$\sim p$	$\vee q$
T	T	T	F	T
T	F	F	F	F
F	T	T	T	T
F	F	T	T	T
		(1)	(2)	(3)

∎

The structure [propositions, ∧, ∨, ~] has many of the same properties as the structure [sets, ∪, ∩, ⁻].

Theorem 1 The operations for propositions have the following properties.
Commutative Properties

    1. $p \vee q \equiv q \vee p$
    2. $p \wedge q \equiv q \wedge p$

Associative Properties

    3. $p \vee (q \vee r) \equiv (p \vee q) \vee r$
    4. $p \wedge (q \wedge r) \equiv (p \wedge q) \wedge r$

Distributive Properties

    5. $p \vee q \wedge r) \equiv (p \vee q) \wedge (p \vee r)$
    6. $p \wedge (q \vee r) \equiv (p \wedge q) \vee (p \wedge r)$

Idempotent Properties

    7. $p \vee p \equiv p$
    8. $p \wedge p \equiv p$

Properties of Negation

    9. $\sim(\sim p) \equiv p$
    10. $\sim(p \vee q) \equiv (\sim p) \wedge (\sim q)$
    11. $\sim(p \wedge q) \equiv (\sim p) \vee (\sim q)$ 10 and 11 are De Morgan's laws   ●

Proof We have proved Property 1 in Example 6. The remaining properties may be proved the same way and are left for the reader as exercises.   ▼

Truth tables can be used to prove statements about propositions, because in a truth table all possible cases are examined.

The implication operation also has a number of important properties.

Theorem 2 (a) $(p \Rightarrow q) \equiv ((\sim p) \vee q)$
    (b) $(p \Rightarrow q) \equiv (\sim q \Rightarrow \sim p)$
    (c) $(p \Leftrightarrow q) \equiv ((p \Rightarrow q) \wedge (q \Rightarrow p))$
    (d) $\sim(p \Rightarrow q) \equiv (p \wedge \sim q)$
    (e) $\sim(p \Leftrightarrow q) \equiv ((p \wedge \sim q) \vee (q \wedge \sim p))$ ●

Proof

    (a) was proved in Example 7 and (b) was proved in Example 4. Notice that (b) says a conditional statement is equivalent to its contrapositive.

    (d) gives an alternate version for the negation of a conditional statement. This could be proved using truth tables, but it can also be proved by using previously proven facts. Since $(p \Rightarrow q) \equiv ((\sim p) \vee q)$, the negation of $p \Rightarrow q$ must be equivalent to $\sim((\sim p) \vee q)$. By De Morgan's laws, $\sim((\sim p) \vee q) \equiv (\sim(\sim p)) \wedge (\sim q)$ or $p \wedge (\sim q)$. Thus, $\sim(p \Rightarrow q) \equiv (p \wedge \sim q)$.

The remaining parts of Theorem 2 are left as exercises.   ▼

Theorem 3 states two results from Section 2.1, and several other properties for the universal and existential quantifiers.

Theorem 3   (a) $\sim(\forall x \, P(x)) \equiv \exists x \, \sim P(x)$

(b) $\sim(\exists x \, P(x)) \equiv \forall x \, (\sim P(x))$

(c) $\exists x \, (P(x) \Rightarrow Q(x)) \equiv \forall x \, P(x) \Rightarrow \exists x \, Q(x)$

(d) $\exists x \, P(x) \Rightarrow \forall x \, Q(x) \equiv \forall x \, (P(x) \Rightarrow Q(x))$

(e) $\exists x \, (P(x) \vee Q(x)) \equiv \exists x \, P(x) \vee \exists x \, Q(x)$

(f) $\forall x \, (P(x) \wedge Q(x)) \equiv \forall x \, P(x) \wedge \forall x \, Q(x)$

(g) $((\forall x \, P(x)) \vee (\forall x \, Q(x))) \Rightarrow \forall x \, (P(x) \vee Q(x))$ is a tautology.

(h) $\exists x \, (P(x) \wedge Q(x)) \Rightarrow \exists x \, P(x) \wedge \exists x \, Q(x)$ is a tautology.   ●

The following theorem gives several important tautologies that are implications. These are used extensively in proving results in mathematics and computer science and we will illustrate them in Section 2.3.

Theorem 4   Each of the following is a tautology.

(a) $(p \wedge q) \Rightarrow p$ (b) $(p \wedge q) \Rightarrow q$

(c) $p \Rightarrow (p \vee q)$ (d) $q \Rightarrow (p \vee q)$

(e) $\sim p \Rightarrow (p \Rightarrow q)$ (f) $\sim(p \Rightarrow q) \Rightarrow p$

(g) $(p \wedge (p \Rightarrow q)) \Rightarrow q$ (h) $(\sim p \wedge (p \vee q)) \Rightarrow q$

(i) $(\sim q \wedge (p \Rightarrow q)) \Rightarrow \sim p$ (j) $((p \Rightarrow q) \wedge (q \Rightarrow r)) \Rightarrow (p \Rightarrow r)$ ●

2.2 Exercises

In Exercises 1 and 2 use the following: p: I am awake; q: I work hard; r: I dream of home.

1. Write each of the following statements in terms of p, q, r, and logical connectives.

(a) I am awake implies that I work hard.

(b) I dream of home only if I am awake.

(c) Working hard is sufficient for me to be awake.

(d) Being awake is necessary for me not to dream of home.

2. Write each of the following statements in terms of p, q, r, and logical connectives.

(a) I am not awake if and only if I dream of home.

(b) If I dream of home, then I am awake and I work hard.

(c) I do not work hard only if I am awake and I do not dream of home.

(d) Not being awake and dreaming of home is sufficient for me to work hard.

3. State the converse of each of the following implications.

(a) If $2 + 2 = 4$, then I am not the Queen of England.

(b) If I am not President of the United Sates, then I will walk to work.

(c) If I am late, then I did not take the train to work.

(d) If I have time and I am not too tired, then I will go to the store.

(e) If I have enough money, then I will buy a car and I will buy a house.

4. State the contrapositive of each implication in Exercise 3.

5. Determine the truth value for each of the following statements.

(a) If 2 is even, then New York has a large population.

(b) If 2 is even, then New York has a small population.

(c) If 2 is odd, then New York has a large population.

(d) If 2 is odd, then New York has a small population.

In Exercises 6 and 7, let p, q, and r be the following statements: p: I will study discrete structure; q: I will go to a movie; r: I am in a good mood.

6. Write the following statements in terms of p, q, r, and logical connectives.

(a) If I am not in a good mood, then I will go to a movie.

(b) I will not go to a movie and I will study discrete structures.

(c) I will go to a movie only if I will not study discrete structures.

(d) If I will not study discrete structures, then I am not in a good mood.

7. Write English sentences corresponding to the following statements.

(a) $((\sim p) \wedge q) \Rightarrow r$

(b) $r \Rightarrow (p \vee q)$

(c) $(\sim r) \Rightarrow ((\sim q \vee p)$

(d) $(q \wedge (\sim p)) \Leftrightarrow r$

In Exercises 8 and 9, let p, q, r, and s be the following statements: p: 4 > 1; q: 4 < 5; r: 3 ≤ 3; s: 2 > 2.

8. Write the following statements in terms of p, q, r, and logical connectives.

(a) Either $4 > 1$ or $4 < 5$.

(b) If $3 \leq 4$, then $2 > 2$.

(c) It is not the case that $2 > 2$ or $4 > 1$.

9. Write English sentences corresponding to the following statements.

(a) $(p \wedge s) \Rightarrow q$ (b) $\sim(r \wedge q)$ (c) $(\sim r) \Rightarrow p$

In Exercises 10 through 12, construct truth tables to determine whether the given statement is a tautology, a contingency, or an absurdity.

10. (a) $p \wedge \sim p$ (b) $q \vee (\sim q \wedge p)$

11. (a) $p \Rightarrow (q \Rightarrow p)$ (b) $q \Rightarrow (q \Rightarrow p)$

12. (a) $(q \wedge p) \vee (q \wedge \sim p)$

(b) $(p \wedge q) \Rightarrow p$ (c) $p \Rightarrow (q \wedge p)$

13. If $p \Rightarrow q$ is false, can you determine the truth value of $(\sim(p \wedge q)) \Rightarrow q$? Explain your answer.

14. If $p \Rightarrow q$ is false, can you determine the truth value of $(\sim p) \vee (p \Leftrightarrow q)$? Explain your answer.

In Exercises 15 and 16, find the truth value of each statement if p and q are true and r, s, and t are false.

15. (a) $\sim(p \Rightarrow q)$ (b) $(\sim p) \Rightarrow r$

(c) $(p \Rightarrow s) \wedge (s \Rightarrow t)$ (d) $t \Rightarrow \sim q$

16. (a) $(\sim q) \Rightarrow (r \Rightarrow (r \Rightarrow (p \vee s))$

(b) $p \Rightarrow (r \Rightarrow q)$

(c) $(q \Rightarrow (r \Rightarrow s)) \wedge ((p \Rightarrow s) \Rightarrow (\sim t)) \Rightarrow$

(d) $(r \wedge s \wedge t) \Rightarrow (p \vee q)$

17. Use the definition of $p \downarrow q$ given for Exercise 24 in Section 2.1 and show that $((p \downarrow p) \downarrow (q \downarrow q))$ is equivalent to $p \wedge q$.

18. Write the negation of each of the following in good English.

(a) The weather is bad and I will not go to work.

(b) If Carol is not sick, then if she goes to the picnic, she will have a good time.

(c) I will not win the game or I will not enter the contest.

19. Consider the following conditional statement:

   p: If the flood destroys my house or the fire destroys my house, then my insurance company will pay me.

(a) Which of the following is the converse of p?

(b) Which of the following is the contrapositive of p?

(i) If my insurance company pays me, then the flood destroys my house or the fire destroys my house.

(ii) If my insurance company pays me, then the flood destroys my house and the fire destroys my house.

(iii) If my insurance company does not pay me, then the flood does not destroy my house or the fire does not destroy my house.

(iv) If my insurance company does not pay me, then the flood does not destroy my house and the fire does not destroy my house.

20. Prove Theorem 1 part 6.

21. Prove Theorem 1 part 11.

22. Prove Theorem 2 part e.

23. Prove Theorem 3 part e.

24. Prove Theorem 3 part f.

25. Prove Theorem 4 part a.

26. Prove Theorem 4 part d.

27. Prove Theorem 4 part g.

28. Prove Theorem 4 part j.

2.3   METHODS OF PROOF

Some methods of proof we have already used are direct proofs using generic elements, definitions, and previously proven facts, and proofs by cases, such as examining all possible truth value situations in a truth table. Here we look at proofs in more detail.

If an implication $p \Rightarrow q$ is a tautology, where p and q may be compound statements involving any number of propositional variables, we say that q **logically follows** from p. Suppose that an implication of the form $(p_1 \wedge p_2 \wedge \cdots \wedge p_n) \Rightarrow q$ is a tautology. Then this implication is true regardless of the truth values of any of its components. In this case, we say that q **logically follows** from $p_1, p_2, \ldots, p_n$. When q logically follows from $p_1, p_2, \ldots, p_n$, we write

$$
\begin{array}{c}
p_1 \\
p_2 \\
\vdots \\
p_n \\
\hline
\therefore \quad q
\end{array}
$$

where the symbol $\therefore$ means therefore. This means if we know that p_1 is true, p_2 is true, $\ldots$, and p_n is true, then we know q is true.

Virtually all mathematical theorems are composed of implications of the type

$$(p_1 \wedge p_2 \wedge \cdots \wedge p_n) \Rightarrow q.$$

The p_i's are called the **hypotheses** or **premises**, and q is called the **conclusion**. To "prove the theorem" means to show that the *implication* is a tautology. Note that we are not trying to show that q (the conclusion) is true, but only that q will be true if all the p_i are true. For this reason, mathematical proofs often begin with the statement "suppose that $p_1, p_2, \ldots,$ and p_n are true" and conclude with the statement "therefore, q is true." The proof does not show that q is true, but simply shows that q has to be true if the p_i are all true.

Arguments based on tautologies represent universally correct methods of reasoning. Their validity depends only on the form of the statements involved and not on the truth values of the variables they contain. Such arguments are called **rules of inference**. The various steps in a mathematical proof of a theorem must follow from the use of various rules of inference, and a mathematical proof of a theorem must begin with the hypotheses, proceed through various steps, each justified by some rule of inference, and arrive at the conclusion.

EXAMPLE 1 According to Theorem 3(j) of the last section, $((p \Rightarrow q) \wedge (q \Rightarrow r)) \Rightarrow (p \Rightarrow r)$ is a tautology. Thus the argument

$$p \Rightarrow q$$
$$\underline{q \Rightarrow r}$$
$$\therefore \quad p \Rightarrow r$$

is universally valid, and so is a rule of inference.   ■

EXAMPLE 2 Is the following argument valid?

If you invest in the stock market, then you will get rich.
If you get rich, then you will be happy.
$\therefore$ If you invest in the stock market, then you will be happy.

Solution   The argument is of the form given in Example 1, hence the argument is valid, although the conclusion may be false.   ■

EXAMPLE 3 The tautology $(p \Leftrightarrow q) \Leftrightarrow ((p \Rightarrow q) \wedge (q \Rightarrow p))$ is Theorem 2(c), Section 2.2. Thus both of the following arguments are valid.

$$\frac{p \Leftrightarrow q}{\therefore \quad (p \Rightarrow q) \wedge (q \Rightarrow p)} \qquad\qquad \begin{array}{c} p \Rightarrow q \\ \underline{q \Rightarrow p} \\ \therefore \quad p \Leftrightarrow q \end{array}$$ ■

Some mathematical theorems are equivalences; that is, they are of the form $p \Leftrightarrow q$. They are usually stated p if and only if q. By Example 3, the proof of such a theorem is logically equivalent with proving both $p \Rightarrow q$ and $q \Rightarrow p$, and this is almost always the way in which equivalences are proved. We first assume that p is true, and show that q must then be true; next we assume that q is true and show that p must then be true.

A very important rule of inference is

$$p$$
$$\underline{p \Rightarrow q}$$
$$\therefore \quad q.$$

That is, p is true, and $p \Rightarrow q$ is true, so q is true. This follows from Theorem 4(g), Section 2.2.

Some rules of inference were given Latin names by classical scholars. Theorem 4(g) is referred to as **modus ponens**, or loosely, the method of asserting.

EXAMPLE 4 Is the following argument valid?

Smoking is healthy.
If smoking is healthy, then cigarettes are prescribed by physicians.
$\therefore$ Cigarettes are prescribed by physicians.

Solution   The argument is valid since it is of the form modus ponens. However, the conclusion is false. Observe that the first premise p: smoking is healthy is false. The second premise $p \Rightarrow q$ is then true and $(p \wedge (p \Rightarrow q))$, the conjunction of the two premises, is false.   ■

EXAMPLE 5   Is the following argument valid?

> If taxes are lowered, then income rises.
> Income rises.
> ∴   Taxes are lowered.

Solution   Let p: taxes are lowered and q: income rises. Then the argument is of the form

$$p \Rightarrow q$$
$$\underline{q}$$
$$\therefore \quad p.$$

Assume that $p \Rightarrow q$ and q are both true. Now $p \Rightarrow q$ may be true with p being false. Then the conclusion p is false. Hence the argument is not valid. Another approach to answering this question is to verify whether the statement $((p \Rightarrow q) \wedge q)$ logically implies the statement p. A truth table shows this is not the case. (Verify this.) ■

An important proof technique, called an **indirect method** of proof, follows from the tautology $(p \Rightarrow q) \Leftrightarrow ((\sim q) \Rightarrow (\sim p))$. This states, as we previously mentioned, that an implication is equivalent to its contrapositive. Thus to prove $p \Rightarrow q$ indirectly, we assume q is false (the statement $\sim q$) and show that p is then false (the statement $\sim p$).

EXAMPLE 6   Let n be an integer. Prove that if n^2 is odd, then n is odd.

Solution   Let p: n^2 is odd and q: n is odd. We have to prove that $p \Rightarrow q$ is true. Instead, we prove the contrapositive $\sim q \Rightarrow \sim p$. Thus suppose that n is not odd, so that n is even. Then $n = 2k$, where k is an integer. We have $n^2 = (2k)^2 = 4k^2 = 2(2k^2)$, so n^2 is even. We thus show that if n is even, then n^2 is even, which is the contrapositive of the given statement. Hence the given statement has been proved. ■

Another important proof technique is **proof by contradiction**. This method is based on the tautology $((p \Rightarrow q) \wedge (\sim q)) \Rightarrow (\sim p)$. Thus the rule of inference

$$p \Rightarrow q$$
$$\underline{\sim q}$$
$$\therefore \quad \sim p$$

is valid. Informally, this states that if a statement p implies a false statement q, then p must be false. This is often applied to the case where q is an absurdity or contradiction, that is, a statement that is always false. An example is given by taking q as the contradiction $r \wedge (\sim r)$. Thus any statement that implies a contradiction must be false. In order to use proof by contradiction, suppose we wish to show that a statement q logically follows from statements $p_1, p_2, \ldots, p_n$. Assume that $\sim q$ is true (that is, q is false) as an extra hypothesis, and that $p_1, p_2, \ldots, p_n$ are also true. If this enlarged hypothesis $p_1 \wedge p_2 \wedge \cdots \wedge p_n \wedge (\sim q)$ implies a contradiction, then at least one of the statements $p_1, p_2, \ldots, p_n, \sim q$ must be false. This means that if all the p_i's are true, then $\sim q$ must be false, so q must be true. Thus q follows from $p_1, p_2, \ldots, p_n$. This is proof by contradiction.

EXAMPLE 7

Prove there is no rational number p/q whose square is 2. In other words, show $\sqrt{2}$ is irrational.

Solution This statement is a good candidate for proof by contradiction, because we could not check all possible rational numbers to demonstrate that none had a square equal to 2. Assume $(p/q)^2 = 2$ for some integers p and q, which have no common factors. If the original choice of p/q is not in lowest terms, we can replace it with its equivalent lowest-term form. Then $p^2 = 2q^2$, so p^2 is even. This implies p is even, since the square of an odd number is odd. Thus, $p = 2n$ for some integer n. We see that $2q^2 = p^2 = (2n)^2 = 4n^2$, so $q^2 = 2n^2$. Thus q^2 is even, and so q is even. We now have that both p and q are even, and therefore have a common factor 2. This is a contradiction to the assumption. Thus the assumption must be false. ∎

We have presented several rules of inference and logical equivalences that correspond to valid proof techniques. In order to prove a theorem of the (typical) form $(p_1 \wedge p_2 \wedge \cdots \wedge p_n) \Rightarrow q$, we begin with the hypothesis $p_1, p_2, \ldots, p_n$ and show that some result r_1 logically follows. Then, using $p_1, p_2, \ldots, p_n, r_1$, we show that some other statement r_2 logically follows. We continue this process, producing intermediate statements $r_1, r_2, \ldots, r_k$, called **steps in the proof**, until we can finally shows that the conclusion q logically follows from $p_1, p_2, \ldots, p_n, r_1, r_2, \ldots, r_k$. Each logical step must be justified by some valid proof technique, based on the rules of inference we have developed, or on some other rules that come from tautological implications we have not discussed. At any stage, we can replace a statement that needs to be derived by its contrapositive statement, or any other equivalent form.

In practice, the construction of proofs is an art and must be learned in part from observation and experience. The choice of intermediate steps and methods of deriving them is a creative activity that cannot be precisely described. But a few simple techniques are applicable to a wide variety of settings. We will focus on these techniques throughout the book. The "Tips for Proofs" notes at the end of each chapter highlights the methods most useful for that chapter's material.

EXAMPLE 8

Let m and n be integers. Prove that $n^2 = m^2$ if and only if $m = n$ or $m = -n$.

Solution Let us analyze the proof as we present it. Suppose p is the statement $n^2 = m^2$, q is the statement $m = n$, and r is the statement $m = -n$. Then we wish to prove the theorem $p \Leftrightarrow (q \vee r)$. We know from previous discussion that we may instead prove $s: p \Rightarrow (q \vee r)$ and $t: (q \vee r) \Rightarrow p$ are true. Thus we assume that either $q: m = n$ or $r: m = -n$ is true. If q is true, then $m^2 = n^2$, and if r is true, then $m^2 = (-n)^2 = n^2$, so in either case p is true. We have therefore shown that the implication $t: (q \vee r) \Rightarrow p$ is true.

Now we must prove that $s: p \Rightarrow (q \vee r)$ is true; that is, we assume p and try to prove either q or r. If p is true, then $n^2 = m^2$, so $m^2 - n^2 = 0$. But $m^2 - n^2 = (m - n)(m + n)$. If r_1 is the intermediate statement $(m - n)(m + n) = 0$, we have shown $p \Rightarrow r_1$ is true. We now show that $r_1 \Rightarrow (q \vee r)$ is true, by showing that the contrapositive $\sim(q \vee r) \Rightarrow (\sim r_1)$ is true. Now $\sim(q \vee r)$ is equivalent to $(\sim q) \wedge (\sim r)$, so we show that $(\sim q) \wedge (\sim r) \Rightarrow (\sim r_1)$. Thus, if $(\sim q): m \neq n$ and $(\sim r): m \neq -n$ are true, then $(m - n) \neq 0$ and $(m + n) \neq 0$, so $(m - n)(m + n) \neq 0$ and r_1 is false. We have therefore shown that $r_1 \Rightarrow (q \vee r)$ is true. Finally, from the truth of $p \Rightarrow r_1$ and $r_1 \Rightarrow (q \vee r)$, we can conclude that $p \Rightarrow (q \vee r)$ is true, and we are done. ∎

We do not usually analyze proofs in this detailed manner. We have done so only to illustrate that proofs are devised by piecing together equivalences and valid steps resulting from rules of inference. The amount of detail given in a proof depends on who the reader is likely to be.

As a final remark, we remind the reader that many mathematical theorems actually mean that the statement is true for all objects of a certain type. Sometimes this is not evident. Thus the theorem in Example 8 really states that for all integers m and n, $m^2 = n^2$ if and only if $m = n$ or $m = -n$. Similarly, the statement "If x and y are real numbers, and $x \neq y$, then $x < y$ or $y < x$" is a statement about all real numbers x and y. To prove such a theorem, we must make sure that the steps in the proof are valid for every real number. We could not assume, for example, that x is 2, or that y is π or $\sqrt{3}$. This is why proofs often begin by selecting a generic element, denoted by a variable. On the other hand, we know from Section 2.2 that the negation of a statement of the form $\forall x\, P(x)$ is $\exists x \sim P(x)$, so we need only find a single example where the statement is false.

EXAMPLE 9   Prove or disprove the statement that if x and y are real numbers, $(x^2 = y^2) \Leftrightarrow (x = y)$.

Solution   The statement can be restated in the form $\forall x\, \forall y\, R(x, y)$. Thus, to prove this result, we would need to provide steps, each of which would be true for all x and y. To disprove the result, we need only find one example for which the implication is false.

Since $(-3)^2 = 3^2$, but $-3 \neq 3$, the result is false. Our example is called a **counterexample**, and any other counterexample would do just as well. ∎

In summary, if a statement claims that a property holds for all objects of a certain type, then to prove it, we must use steps that are valid for all objects of that type and that do not make references to any particular object. To disprove such a statement, we need only show one counterexample, that is, one particular object or set of objects for which the claim fails.

2.3 Exercises

In Exercises 1 through 11, state whether the argument given is valid or not. If it is valid, identify the tautology or tautologies on which it is based.

1.   If I drive to work, then I will arrive tired.
I am not tried when I arrive at work.
∴   I do not drive to work.

2.   If I drive to work, then I will arrive tired.
I arrive at work tired.
∴   I drive to work.

3.   If I drive to work, then I will arrive tired.
I do not drive to work.
∴   I will not arrive tired.

4.   If I drive to work, then I will arrive tired.
I drive to work.
∴   I will arrive tired.

5.   I will become famous or I will not become a writer.
I will become a writer.
∴   I will become famous.

6.   I will become famous or I will be a writer.
   I will not be a writer.
   ∴ I will become famous.

7.   If I try hard and I have talent, then I will become
      a musician.
   If I become a musician, then I will be happy.
   ∴ If I will not be happy, then I did not try hard or
      I do not have talent.

8.   If I graduate this semester, then I will have
      passed the physics course.
   If I do not study physics for 10 hours a week,
      then I will not pass physics.
   If I study physics for 10 hours a week, then I
      cannot play volleyball.
   ∴ If I play volleyball, I will not graduate this
      semester.

9.   If my plumbing plans do not meet the construc-
      tion code, then I cannot build my house.
   If I hire a licensed contractor, then my plumbing
      plans will meet the construction code.
   I hire a licensed contractor.
   ∴ I can build my house.

10. (a) $p \vee q$ (b) $p \Rightarrow q$
            $\sim q$ $\sim p$
         ∴ p ∴ $\sim q$

11. (a) $(p \Rightarrow q) \wedge (q \Rightarrow r)$
            $(\sim q) \wedge r$
         ∴ p

   (b) $\sim(p \Rightarrow q)$
            p
         ∴ $\sim q$

12. Prove that the sum of two even numbers is even.

13. Prove that the sum of two odd numbers is even.

14. Prove that the structure [even integers, $+$, $*$] is closed with
   respect to $*$.

15. Prove that the structure [odd integers, $+$, $*$] is closed with
   respect to $*$.

16. Prove that n^2 is even if and only if n is even.

17. Prove that $A = B$ if and only if $A \subseteq B$ and $B \subseteq A$.

18. Let A and B be subsets of a universal set U. Prove that
   $A \subseteq B$ if and only if $\overline{B} \subseteq \overline{A}$.

19. Show that

   (a) $A \subseteq B$ is a necessary and sufficient condition for
       $A \cup B = B$.

   (b) $A \subseteq B$ is a necessary and sufficient condition for
       $A \cap B = A$.

20. Prove or disprove: $n^2 + 41n + 41$ is a prime number for
   every integer n.

21. Prove or disprove: the sum of any five consecutive integers is divisible by 5.

22. Prove or disprove that $3 \mid (n^3 - n)$ for every positive
   integer n.

23. Determine if the following is a valid argument. Explain
   your conclusion.

   Prove: $\forall x \ x^3 > x^2$.

   Proof: $\forall x \ x^2 > 0$ so $\forall x \ x^2(x - 1) > 0(x - 1)$ and
   $\forall x \ x^3 - x^2 > 0$. Hence $\forall x \ x^3 > x^2$.

24. Determine if the following is a valid argument. Explain
   your conclusion.

   Prove: If $\mathbf{A}$ and $\mathbf{B}$ are matrices such that $\mathbf{AB} = \mathbf{0}$, then
   either $\mathbf{A} = \mathbf{0}$ or $\mathbf{B} = \mathbf{0}$.

   Proof: There are two cases to consider: $\mathbf{A} = \mathbf{0}$ or
   $\mathbf{A} \neq \mathbf{0}$. If $\mathbf{A} = \mathbf{0}$, then we are done. If $\mathbf{A} \neq \mathbf{0}$, then
   $\mathbf{A}^{-1}(\mathbf{AB}) = \mathbf{A}^{-1}\mathbf{0}$ and $(\mathbf{A}^{-1}\mathbf{A})\mathbf{B} = \mathbf{0}$ and $\mathbf{B} = \mathbf{0}$.

25. Determine if the following is a valid argument. Explain
   your conclusion.
   Let m and n be two relatively prime integers. Prove that if
   mn is a cube, then m and n are each cubes.
   Proof: We first note that in the factorization of any cube
   into prime factors, each prime must have an exponent
   that is a multiple of 3. Write m and n each as a product
   of primes; $m = p_1^{a_1} p_2^{a_2} \cdots p_k^{a_k}$ and $n = q_1^{b_1} q_2^{b_2} \cdots q_j^{b_j}$.
   Suppose m is not a cube. Then at least one a_i is not a
   multiple of 3. Since each prime factor of mn must have an
   exponent that is a multiple of 3, n must have a factor $p_i^{b_i}$
   such that $b_i \neq 0$ and $a_i + b_i$ is a multiple of 3. But this
   means that m and n share a factor, p_i. This contradicts the
   fact that m and n are relatively prime.

26. Determine if the following is a valid argument. Explain
   your conclusion.

   Prove: If x is an irrational number, then $1 - x$ is also an
   irrational number.

   Proof: Suppose $1 - x$ is rational. Then we can write
   $1 - x$ as $\dfrac{a}{b}$, with $a, b \in Z$. Now we have $1 - \dfrac{a}{b} = x$ and
   $x = \dfrac{b - a}{b}$, a rational number. This is a contradiction.
   Hence, if x is irrational, so is $1 - x$.

27. Prove that the sum of two prime numbers, each larger than
   2, is not a prime number.

28. Prove that if two lines are each perpendicular to a third
   line in the plane, then the two lines are parallel.

29. Prove that if x is a rational number and y is an irrational
   number, then $x + y$ is an irrational number.

2.4 MATHEMATICAL INDUCTION

Here we discuss another proof technique. Suppose the statement to be proved can be put in the form $\forall n \geq n_0 \; P(n)$, where n_0 is some fixed integer. That is, suppose we wish to show that $P(n)$ is true for all integers $n \geq n_0$. The following result shows how this can be done. Suppose that (a) $P(n_0)$ is true and (b) If $P(k)$ is true for some $k \geq n_0$, then $P(k+1)$ must also be true. Then $P(n)$ is true for all $n \geq n_0$. This result is called the **principle of mathematical induction**. Thus to prove the truth of a statement $\forall n \geq n_0 \; P(n)$, using the principle of mathematical induction, we must begin by proving directly that the first proposition $P(n_0)$ is true. This is called the **basis step** of the induction and is generally very easy.

Then we must prove that $P(k) \Rightarrow P(k+1)$ is a tautology for any choice of $k \geq n_0$. Since the only case where an implication is false is if the antecedent is true and the consequent is false, this step is usually done by showing that if $P(k)$ were true, then $P(k+1)$ would also have to be true. Note that this is not the same as assuming that $P(k)$ is true for some value of k. This step is called the **induction step**, and some work will usually be required to show that the implication is always true.

EXAMPLE 1   Show, by mathematical induction, that for all $n \geq 1$,

$$1 + 2 + 3 + \cdots + n = \frac{n(n+1)}{2}.$$

Solution   Let $P(n)$ be the predicate $1 + 2 + 3 + \cdots + n = \dfrac{n(n+1)}{2}$. In this example, $n_0 = 1$.

Basis Step: We must first show that $P(1)$ is true. $P(1)$ is the statement

$$1 = \frac{1(1+1)}{2},$$

which is clearly true.

Induction Step: We must now show that for $k \geq 1$, if $P(k)$ is true, then $P(k+1)$ must also be true. We assume that for some fixed $k \geq 1$,

$$1 + 2 + 3 + \cdots + k = \frac{k(k+1)}{2}. \tag{1}$$

We now wish to show the truth of $P(k+1)$:

$$1 + 2 + 3 + \cdots + (k+1) = \frac{(k+1)((k+1)+1)}{2}.$$

The left-hand side of $P(k+1)$ can be written as $1 + 2 + 3 + \cdots + k + (k+1)$

and we have

$$(1 + 2 + 3 + \cdots + k) + (k + 1)$$

$$= \frac{k(k + 1)}{2} + (k + 1) \qquad \text{using (1) to replace } 1 + 2 + \cdots + k$$

$$= (k + 1)\left[\frac{k}{2} + 1\right] \qquad \text{factoring}$$

$$= \frac{(k + 1)(k + 2)}{2}$$

$$= \frac{(k + 1)((k + 1) + 1)}{2} \qquad \text{the right-hand side of P}(k + 1)$$

Thus, we have shown the left-hand side of P$(k + 1)$ equals the right-hand side of P$(k + 1)$. By the principle of mathematical induction, it follows that P(n) is true for all $n \geq 1$. ■

EXAMPLE 2   Let $A_1, A_2, A_3, \ldots, A_n$ be any n sets. We show by mathematical induction that

$$\overline{\left(\bigcup_{i=1}^{n} A_i\right)} = \bigcap_{i=1}^{n} \overline{A_i}.$$

(This is an extended version of one of De Morgan's laws.) Let P(n) be the predicate that the equality holds for any n sets. We prove by mathematical induction that for all $n \geq 1$, P(n) is true.

Basis Step: P(1) is the statement $\overline{A_1} = \overline{A_1}$, which is obviously true.

Induction Step: We use P(k) to show P$(k+1)$. The left-hand side of P$(k+1)$ is

$$\overline{\left(\bigcup_{i=1}^{k+1} A_i\right)} = \overline{A_1 \cup A_2 \cup \cdots \cup A_k \cup A_{k+1}}$$

$$= \overline{(A_1 \cup A_2 \cup \cdots \cup A_k) \cup A_{k+1}} \qquad \text{associative property of } \cup$$

$$= \overline{(A_1 \cup A_2 \cup \cdots \cup A_k)} \cap \overline{A_{k+1}} \qquad \text{by De Morgan's law for two sets}$$

$$= \left(\bigcap_{i=1}^{k} \overline{A_i}\right) \cap \overline{A_{k+1}} \qquad \text{using P}(k)$$

$$= \bigcap_{i=1}^{k+1} \overline{A_i} \qquad \text{right-hand side of P}(k + 1)$$

Thus, the implication P$(k) \Rightarrow$ P$(k + 1)$ is a tautology, and by the principle of mathematical induction P(n) is true for all $n \geq 1$. ■

EXAMPLE 3

We show by mathematical induction that any finite, nonempty set is countable; that is, it can be arranged in a list.

Let $P(n)$ be the predicate that if A is any set with $|A| = n \geq 1$, then A is countable. (See Chapter 1 for definitions.)

Basis Step: Here n_0 is 1, so we let A be any set with one element, say $A = \{x\}$. In this case x forms a sequence all by itself whose set is A, so $P(1)$ is true.

Induction Step: We want to use the statement $P(k)$ that if A is any set with k elements, then A is countable. Now choose any set B with $k + 1$ elements and pick any element x in B. Since $B - \{x\}$ is a set with k elements, the induction hypothesis $P(k)$ tells us there is a sequence $x_1, x_2, \ldots, x_k$ with $B - \{x\}$ as its corresponding set. The sequence $x_1, x_2, \ldots, x_k, x$ then has B as the corresponding set so B is countable. Since B can be any set with $k + 1$ elements, $P(k + 1)$ is true if $P(k)$ is. Thus, by the principle of mathematical induction, $P(n)$ is true for all $n \geq 1$. ∎

In proving results by induction, you should not start by assuming that $P(k+1)$ is true and attempting to manipulate this result until you arrive at a true statement. This common mistake is always an incorrect use of the principle of mathematical induction.

A natural connection exists between recursion and induction, because objects that are recursively defined often use a natural sequence in their definition. Induction is frequently the best, maybe the only, way to prove results about recursively defined objects.

EXAMPLE 4

Consider the following recursive definition of the factorial function: $1! = 1$, $n! = n(n - 1)!$, $n > 1$. Suppose we wish to prove for all $n \geq 1$, $n! \geq 2^{n-1}$. We proceed by mathematical induction. Let $P(n): n! \geq 2^{n-1}$. Here n_0 is 1.

Basis Step: $P(1)$ is the statement $1! \geq 2^0$. Since $1!$ is 1, this statement is true.

Induction Step: We want to show $P(k) \Rightarrow P(k + 1)$ is a tautology. It will be a tautology if $P(k)$ true guarantees $P(k + 1)$ is true. Suppose $k! \geq 2^{k-1}$ for some $k \geq 1$. Then by the recursive definition, the left side of $P(k + 1)$ is

$$
\begin{aligned}
(k + 1)! = (k + 1)k! & \\
\geq (k + 1)2^{k-1} \quad & \text{using } P(k) \\
\geq 2 \times 2^{k-1} \quad & k + 1 \geq 2, \text{ since } k \geq 1 \\
= 2^k \quad & \text{right-hand side of } P(k + 1)
\end{aligned}
$$

Thus, $P(k + 1)$ is true. By the principle of mathematical induction, it follows that $P(n)$ is true for all $n \geq 1$. ∎

The following example shows one way in which induction can be useful in computer programming. The pseudocode used in this and following examples is described in Appendix A.

EXAMPLE 5   Consider the following function given in pseudocode.

FUNCTION SQ(A)
1. $C \leftarrow 0$
2. $D \leftarrow 0$
3.   **WHILE** ($D \neq A$)
     a. $C \leftarrow C + A$
     b. $D \leftarrow D + 1$
4.   **RETURN** (C)
END OF FUNCTION SQ

   The name of the function, SQ, suggests that it computes the square of A. Step 3b shows A must be a positive integer if the looping is to end. A few trials with particular values of A will provide evidence that the function does carry out this task. However, suppose we now want to prove that SQ always computes the square of the positive integer A, no matter how large A might be. We shall give a proof by mathematical induction. For each integer $n \geq 0$, let C_n and D_n be the values of the variables C and D, respectively, after passing through the **WHILE** loop n times. In particular, C_0 and D_0 represent the values of the variables before looping starts. Let $P(n)$ be the predicate $C_n = A \times D_n$. We shall prove by induction that $\forall n \geq 0$ $P(n)$ is true. Here n_0 is 0.

   Basis Step: P(0) is the statement $C_0 = A \times D_0$, which is true since the value of both C and D is zero "after" zero passes through the **WHILE** loop.

   Induction Step: We must now use

$$P(k): C_k = A \times D_k \tag{2}$$

to show that P($k + 1$): $C_{k+1} = A \times D_{k+1}$. After a pass through the loop, C is increased by A, and D is increased by 1, so $C_{k+1} = C_k + A$ and $D_{k+1} = D_k + 1$.

left-hand side of P($k + 1$): $C_{k+1} = C_k + A$
$\qquad\qquad\qquad = A \times D_k + A$ using (2) to replace C_k
$\qquad\qquad\qquad = A \times (D_k + 1)$ factoring
$\qquad\qquad\qquad = A \times D_{k+1}$ right-hand side of
$\qquad\qquad\qquad\qquad\qquad\qquad$ P($k + 1$)

   By the principle of mathematical induction, it follows that as long as looping occurs, $C_n = A \times D_n$. The loop must terminate. (Why?) When the loop terminates, $D = A$, so $C = A \times A$, or A^2, and this is the value returned by the function SQ.   ∎

   Example 5 illustrates the use of a **loop invariant**, a relationship between variables that persists through all iterations of the loop. This technique for proving that loops and programs do what is claimed they do is an important part of the theory of algorithm verification. In Example 5 it is clear that the looping stops if A is a positive integer, but for more complex cases, this may also be proved by induction.

EXAMPLE 6   Use the technique of Example 5 to prove that the pseudocode program given in Section 1.4 does compute the greatest common divisor of two positive integers.

Solution   Here is the pseudocode given earlier.

FUNCTION GCD(X, Y)
1.   **WHILE** ($X \neq Y$)
    a.   **IF** ($X > Y$) **THEN**
        1. $X \leftarrow X - Y$
    b.   **ELSE**
        1. $Y \leftarrow Y - X$
2.   **RETURN** (X)
END OF FUNCTION GCD

We claim that if X and Y are positive integers, then GCD returns GCD(X, Y). To prove this, let X_n and Y_n be the values of X and Y after $n \geq 0$ passes through the **WHILE** loop. We claim that P(n): GCD(X_n, Y_n) = GCD(X, Y) is true for all $n \geq 0$, and we prove this by mathematical induction. Here n_0 is 0.

Basis Step: $X_0 = X$, $Y_0 = Y$, since these are the values of the variables before looping begins; thus P(0) is the statement GCD(X_0, Y_0) = GCD(X, Y), which is true.

Induction Step: Consider the left-hand side of P($k + 1$), that is, GCD(X_{k+1}, Y_{k+1}). After the $k + 1$ pass through the loop, either $X_{k+1} = X_k$ and $Y_{k+1} = Y_k - X_k$ or $X_{k+1} = X_k - Y_k$ and $Y_{k+1} = Y_k$. Then if P(k): GCD(X_k, Y_k) = GCD(X, Y) is true, we have, by Theorem 5, Section 1.4, that GCD(X_{k+1}, Y_{k+1}) = GCD(X_k, Y_k) = GCD(X, Y). Thus, by the principle of mathematical induction, P(n) is true for all $n \geq 0$. The exit condition for the loop is $X_n = Y_n$ and we have GCD(X_n, Y_n) = X_n. Hence the function always returns the value GCD(X, Y). ∎

Strong Induction

A slightly different form of mathematical induction is easier to use in some proofs. In the **strong form of mathematical induction**, or strong induction, the induction step is to show that

$$\text{P}(n_0) \wedge \text{P}(n_0 + 1) \wedge \text{P}(n_0 + 2) \wedge \cdots \wedge \text{P}(k) \Rightarrow \text{P}(k + 1)$$

is a tautology. As before, the only case we need to check is that if each P(j), $j = n_0, \ldots, k$ is true, then P($k + 1$) is true. The strong form of induction is equivalent to the form we first presented so it is a matter of convenience which we use in a proof.

> **EXAMPLE 7**

Prove that every positive integer $n > 1$ can be written uniquely as $p_1^{a_1} p_2^{a_2} \cdots p_s^{a_s}$, where the p_i are primes and $p_1 < p_2 < \cdots < p_s$. (Theorem 3, Section 1.4)

Proof (by strong induction)

Basis Step: Here n_0 is 2. P(2) is clearly true, since 2 is prime.

Induction Step: We use P(2), P(3), ..., P(k) to show P($k + 1$): $k + 1$ can be written uniquely as $p_1^{a_1} p_2^{a_2} \cdots p_s^{a_s}$, where the p_i are primes and $p_1 < p_2 < \cdots < p_s$. There are two cases to consider. If $k + 1$ is a prime, then P($k + 1$) is true. If $k+1$ is not prime, then $k+1 = lm$, $2 \leq l \leq k$, $2 \leq m \leq k$. Using $P(l)$ and $P(m)$, we have $k = lm = q_1^{b_1} q_2^{b_2} \cdots q_t^{b_t} r_1^{c_1} r_2^{c_2} \cdots r_u^{c_u} = p_1^{a_1} p_2^{a_2} \cdots p_s^{a_s}$, where each $p_i = q_j$ or r_k, $p_1 < p_2 < \cdots < p_s$, and if $q_j = r_k = p_i$, then $a_i = b_j + c_k$, otherwise $p_i = q_j$ and $a_i = b_j$ or $p_i = r_k$ and $a_i = c_k$. Since the factorization of l and m are unique, so is the factorization of $k+1$. ∎

2.4 Exercises

In Exercises 1 through 7, prove the statement is true by using mathematical induction.

1. $2 + 4 + 6 + \cdots + 2n = n(n+1)$

2. $1^2 + 3^2 + 5^2 + \cdots + (2n-1)^2 = \dfrac{n(2n+1)(2n-1)}{3}$

3. $1 + 2^1 + 2^2 + \cdots + 2^n = 2^{n+1} - 1$

4. $5 + 10 + 15 + \cdots + 5n = \dfrac{5n(n+1)}{2}$

5. $1^2 + 2^2 + 3^2 + \cdots + n^2 = \dfrac{n(n+1)(2n+1)}{6}$

6. $1 + a + a^2 + \cdots + a^{n-1} = \dfrac{a^n - 1}{a - 1}$

7. $a + ar + ar^2 + \cdots + ar^{n-1} = \dfrac{a(1-r^n)}{1-r}$ for $r \neq 1$

8. Let $P(n)$: $1^3 + 2^3 + 3^3 + \cdots + n^3 = \dfrac{n^2(n+1)^2 + 4}{4}$.
   (a) Use $P(k)$ to show $P(k+1)$.
   (b) Is $P(n)$ true for all $n \geq 1$?

9. Let $P(n)$: $1 + 5 + 9 + \cdots + (4n-3) = (2n+1)(n-1)$.
   (a) Use $P(k)$ to show $P(k+1)$.
   (b) Is $P(n)$ true for all $n \geq 1$?

10. Prove $1 + 2^n < 3^n$ for $n \geq 2$

11. Prove $n < 2^n$ for $n > 1$

12. Prove $1 + 2 + 3 + \cdots + n < \dfrac{(2n+1)^2}{8}$

13. Find the least n for which the statement is true and then prove that $(1 + n^2) < 2^n$.

14. Find the least n for which the statement is true and then prove that $10n < 3^n$.

15. Prove by mathematical induction that if a set A has n elements, then $P(A)$ has 2^n elements.

16. Prove by mathematical induction that $3 \mid (n^3 - n)$ for every positive integer n.

17. Prove by mathematical induction that if $A_1, A_2, \ldots, A_n$ are any n sets, then
$$\overline{\left(\bigcap_{i=1}^{n} A_i \right)} = \bigcup_{i=1}^{n} \overline{A_i}.$$

18. Prove by mathematical induction that if $A_1, A_2, \ldots, A_n$ and B are any $n+1$ sets, then
$$\left(\bigcup_{i=1}^{n} A_i \right) \cap B = \bigcup_{i=1}^{n} (A_i \cap B).$$

19. Prove by mathematical induction that if $A_1, A_2, \ldots, A_n$ and B are any $n+1$ sets, then
$$\left(\bigcap_{i=1}^{n} A_i \right) \cup B = \bigcap_{i=1}^{n} (A_i \cup B).$$

20. Let $P(n)$ be the statement $2 \mid (2n-1)$.
   (a) Prove that $P(k) \Rightarrow P(k+1)$ is a tautology.
   (b) Show that $P(n)$ is not true for any integer n.
   (c) Do the results in (a) and (b) contradict the principle of mathematical induction? Explain.

In Exercises 21 through 23, prove the given statement about matrices. Assume $\mathbf{A}$ is $n \times n$.

21. $(\mathbf{A}_1 + \mathbf{A}_2 + \cdots + \mathbf{A}_n)^T = \mathbf{A}_1^T + \mathbf{A}_2^T + \cdots + \mathbf{A}_n^T$

22. $\mathbf{A}^2 \mathbf{A}^n = \mathbf{A}^{2+n}$

23. Let $\mathbf{A}$ and $\mathbf{B}$ be square matrices. If $\mathbf{AB} = \mathbf{BA}$, then $(\mathbf{AB})^n = \mathbf{A}^n \mathbf{B}^n$, for $n \geq 1$.

24. Prove that any restaurant bill of $\$n$, $n \geq 5$, can be paid exactly using only $\$2$ and $\$5$ bills.

25. Prove that every integer greater than 27 can be written as $5a + 8b$, where $a, b \in Z^+$.

26. Use induction to show that if p is a prime and $p \mid a^n$ for $n > 1$, then $p \mid a$.

27. Prove that if $\text{GCD}(a, b) = 1$, then $\text{GCD}(a^n, b^n) = 1$ for all $n \geq 1$. (*Hint:* Use Exercise 26.)

28. (a) Find the smallest positive integer n_0 such that $2^{n_0} > n_0^2$.

   (b) Prove $2^n > n^2$ for all $n \geq n_0$.

In Exercises 29 through 34, show that the given algorithm, correctly used, produces the output stated, by using mathematical induction to prove the relationship indicated is a loop invariant and checking values when the looping stops. All variables represent nonnegative integers.

29. SUBROUTINE COMP $(X, Y; Z)$
1. $Z \leftarrow X$
2. $W \leftarrow Y$
3. **WHILE** $(W > 0)$
   a. $Z \leftarrow Z + Y$
   b. $W \leftarrow W - 1$
4. **RETURN**
END OF SUBROUTINE COMP
COMPUTES: $Z = X + Y^2$
LOOP INVARIANT: $(Y \times W) + Z = X + Y^2$

30. SUBROUTINE DIFF $(X, Y; Z)$
1. $Z \leftarrow X$
2. $W \leftarrow Y$
3. **WHILE** $(W > 0)$
   a. $Z \leftarrow Z - 1$
   b. $W \leftarrow W - 1$
4. **RETURN**
END OF SUBROUTINE DIFF
COMPUTES: $Z = X - Y$
LOOP INVARIANT: $X - Z + W = Y$

31. SUBROUTINE EXP2 $(N, M; R)$
1. $R \leftarrow 1$
2. $K \leftarrow 2M$
3. **WHILE** $(K > 0)$
   a. $R \leftarrow R \times N$
   b. $K \leftarrow K - 1$
4. **RETURN**
END OF SUBROUTINE EXP2
COMPUTES: $R = N^{2M}$
LOOP INVARIANT: $R \times N^K = N^{2M}$

32. SUBROUTINE POWER $(X, Y; Z)$
1. $Z \leftarrow 0$
2. $W \leftarrow Y$
3. **WHILE** $(W > 0)$
   a. $Z \leftarrow Z + X$
   b. $W \leftarrow W - 1$
4. $W \leftarrow Y - 1$

5. $U \leftarrow Z$
6. **WHILE** $(W > 0)$
   a. $Z \leftarrow Z + U$
   b. $W \leftarrow W - 1$
7. **RETURN**
END OF SUBROUTINE POWER
COMPUTES: $Z = X \times Y^2$
LOOP INVARIANT (first loop):
$X + (X \times W) = X \times Y$
LOOP INVARIANT (second loop):
$X + (X \times Y \times W) = X \times Y^2$
(*Hint*: Use the value of Z at the end of the first loop in loop 2.)

33. SUBROUTINE DIV (X, Y)
1. **IF** $(Y = 0)$ **THEN**
   a. **PRINT** ('error $Y = 0$')
2. **ELSE**
   a. $R \leftarrow X$
   b. $K \leftarrow 0$
   c. **WHILE** $(K \geq Y)$
      i. $R \leftarrow R - Y$
      ii. $K \leftarrow K + 1$
   d. **IF** $(R = 0)$ **THEN**
      i. **PRINT** ('true')
   e. **ELSE**
      i. **PRINT** ('false')
3. **RETURN**
END OF SUBROUTINE DIV
COMPUTES: TRUTH VALUE OF $Y \mid X$.
LOOP INVARIANT: $R + K \times Y = X$

34. SUBROUTINE SQS $(X, Y; Z)$
1. $Z \leftarrow Y$
2. $W \leftarrow X$
3. **WHILE** $(W > 0)$
   a. $Z \leftarrow Z + X$
   b. $W \leftarrow W - 1$
4. $W \leftarrow Y - 1$
5. **WHILE** $(W > 0)$
   a. $Z \leftarrow Z + X$
   b. $W \leftarrow W - 1$
6. **RETURN**
END OF SUBROUTINE SQS
COMPUTES: $Z = X^2 \times Y^2$
LOOP INVARIANT (first loop):
$Z + (X \times W) = Y + X^2$
LOOP INVARIANT (second loop):
$Z + (Y \times W) = X^2 + Y^2$

TIPS FOR PROOFS

This chapter provides the formal basis for our proofs, although most proofs are not so formal as the patterns given in Section 2.3. Two new types of proofs are presented: indirect proofs and induction proofs. Indirect proofs are based either on the pattern $(p \Rightarrow q) \wedge \sim q$ (proof by contradiction) or on the fact that $(p \Rightarrow q) \equiv (\sim q \Rightarrow \sim p)$ (prove the contrapositive). There are no hard and fast rules about when to use a direct or indirect proof. One strategy is to proceed optimistically with a direct proof. If that does not lead to anything useful, you may be able to identify a counterexample if the statement is in fact false or start a new proof based on one of the indirect models. Where the difficulty occurs in the attempted direct proof can often point the way to go next. Remember that a certain amount of creativity is required for any proof.

Statements that are good candidates for proof by induction are ones that involve the counting or whole numbers in some way, either to count something or to describe a pattern. Examples of these are in Section 2.4, Exercises 11 and 15. Notice that for most of the induction proofs in Section 2.4, $P(k)$ is used early and then properties of operations and arithmetic are used, but in proving loop invariants, the "arithmetic" comes first, then the use of $P(k)$.

In proving statements about propositions, try to use the properties of logical operations (see Section 2.2, Theorem 1 for some of these). Building truth tables should be your second-choice strategy.

KEY IDEAS FOR REVIEW

- Statement: declarative sentence that is either true or false, but not both
- Propositional variable: letter denoting a statement
- Compound statement: statement obtained by combining two or more statements by a logical connective
- Logical connectives: not ($\sim$), and ($\wedge$), or ($\vee$), if then ($\Rightarrow$), if and only if ($\Leftrightarrow$)
- Conjunction: $p \wedge q$ (p and q)
- Disjunction: $p \vee q$ (p or q)
- Predicate (propositional function): a sentence of the form $P(x)$
- Universal quantification: $\forall x\ P(x)$ [For all values of x, $P(x)$ is true.]
- Existential quantification: $\exists x\ P(x)$ [There exists an x such that $P(x)$ is true.]
- Conditional statement or implication: $p \Rightarrow q$ (if p then q); p is the antecedent or hypothesis and q is the consequent or conclusion
- Converse of $p \Rightarrow q$: $q \Rightarrow p$
- Contrapositive of $p \Rightarrow q$: $\sim q \Rightarrow \sim p$
- Equivalence: $p \Leftrightarrow q$
- Tautology: a statement that is true for all possible values of its propositional variables

- Absurdity: a statement that is false for all possible values of its propositional variables
- Contingency: a statement that may be true or false, depending on the truth values of its propositional variables
- $p \equiv q$ (Logically equivalent statements p and q): $p \Leftrightarrow q$ is a tautology
- Methods of proof:
    q logically follows from p: see page 58
    Rules of inference: see page 58
    Modus ponens: see page 59
    Indirect method: see page 60
    Proof by contradiction: see page 60
- Counterexample: single instance that disproves a theorem or proposition
- Principle of mathematical induction: Let n_0 be a fixed integer. Suppose that for each integer $n \geq n_0$ we have a proposition $P(n)$. Suppose that (a) $P(n_0)$ is true and (b) If $P(k)$, then $P(k + 1)$ is a tautology for every $k \geq n_0$. Then the principle of mathematical induction states that $P(n)$ is true for all $n \geq n_0$.
- Loop invariant: a statement that is true before and after every pass through a programming loop
- Strong form of mathematical induction: see page 68

CODING EXERCISES

For each of the following, write the requested program or subroutine in pseudocode (as described in Appendix A) or in a programming language that you know. Test your code either with a paper-and-pencil trace or with a computer run.

1. Write a program that will print a truth table for $p \wedge \sim q$.

2. Write a program that will print a truth table for $(p \vee q) \Rightarrow r$.

3. Write a program that will print a truth table for any two-variable propositional function.

4. Write a subroutine EQUIVALENT that determines if two logical expressions are equivalent.

5. Write a subroutine that determines if a logical expression is a tautology, a contingency, or an absurdity.

CHAPTER 2 SELF-TEST

1. Determine the truth value of the given statements if p is true and q is false.

    (a) $\sim p \wedge q$ (b) $\sim p \vee \sim q$

2. Determine the truth value for each of the following statements. Assume $x, y \in Z$.

    (a) $\forall x, y \; x + y$ is even.

    (b) $\exists x \, \forall y \; x + y$ is even.

3. Make a truth table for $(p \wedge \sim p) \vee (\sim (q \wedge r))$.

For questions 4 through 6, let p: $1 < -1$, q: $|2| = |-2|$, r: $-3 < -1$, and s: $1 < 3$.

4. Write the symbolic version of the converse and of the contrapositive for each of the following propositions.

    (a) $p \Rightarrow q$

    (b) $(\sim r) \vee (\sim s) \Rightarrow q$

    (c) $q \Rightarrow p \vee s$

5. Write the converse and the contrapositive of these propositions as English sentences.

6. Give the truth value of each proposition in Problem 4.

7. The English word "or" is sometimes used in the exclusive sense meaning that either p or q, but not both, is true. Make a truth table for this exclusive or, *xor*.

8. Let p: An Internet business is cheaper to start, q: I will start an Internet business, and r: An Internet business makes less money. For each of the following write the argument in English sentences and also determine the validity of the argument.

    (a) $\begin{array}{l} r \Rightarrow (q \Rightarrow p) \\ \underline{\sim p} \\ \therefore \; (\sim r) \vee (\sim q) \end{array}$ (b) $\begin{array}{l} p \Rightarrow q \\ q \Rightarrow r \\ \underline{p} \\ \therefore \; r \end{array}$

9. Suppose that m and n are integers such that $n \mid m$ and $m \mid n$. Are these hypotheses sufficient to prove that $m = n$? If so, give a proof. If not, supply a simple additional hypothesis that will guarantee $m = n$ and provide a proof.

10. Prove or disprove by giving a counterexample that the sum of any three consecutive odd integers is divisible by 6.

11. Use mathematical induction to prove that $4^n - 1$ is divisible by 3.

12. Use mathematical induction to prove that $1 + 2 + 3 + \cdots + n < \dfrac{(n + 1)^2}{2}$.

3

COUNTING

Prerequisites: Chapter 1

Techniques for counting are important in mathematics and in computer science, especially in the analysis of algorithms. In Section 1.2, the addition principle was introduced. In this chapter, we present other counting techniques, in particular those for permutations and combinations, and we look at two applications of counting, the pigeonhole principle and probability. In addition, recurrence relations, another tool for the analysis of computer programs, are discussed.

3.1 PERMUTATIONS

We begin with a simple but general result we will use frequently in this section and elsewhere.

Theorem 1

Suppose that two tasks T_1 and T_2 are to be performed in sequence. If T_1 can be performed in n_1 ways, and for each of these ways T_2 can be performed in n_2 ways, then the sequence $T_1 T_2$ can be performed in $n_1 n_2$ ways. ●

Proof Each choice of a method of performing T_1 will result in a different way of performing the task sequence. There are n_1 such methods, and for each of these we may choose n_2 ways of performing T_2. Thus, in all, there will be $n_1 n_2$ ways of performing the sequence $T_1 T_2$. See Figure 3.1 for the case where n_1 is 3 and n_2 is 4. ▼

Theorem 1 is sometimes called the **multiplication principle of counting**. (You should compare it carefully with the addition principle of counting from Section 1.2.) It is an easy matter to extend the multiplication principle as follows.

Theorem 2

Suppose that tasks $T_1, T_2, \ldots, T_k$ are to be performed in sequence. If T_1 can be performed in n_1 ways, and for each of these ways T_2 can be performed in n_2 ways, and

73

Possible ways of performing task 1   Possible ways of performing task 2

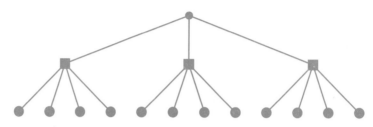

Possible ways of performing task 1, then task 2 in sequence

Figure 3.1

for each of these $n_1 n_2$ ways of performing $T_1 T_2$ in sequence, T_3 can be performed in n_3 ways, and so on, then the sequence $T_1 T_2 \cdots T_k$ can be performed in exactly $n_1 n_2 \cdots n_k$ ways.   ●

Proof   This result can be proved by using the principle of mathematical induction on k.   ▼

EXAMPLE 1   A label identifier, for a computer system, consists of one letter followed by three digits. If repetitions are allowed, how many distinct label identifiers are possible?

Solution   There are 26 possibilities for the beginning letter and there are 10 possibilities for each of the three digits. Thus, by the extended multiplication principle, there are $26 \times 10 \times 10 \times 10$ or 26,000 possible label identifiers.   ■

EXAMPLE 2   Let A be a set with n elements. How many subsets does A have?

Solution   We know from Section 1.3 that each subset of A is determined by its characteristic function, and if A has n elements, this function may be described as an array of 0's and 1's having length n. The first element of the array can be filled in two ways (with a 0 or a 1), and this is true for all succeeding elements as well. Thus, by the extended multiplication principle, there are

$$\underbrace{2 \cdot 2 \cdot \cdots \cdot 2}_{n \text{ factors}} = 2^n$$

ways of filling the array, and therefore 2^n subsets of A.   ■

We now turn our attention to the following counting problem. Let A be any set with n elements, and suppose that $1 \le r \le n$.

Problem 1 How many different sequences, each of length r, can be formed using elements from A if

(a) elements in the sequence may be repeated?

(b) all elements in the sequence must be distinct?

First we note that any sequence of length r can be formed by filling r boxes in order from left to right with elements of A. In case (a) we may use copies of elements of A.

box 1   box 2   box 3   box $r-1$ box r

Let T_1 be the task "fill box 1," let T_2 be the task "fill box 2," and so on. Then the combined task $T_1 T_2 \cdots T_r$ represents the formation of the sequence.

Case (a). T_1 can be accomplished in n ways, since we may copy any element of A for the first position of the sequence. The same is true for each of the tasks $T_2, T_3, \ldots, T_r$. Then by the extended multiplication principle, the number of sequences that can be formed is

$$\underbrace{n \cdot n \cdot \cdots \cdot n}_{r \text{ factors}} = n^r.$$

We have therefore proved the following result.

Theorem 3 Let A be a set with n elements and $1 \leq r \leq n$. Then the number of sequences of length r that can be formed from elements of A, allowing repetitions, is n^r. ●

EXAMPLE 3 How many three-letter "words" can be formed from letters in the set $\{a, b, y, z\}$ if repeated letters are allowed?

Solution Here n is 4 and r is 3, so the number of such words is 4^3 or 64, by Theorem 3. ■

Now we consider case (b) of Problem 1. Here also T_1 can be performed in n ways, since any element of A can be chosen for the first position. Whichever element is chosen, only $(n-1)$ elements remain, so that T_2 can be performed in $(n-1)$ ways, and so on, until finally T_r can be performed in $n - (r-1))$ or $(n-r+1)$ ways. Thus, by the extended principle of multiplication, a sequence of r distinct elements from A can be formed in $n(n-1)(n-2)\cdots(n-r+1)$ ways.

A sequence of r distinct elements of A is often called a permutation of A taken r at a time. This terminology is standard, and therefore we adopt it, but it is confusing. A better terminology might be a "permutation of r elements chosen from A." Many sequences of interest are permutations of some set of n objects taken r at a time. The preceding discussion shows that the number of such sequences depends only on n and r, not on A. This number is often written $_nP_r$ and is called the **number of permutations of n objects taken r at a time**. We have just proved the following result.

Theorem 4   If $1 \le r \le n$, then $_nP_r$, the number of permutations of n objects taken r at a time, is $n \cdot (n-1) \cdot (n-2) \cdot \cdots \cdot (n-r+1)$.   ●

EXAMPLE 4   Let A be $\{1, 2, 3, 4\}$. Then the sequences 124, 421, 341, and 243 are some permutations of A taken 3 at a time. The sequences 12, 43, 31, 24, and 21 are examples of different permutations of A taken two at a time. By Theorem 4, the total number of permutations of A taken three at a time is $_4P_3$ or $4 \cdot 3 \cdot 2$ or 24. The total; number of permutations of A taken two at a time is $_4P_2$ or $4 \cdot 3$ or 12.   ■

When $r = n$, we are counting the distinct arrangements of the elements of A, with $|A| = n$, into sequences of length n. Such a sequence is simply called a **permutation** of A. In Chapter 5 we use the term "permutation" in a slightly different way to increase its utility. The number of permutations of A is thus $_nP_n$ or $n \cdot (n-1) \cdot (n-2) \cdot \cdots \cdot 2 \cdot 1$, if $n \ge 1$. This number is also written $n!$ and is read ***n* factorial**. Both $_nP_r$ and $n!$ are built-in functions on many calculators.

EXAMPLE 5   Let A be $\{a, b, c\}$. Then the possible permutations of A are the sequences abc, acb, bac, bca, cab, and cba.   ■

For convenience, we define $0!$ to be 1. Then for every $n \ge 0$ the number of permutations of n objects is $n!$. If $n \ge 1$ and $1 \le r \le n$, we can now give a more compact form for $_nP_r$ as follows:

$$_nP_r = n \cdot (n-1) \cdot (n-2) \cdot \cdots \cdot (n-r+1)$$

$$= \frac{n \cdot (n-1) \cdot \cdots \cdot (n-r+1) \cdot (n-r) \cdot (n-r-1) \cdot \cdots \cdot 2 \cdot 1}{(n-r) \cdot (n-r-1) \cdot \cdots \cdot 2 \cdot 1}$$

$$= \frac{n!}{(n-r)!}.$$

EXAMPLE 6   Let A consist of all 52 cards in an ordinary deck of playing cards. Suppose that this deck is shuffled and a hand of five cards is dealt. A list of cards in this hand, in the order in which they were dealt, is a permutation of A taken five at a time. Examples would include AH, 3D, 5C, 2H, JS; 2H, 3H, 5H, QH, KD; JH, JD, JS, 4H, 4C; and 3D, 2H, AH, JS, 5C. Note that the first and last hands are the same, but they represent different permutations since they were dealt in a different order. The number of permutations of A taken five at a time is $_{52}P_5 = \frac{52!}{47!}$ or $52 \cdot 51 \cdot 50 \cdot 49 \cdot 48$ or 311,875,200. This is the number of five-card hands that can be dealt if we consider the order in which they were dealt.   ■

EXAMPLE 7   If A is the set in Example 5, then n is 3 and the number of permutations of A is $3!$ or 6. Thus, all the permutations of A are listed in Example 5, as claimed.   ■

EXAMPLE 8   How many "words" of three distinct letters can be formed from the letters of the word MAST?

Solution   The number is $_4P_3 = \dfrac{4!}{(4-3)!}$ or $\dfrac{4!}{1!}$ or 24.   ■

In Example 8 if the word had been MASS, $_4P_3$ would count as distinct some permutations that cannot be distinguished. For example, if we tag the two S's as S_1 and S_2, then S_1AS_2 and S_2AS_1 are two of the 24 permutations counted, but without the tags, these are the same "word." We have one more case to consider, permutations with limited repeats.

EXAMPLE 9   How many distinguishable permutations of the letters in the word BANANA are there?

Solution   We begin by tagging the A's and N's in order to distinguish between them temporarily. For the letters B, A_1, N_1, A_2, N_2, A_3, there are 6! or 720 permutations. Some of these permutations are identical except for the order in which the N's appear; for example, $A_1A_2A_3BN_1N_2$ and $A_1A_2A_3BN_2N_1$. In fact, the 720 permutations can be listed in pairs whose members differ only in the order of the two N's. This means that if the tags are dropped from the N's only $\frac{720}{2}$ or 360 distinguishable permutations remain. Reasoning in a similar way we see that these can be grouped in groups of 3! or 6 that differ only in the order of the three A's. For example, one group of 6 consists of $BNNA_1A_2A_3$, $BNNA_1A_3A_2$, $BNNA_2A_1A_3$, $BNNA_2A_3A_1$, $BNNA_3A_1A_2$, $BNNA_3A_2A_1$. Dropping the tags would change these 6 into the single permutation BNNAAA. Thus, there are $\frac{360}{6}$ or 60 distinguishable permutations of the letters of BANANA. ∎

The following theorem describes the general situation for permutations with limited repeats.

Theorem 5   The number of distinguishable permutations that can be formed from a collection of n objects where the first object appears k_1 times, the second object k_2 times, and so on, is

$$\frac{n!}{k_1!\,k_2!\cdots k_t!}.$$

EXAMPLE 10   The number of distinguishable "words" that can be formed from the letters of MISSISSIPPI is $\frac{11!}{1!\,4!\,4!\,2!}$ or 34,650. ∎

3.1 Exercises

1. A bank password consists of two letters of the English alphabet followed by two digits. How many different passwords are there?

2. In a psychological experiment, a person must arrange a square, a cube, a circle, a triangle, and a pentagon in a row. How many different arrangements are possible?

3. A coin is tossed four times and the result of each toss is recorded. How many different sequences of heads and tails are possible?

4. A catered menu is to include a soup, a main course, a dessert, and a beverage. Suppose a customer can select from four soups, five main courses, three desserts, and two beverages. How many different menus can be selected?

5. A fair six-sided die is tossed four times and the numbers shown are recorded in a sequence. How many different sequences are there?

6. Let $A = \{0, 1\}$.

   (a) How many strings of length three are there in A^*?

   (b) How many strings of length seven are there in A^*?

7. (a) Compute the number of strings of length four in the set corresponding to the regular expression $(01)^*1$.

   (b) Compute the number of strings of length five in the set corresponding to the regular expression $(01)^*1$.

8. Compute each of the following.

   (a) $_4P_4$ (b) $_6P_5$ (c) $_7P_2$

9. Compute each of the following.

   (a) $_nP_{n-1}$ (b) $_nP_{n-2}$ (c) $_{n+1}P_{n-1}$

In Exercises 10 through 13, compute the number of permutations of the given set.

10. $\{r, s, t, u\}$ $4!$

11. $\{1, 2, 3, 4, 5\}$ $5!$

12. $\{a, b, 1, 2, 3, c\}$ $6!$

13. $\{4, 7, 10, 13\}$ $4!$

In Exercises 14 through 16, find the number of permutations of A taken r at a time.

14. $A = \{1, 2, 3, 4, 5, 6, 7\}, r = 3$

15. $A = \{a, b, c, d, e, f\}, r = 2$

16. $A = \{x \mid x \text{ is an integer and } x^2 < 16\}, r = 4$

17. In how many ways can six men and six women be seated in a row if

   (a) any person may sit next to any other?

   (b) men and women must occupy alternate seats?

18. Find the number of different permutations of the letters in the word GROUP.

19. How many different arrangements of the letters in the word BOUGHT can be formed if the vowels must be kept next to each other?

20. Find the number of distinguishable permutations of the letters in BOOLEAN.

21. Find the number of distinguishable permutations of the letters in PASCAL.

22. Find the number of distinguishable permutations of the letters in ASSOCIATIVE.

23. Find the number of distinguishable permutations of the letters in REQUIREMENTS.

24. In how many ways can seven people be seated in a circle?

25. A bookshelf is to be used to display six new books. Suppose there are eight computer science books and five French books from which to choose. If we decide to show four computer science books and two French books and we are required to keep the books in each subject together, how many different displays are possible?

26. Three fair six-sided dice are tossed and the numbers showing on the top faces are recorded as a triple. How many different records are possible?

27. Prove that $n \cdot {_{n-1}P_{n-1}} = {_nP_n}$.

28. Most versions of Pascal allow variable names to consist of eight letters or digits with the requirement that the first character must be a letter. How many eight-character variable names are possible?

29. Until recently, U.S. telephone area codes were three-digit numbers whose middle digit was 0 or 1. Codes whose last two digits are 1's are used for other purposes (for example, 911). With these conditions how many area codes were available?

30. How many Social Security numbers can be assigned at any one time? Identify any assumptions you have made.

31. How many zeros are there at the end of 12!? at the end of 26!? at the end of 53!?

32. Give a procedure for determining the number of zeros at the end of $n!$. Justify your procedure.

3.2 COMBINATIONS

The multiplication principle and the counting methods for permutations all apply to situations where order matters. In this section we look at some counting problems where order does not matter.

Problem 2   Let A be any set with n elements and $1 \leq r \leq n$. How many different subsets of A are there, each with r elements?

The traditional name for an r-element subset of an n-element set A is a **combination of A, taken r at a time**.

EXAMPLE 1

Let $A = \{1, 2, 3, 4\}$. The following are all distinct combinations of A, taken three at a time: $A_1 = \{1, 2, 3\}$, $A_2 = \{1, 2, 4\}$, $A_3 = \{1, 3, 4\}$, $A_4 = \{2, 3, 4\}$. Note that these are subsets, not sequences. Thus $A_1 = \{2, 1, 3\} = \{2, 3, 1\} = \{1, 3, 2\} = \{3, 1, 2\} = \{3, 2, 1\}$. In other words, when it comes to combinations, unlike permutations, the order of the elements is irrelevant. ∎

EXAMPLE 2

Let A be the set of all 52 cards in an ordinary deck of playing cards. Then a combination of A, taken five at a time, is just a hand of five cards regardless of how these cards were dealt. ∎

We now want to count the number of r-element subsets of an n-element set A. This is most easily accomplished by using what we already know about permutations. Observe that each permutation of the elements of A, taken r at a time, can be produced by performing the following two tasks in sequence.

TASK 1:   Choose a subset B of A containing r elements.

TASK 2:   Choose a particular permutation of B.

We are trying to compute the number of ways to choose B. Call this number C. Then task 1 can be performed in C ways, and task 2 can be performed in $r!$ ways. Thus the total number of ways of performing both tasks is, by the multiplication principle, $C \cdot r!$. But it is also $_nP_r$. Hence,

$$C \cdot r! = {}_nP_r = \frac{n!}{(n-r)!}.$$

Therefore,

$$C = \frac{n!}{r!\,(n-r)!}.$$

We have proved the following result.

Theorem 1   Let A be a set with $|A| = n$, and let $1 \le r \le n$. Then the number of combinations of the elements of A, taken r at a time, that is the number of r-element subsets of A is

$$\frac{n!}{r!\,(n-r)!}.$$
●

Note again that the number of combinations of A, taken r at a time, does not depend on A, but only on n and r. This number is often written $_nC_r$ and is called the **number of combinations of n objects taken r at a time**. We have

$$_nC_r = \frac{n!}{r!\,(n-r)!}.$$

This computation is a built-in function on many calculators.

EXAMPLE 3   Compute the number of distinct five-card hands that can be dealt from a deck of 52 cards.

> *Solution*   This number is $_{52}C_5$ because the order in which the cards were dealt is irrelevant. $_{52}C_5 \doteq \frac{52!}{5!\,47!}$ or 2,598,960. Compare this number with the number computed in Section 3.1, Example 6. ∎

In the discussion of permutations, we considered cases where repetitions are allowed. We now look at one such case for combinations.

Consider the following situation. A radio station offers a prize of three CDs from the Top Ten list. The choice of CDs is left to the winner, and repeats are allowed. The order in which the choices are made is irrelevant. To determine the number of ways in which prize winners can make their choices, we use a problem-solving technique we have used before; we model the situation with one we already know how to handle.

Suppose choices are recorded by the station's voice mail system. After properly identifying herself, a winner is asked to press 1 if she wants CD number n and to press 2 if she does not. If 1 is pressed, the system asks again about CD number n. When 2 is pressed, the system asks about the next CD on the list. When three 1's have been recorded, the system tells the caller the selected CDs will be shipped. A record must be created for each of these calls. A record will be a sequence of 1's and 2's. Clearly there will be three 1's in the sequence. A sequence may contain as many as nine 2's, for example, if the winner refuses the first nine CDs and chooses three copies of CD number 10. Our model for counting the number of ways a prize winner can choose her three CDs is the following. Each three-CD selection can be represented by an array containing three 1's and nine 2's or blanks, or a total of twelve cells. Some possible records are 222122122221 (selecting numbers 4, 6, 10), 1211*bbbbbbbbb* (selecting number 1 and two copies of number 2), and 222222222111 (selecting three copies of number 10). The number of ways to select three cells of the array to hold 1's is $_{12}C_3$ since the array has $3 + 9$ or 12 cells and the order in which this selection is made does not matter. The following theorem generalizes this discussion.

Theorem 2   Suppose k selections are to be made from n items without regard to order and that repeats are allowed, assuming at least k copies of each of the n items. The number of ways these selections can be made is $_{(n+k-1)}C_k$. ●

EXAMPLE 4   In how many ways can a prize winner choose three CDs from the Top Ten list if repeats are allowed?

> *Solution*   Here n is 10 and k is 3. By Theorem 2, there are $_{10+3-1}C_3$ or $_{12}C_3$ ways to make the selections. The prize winner can make the selection in 220 ways. ∎

In general, when order matters, we count the number of sequences or permutations; when order does not matter, we count the number of subsets or combinations.

Some problems require that the counting of permutations and combinations be combined or supplemented by the direct use of the addition or the multiplication principle.

EXAMPLE 5 Suppose that a valid computer password consists of seven characters, the first of which is a letter chosen from the set {A, B, C, D, E, F, G} and the remaining six characters are letters chosen from the English alphabet or a digit. How many different passwords are possible?

Solution A password can be constructed by performing the tasks T_1 and T_2 in sequence.

TASK 1: Choose a starting letter from the set given.

TASK 2: Choose a sequence of letter and digits. Repeats are allowed.

Task T_1 can be performed in $_7C_1$ or 7 ways. Since there are 26 letters and 10 digits that can be chosen for each of the remaining six characters, and since repeats are allowed, task T_2 can be performed in 36^6 or 2,176,782,336 ways. By the multiplication principle, there are $7 \cdot 2176782336$ or 15,237,476,352 different passwords. ∎

EXAMPLE 6 How many different seven-person committees can be formed each containing three women from an available set of 20 women and four men from an available set of 30 men?

Solution In this case a committee can be formed by performing the following two tasks in succession:

TASK 1: Choose three women from the set of 20 women.

TASK 2: Choose four men from the set of 30 men.

Here order does not matter in the individual choices, so we are merely counting the number of possible subsets. Thus task 1 can be performed in $_{20}C_3$ or 1140 ways and task 2 can be performed in $_{30}C_4$ or 27,405 ways. By the multiplication principle, there are $(1140)(27405)$ or 31,241,700 different committees. ∎

3.2 Exercises

1. Compute each of the following.

   (a) $_7C_7$ (b) $_7C_4$ (c) $_{16}C_5$

2. Compute each of the following.

   (a) $_nC_{n-1}$ (b) $_nC_{n-2}$ (c) $_{n+1}C_{n-1}$

3. Show that $_nC_r = {_nC_{n-r}}$.

4. In how many ways can a committee of three faculty members and two students be selected from seven faculty members and eight students?

5. In how many ways can a six-card hand be dealt from a deck of 52 cards?

6. At a certain college, the housing office has decided to appoint, for each floor, one male and one female residential advisor. How many different pairs of advisors can be selected for a seven-story building from 12 male candidates and 15 female candidates?

7. A microcomputer manufacturer who is designing an advertising campaign is considering six magazines, three newspapers, two television stations, and four radio stations. In how many ways can six advertisements be run if

   (a) all six are to be in magazines?

   (b) two are to be in magazines, two are to be in newspapers, one is to be on television, and one is to be on radio?

8. How many different eight-card hands with five red cards and three black cards can be dealt from a deck of 52 cards?

9. (a) Find the number of subsets of each possible size for a set containing four elements.

(b) Find the number of subsets of each possible size for a set containing n elements.

For Exercises 10 through 13, suppose that an urn contains 15 balls, of which eight are red and seven are black.

10. In how many ways can five balls be chosen so that

(a) all five are red?

(b) all five are black?

11. In how many ways can five balls be chosen so that

(a) two are red and three are black?

(b) three are red and two are black?

12. In how many ways can five balls be chosen so that at most three are black?

13. In how many ways can five balls be chosen so that at least two are red?

14. A committee of six people with one person designated as chair of the committee is to be chosen. How many different committees of this type can be chosen from a group of 10 people?

15. A gift certificate at a local bookstore allows the recipient to choose six books from the combined list of ten best-selling fiction books and ten best-selling non-fiction books. In how many different ways can the selection of six books be made?

16. The college food plan allows a student to chose three pieces of fruit each day. The fruits available are apples, bananas, peaches, pears, and plums. For how many days can a student make a different selection?

17. Show that $_{n+1}C_r = {_n}C_{r-1} + {_n}C_r$.

18. (a) How many ways can a student choose eight out of ten questions to answer on an exam?

(b) How many ways can a student choose eight out of ten questions to answer on an exam if the first three questions must be answered?

19. Five fair coins are tossed and the results are recorded.

(a) How many different sequences of heads and tails are possible?

(b) How many of the sequences in part (a) have exactly one head recorded?

(c) How many of the sequences in part (a) have exactly three heads recorded?

20. Three fair six-sided dice are tossed and the numbers showing on top are recorded.

(a) How many different record sequences are possible?

(b) How many of the records in part (a) contain exactly one six?

(c) How many of the records in part (a) contain exactly two fours?

21. If n fair coins are tossed and the results recorded, how many

(a) record sequences are possible?

(b) sequences contain exactly three tails, assuming $n \geq 3$?

(c) sequences contain exactly k heads, assuming $n \geq k$?

22. If n fair six-sided dice are tossed and the numbers showing on top are recorded, how many

(a) record sequences are possible?

(b) sequences contain exactly one six?

(c) sequences contain exactly four twos, assuming $n \geq 4$?

23. How many ways can you choose three of seven fiction books and two of six non-fiction books to take with you on your vacation?

24. For the driving part of your vacation you will take 6 of the 35 rock cassettes in your collection, 3 of the 22 classical cassettes, and 1 of the 8 comedy cassettes. In how ways can you make your choice(s)?

25. The array commonly called Pascal's triangle can be defined by giving enough information to establish its pattern.

(a) Write the next three rows of Pascal's triangle.

(b) Give a rule for building the next row from the previous row(s).

26. Pascal's triangle can also be defined by an explicit pattern. Use the results of Exercises 9 and 25 to give an explicit rule for building the nth row of Pascal's triangle.

27. Explain the connections between Exercises 17, 25, and 26.

28. The list of numbers in any row of Pascal's triangle reads the same from left to right as it does from right to left. Such a sequence is called a *palindrome*. Use the results of Exercise 26 to prove that each row of Pascal's triangle is a palindrome.

3.3 PIGEONHOLE PRINCIPLE

In this section we introduce another proof technique, one that makes use of the counting methods we have discussed.

Theorem 1
The Pigeonhole Principle

If n pigeons are assigned to m pigeonholes, and $m < n$, then at least one pigeonhole contains two or more pigeons.   ●

Proof Suppose each pigeonhole contains at most 1 pigeon. Then at most m pigeons have been assigned. But since $m < n$, not all pigeons have been assigned pigeonholes. This is a contradiction. At least one pigeonhole contains two or more pigeons.   ▽

This informal and almost trivial sounding theorem is easy to use and has unexpected power in proving interesting consequences.

EXAMPLE 1

If eight people are chosen in any way from some group, at least two of them will have been born on the same day of the week. Here each person (pigeon) is assigned to the day of the week (pigeonhole) on which he or she was born. Since there are eight people and only seven days of the week, the pigeonhole principle tells us that at least two people must be assigned to the same day of the week.   ■

Note that the pigeonhole principle provides an existence proof; there must be an object or objects with a certain characteristic. In Example 1, this characteristic is having been born on the same day of the week. The pigeonhole principle guarantees that there are at least two people with this characteristic but gives no information on identifying these people. Only their existence is guaranteed. In contrast, a constructive proof guarantees the existence of an object or objects with a certain characteristic by actually constructing such an object or objects. For example, we could prove that given two rational numbers p and q there is a rational number between them by showing that $\frac{p+q}{2}$ is between p and q.

In order to use the pigeonhole principle we must identify pigeons (objects) and pigeonholes (categories of the desired characteristic) and be able to count the number of pigeons and the number of pigeonholes.

EXAMPLE 2

Show that if any five numbers from 1 to 8 are chosen, then two of them will add to 9.

Solution Construct four different sets, each containing two numbers that add up to 9 as follows: $A_1 = \{1, 8\}$, $A_2 = \{2, 7\}$, $A_3 = \{3, 6\}$, $A_4 = \{4, 5\}$. Each of the five numbers chosen must belong to one of these sets. Since there are only four sets, the pigeonhole principle tells us that two of the chosen numbers belong to the same set. These numbers add up to 9.   ■

EXAMPLE 3

Show that if any 11 numbers are chosen from the set $\{1, 2, \ldots, 20\}$, then one of them will be a multiple of another.

Solution   The key to solving this problem is to create 10 or fewer pigeonholes in such a way that each number chosen can be assigned to only one pigeonhole, and when x and y are assigned to the same pigeonhole we are guaranteed that either $x \mid y$ or $y \mid x$. Factors are a natural feature to explore. There are eight prime numbers between 1 and 20, but knowing that x and y are multiples of the same prime will not guarantee that either $x \mid y$ or $y \mid x$. We try again. There are ten odd numbers between 1 and 20. Every positive integer n can be written as $n = 2^k m$, where m is odd and $k \geq 0$. This can be seen by simply factoring all powers of 2 (if any) out of n. In this case let us call m the odd part of n. If 11 numbers are chosen from the set $\{1, 2, \ldots, 20\}$, then two of them must have the same odd part. This follows from the pigeonhole principle since there are 11 numbers (pigeons), but only 10 odd numbers between 1 and 20 (pigeonholes) that can be odd parts of these numbers.

Let n_1 and n_2 be two chosen numbers with the same odd part. We must have $n_1 = 2^{k_1} m$ and $n_2 = 2^{k_2} m$, for some k_1 and k_2. If $k_1 \geq k_2$, then n_1 is a multiple of n_2; otherwise, n_2 is a multiple of n_1.   ∎

Figure 3.2

Figure 3.3

EXAMPLE 4   Consider the region shown in Figure 3.2. It is bounded by a regular hexagon whose sides are of length 1 unit. Show that if any seven points are chosen in this region, then two of them must be no farther apart than 1 unit.

Solution   Divide the region into six equilateral triangles, as shown in Figure 3.3. If seven points are chosen in the region, we can assign each of them to a triangle that contains it. If the point belongs to several triangles, arbitrarily assign it to one of them. Then the seven points are assigned to six triangular regions, so by the pigeonhole principle, at least two points must belong to the same region. These two cannot be more than one unit apart. (Why?)   ∎

EXAMPLE 5   Shirts numbered consecutively from 1 to 20 are worn by the 20 members of a bowling league. When any three of these members are chosen to be a team, the league proposes to use the sum of their shirt numbers as a code number for the team. Show that if any eight of the 20 are selected, then from these eight one may form at least two different teams having the same code number.

Solution   From the eight selected bowlers, we can form a total of $_8C_3$ or 56 different teams. These will play the role of pigeons. The largest possible team code number is $18 + 19 + 20$ or 57, and the smallest possible is $1 + 2 + 3$ or 6. Thus only the 52 code numbers (pigeonholes) between 6 and 57 inclusive are available for the 56 possible teams. By the pigeonhole principle, at least two teams will have the same code number. The league should use another way to assign team numbers.   ∎

The Extended Pigeonhole Principle

Note that if there are m pigeonholes and more than $2m$ pigeons, three or more pigeons will have to be assigned to at least one of the pigeonholes. (Consider the most even distribution of pigeons you can make.) In general, if the number of pigeons is much larger than the number of pigeonholes, Theorem 1 can be restated to give a stronger conclusion.

First a word about notation. If n and m are positive integers, then $\lfloor n/m \rfloor$ stands for the largest integer less than or equal to the rational number n/m. Thus $\lfloor 3/2 \rfloor$ is 1, $\lfloor 9/4 \rfloor$ is 2, and $\lfloor 6/3 \rfloor$ is 2.

Theorem 2
The Extended
Pigeonhole Principle

If n pigeons are assigned to m pigeonholes, then one of the pigeonholes must contain at least $\lfloor (n-1)/m \rfloor + 1$ pigeons.   ●

Proof (*by contradiction*) If each pigeonhole contains no more than $\lfloor (n-1)/m \rfloor$ pigeons, then there are at most $m \cdot \lfloor (n-1)/m \rfloor \leq m \cdot (n-1)/m = n-1$ pigeons in all. This contradicts our hypothesis, so one of the pigeonholes must contain at least $\lfloor (n-1)/m \rfloor + 1$ pigeons.   ▼

This proof by contradiction uses the fact that there are two ways to count the total number of pigeons, the original count n and as the product of the number of pigeonholes times the number of pigeons per pigeonhole.

EXAMPLE 6

We give an extension of Example 1. Show that if any 30 people are selected, then one may choose a subset of five so that all five were born on the same day of the week.

Solution   Assign each person to the day of the week on which she or he was born. Then 30 pigeons are being assigned to 7 pigeonholes. By the extended pigeonhole principle with $n = 30$ and $m = 7$, at least $\lfloor (30-1)/7 \rfloor + 1$ or 5 of the people must have been born on the same day of the week.   ■

EXAMPLE 7

Show that if 30 dictionaries in a library contain a total of 61,327 pages, then one of the dictionaries must have at least 2045 pages.

Solution   Let the pages be the pigeons and the dictionaries the pigeonholes. Assign each page to the dictionary in which it appears. Then by the extended pigeonhole principle, one dictionary must contain at least $\lfloor 61{,}326/30 \rfloor + 1$ or 2045 pages.   ■

3.3 Exercises

1. If thirteen people are assembled in a room, show that at least two of them must have their birthday in the same month.

2. Show that if seven numbers from 1 to 12 are chosen, then two of them will add up to 13.

3. Let T be an equilateral triangle whose sides are of length 1 unit. Show that if any five points are chosen lying on or inside the triangle, then two of them must be no more than $\frac{1}{2}$ unit apart.

4. Show that if any eight positive integers are chosen, two of them will have the same remainder when divided by 7.

5. Show that if seven colors are used to paint 50 bicycles, at least eight bicycles will be the same color.

6. Ten people volunteer for a three-person committee. Every possible committee of three that can be formed from these ten names is written on a slip of paper, one slip for each possible committee, and the slips are put in ten hats. Show that at least one hat contains 12 or more slips of paper.

7. Six friends discover that they have a total of $21.61 with them on a trip to the movies. Show that one or more of them must have at least $3.61.

8. A store has an introductory sale on 12 types of candy bars. A customer may choose one bar of any five different types and will be charged no more than $1.75. Show that although different choices may cost different amounts, there must be at least two different ways to choose so that the cost will be the same for both choices.

9. If the store in Exercise 8 allows repetitions in the choices, show that there must be at least ten ways to make different choices that have the same cost.

10. Show that there must be at least 90 ways to choose six numbers from 1 to 15 so that all the choices have the same sum.

11. How many friends must you have to guarantee at least five of them will have birthdays in the same month?

12. Show that if five points are selected in a square whose sides have length 1 inch, at least two of the points must be no more than $\sqrt{2}$ inches apart.

13. Let $\mathbf{A}$ be an 8×8 Boolean matrix. If the sum of the entries in $\mathbf{A}$ is 51, prove that there is a row i and a column j in $\mathbf{A}$ such that the entries in row i and in column j add up to more than 13.

14. Write an exercise similar to Exercise 13 for a 12×12 Boolean matrix.

15. Prove that if any 14 numbers from 1 to 25 are chosen, then one of them is a multiple of another.

16. Twenty disks numbered 1 through 20 are placed face down on a table. Disks are selected one at a time and turned over until 10 disks have been chosen. If two of the disks add up to 21, the player loses. Is it possible to win this game?

17. Suppose the game in Exercise 16 has been changed so that 12 disks are chosen. Is it possible to win this game?

18. Complete the following proof. It is not possible to arrange the numbers $1, 2, 3, \ldots, 10$ in a circle so that every triple of consecutively placed numbers has a sum less than 15. Proof: In any arrangement of $1, 2, 3, \ldots, 10$ in a circle, there are _____ triples of consecutively placed numbers, because _____. Each number appears in _____ of these triples. If the sum of each triple were less than 15, then the total sum of all triples would be less than _____ times 15 or _____. But $1 + 2 + 3 + \cdots + 10$ is 55 and since each number appears in _____ triples, the total sum should be _____ times 55. This is a contradiction so not all triples can have a sum less than 15.

19. Prove that any sequence of six numbers must contain a subsequence whose sum is divisible by six. (*Hint*: Consider the sums $c_1, c_1 + c_2, c_1 + c_2 + c_3, \ldots$ and the possible remainders when dividing by six.)

20. Prove that any sequence of numbers must contain a subsequence whose sum is divisible by n.

3.4 ELEMENTS OF PROBABILITY

Another area where counting techniques are important is probability theory. In this section we present a brief introduction to probability.

Many experiments do not yield exactly the same results when performed repeatedly. For example, if we toss a coin, we are not sure if we will get heads or tails, and if we toss a die, we have no way of knowing which of the six possible numbers will turn up. Experiments of this type are called **probabilistic**, in contrast to **deterministic** experiments, whose outcome is always the same.

Sample Spaces

A set A consisting of all the outcomes of an experiment is called a **sample space** of the experiment. With a given experiment, we can often associate more than one sample space, depending on what the observer chooses to record as an outcome.

EXAMPLE 1   Suppose that a nickel and a quarter are tossed in the air. We describe three possible sample spaces that can be associated with this experiment.

1. If the observer decides to record as an outcome the number of heads observed, the sample space is $A = \{0, 1, 2\}$.

2. If the observer decides to record the sequence of heads (H) and tails (T) observed, listing the condition of the nickel first and then that of the quarter, then the sample space is $A = \{HH, HT, TH, TT\}$.

3. If the observer decides to record the fact that the coins match (M) (both heads or both tails) or do not match (N), then the sample space is $A = \{M, N\}$. ∎

We thus see that in addition to describing the experiment, we must indicate exactly what the observer wishes to record. Then the set of all outcomes of this type become the sample space for the experiment.

A sample space may contain a finite or an infinite number of outcomes, but in this chapter, we need only finite sample spaces.

EXAMPLE 2   Determine the sample space for an experiment consisting of tossing a six-sided die twice and recording the sequence of numbers showing on the top face of the die after each toss.

Solution   An outcome of the experiment can be represented by an ordered pair of numbers (n, m), where n and m can be 1, 2, 3, 4, 5, or 6. Thus the sample space A contains 6×6 or 36 elements (by the multiplication principle). ∎

EXAMPLE 3   An experiment consists of drawing three coins in succession from a box containing four pennies and five dimes, and recording the sequence of results. Determine the sample space of this experiment.

Solution   An outcome can be recorded as a sequence of length 3 constructed from the letters P (penny) and D (dime). Thus the sample space A is {PPP, PPD, PDP, PDD, DPP, DPD, DDP, DDD}. ∎

Events

A statement about the outcome of an experiment, which for a particular outcome will be either true or false, is said to describe an **event**. Thus for Example 2, the statements, "Each of the numbers recorded is less than 3" and "The sum of the numbers recorded is 4" would describe events. The event described by a statement is taken to be the set of all outcomes for which the statement is true. With this interpretation, any event can be considered a subset of the sample space. Thus the event E described by the first statement is $E = \{(1, 1), (1, 2), (2, 1), (2, 2)\}$. Similarly, the event F described by the second statement is $F = \{(1, 3), (2, 2), (3, 1)\}$.

EXAMPLE 4   Consider the experiment in Example 2. Determine the events described by each of the following statements.

(a) The sum of the numbers showing on the top faces is 8.

(b) The sum of the numbers showing on the top faces is at least 10.

Solution

(a) The event consists of all ordered pairs whose sum is 8. Thus the event is $\{(2, 6), (3, 5), (4, 4), (5, 3), (6, 2)\}$.

(b) The event consists of all ordered pairs whose sum is 10, 11, or 12. Thus the event is $\{(4, 6), (5, 5), (5, 6), (6, 4), (6, 5), (6, 6)\}$. ∎

If A is a sample space of an experiment, then A itself is an event called the **certain event** and the empty subset of A is called the **impossible event**.

Since events are sets, we can combine them by applying the operations of union, intersection, and complementation to form new events. The sample space A is the universal set for these events. Thus if E and F are events, we can form the new events $E \cup F$, $E \cap F$, and $\overline{E}$. What do these new events mean in terms of the experiment? An outcome of the experiment belongs to $E \cup F$ when it belongs to E or F (or both). In other words, the event $E \cup F$ occurs exactly when E or F occurs. Similarly, the event $E \cap F$ occurs if and only if both E and F occur. Finally, $\overline{E}$ occurs if and only if E does not occur.

EXAMPLE 5 Consider the experiment of tossing a die and recording the number on the top face. Let E be the event that the number is even and let F be the event that the number is prime. Then $E = \{2, 4, 6\}$ and $F = \{2, 3, 5\}$. The event that the number showing is either even or prime is $E \cup F = \{2, 3, 4, 5, 6\}$. The event that the number showing is an even prime is $E \cap F = \{2\}$. Finally, the event that the number showing is not even is $\overline{E} = \{1, 3, 5\}$ and the event that the number showing is not prime is $\overline{F} = \{1, 4, 6\}$. ∎

Events E and F are said to be **mutually exclusive** or **disjoint** if $E \cap F = \{\ \}$. If E and F are mutually exclusive events, then E and F cannot both occur at the same time; if E occurs, then F does not occur, and if F occurs, then E does not. If $E_1, E_2, \ldots, E_n$ are all events, then we say that these sets are **mutually exclusive**, or **disjoint**, if each pair of them is mutually exclusive. Again, this means that at most one of the events can occur on any given outcome of the experiment.

Assigning Probabilities to Events

In probability theory, we assume that each event E has been assigned a number $p(E)$ called the **probability of the event E**. We now look at probabilities. We will investigate ways in which they can be assigned, properties they must satisfy, and the meaning that can be given to them.

The number $p(E)$ reflects our assessment of the likelihood that the event E will occur. More precisely, suppose the underlying experiment is performed repeatedly, and that after n such performances, the event E has occurred n_E times. Then the fraction $f_E = n_E / n$, called the **frequency of occurrence of E in n trials**, is a measure of the likelihood that E will occur. When we assign the probability $p(E)$ to the event E, it means that in our judgment or experience, we believe that the fraction f_E will tend ever closer to a certain number as n becomes larger, and that $p(E)$ is this number. Thus probabilities can be thought of as idealized frequencies of occurrence of events, to which actual frequencies of occurrence will tend when the experiment is performed repeatedly.

EXAMPLE 6 Suppose an experiment is performed 2000 times, and the frequency of occurrence f_E of an event E is recorded after 100, 500, 1000, and 2000 trials. Table 3.1 summarizes the results.

Table 3.1

Number of Repetitions of the Experiment	n_E	$f_E = n_E/n$
100	48	0.48
500	259	0.518
1000	496	0.496
2000	1002	0.501

Based on this table, it appears that the frequency f_E approaches $\frac{1}{2}$ as n becomes larger. It could therefore be argued that $p(E)$ should be set equal to $\frac{1}{2}$. On the other hand, one might require more extensive evidence before assigning $\frac{1}{2}$ as the value of $p(E)$. In any case, this sort of evidence can never "prove" that $p(E)$ is $\frac{1}{2}$. It only serves to make this a plausible assumption. ∎

If probabilities assigned to various events are to represent meaningfully frequencies of occurrence of the events, as explained previously, then they cannot be assigned in a totally arbitrary way. They must satisfy certain conditions. In the first place, since every frequency f_E must satisfy the inequalities $0 \leq f_E \leq 1$, it is only reasonable to assume that

P1: $0 \leq p(E) \leq 1$ for every event E in A.

Also, since the event A must occur every time (every outcome belongs to A), and the event $\varnothing$ cannot occur, we assume that

P2: $p(A) = 1$ and $p(\varnothing) = 0$.

Finally, if $E_1, E_2, \ldots, E_k$ are mutually exclusive events, then

$$n_{(E_1 \cup E_2 \cup \cdots \cup E_k)} = n_{E_1} + n_{E_2} + \cdots + n_{E_k},$$

since only one of these events can occur at a time. If we divide both sides of this equation by n, we see that the frequencies of occurrence must satisfy a similar equation. We therefore assume

P3: $p(E_1 \cup E_2 \cup \cdots \cup E_k) = p(E_1) + p(E_2) + \cdots + p(E_k)$

whenever the events are mutually exclusive. If the probabilities are assigned to all events in such a way that P1, P2, and P3 are always satisfied, then we have a **probability space**. We call P1, P2, and P3 the **axioms for a probability space**.

It is important to realize that mathematically, no demands are made on a probability space except those given by the probability axioms P1, P2, and P3. Probability theory begins with all probabilities assigned, and then investigates consequences of any relations between these probabilities. No mention is made of how the probabilities were assigned. However, the mathematical conclusions will be useful in an actual situation only if the probabilities assigned reflect what actually occurs in that situation.

Experimentation is not the only way to determine reasonable probabilities for events. The probability axioms can sometimes provide logical arguments for choosing certain probabilities.

EXAMPLE 7   Consider the experiment of tossing a coin and recording whether heads or tails results. Consider the events E: heads turns up and F: tails turns up. The mechanics of the toss are not controllable in detail. Thus in the absence of any defect in the coin that might unbalance it, one may argue that E and F are equally likely to occur. There is a symmetry in the situation that makes it impossible to prefer one outcome over the other. This argument lets us compute what the probabilities of E and F must be.

We have assumed that $p(E) = p(F)$, and it is clear that E and F are mutually exclusive events and $A = E \cup F$. Thus, using the properties P2 and P3, we see that $1 = p(A) = p(E) + p(F) = 2p(E)$ since $p(E) = p(F)$. This shows that $p(E) = \frac{1}{2} = p(F)$. One may often assign appropriate probabilities to events by combining the symmetry of situations with the axioms of probability. ∎

Finally, we will show that the problem of assigning probabilities to events can be reduced to the consideration of the simplest cases. Let A be a probability space. We assume that A is finite, that is, $A = \{x_1, x_2, \ldots, x_n\}$. Then each event $\{x_k\}$, consisting of just one outcome, is called an **elementary event**. For simplicity, let us write $p_k = p(\{x_k\})$. Then p_k is called the **elementary probability corresponding to the outcome** x_k. Since the elementary events are mutually exclusive and their union is A, the axioms of probability tell us that

EP1: $0 \le p_k \le 1$ for all k

EP2: $p_1 + p_2 + \cdots + p_n = 1$.

If E is any event in A, say $E = \{x_{i_1}, x_{i_2}, \ldots, x_{i_m}\}$, then we can write $E = \{x_{i_1}\} \cup \{x_{i_2}\} \cup \cdots \cup \{x_{i_m}\}$. This means, by axiom P2, that $p(E) = p_{i_1} + p_{i_2} + \cdots + p_{i_m}$. Thus if we know the elementary probabilities, then we can compute the probability of any event E.

EXAMPLE 8   Suppose that an experiment has a sample space $A = \{1, 2, 3, 4, 5, 6\}$ and that the elementary probabilities have been determined as follows:

$$p_1 = \frac{1}{12}, \quad p_2 = \frac{1}{12}, \quad p_3 = \frac{1}{3}, \quad p_4 = \frac{1}{6}, \quad p_5 = \frac{1}{4}, \quad p_6 = \frac{1}{12}.$$

Let E be the event "The outcome is an even number." Compute $p(E)$.

Solution   Since $E = \{2, 4, 6\}$, we see that

$$p(E) = p_2 + p_4 + p_6 = \frac{1}{12} + \frac{1}{6} + \frac{1}{12} \quad \text{or} \quad \frac{1}{3}.$$

In a similar way we can determine the probability of any event in A. ∎

Thus we see that the problem of assigning probabilities to all events in a consistent way can be reduced to the problem of finding numbers $p_1, p_2, \ldots, p_n$ that satisfy EP1 and EP2. Again, mathematically speaking, there are no other restrictions on the p_k's. However, if the mathematical structure that results is to be useful in a particular situation, then the p_k's must reflect the actual behavior occurring in that situation.

Equally Likely Outcomes

Let us assume that all outcomes in a sample space A are equally likely to occur. This is, of course, an assumption, and so cannot be proved. We would make such an assumption if experimental evidence or symmetry indicated that it was appropriate in a particular situation (see Example 7). Actually these situations arise commonly. One additional piece of terminology is customary. Sometimes experiments involve choosing an object, in a nondeterministic way, from some collection. If the selection is made in such a way that all objects have an equal probability of being chosen, we say that we have made a **random selection** or **chosen an object at random** from the collection. We will often use this terminology to specify examples of experiments with equally likely outcomes.

Suppose that $|A| = n$ and these n outcomes are equally likely. Then the elementary probabilities are all equal, and since they must add up to 1, this means that each elementary probability is $1/n$. Now let E be an event that contains k outcomes, say $E = \{x_1, x_2, \ldots, x_k\}$. Since all elementary probabilities are $1/n$, we must have

$$p(E) = \underbrace{\frac{1}{n} + \frac{1}{n} + \cdots + \frac{1}{n}}_{k \text{ summands}} = \frac{k}{n}.$$

Since $k = |E|$, we have the following principle: If all outcomes are equally likely, then for every event E

$$p(E) = \frac{|E|}{|A|} = \frac{\text{total number of outcomes in } E}{\text{total number of outcomes}}.$$

In this case, the computation of probabilities reduces to counting numbers of elements in sets. For this reason, the methods of counting discussed in the earlier sections of this chapter are quite useful.

EXAMPLE 9

Choose four cards at random from a standard 52-card deck. What is the probability that four kings will be chosen?

Solution   The outcomes of this experiment are four-card hands; each is equally likely to be chosen. The number of four-card hands is $_{52}C_4$ or 270,725. Let E be the event that all four cards are kings. The event E contains only one outcome. Thus $p(E) = \frac{1}{270,725}$ or approximately 0.000003694. This is an extremely unlikely event. ∎

EXAMPLE 10

A box contains six red balls and four green balls. Four balls are selected at random from the box. What is the probability that two of the selected balls will be red and two will be green?

Solution   The total number of outcomes is the number of ways to select four objects out of ten, without regard to order. This is $_{10}C_4$ or 210. Now the event E, that two of the balls are red and two of them are green, can be thought of as the result of performing two tasks in succession.

TASK 1:   Choose two red balls from the six red balls in the box.
TASK 2:   Choose two green balls from the four green balls in the box.

Task 1 can be done in $_6C_2$ or 15 ways and task 2 can be done in $_4C_2$ or 6 ways. Thus, event E can occur in $15 \cdot 6$ or 90 ways, and therefore $p(E) = \frac{90}{210}$ or $\frac{3}{7}$. ∎

EXAMPLE 11   A fair six-sided die is tossed three times and the resulting sequence of numbers is recorded. What is the probability of the event E that either all three numbers are equal or none of them is a 4?

Solution   Since the die is assumed to be fair, all outcomes are equally likely. First, we compute the total number of outcomes of the experiment. This is the number of sequences of length 3, allowing repetitions, that can be constructed from the set $\{1, 2, 3, 4, 5, 6\}$. This number is 6^3 or 216.

Event E cannot be described as the result of performing two successive tasks as in Example 10. We can, however, write E as the union of two simpler events. Let F be the event that all three numbers recorded are equal, and let G be the event that none of the numbers recorded is a 4. Then $E = F \cup G$. By the addition principle (Theorem 2, Section 1.2), $|F \cup G| = |F| + |G| - |F \cap G|$.

There are only six outcomes in which the numbers are equal, so $|F|$ is 6. The event G consists of all sequences of length 3 that can be formed from the set $\{1, 2, 3, 5, 6\}$. Thus $|G|$ is 5^3 or 125. Finally, the event $F \cap G$ consists of all sequences for which the three numbers are equal and none is a 4. Clearly, there are five ways for this to happen, so $|F \cap G|$ is 5. Using the addition principle, $|E| = |F \cup G| = 6 + 125 - 5$ or 126. Thus, we have $p(E) = \frac{126}{216}$ or $\frac{7}{12}$. ∎

EXAMPLE 12   Consider again the experiment in Example 10, in which four balls are selected at random from a box containing six red balls and four green balls.

(a) If E is the event that no more than two of the balls are red, compute the probability of E.

(b) If F is the event that no more than three of the balls are red, compute the probability of F.

Solution

(a) Here E can be decomposed as the union of mutually exclusive events. Let E_0 be the event that none of the chosen balls are red, let E_1 be the event that exactly one of the chosen balls is red, and let E_2 be the event that exactly two of the chosen balls are red. Then E_0, E_1, and E_2 are mutually exclusive and $E = E_0 \cup E_1 \cup E_2$. Using the addition principle twice, $|E| = |E_0| + |E_1| + |E_2|$. If none of the balls is red, then all four must be green. Since there are only four green balls in the box, there is only one way for event E_0 to occur. Thus $|E_0| = 1$. If one ball is red, then the other three must be green. To make such a choice, we must choose one red ball from a set of six, and then three green balls from a set of four. Thus, the number of outcomes in E_1 is $(_6C_1)(_4C_3)$ or 24.

In exactly the same way, we can show that the number of outcomes in E_2 is $(_6C_2)(_4C_2)$ or 90. Thus, $|E| = 1 + 24 + 90$ or 115. On the other hand, the total number of ways of choosing four balls from the box is $_{10}C_4$ or 210, so $p(E) = \frac{115}{210}$ or $\frac{23}{42}$.

(b) We could compute $|F|$ in the same way we computed $|E|$ in part (a), by decomposing F into four mutually exclusive events. The analysis would, however, be even longer than that of part (a). We choose instead to illustrate another approach that is frequently useful.

Let $\overline{F}$ be the complementary event to F. Since F and $\overline{F}$ are mutually exclusive and their union is the sample space, we must have $p(F) + p(\overline{F}) = 1$. This formula holds for any event F and is used when the complementary event is easier to analyze. This is the case here, since $\overline{F}$ is the event that all four balls chosen are red. These four red balls can be chosen from the six red balls in $_6C_4$ or 15 ways, so $p(\overline{F}) = \frac{15}{210}$ or $\frac{1}{14}$. This means that $p(F) = 1 - \frac{1}{14}$ or $\frac{13}{14}$. ∎

A common use of probability in computer science is in analyzing the efficiency of algorithms. For example, this may be done by considering the number of steps we "expect" the algorithm to execute on an "average" run. Here is a simple case to consider. If a fair coin is tossed 500 times, we expect $250 \left(\frac{1}{2} \cdot 500 \right)$ heads to occur. Of course, we would not be surprised if the number of heads were not exactly 250. This idea leads to the following definition. The **expected value** of an experiment is the sum of the value of each outcome times its probability. Roughly speaking, the expected value describes the "average" value for a large number of trials.

EXAMPLE 13   An array of length 10 is searched for a key word. The number of steps needed to find it is recorded. Assuming that the key is equally likely to be in any position of the array, the expected value of this experiment is $1 \cdot \frac{1}{10} + 2 \cdot \frac{1}{10} + \cdots + 10 \cdot \frac{1}{10}$ or $\frac{55}{10}$. On the average, we can expect to find a key word in 5.5 steps. ∎

3.4 Exercises

In Exercises 1 through 4, describe the associated sample space.

1. A coin is tossed three times and the sequence of heads and tails is recorded.

2. Two letters are selected simultaneously at random from the letters a, b, c, d.

3. A silver urn and a copper urn contain blue, red, and green balls. An urn is chosen at random and then a ball is selected at random from this urn.

4. A box contains 12 items, four of which are defective. An item is chosen at random and not replaced. This is continued until all four defective items have been selected. The total number of items selected is recorded.

5. (a) Suppose that the sample space of an experiment is $\{1, 2, 3\}$. Determine all possible events.

(b) Let S be a sample space containing n elements. How many events are there for the associated experiment?

In Exercises 6 through 8, use the following assumptions. A card is selected at random from a standard deck. Let E, F, and G be the following events.

> E: The card is black.
>
> F: The card is a diamond.
>
> G: The card is an ace.

Describe the following events in complete sentences.

6. (a) $E \cup G$ (b) $E \cap G$

7. (a) $\overline{E} \cap G$ (b) $E \cup F \cup G$

8. (a) $E \cup \overline{F} \cup G$ (b) $(F \cap \overline{G}) \cup E$

In Exercises 9 and 10, assume that a die is tossed twice and the numbers showing on the top faces are recorded in sequence. Determine the elements in each of the given events.

9. (a) At least one of the numbers is a 5.

   (b) At least one of the numbers is an 8.

10. (a) The sum of the numbers is less than 7.

   (b) The sum of the numbers is greater than 8.

11. A die is tossed and the number showing on the top face is recorded. Let E, F, and G be the following events.

   E: The number is at least 3.

   F: The number is at most 3.

   G: The number is divisible by 2.

   (a) Are E and F mutually exclusive? Justify your answer.

   (b) Are F and G mutually exclusive? Justify your answer.

   (c) Is $E \cup F$ the certain event? Justify your answer.

   (d) Is $E \cap F$ the impossible event? Justify your answer.

12. Let E be an event for an experiment with sample space A. Show that

   (a) $E \cup \overline{E}$ is the certain event.

   (b) $E \cap \overline{E}$ is the impossible event.

13. A medical team classifies people according to the following characteristics.

   Drinking habits: drinks (d), abstains (a)

   Income level: low (l), middle (m), upper (u)

   Smoking habits: smoker (s), nonsmoker (n)

   Let E, F, and G be the following events.

   E: A person drinks.

   F: A person's income level is low.

   G: A person smokes.

   List the elements in each of the following events.

   (a) $E \cup F$ (b) $\overline{E} \cap F$ (c) $(E \cup G) \cap F$

In Exercises 14 and 15, let $S = \{1, 2, 3, 4, 5, 6\}$ be the sample space of an experiment and let

$$E = \{1, 3, 4, 5\}, \quad F = \{2, 3\}, \quad G = \{4\}.$$

14. Compute the events $E \cup F$, $E \cap F$, and $\overline{F}$.

15. Compute the following events: $\overline{E} \cup F$ and $\overline{F} \cap G$.

In Exercises 16 and 17, list the elementary events for the given experiments.

16. A vowel is selected at random from the set of all vowels a, e, i, o, u.

17. A card is selected at random from a standard deck and it is recorded whether the card is a club, spade, diamond, or heart.

18. (a) What is the probability of correctly guessing a person's four-digit PIN?

   (b) People often use the four digits of their birthday (MM-DD) to create a PIN. What is the probability of correctly guessing a PIN created this way, if the birthday is known?

19. When a certain defective die is tossed, the numbers from 1 to 6 will be on the top face with the following probabilities.

$$p_1 = \frac{2}{18}, \quad p_2 = \frac{3}{18}, \quad p_3 = \frac{4}{18}, \quad p_4 = \frac{3}{18}$$

$$p_5 = \frac{4}{18}, \quad p_6 = \frac{2}{18}$$

   Find the probability that

   (a) an odd number is on top.

   (b) a prime number is on top.

   (c) a number less than 5 is on top.

   (d) a number greater than 3 is on top.

20. Repeat Exercise 19, assuming that the die is not defective.

21. Suppose that E and F are mutually exclusive events such that $p(E) = 0.3$ and $p(F) = 0.4$. Find the probability that

   (a) E does not occur.   (b) E and F occur.

   (c) E or F occurs.

   (d) E does not occur or F does not occur.

22. Consider an experiment with sample space $A = \{x_1, x_2, x_3, x_4\}$ for which

$$p_1 = \frac{2}{7}, \quad p_2 = \frac{3}{7}, \quad p_3 = \frac{1}{7}, \quad p_4 = \frac{1}{7}.$$

   Find the probability of the given event.

   (a) $E = \{x_1, x_2\}$ (b) $F = \{x_1, x_3, x_4\}$

23. There are four candidates for president, A, B, C, and D. Suppose A is twice as likely to be elected as B, B is three times as likely as C, and C and D are equally likely to be elected. What is the probability of being elected for each candidate?

24. The outcome of a particular game of chance is an integer from 1 to 5. Integers 1, 2, and 3 are equally likely to occur, and integers 4 and 5 are equally likely to occur. The probability that the outcome is greater than 2 is $\frac{1}{2}$. Find the probability of each possible outcome.

25. A fair coin is tossed five times. What is the probability of obtaining three heads and two tails?

In Exercises 26 through 28, suppose a fair die is tossed and the number showing on the top face is recorded. Let E, F, and G be the following events.

$$E: \{1, 2, 3, 5\}, \quad F: \{2, 4\}, \quad G: \{1, 4, 6\}$$

Compute the probability of the event indicated.

26. (a) $E \cup F$ (b) $E \cap F$

27. (a) $\overline{E} \cap F$ (b) $E \cup G$

28. (a) $\overline{E} \cup \overline{G}$ (b) $\overline{E} \cap \overline{G}$

29. Suppose two dice are tossed and the numbers on the top faces recorded. What is the probability that

    (a) a 4 was tossed?

    (b) a prime number was tossed?

    (c) the sum of the numbers is less than 5?

    (d) the sum of the numbers is at least 7?

30. Suppose that two cards are selected at random from a standard 52-card deck. What is the probability that both cards are less than 10 and neither of them is red?

31. Suppose that three balls are selected at random from an urn containing seven red balls and five black balls. Compute the probability that

    (a) all three balls are red.

    (b) at least two balls are black.

    (c) at most two balls are black.

    (d) at least one ball is red.

32. A fair die is tossed three times in succession. Find the probability that the three resulting numbers

    (a) include exactly two 3's.

    (b) form an increasing sequence.

    (c) include at least one 3.

    (d) include at most one 3.

    (e) include no 3's.

33. An array of length n is searched for a key word. On the average, how many steps will it take to find the key?

34. How should the analysis in Exercise 33 be changed if we do not assume that the key word is in the array?

35. A game is played by rolling two dice and paying the player an amount (in dollars) equal to the sum of the numbers on top if this is 10 or greater. The player must pay $3 for each game. What is the expected value of this game?

36. For the game described in Exercise 35, what would be a "fair" cost to play the game? Justify your answer.

3.5 RECURRENCE RELATIONS

The recursive definitions of sequences in Section 1.3 are examples of recurrence relations. When the problem is to find an explicit formula for a recursively defined sequence, the recursive formula is called a **recurrence relation**. Remember that to define a sequence recursively, a recursive formula must be accompanied by information about the beginning of the sequence. This information is called the **initial condition** or **conditions** for the sequence.

EXAMPLE 1

(a) The recurrence relation $a_n = a_{n-1} + 3$ with $a_1 = 4$ recursively defines the sequence 4, 7, 10, 13, ...

(b) The recurrence relation $f_n = f_{n-1} + f_{n-2}$, $f_1 = f_2 = 1$, defines the **Fibonacci sequence** 1, 1, 2, 3, 5, 8, 13, 21, ... The initial conditions are $f_1 = 1$ and $f_2 = 1$. ∎

Recurrence relations arise naturally in many counting problems and in analyzing programming problems.

EXAMPLE 2   Let $A = \{0, 1\}$. Give a recurrence relation for c_n, the number of strings of length n in A^* that do not contain adjacent 0's.

Solution   Since 0 and 1 are the only strings of length 1, $c_1 = 2$. Also, $c_2 = 3$; the only such strings are 01, 10, 11. In general, any string w of length $n - 1$ that does not contain 00 can be catenated with 1 to form a string $1 \cdot w$, a string of length n that does not contain 00. The only other possible beginning for a "good" string of length n is 01. But any of these strings must be of the form $01 \cdot v$, where v is a "good" string of length $n - 2$. Hence, $c_n = c_{n-1} + c_{n-2}$ with the initial conditions $c_1 = 2$ and $c_2 = 3$.   ∎

EXAMPLE 3   Suppose we wish to print out all n-element sequences without repeats that can be made from the set $\{1, 2, 3, \ldots, n\}$. One approach to this problem is to proceed recursively as follows.

STEP 1:   Produce a list of all sequences that can be made from $\{1, 2, 3, \ldots, n - 1\}$.

STEP 2:   For each sequence in step 1, insert n in turn in each of the n available places (at the front, at the end, and between every pair of numbers in the sequence), print the result, and remove n.

The number of insert-print-remove actions is the number of n-element sequences. It is also clearly n times the number of sequences produced in step 1. Thus we have

number of n-element sequences $= n \times$ (number of $(n - 1)$-sequences).

This gives a recursive formula for the number of n-element sequences. What is the initial condition?   ∎

One technique for finding an explicit formula for the sequence defined by a recurrence relation is **backtracking**, as illustrated in the following example.

EXAMPLE 4   The recurrence relation $a_n = a_{n-1} + 3$ with $a_1 = 2$ defines the sequence $2, 5, 8, \ldots$. We backtrack the value of a_n by substituting the definition of a_{n-1}, a_{n-2}, and so on until a pattern is clear.

$$a_n = a_{n-1} + 3 \qquad \text{or} \qquad a_n = a_{n-1} + 3$$
$$= (a_{n-2} + 3) + 3 \qquad\qquad\qquad = a_{n-2} + 2 \cdot 3$$
$$= ((a_{n-3} + 3) + 3) + 3 \qquad\qquad = a_{n-3} + 3 \cdot 3$$

Eventually this process will produce

$$a_n = a_{n-(n-1)} + (n - 1) \cdot 3$$
$$= a_1 + (n - 1) \cdot 3$$
$$= 2 + (n - 1) \cdot 3.$$

An explicit formula for the sequence is $a_n = 2 + (n - 1)3$.   ∎

EXAMPLE 5   Backtrack to find an explicit formula for the sequence defined by the recurrence relation $b_n = 2b_{n-1} + 1$ with initial condition $b_1 = 7$.

Solution We begin by substituting the definition of the previous term in the defining formula.

$$b_n = 2b_{n-1} + 1$$
$$= 2(2b_{n-2} + 1) + 1$$
$$= 2[2(2b_{n-3} + 1) + 1] + 1$$
$$= 2^3 b_{n-3} + 4 + 2 + 1$$
$$= 2^3 b_{n-3} + 2^2 + 2^1 + 1.$$

A pattern is emerging with these rewriting of b_n. (*Note:* There are no set rules for how to rewrite these expressions and a certain amount of experimentation may be necessary.) The backtracking will end at

$$b_n = 2^{n-1} b_{n-(n-1)} + 2^{n-2} + 2^{n-3} + \cdots + 2^2 + 2^1 + 1$$
$$= 2^{n-1} b_1 + 2^{n-1} - 1 \qquad \text{Using Exercise 3, Section 2.4}$$
$$= 7 \cdot 2^{n-1} + 2^{n-1} - 1 \qquad \text{Using } b_1 = 7$$
$$= 8 \cdot 2^{n-1} - 1 \quad \text{or} \quad 2^{n+2} - 1. \qquad \blacksquare$$

Two useful summing rules were proved in Section 2.4. We record them again for use in this section.

S1. $1 + a + a^2 + a^3 + \cdots + a^{n-1} = \dfrac{a^n - 1}{a - 1}$.

S2. $1 + 2 + 3 + \cdots + n = \dfrac{n(n+1)}{2}$.

Backtracking may not reveal an explicit pattern for the sequence defined by a recurrence relation. We now introduce a more general technique for solving a recurrence relation. First we give a definition. A recurrence relation is a **linear homogeneous relation of degree k** if it is of the form

$$a_n = r_1 a_{n-1} + r_2 a_{n-2} + \cdots + r_k a_{n-k} \qquad \text{with the } r_i\text{'s constants.}$$

Note that on the right-hand side, the summands are each built the same (homogeneous) way as a multiple of one of the k (degree k) previous terms (linear).

EXAMPLE 6

(a) The relation $c_n = (-2)c_{n-1}$ is a linear homogeneous recurrence relation of degree 1.

(b) The relation $a_n = a_{n-1} + 3$ is not a linear homogeneous recurrence relation.

(c) The recurrence relation $f_n = f_{n-1} + f_{n-2}$ is a linear homogeneous relation of degree 2.

(d) The recurrence relation $g_n = g_{n-1}^2 + g_{n-2}$ is not a linear homogeneous relation. $\blacksquare$

For a linear homogeneous recurrence relation of degree k, $a_n = r_1 a_{n-1} + r_2 a_{n-2} + \cdots + r_k a_{n-k}$, we call the associated polynomial of degree k, $x^k = r_1 x^{k-1} + r_2 x^{k-2} + \cdots + r_k$, its **characteristic equation**. The roots of the characteristic equation play a key role in the explicit formula for the sequence defined by the recurrence relation and the initial conditions. While the problem can be solved in general, we give a theorem for degree 2 only. Here it is common to write the characteristic equation as $x^2 - r_1 x - r_2 = 0$.

Theorem 1   (a) If the characteristic equation $x^2 - r_1 x - r_2 = 0$ of the recurrence relation $a_n = r_1 a_{n-1} + r_2 a_{n-2}$ has two distinct roots, s_1 and s_2, then $a_n = u s_1^n + v s_2^n$, where u and v depend on the initial conditions, is the explicit formula for the sequence.

(b) If the characteristic equation $x^2 - r_1 x - r_2 = 0$ has a single root s, the explicit formula is $a_n = u s^n + v n s^n$, where u and v depend on the initial conditions.   ●

Proof

(a) Suppose that s_1 and s_2 are roots of $x^2 - r_1 x - r_2 = 0$, so $s_1^2 - r_1 s_1 - r_2 = 0$, $s_2^2 - r_1 s_2 - r_2 = 0$, and $a_n = u s_1^n + v s_2^n$, for $n \geq 1$. We show that this definition of a_n defines the same sequence as $a_n = r_1 a_{n-1} + r_2 a_{n-2}$. First we note that u and v are chosen so that $a_1 = u s_1 + v s_2$ and $a_2 = u s_1^2 + v s_2^2$ and so the initial conditions are satisfied. Then

$$a_n = u s_1^n + v s_2^n \qquad\qquad \text{Split out } s_1^2 \text{ and } s_2^2.$$

$$= u s_1^{n-2} s_1^2 + v s_2^{n-2} s_2^2 \qquad\qquad \text{Substitute for } s_1^2 \text{ and } s_2^2.$$

$$= u s_1^{n-2}(r_1 s_1 + r_2) + v s_2^{n-2}(r_1 s_2 + r_2)$$

$$= r_1 u s_1^{n-1} + r_2 u s_1^{n-2} + r_1 v s_2^{n-1} + r_2 v s_2^{n-2}$$

$$= r_1(u s_1^{n-1} + v s_2^{n-1}) + r_2(u s_1^{n-2} + v s_2^{n-2})$$

$$= r_1 a_{n-1} + r_2 a_{n-2} \qquad\qquad \text{Use definitions of } a_{n-1}$$
$$\text{and } a_{n-2}.$$

(b) This part may be proved in a similar way.   ▼

This direct proof requires that we find a way to use what is known about s_1 and s_2. We know something about s_1^2 and s_2^2, and this suggests the first step of the algebraic rewriting. Finding a useful first step in a proof may involve some false starts. Be persistent.

EXAMPLE 7

Find an explicit formula for the sequence defined by $c_n = 3c_{n-1} - 2c_{n-2}$ with initial conditions $c_1 = 5$ and $c_2 = 3$.

Solution   The recurrence relation $c_n = 3c_{n-1} - 2c_{n-2}$ is a linear homogeneous relation of degree 2. Its associated equation is $x^2 = 3x - 2$. Rewriting this as $x^2 - 3x + 2 = 0$, we see there are two roots, 1 and 2. Theorem 1(a) says we can find u and v so that $c_1 = u(1) + v(2)$ and $c_2 = u(1)^2 + v(2)^2$. Solving this 2 × 2 system yields u is 7 and v is -1.

By Theorem 1, we have $c_n = 7 \cdot 1^n + (-1) \cdot 2^n$ or $c_n = 7 - 2^n$. Note that using $c_n = 3c_{n-1} - 2c_{n-2}$ with initial conditions $c_1 = 5$ and $c_2 = 3$, gives 5, 3, -1, -9 as the first four terms of the sequence. The formula $c_n = 7 - 2^n$ also produces 5, 3, -1, -9 as the first four terms.   ■

EXAMPLE 8

Solve the recurrence relation $d_n = 2d_{n-1} - d_{n-2}$ with initial conditions $d_1 = 1.5$ and $d_2 = 3$.

Solution The associated equation for this linear homogeneous relation is $x^2 - 2x + 1 = 0$. This equation has one (multiple) root, 1. Thus, by Theorem 1(b), $d_n = u(1)^n + vn(1)^n$. Using this formula and the initial conditions, $d_1 = 1.5 = u + v(1)$ and $d_2 = 3 = u + v(2)$, we find that u is 0 and v is 1.5. Then $d_n = 1.5n$. ∎

The Fibonacci sequence in Example 1(b) is a well-known sequence whose explicit formula took over two hundred years to find.

EXAMPLE 9

The Fibonacci sequence is defined by a linear homogeneous recurrence relation of degree 2, so by Theorem 1, the roots of the associated equation are needed to describe the explicit formula for the sequence. From $f_n = f_{n-1} + f_{n-2}$ and $f_1 = f_2 = 1$, we have $x^2 - x - 1 = 0$. Using the quadratic formula to obtain the roots, we find

$$s_1 = \frac{1 + \sqrt{5}}{2} \quad \text{and} \quad s_2 = \frac{1 - \sqrt{5}}{2}.$$

It remains to determine the u and v of Theorem 1. We solve

$$1 = u\left(\frac{1 + \sqrt{5}}{2}\right) + v\left(\frac{1 - \sqrt{5}}{2}\right) \quad \text{and} \quad 1 = u\left(\frac{1 + \sqrt{5}}{2}\right)^2 + v\left(\frac{1 - \sqrt{5}}{2}\right)^2.$$

For the given initial conditions, u is $\frac{1}{\sqrt{5}}$ and v is $-\frac{1}{\sqrt{5}}$. The explicit formula for the Fibonacci sequence is

$$f_n = \frac{1}{\sqrt{5}}\left(\frac{1 + \sqrt{5}}{2}\right)^n - \frac{1}{\sqrt{5}}\left(\frac{1 - \sqrt{5}}{2}\right)^n. \quad ∎$$

Sometimes properties of a recurrence relation are useful to know. Because of the close connection between recurrence (recursion) and mathematical induction, proofs of these properties by induction are common.

EXAMPLE 10

For the Fibonacci numbers in Example 1(b), $f_n \leq \left(\frac{5}{3}\right)^n$. This gives a bound on how fast the Fibonacci numbers grow.

Proof (by strong induction)

Basis Step: Here n_0 is 1. $P(1)$ is $1 \leq \frac{5}{3}$ and this is clearly true.

Induction Step: We use $P(j)$, $j \leq k$ to show $P(k+1)$: $f_{k+1} \leq \left(\frac{5}{3}\right)^{k+1}$. Consider the left-hand side of $P(k+1)$:

$$
\begin{aligned}
f_{k+1} = f_k + f_{k-1} &\leq \left(\frac{5}{3}\right)^k + \left(\frac{5}{3}\right)^{k-1} \\
&= \left(\frac{5}{3}\right)^{k-1} \left(\frac{5}{3} + 1\right) \\
&= \left(\frac{5}{3}\right)^{k-1} \left(\frac{8}{3}\right) \\
&< \left(\frac{5}{3}\right)^{k-1} \left(\frac{5}{3}\right)^2 \\
&= \left(\frac{5}{3}\right)^{k+1}, \quad \text{the right-hand side of } P(k+1). \qquad \blacksquare
\end{aligned}
$$

3.5 Exercises

In Exercises 1 through 6, give the first four terms and identify the given recurrence relation as linear homogeneous or not. If the relation is a linear homogeneous relation, give its degree.

1. $a_n = 2.5a_{n-1}, a_1 = 4$

2. $b_n = -3b_{n-1} - 2b_{n-2}, b_1 = -2, b_2 = 4$

3. $c_n = 2^n c_{n-1}, c_1 = 3$

4. $d_n = nd_{n-1}, d_1 = 2$

5. $e_n = 5e_{n-1} + 3, e_1 = 1$

6. $g_n = \sqrt{g_{n-1} + g_{n-2}}, g_1 = 1, g_2 = 3$

7. Let $A = \{0, 1\}$. Give a recurrence relation for the number of strings of length n in A^* that do not contain 01.

8. Let $A = \{0, 1\}$. Give a recurrence relation for the number of strings of length n in A^* that do not contain 111.

9. On the first of each month Mr. Martinez deposits $100 in a savings account that pays 6% compounded monthly. Assuming that no withdrawals are made, give a recurrence relation for the total amount of money in the account at the end of n months.

10. An annuity of $10,000 earns 8% compounded monthly. Each month $250 is withdrawn from the annuity. Write a recurrence relation for the monthly balance at the end of n months.

In Exercises 11 through 16, use the technique of backtracking to find an explicit formula for the sequence defined by the recurrence relation and initial condition(s).

11. $a_n = 2.5a_{n-1}, a_1 = 4$

12. $b_n = 5b_{n-1} + 3, b_1 = 3$

13. $c_n = c_{n-1} + n, c_1 = 4$

14. $d_n = -1.1d_{n-1}, d_1 = 5$

15. $e_n = e_{n-1} - 2, e_1 = 0$

16. $g_n = ng_{n-1}, g_1 = 6$

In Exercises 17 through 22, solve each of the recurrence relations.

17. $a_n = 4a_{n-1} + 5a_{n-2}, a_1 = 2, a_2 = 6$

18. $b_n = -3b_{n-1} - 2b_{n-2}, b_1 = -2, b_2 = 4$

19. $c_n = -6c_{n-1} - 9c_{n-2}, c_1 = 2.5, c_2 = 4.7$

20. $d_n = 4d_{n-1} - 4d_{n-2}, d_1 = 1, d_2 = 7$

21. $e_n = 2e_{n-2}, e_1 = \sqrt{2}, e_2 = 6$

22. $g_n = 2g_{n-1} - 2g_{n-2}, g_1 = 1, g_2 = 4$

23. Develop a general explicit formula for a nonhomogeneous recurrence relation of the form $a_n = ra_{n-1} + s$, where r and s are constants.

24. Test the results of Exercise 23 on Exercises 12 and 15.

25. Prove Theorem 1(b). (*Hint:* Find the condition on r_1 and r_2 that guarantees that there is one solution s.)

26. Solve the recurrence relation of Example 2.

27. Using the argument in Example 3 for $_nP_r$ would produce $_nP_r = r \cdot {_nP_{r-1}}$. But this is easily shown to be false for nearly all choices of n and r. Explain why the argument is not valid.

28. For the Fibonacci sequence, prove that for $n \geq 2$, $f_{n+1}^2 - f_n^2 = f_{n-1}f_{n+2}$.

29. Solve the recurrence relation of Exercise 7.

30. Solve the recurrence relation of Exercise 9.

31. Use mathematical induction to prove that for the recurrence relation $b_n = b_{n-1} + 2b_{n-2}$, $b_1 = 1$, $b_2 = 3$, $b_n < \left(\frac{5}{2}\right)^n$.

32. Use mathematical induction to prove that for the recurrence relation $a_n = 2a_{n-1} + a_{n-2}$, $a_1 = 10$, $a_2 = 12$, $5 \mid a_{3n+1}$, $n \geq 0$.

33. Let $\mathbf{A}_1, \mathbf{A}_2, \mathbf{A}_3, \ldots, \mathbf{A}_{n+1}$ each be a $k \times k$ matrix. Let C_n be the number of ways to evaluate the product $\mathbf{A}_1 \times \mathbf{A}_2 \times \mathbf{A}_3 \times \cdots \times \mathbf{A}_{n+1}$ by choosing different orders in which to do the n multiplications. Compute C_1, C_2, C_3, C_4, C_5.

34. Give a recurrence relation for C_n (defined in Exercise 33).

35. Verify that $C_n = \frac{_{2n}C_n}{n+1}$ is a possible solution to the recurrence relation of Exercise 34 by showing that this formula produces the first five values as found in Exercise 33. (The terms of this sequence are called the **Catalan numbers**.)

TIPS FOR PROOFS

Proofs based on the pigeonhole principle are introduced in this chapter. Two situations are possible; the pigeons and pigeonholes are implicitly defined in the statement of the problem (Section 3.3, Exercise 5) or you must create pigeons and pigeonholes by defining categories into which the objects must fall (Section 3.3, Exercises 12 and 13). In the first case, the phrases "at least k objects" "have the same property" identify the pigeons (objects) and the labels on the pigeonholes (possible properties).

Proofs of statements about $_nC_r$ and $_nP_r$ are usually direct proofs based on the definitions and elementary algebra. Remember that a direct proof is generally the first approach to try.

KEY IDEAS FOR REVIEW

- Theorem (The Multiplication Principle): Suppose two tasks T_1 and T_2 are to be performed in sequence. If T_1 can be performed in n_1 ways and for each of these ways T_2 can be performed in n_2 ways, then the sequence $T_1 T_2$ can be performed in $n_1 n_2$ ways.

- Theorem (The Extended Multiplication Principle): see pages 73–74

- Theorem: Let A be a set with n elements and $1 \leq r \leq n$. Then the number of sequences of length r that can be formed from elements of A, allowing repetitions, is n^r.

- Permutation of n objects taken r at a time ($1 \leq r \leq n$): a sequence of length r formed from distinct elements

- Theorem: If $1 \leq r \leq n$, then $_nP_r$, the number of permutations of n objects taken r at a time, is $n \cdot (n-1) \cdot (n-2) \cdot \cdots \cdot (n-r+1)$ or $\frac{n!}{(n-r)!}$.

- Permutation: an arrangement of n elements of a set A into a sequence of length n

- Theorem: The number of distinguishable permutations that can be formed from a collection of n objects where the first object appears k_1 times, the second object k_2 times, and so on, is $\frac{n!}{k_1! k_2! \cdots k_t!}$.

- Combination of n objects taken r at a time: a subset of r elements taken from a set with n elements

- Theorem: Let A be a set with $|A| = n$ and let $1 \leq r \leq n$. Then $_nC_r$, the number of combinations of the elements of A, taken r at a time, is $\frac{n!}{r!(n-r)!}$.

- Theorem: Suppose k selections are to be made from n items without regard to order and that repeats are allowed, assuming at least k copies of each of the n items. The number of ways these selections can be made is $_{(n+k-1)}C_k$.

- The pigeonhole principle: see page 83

- The extended pigeonhole principle: see page 85

- Sample space: the set of all outcomes of an experiment

- Event: a subset of the sample space

- Certain event: an event certain to occur

- Impossible event: the empty subset of the sample space

- Mutually exclusive events: any two events E and F with $E \cap F = \{ \}$

- f_E: the frequency of occurrence of the event E in n trials
- $p(E)$: the probability of event E
- Probability space: see page 89
- Elementary event: an event consisting of just one outcome
- Random selection: see page 91
- Expected value: the sum of the products (value of a_i) · $(p(a_i))$ for all outcomes a_i of an experiment

- Recurrence relation: a recursive formula for a sequence
- Initial conditions: information about the beginning of a recursively defined sequence
- Linear homogeneous relation of degree k: a recurrence relation of the form $a_n = r_1 a_{n-1} + r_2 a_{n-2} + \cdots + r_k a_{n-k}$ with the r_i's constants
- Characteristic equation: see page 97
- Catalan numbers: see page 101

CODING EXERCISES

For each of the following, write the requested program or subroutine in pseudocode (as described in Appendix A) or in a programming language that you know. Test your code either with a paper-and-pencil trace or with a computer run.

1. Write a subroutine that accepts two positive integers n and r and if $r \leq n$, returns the number of permutations of n objects taken r at a time.

2. Write a program that has as input positive integers n and r and if $r \leq n$, prints the permutations of $1, 2, 3, \ldots, n$ taken r at a time.

3. Write a subroutine that accepts two positive integers n and r and if $r \leq n$, returns the number of combinations of n objects taken r at a time.

4. Write a program that has as input positive integers n and r and if $r \leq n$, prints the combinations of $1, 2, 3, \ldots, n$ taken r at a time.

5. (a) Write a recursive subroutine that with input k prints the first k Fibonacci numbers.

   (b) Write a nonrecursive subroutine that with input k prints the kth Fibonacci number.

CHAPTER 3 SELF-TEST

1. Compute the number of
   (a) five-digit binary numbers.
   (b) five-card hands from a deck of 52 cards.
   (c) distinct arrangements of the letters of DISCRETE.

2. A computer program is used to generate all possible six-letter names for a new medication. Suppose that all the letters of the English alphabet may be used. How many possible names can be formed
   (a) if the letters are to be distinct?
   (b) if exactly two letters are repeated?

3. A fair six-sided die is rolled five times. If the results of each roll are recorded, how many
   (a) record sequences are possible?
   (b) record sequences begin 1, 2?

4. The IC Shoppe has 14 flavors of ice cream today. If you allow repeats, how many different triple-scoop ice cream cones can be chosen? (The order in which the scooping is done does not matter.)

5. The Spring Dance Committee must have 3 freshman and 5 sophomore members. If there are 23 eligible freshman and 18 eligible sophomores, how many different committees are possible?

6. Show that $_{2n}C_2 = 2 \cdot {}_nC_2 + n^2$.

7. Pizza Quik always puts 50 pepperoni slices on a pepperoni pizza. If you cut a pepperoni pizza into eight equal size pieces, at least one piece must have _____ pepperoni slices. Justify your answer.

8. Complete and prove the following statement. At least _____ months of the year must begin on the same day of the week.

9. What is the probability that exactly two coins will land heads up when five fair coins are tossed?

10. Let $p(A) = 0.29$, $p(B) = 0.41$, and $p(A \cup B) = 0.65$. Are A and B mutually exclusive events? Justify your answer.

11. Solve the recurrence relation $b_n = 7b_{n-1} - 12b_{n-2}$, $b_1 = 1$, $b_2 = 7$.

12. Develop a formula for the solution of a recurrence relation of the form $a_n = ma_{n-1} - 1$, $a_1 = m$.

4

RELATIONS AND DIGRAPHS

Prerequisites: Chapters 1 and 2

Relationships between people, numbers, sets, and many other entities can be formalized in the idea of a binary relation. In this chapter we develop the concept of binary relation, and we give several geometric and algebraic methods of representing such objects. We also discuss a variety of different properties that a binary relation may possess, and we introduce important examples such as equivalence relations. Finally, we introduce several useful types of operations that may be performed on binary relations. We discuss these operations from both a theoretical and computational point of view.

4.1 PRODUCT SETS AND PARTITIONS

Product Sets

An **ordered pair** (a, b) is a listing of the objects a and b in a prescribed order, with a appearing first and b appearing second. Thus an ordered pair is merely a sequence of length 2. From our earlier discussion of sequences (see Section 1.3) it follows that the ordered pairs (a_1, b_1) and (a_2, b_2) are equal if and only if $a_1 = a_2$ and $b_1 = b_2$.

If A and B are two nonempty sets, we define the **product set** or **Cartesian product** $A \times B$ as the set of all ordered pairs (a, b) with $a \in A$ and $b \in B$. Thus

$$A \times B = \{(a, b) \mid a \in A \text{ and } b \in B\}.$$

EXAMPLE 1

Let

$$A = \{1, 2, 3\} \quad \text{and} \quad B = \{r, s\};$$

then

$$A \times B = \{(1, r), (1, s), (2, r), (2, s), (3, r), (3, s)\}.$$

Observe that the elements of $A \times B$ can be arranged in a convenient tabular array as shown in Figure 4.1. ∎

A \ B	r	s
1	$(1, r)$	$(1, s)$
2	$(2, r)$	$(2, s)$
3	$(3, r)$	$(3, s)$

Figure 4.1

EXAMPLE 2   If A and B are as in Example 1, then

$$B \times A = \{(r, 1), (s, 1), (r, 2), (s, 2), (r, 3), (s, 3)\}.$$ ■

Theorem 1   For any two finite, nonempty sets A and B, $|A \times B| = |A| |B|$.   ●

Proof   Suppose that $|A| = m$ and $|B| = n$. To form an ordered pair (a, b), $a \in A$ and $b \in B$, we must perform two successive tasks. Task 1 is to choose a first element from A, and task 2 is to choose a second element from B. There are m ways to perform task 1 and n ways to perform task 2; so, by the multiplication principle (see Section 3.1), there are $m \times n$ ways to form an ordered pair (a, b). In other words, $|A \times B| = m \cdot n = |A| |B|$.   ▼

EXAMPLE 3   If $A = B = \mathbb{R}$, the set of all real numbers, then $\mathbb{R} \times \mathbb{R}$, also denoted by $\mathbb{R}^2$, is the set of all points in the plane. The ordered pair (a, b) gives the coordinates of a point in the plane.   ■

EXAMPLE 4   A marketing research firm classifies a person according to the following two criteria:

Gender: male (m); female (f)

Highest level of education completed: elementary school (e); high school (h); college (c); graduate school (g)

Let $S = \{m, f\}$ and $L = \{e, h, c, g\}$. The product set $S \times L$ contains all the categories into which the population is classified. Thus the classification (f, g) represents a female who has completed graduate school. There are eight categories in this classification scheme.   ■

We now define the Cartesian product of three or more nonempty sets by generalizing the earlier definition of the Cartesian product of two sets. That is, the **Cartesian product** $A_1 \times A_2 \times \cdots \times A_m$ of the nonempty sets $A_1, A_2, \ldots, A_m$ is the set of all ordered m-tuples $(a_1, a_2, \ldots, a_m)$, where $a_i \in A_i$, $i = 1, 2, \ldots, m$. Thus

$$A_1 \times A_2 \times \cdots \times A_m = \{(a_1, a_2, \ldots, a_m) \mid a_i \in A_i, \ i = 1, 2, \ldots, m\}.$$

EXAMPLE 5   A software firm provides the following three characteristics for each program that it sells:

Language: FORTRAN (f); PASCAL (p); LISP (l)

Memory: 2 meg (2); 4 meg (4); 8 meg (8)

Operating system: UNIX (*u*); DOS (*d*)

Let $L = \{f, p, l\}$, $M = \{2, 4, 8\}$, and $O = \{u, d\}$. Then the Cartesian product $L \times M \times O$ contains all the categories that describe a program. There are $3 \cdot 3 \cdot 2$ or 18 categories in this classification scheme. ∎

Proceeding in a manner similar to that used to prove Theorem 1, using the extended multiplication principle, we can show that if A_1 has n_1 elements, A_2 has n_2 elements, ..., and A_m has n_m elements, then $A_1 \times A_2 \times \cdots \times A_m$ has $n_1 \cdot n_2 \cdot \cdots \cdot n_m$ elements.

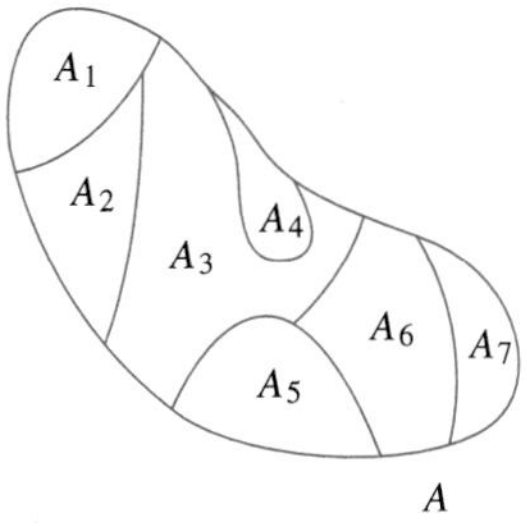

Partitions

A **partition** or **quotient set** of a nonempty set A is a collection $\mathcal{P}$ of nonempty subsets of A such that

1. Each element of A belongs to one of the sets in $\mathcal{P}$.
2. If A_1 and A_2 are distinct elements of $\mathcal{P}$, then $A_1 \cap A_2 = \varnothing$.

The sets in $\mathcal{P}$ are called the **blocks** or **cells** of the partition. Figure 4.2 shows a partition $\mathcal{P} = \{A_1, A_2, A_3, A_4, A_5, A_6, A_7\}$ into seven blocks.

Figure 4.2

EXAMPLE 6

Let $A = \{a, b, c, d, e, f, g, h\}$. Consider the following subsets of A:

$$A_1 = \{a, b, c, d\}, \quad A_2 = \{a, c, e, f, g, h\}, \quad A_3 = \{a, c, e, g\},$$
$$A_4 = \{b, d\}, \quad A_5 = \{f, h\}.$$

Then $\{A_1, A_2\}$ is not a partition since $A_1 \cap A_2 \neq \varnothing$. Also, $\{A_1, A_5\}$ is not a partition since $e \notin A_1$ and $e \notin A_5$. The collection $\mathcal{P} = \{A_3, A_4, A_5\}$ is a partition of A. ∎

EXAMPLE 7

Let

$$Z = \text{set of all integers,}$$
$$A_1 = \text{set of all even integers, and}$$
$$A_2 = \text{set of all odd integers.}$$

Then $\{A_1, A_2\}$ is a partition of Z. ∎

Since the members of a partition of a set A are subsets of A, we see that the partition is a subset of $P(A)$, the power set of A. That is, partitions can be considered as particular kinds of subsets of $P(A)$.

4.1 Exercises

In Exercises 1 through 4, find x or y so that the statement is true.

1. (a) $(x, 3) = (4, 3)$ (b) $(a, 3y) = (a, 9)$

2. (a) $(3x + 1, 2) = (7, 2)$

(b) $(C^{++}, \text{PASCAL}) = (y, x)$

3. (a) $(4x, 6) = (16, y)$

(b) $(2x - 3, 3y - 1) = (5, 5)$

4. (a) $(x^2, 25) = (49, y)$ (b) $(x, y) = (x^2, y^2)$

In Exercises 5 and 6, let $A = \{a, b\}$ and $B = \{4, 5, 6\}$.

5. List the elements in

   (a) $A \times B$ (b) $B \times A$

6. List the elements in

   (a) $A \times A$ (b) $B \times B$

7. Let $A = \{\text{Fine, Yang}\}$ and $B = \{\text{president, vice-president, secretary, treasurer}\}$. Give each of the following.

   (a) $A \times B$ (b) $B \times A$ (c) $A \times A$

8. A genetics experiment classifies fruit flies according to the following two criteria:

   Gender: male (m), female (f)

   Wing span: short (s), medium (m), long (l)

   (a) How many categories are there in this classification?

   (b) List all the categories in this classification scheme.

9. A car manufacturer makes three different types of car frames and two types of engines.

   Frame type: sedan (s), coupe (c), van (v)

   Engine type: gas (g), diesel (d)

   List all possible models of cars.

10. If $A = \{a, b, c\}$, $B = \{1, 2\}$, and $C = \{\#, *\}$, list all the elements of $A \times B \times C$.

11. If A has three elements and B has $n \geq 1$ elements, use mathematical induction to prove that $|A \times B| = 3n$.

In Exercises 12 and 13, let $A = \{a \mid a \text{ is a real number}\}$ and $B = \{1, 2, 3\}$. Sketch the given set in the Cartesian plane.

12. $A \times B$ 13. $B \times A$

In Exercises 14 and 15, let $A = \{a \mid a \text{ is a real number and } -2 \leq a \leq 3\}$ and $B = \{b \mid b \text{ is a real number and } 1 \leq b \leq 5\}$. Sketch the given set in the Cartesian plane.

14. $A \times B$ 15. $B \times A$

16. Show that if A_1 has n_1 elements, A_2 has n_2 elements, and A_3 has n_3 elements, then $A_1 \times A_2 \times A_3$ has $n_1 \cdot n_2 \cdot n_3$ elements.

17. If $A \subseteq C$ and $B \subseteq D$, prove that $A \times B \subseteq C \times D$.

In Exercises 18 and 19, let $A = \{1, 2, 3, 4, 5, 6, 7, 8, 9, 10\}$ and

$$A_1 = \{1, 2, 3, 4\}, \quad A_2 = \{5, 6, 7\}$$
$$A_3 = \{4, 5, 7, 9\}, \quad A_4 = \{4, 8, 10\}$$
$$A_5 = \{8, 9, 10\}, \quad A_6 = \{1, 2, 3, 6, 8, 10\}.$$

18. Which of the following are partitions of A?

    (a) $\{A_1, A_2, A_5\}$ (b) $\{A_1, A_3, A_5\}$

19. Which of the following are partitions of A?

    (a) $\{A_3, A_6\}$ (b) $\{A_2, A_3, A_4\}$

20. If A_1 is the set of positive integers and A_2 is the set of all negative integers, is $\{A_1, A_2\}$ a partition of Z? Explain your conclusion.

For Exercises 21 through 23, use $A = \{a, b, c \ldots, z\}$.

21. Give a partition $\mathcal{P}$ of A such that $|\mathcal{P}| = 4$ and one element of $\mathcal{P}$ contains only the letters needed to spell your first name.

22. Give a partition $\mathcal{P}$ of A such that $|\mathcal{P}| = 3$ and each element of $\mathcal{P}$ contains at least five elements.

23. Is it possible to have a partition $\mathcal{P}$ of A such that $\mathcal{P} = \{A_1, A_2, \ldots, A_{10}\}$ and $\forall i \; |A_i| \geq 3$?

24. If $B = \{0, 3, 6, 9, \ldots\}$, give a partition of B containing

    (a) two infinite subsets.

    (b) three infinite subsets.

25. List all partitions of $A = \{1, 2, 3\}$.

26. List all partitions of $B = \{a, b, c, d\}$.

27. The number of partitions of a set with n elements into k subsets satisfies the recurrence relation

$$S(n, k) = S(n - 1, k - 1) + k \cdot S(n - 1, k)$$

with initial conditions $S(n, 1) = S(n, n) = 1$. Find the number of partitions of a set with three elements into two subsets, that is, $S(3, 2)$. Compare your result with the results of Exercise 25.

28. Find the number of partitions of a set with four elements into two subsets using the recurrence relation in Exercise 27. Compare the result with the results of Exercise 26.

29. Let A, B, and C be subsets of U. Prove that $A \times (B \cup C) = (A \times B) \cup (A \times C)$.

30. Use the sets $A = \{1, 2, 4\}$, $B = \{2, 5, 7\}$, and $C = \{1, 3, 7\}$ to investigate whether $A \times (B \cap C) = (A \times B) \cap (A \times C)$. Explain your conclusions.

4.2 RELATIONS AND DIGRAPHS

The notion of a relation between two sets of objects is quite common and intuitively clear (a formal definition will be given later). If A is the set of all living humans males and B is the set of all living human females, then the relation F (father) can be defined between A and B. Thus, if $x \in A$ and $y \in B$, then x is related to y by the relation F if x is the father of y, and we write $x \ F \ y$. Because order matters here, we refer to F as a relation from A to B. We could also consider the relations S and H from A to B by letting $x \ S \ y$ mean that x is a son of y, and $x \ H \ y$ mean that x is the husband of y.

If A is the set of all real numbers, there are many commonly used relations from A to A. An example is the relation "less than," which is usually denoted by $<$, so that X is related to y if $x < y$, and the other order relations $>$, $\geq$, and $\leq$. We see that a relation is often described verbally and may be denoted by a familiar name or symbol. The problem with this approach is that we will need to discuss *any possible* relation from one abstract set to another. Most of these relations have no simple verbal description and no familiar name or symbol to remind us of their nature or properties. Furthermore, it is usually awkward, and sometimes nearly impossible, to give any precise proofs of the properties that a relation satisfies if we must deal with a verbal description of it.

To solve this problem, observe that the only thing that really matters about a relation is that we know precisely which elements in A are related to which elements in B. Thus suppose that $A = \{1, 2, 3, 4\}$ and R is a relation from A to A. If we know that $1 \ R \ 2, 1 \ R \ 3, 1 \ R \ 4, 2 \ R \ 3, 2 \ R \ 4$, and $3 \ R \ 4$, then we know everything we need to know about R. Actually, R is the familiar relation $<$, "less than," but we need not know this. It would be enough to be given the foregoing list of related pairs. Thus we may say that R is completely known if we know all R-related pairs. We could then write $R = \{(1, 2), (1, 3), (1, 4), (2, 3), (2, 4), (3, 4)\}$, since R is essentially equal to or completely specified by this set of ordered pairs. Each ordered pair specifies that its first element is related to its second element, and all possible related pairs are assumed to be given, at least in principle. This method of specifying a relation does not require any special symbol or description and so is suitable for any relation between any two sets. Note that from this point of view a relation from A to B is simply a subset of $A \times B$ (giving the related pairs), and, conversely, any subset of $A \times B$ can be considered a relation, even if it is an unfamiliar relation for which we have no name or alternative description. We choose this approach for defining relations.

Let A and B be nonempty sets. A **relation** R **from** A **to** B is a subset of $A \times B$. If $R \subseteq A \times B$ and $(a, b) \in R$, we say that a **is related to** b **by** R, and we also write $a \ R \ b$. If a is not related to b by R, we write $a \ \not{R} \ b$. Frequently, A and B are equal. In this case, we often say that $R \subseteq A \times A$ **is a relation on** A, instead of a relation from A to A.

Relations are extremely important in mathematics and its applications. It is not an exaggeration to say that 90% of what will be discussed in the remainder of this book will concern some type of object that may be considered a relation. We now give a number of examples.

EXAMPLE 1   Let $A = \{1, 2, 3\}$ and $B = \{r, s\}$. Then $R = \{(1, r), (2, s), (3, r)\}$ is a relation from A to B.   ∎

EXAMPLE 2   Let A and B be sets of real numbers. We define the following relation R (equals) from A to B:

$$a \ R \ b \quad \text{if and only if} \quad a = b.$$ ∎

EXAMPLE 3   Let $A = \{1, 2, 3, 4, 5\}$. Define the following relation R (less than) on A:

$$a \ R \ b \quad \text{if and only if} \quad a < b.$$

Then

$$R = \{(1, 2), (1, 3), (1, 4), (1, 5), (2, 3), (2, 4), (2, 5), (3, 4), (3, 5), (4, 5)\}.$$ ∎

EXAMPLE 4   Let $A = Z^+$, the set of all positive integers. Define the following relation R on A:

$$a \ R \ b \quad \text{if and only if} \quad a \text{ divides } b.$$

Then $4 \ R \ 12$, but $5 \ \not{R} \ 7$.   ∎

EXAMPLE 5   Let A be the set of all people in the world. We define the following relation R on A: $a \ R \ b$ if and only if there is a sequence $a_0, a_1, \ldots, a_n$ of people such that $a_0 = a$, $a_n = b$ and a_{i-1} knows a_i, $i = 1, 2, \ldots, n$ (n will depend on a and b).   ∎

EXAMPLE 6   Let $A = \mathbb{R}$, the set of real numbers. We define the following relation R on A:

$$x \ R \ y \quad \text{if and only if} \quad x \text{ and } y \text{ satisfy the equation } \frac{x^2}{4} + \frac{y^2}{9} = 1.$$

The set R consists of all points on the ellipse shown in Figure 4.3.   ∎

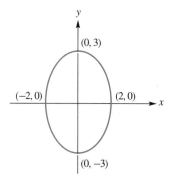

Figure 4.3

EXAMPLE 7   Let A be the set of all possible inputs to a given computer program, and let B be the set of all possible outputs from the same program. Define the following relation R from A to B: $a \ R \ b$ if and only if b is the output produced by the program when input a is used.   ∎

EXAMPLE 8   Let $A =$ the set of all lines in the plane. Define the following relation R on A:

$$l_1 \ R \ l_2 \quad \text{if and only if} \quad l_1 \text{ is parallel to } l_2. \qquad \blacksquare$$

EXAMPLE 9   An airline services the five cities $c_1, c_2, c_3, c_4,$ and c_5. Table 4.1 gives the cost (in dollars) of going from c_i to c_j. Thus the cost of going from c_1 to c_3 is $100, while the cost of going from c_4 to c_2 is $200.

Table 4.1

From \ To	c_1	c_2	c_3	c_4	c_5
c_1		140	100	150	200
c_2	190		200	160	220
c_3	110	180		190	250
c_4	190	200	120		150
c_5	200	100	200	150	

We now define the following relation R on the set of cities $A = \{c_1, c_2, c_3, c_4, c_5\}$: $c_i \ R \ c_j$ if and only if the cost of going from c_i to c_j is defined and less than or equal to $180. Find R.

Solution   The relation R is the subset of $A \times A$ consisting of all cities (c_i, c_j), where the cost of going from c_i to c_j is less than or equal to $180. Hence

$$R = \{(c_1, c_2), (c_1, c_3), (c_1, c_4), (c_2, c_4), (c_3, c_1), (c_3, c_2),$$
$$(c_4, c_3), (c_4, c_5), (c_5, c_2), (c_5, c_4)\}. \qquad \blacksquare$$

Sets Arising from Relations

Let $R \subseteq A \times B$ be a relation from A to B. We now define various important and useful sets related to R.

The **domain** of R, denoted by $\text{Dom}(R)$, is the set of elements in A that are related to some element in B. In other words, $\text{Dom}(R)$, a subset of A, is the set of all first elements in the pairs that make up R. Similarly, we define the **range** of R, denoted by $\text{Ran}(R)$, to be the set of elements in B that are second elements of pairs in R, that is, all elements in B that are related to some element in A.

Elements of A that are not in $\text{Dom}(R)$ are not involved in the relation R in any way. This is also true for elements of B that are not in $\text{Ran}(R)$.

EXAMPLE 10   If R is the relation defined in Example 1, then $\text{Dom}(R) = A$ and $\text{Ran}(R) = B$. $\blacksquare$

EXAMPLE 11   If R is the relation given in Example 3, then $\text{Dom}(R) = \{1, 2, 3, 4\}$ and $\text{Ran}(R) = \{2, 3, 4, 5\}$. $\blacksquare$

EXAMPLE 12   Let R be the relation of Example 6. Then $\text{Dom}(R) = [-2, 2]$ and $\text{Ran}(R) = [-3, 3]$. Note that these sets are given in interval notation. $\blacksquare$

If R is a relation from A to B and $x \in A$, we define $R(x)$, the **R-relative set of x**, to be the set of all y in B with the property that x is R-related to y. Thus, in symbols,

$$R(x) = \{y \in B \mid x \; R \; y\}.$$

Similarly, if $A_1 \subseteq A$, then $R(A_1)$, the **R-relative set of A_1**, is the set of all y in B with the property that x is R-related to y for some x in A_1. That is,

$$R(A_1) = \{y \in B \mid x \; R \; y \text{ for some } x \text{ in } A_1\}.$$

From the preceding definitions, we see that $R(A_1)$ is the union of the sets $R(x)$, where $x \in A_1$. The sets $R(x)$ play an important role in the study of many types of relations.

EXAMPLE 13   Let $A = \{a, b, c, d\}$ and let $R = \{(a, a), (a, b), (b, c), (c, a), (d, c), (c, b)\}$. Then $R(a) = \{a, b\}$, $R(b) = \{c\}$, and if $A_1 = \{c, d\}$, then $R(A_1) = \{a, b, c\}$.   ■

EXAMPLE 14   Let R be the relation of Example 6, and let $x \in \mathbb{R}$. If $x \; R \; y$ for some y, then $x^2/4 + y^2/9 = 1$. We see that if x is not in the interval $(-2, 2)$, then no y can satisfy the preceding equation, since $x^2/4 > 1$. Thus, in this case, $R(x) = \varnothing$. If $x = -2$, then $x^2/4 = 1$, so x can only be related to 0. Thus $R(-2) = \{0\}$. Similarly, $R(2) = \{0\}$. Finally, if $-2 < x < 2$ and $x \; R \; y$, then we must have $y = \sqrt{9 - (9x^2/4)}$ or $y = -\sqrt{9 - (9x^2/4)}$, as we see by solving the equation $x^2/4 + y^2/9 = 1$, so that $R(x) = \{\sqrt{9 - (9x^2/4)}, -\sqrt{9 - (9x^2/4)}\}$. Thus, for example, $R(1) = \{(3\sqrt{3})/2, -(3\sqrt{3})/2\}$.   ■

The following theorem shows the behavior of the R-relative sets with regard to basic set operations.

Theorem 1   Let R be a relation from A to B, and let A_1 and A_2 be subsets of A. Then
  (a) If $A_1 \subseteq A_2$, then $R(A_1) \subseteq R(A_2)$.
  (b) $R(A_1 \cup A_2) = R(A_1) \cup R(A_2)$.
  (c) $R(A_1 \cap A_2) \subseteq R(A_1) \cap R(A_2)$.   ●

Proof
  (a) If $y \in R(A_1)$, then $x \; R \; y$ for some $x \in A_1$. Since $A_1 \subseteq A_2$, $x \in A_2$. Thus, $y \in R(A_2)$, which proves part (a).
  (b) If $y \in R(A_1 \cup A_2)$, then by definition $x \; R \; y$ for some x in $A_1 \cup A_2$. If x is in A_1, then, since $x \; R \; y$, we must have $y \in R(A_1)$. By the same argument, if x is in A_2, then $y \in R(A_2)$. In either case, $y \in R(A_1) \cup R(A_2)$. Thus we have shown that $R(A_1 \cup A_2) \subseteq R(A_1) \cup R(A_2)$.
      Conversely, since $A_1 \subseteq (A_1 \cup A_2)$, part (a) tells us that $R(A_1) \subseteq R(A_1 \cup A_2)$. Similarly, $R(A_2) \subseteq R(A_1 \cup A_2)$. Thus $R(A_1) \cup R(A_2) \subseteq R(A_1 \cup A_2)$, and therefore part (b) is true.
  (c) If $y \in R(A_1 \cap A_2)$, then, for some x in $A_1 \cap A_2$, $x \; R \; y$. Since x is in both A_1 and A_2, it follows that y is in both $R(A_1)$ and $R(A_2)$; that is, $y \in R(A_1) \cap R(A_2)$. Thus part (c) holds.   ▼

The strategy of this proof is one we have seen many times in earlier sections: Apply a relevant definition to a generic object.

Notice that Theorem 1(c) does not claim equality of sets. See Exercise 18 for conditions under which the two sets are equal. In the following example, we will see that equality does not always hold.

EXAMPLE 15   Let $A = Z$, R be "$\leq$," $A_1 = \{0, 1, 2\}$, and $A_2 = \{9, 13\}$. Then $R(A_1)$ consists of all integers n such that $0 \leq n$, or $1 \leq n$, or $2 \leq n$. Thus $R(A_1) = \{0, 1, 2, \ldots\}$. Similarly, $R(A_2) = \{9, 10, 11, \ldots\}$, so $R(A_1) \cap R(A_2) = \{9, 10, 11, \ldots\}$. On the other hand, $A_1 \cap A_2 = \varnothing$; thus $R(A_1 \cap A_2) = \varnothing$. This shows that the containment in Theorem 1(c) is not always an equality.   ∎

EXAMPLE 16   Let $A = \{1, 2, 3\}$ and $B = \{x, y, z, w, p, q\}$, and consider the relation $R = \{(1, x), (1, z), (2, w), (2, p), (2, q), (3, y)\}$. Let $A_1 = \{1, 2\}$ and $A_2 = \{2, 3\}$. Then $R(A_1) = \{x, z, w, p, q\}$ and $R(A_2) = \{w, p, q, y\}$. Thus $R(A_1) \cup R(A_2) = B$. Since $A_1 \cup A_2 = A$, we see that $R(A_1 \cup A_2) = R(A) = B$, as stated in Theorem 1(b). Also, $R(A_1) \cap R(A_2) = \{w, p, q\} = R(\{2\}) = R(A_1 \cap A_2)$, so in this case equality does hold for the containment in Theorem 1(c).   ∎

It is a useful and easily seen fact that the sets $R(a)$, for a in A, completely determine a relation R. We state this fact precisely in the following theorem.

Theorem 2   Let R and S be relations from A to B. If $R(a) = S(a)$ for all a in A, then $R = S$.●

Proof   If $a \, R \, b$, then $b \in R(a)$. Therefore, $b \in S(a)$ and $a \, S \, b$. A completely similar argument shows that, if $a \, S \, b$, then $a \, R \, b$. Thus $R = S$.   ▼

The Matrix of a Relation

We can represent a relation between two finite sets with a matrix as follows. If $A = \{a_1, a_2, \ldots, a_m\}$ and $B = \{b_1, b_2, \ldots, b_n\}$ are finite sets containing m and n elements, respectively, and R is a relation from A to B, we represent R by the $m \times n$ matrix $\mathbf{M}_R = \left[m_{ij} \right]$, which is defined by

$$m_{ij} = \begin{cases} 1 & \text{if } (a_i, b_j) \in R \\ 0 & \text{if } (a_i, b_j) \notin R. \end{cases}$$

The matrix $\mathbf{M}_R$ is called the **matrix of** R. Often $\mathbf{M}_R$ provides an easy way to check whether R has a given property.

EXAMPLE 17   Let R be the relation defined in Example 1. Then the matrix of R is

$$\mathbf{M}_R = \begin{bmatrix} 1 & 0 \\ 0 & 1 \\ 1 & 0 \end{bmatrix}.$$

∎

Conversely, given sets A and B with $|A| = m$ and $|B| = n$, an $m \times n$ matrix whose entries are zeros and ones determines a relation, as is illustrated in the following example.

EXAMPLE 18   Consider the matrix

$$\mathbf{M} = \begin{bmatrix} 1 & 0 & 0 & 1 \\ 0 & 1 & 1 & 0 \\ 1 & 0 & 1 & 0 \end{bmatrix}.$$

Since **M** is 3×4, we let

$$A = \{a_1, a_2, a_3\} \quad \text{and} \quad B = \{b_1, b_2, b_3, b_4\}.$$

Then $(a_i, b_j) \in R$ if and only if $m_{ij} = 1$. Thus

$$R = \{(a_1, b_1), (a_1, b_4), (a_2, b_2), (a_2, b_3), (a_3, b_1), (a_3, b_3)\}.$$ ∎

The Digraph of a Relation

If A is a finite set and R is a relation on A, we can also represent R pictorially as follows. Draw a small circle for each element of A and label the circle with the corresponding element of A. These circles are called **vertices**. Draw an arrow, called an **edge**, from vertex a_i to vertex a_j if and only if $a_i \, R \, a_j$. The resulting pictorial representation of R is called a **directed graph** or **digraph** of R.

Thus, if R is a relation on A, the edges in the digraph of R correspond exactly to the pairs in R, and the vertices correspond exactly to the elements of the set A. Sometimes, when we want to emphasize the geometric nature of some property of R, we may refer to the pairs of R themselves as edges and the elements of A as vertices.

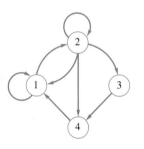

Figure 4.4

EXAMPLE 19   Let

$$A = \{1, 2, 3, 4\}$$
$$R = \{(1, 1), (1, 2), (2, 1), (2, 2), (2, 3), (2, 4), (3, 4), (4, 1)\}.$$

Then the digraph of R is as shown in Figure 4.4. ∎

A collection of vertices with edges between some of the vertices determines a relation in a natural manner.

EXAMPLE 20   Find the relation determined by Figure 4.5.

Solution   Since $a_i \, R \, a_j$ if and only if there is an edge from a_i to a_j, we have

$$R = \{(1, 1), (1, 3), (2, 3), (3, 2), (3, 3), (4, 3)\}.$$ ∎

In this book, digraphs are nothing but geometrical representations of relations, and any statement made about a digraph is actually a statement about the corresponding relation. This is especially important for theorems and their proofs. In some cases, it is easier or clearer to state a result in graphical terms, but a proof will always refer to the underlying relation. The reader should be aware that some authors allow more general objects as digraphs; for example, by permitting several edges in the same direction between the same vertices.

An important concept for relations is inspired by the visual form of digraphs. If R is a relation on a set A and $a \in A$, then the **in-degree** of a (relative to the

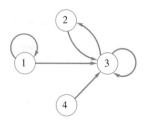

Figure 4.5

relation R) is the number of $b \in A$ such that $(b, a) \in R$. The **out-degree** of a is the number of $b \in A$ such that $(a, b) \in R$.

What this means, in terms of the digraph of R, is that the in-degree of a vertex is the number of edges terminating at the vertex. The out-degree of a vertex is the number of edges leaving the vertex. Note that the out-degree of a is $|R(a)|$.

EXAMPLE 21   Consider the digraph of Figure 4.4. Vertex 1 has in-degree 3 and out-degree 2. Also consider the digraph shown in Figure 4.5. Vertex 3 has in-degree 4 and out-degree 2, while vertex 4 has in-degree 0 and out-degree 1. ∎

EXAMPLE 22   Let $A = \{a, b, c, d\}$, and let R be the relation on A that has the matrix

$$\mathbf{M}_R = \begin{bmatrix} 1 & 0 & 0 & 0 \\ 0 & 1 & 0 & 0 \\ 1 & 1 & 1 & 0 \\ 0 & 1 & 0 & 1 \end{bmatrix}.$$

Construct the digraph of R, and list in-degrees and out-degrees of all vertices.

Solution   The digraph of R is shown in Figure 4.6. The following table gives the in-degrees and out-degrees of all vertices. Note that the sum of all in-degrees must equal the sum of all out-degrees.

	a	b	c	d
In-degree	2	3	1	1
Out-degree	1	1	3	2

∎

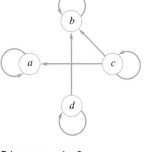

Figure 4.6

EXAMPLE 23   Let $A = \{1, 4, 5\}$, and let R be given by the digraph shown in Figure 4.7. Find $\mathbf{M}_R$ and R.

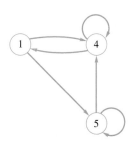

Figure 4.7

Solution

$$\mathbf{M}_R = \begin{bmatrix} 0 & 1 & 1 \\ 1 & 1 & 0 \\ 0 & 1 & 1 \end{bmatrix}, \qquad R = \{(1, 4), (1, 5), (4, 1), (4, 4), (5, 4), (5, 5)\}. ∎$$

If R is a relation on a set A, and B is a subset of A, the **restriction of R to B** is $R \cap (B \times B)$.

EXAMPLE 24   Let $A = \{a, b, c, d, e, f\}$ and $R = \{(a, a), (a, c), (b, c), (a, e), (b, e), (c, e)\}$. Let $B = \{a, b, c\}$. Then

$$B \times B = \{(a, a), (a, b), (a, c), (b, a), (b, b), (b, c), (c, a), (c, b), (c, c)\}$$

and the restriction of R to B is $\{(a, a), (a, c), (b, c)\}$.   ∎

4.2 Exercises

1. For the relation R defined in Example 4, which of the following ordered pairs belong to R?

   (a) $(2, 3)$ (b) $(0, 8)$ (c) $(1, 3)$

   (d) $(6, 18)$ (e) $(-6, 24)$ (f) $(8, 0)$

2. For the relation R defined in Example 6, which of the following ordered pairs belong to R?

   (a) $(2, 0)$ (b) $(0, 2)$ (c) $(0, 3)$

   (d) $(0, 0)$ (e) $(1, 3/2\sqrt{3})$ (f) $(1, 1)$

3. Let $A = Z^+$, the positive integers, and R be the relation defined by $a \, R \, b$ if and only if $2a \le b + 1$. Which of the following ordered pairs belong to R?

   (a) $(2, 2)$ (b) $(3, 2)$ (c) $(6, 15)$

   (d) $(1, 1)$ (e) $(15, 6)$ (f) (n, n)

In Exercises 4 through 12, find the domain, range, matrix, and, when $A = B$, the digraph of the relation R.

4. $A = \{a, b, c, d\}$, $B = \{1, 2, 3\}$,
   $R = \{(a, 1), (a, 2), (b, 1), (c, 2), (d, 1)\}$

5. $A = \{\text{IBM, COMPAQ, Dell, Gateway, Zenith}\}$,
   $B = \{750\text{C}, \text{PS60}, 450\text{SV}, 4/33\text{S}, 525\text{SX}, 466\text{V}, 486\text{SL}\}$
   $R = \{(\text{IBM}, 750\text{C}), (\text{Dell}, 466\text{V}), (\text{COMPAQ}, 450\text{SV}), (\text{Gateway}, \text{PS60})\}$

6. $A = \{1, 2, 3, 4\}$, $B = \{1, 4, 6, 8, 9\}$; $a \, R \, b$ if and only if $b = a^2$.

7. $A = \{1, 2, 3, 4, 8\} = B$; $a \, R \, b$ if and only if $a = b$.

8. $A = \{1, 2, 3, 4, 8\}$, $B = \{1, 4, 6, 9\}$; $a \, R \, b$ if and only if $a \mid b$.

9. $A = \{1, 2, 3, 4, 6\} = B$; $a \, R \, b$ if and only if a is a multiple of b.

10. $A = \{1, 2, 3, 4, 5\} = B$; $a \, R \, b$ if and only if $a \le b$.

11. $A = \{1, 3, 5, 7, 9\}$, $B = \{2, 4, 6, 8\}$; $a \, R \, b$ if and only if $b < a$.

12. $A = \{1, 2, 3, 4, 8\} = B$; $a \, R \, b$ if and only if $a + b \le 9$.

13. Let $A = Z^+$, the positive integers, and R be the relation defined by $a \, R \, b$ if and only if there exists a k in Z^+ so that $a = b^k$ (k depends on a and b). Which of the following belong to R?

   (a) $(4, 16)$ (b) $(1, 7)$ (c) $(8, 2)$

   (d) $(3, 3)$ (e) $(2, 8)$ (f) $(2, 32)$

14. Let $A = \mathbb{R}$. Consider the following relation R on A: $a \, R \, b$ if and only if $2a + 3b = 6$. Find $\text{Dom}(R)$ and $\text{Ran}(R)$.

15. Let $A = \mathbb{R}$. Consider the following relation R on A: $a \, R \, b$ if and only if $a^2 + b^2 = 25$. Find $\text{Dom}(R)$ and $\text{Ran}(R)$.

16. Let R be the relation defined in Example 6. Find $R(A_1)$ for each of the following.

   (a) $A_1 = \{1, 8\}$ (b) $A_1 = \{3, 4, 5\}$

   (c) $A_1 = \{ \ \}$

17. Let R be the relation defined in Exercise 9. Find each of the following.

   (a) $R(3)$ (b) $R(6)$ (c) $R(\{2, 4, 6\})$

18. Let R be a relation from A to B. Prove that for all subsets A_1 and A_2 of A

   $$R(A_1 \cap A_2) = R(A_1) \cap R(A_2) \quad \text{if and only if}$$
   $$R(a) \cap R(b) = \{ \ \} \quad \text{for any distinct } a, b \text{ in } A.$$

19. Let $A = \mathbb{R}$. Give a description of the relation R specified by the shaded region in Figure 4.8.

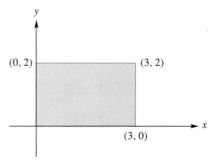

Figure 4.8

20. If A has n elements and B has m elements, how many different relations are there from A to B?

In Exercises 21 and 22, give the relation R defined on A and its digraph.

21. Let $A = \{1, 2, 3, 4\}$ and $\mathbf{M}_R = \begin{bmatrix} 1 & 1 & 0 & 1 \\ 0 & 1 & 1 & 0 \\ 0 & 0 & 1 & 1 \\ 1 & 0 & 0 & 0 \end{bmatrix}$.

22. Let $A = \{a, b, c, d, e\}$ and $\mathbf{M}_R = \begin{bmatrix} 1 & 1 & 0 & 0 & 0 \\ 0 & 0 & 1 & 1 & 0 \\ 0 & 0 & 0 & 1 & 1 \\ 0 & 1 & 1 & 0 & 0 \\ 1 & 0 & 0 & 0 & 0 \end{bmatrix}$.

In Exercises 23 and 24, find the relation determined by the digraph and give its matrix.

23.

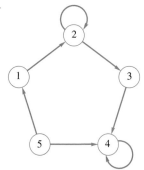

Figure 4.9

24.

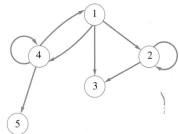

Figure 4.10

25. For the digraph in Exercise 23, give the in-degree and the out-degree of each vertex.

26. For the digraph in Exercise 24, give the in-degree and the out-degree of each vertex.

27. Describe how to find the in-degree and the out-degree of a vertex directly from the matrix of a relation.

In Exercises 28 and 29, let $A = \{1, 2, 3, 4, 5, 6, 7\}$ and $R = \{(1, 2), (1, 4), (2, 3), (2, 5), (3, 6), (4, 7)\}$. Compute the restriction of R to B for the given subset of A.

28. $B = \{1, 2, 4, 5\}$

29. $B = \{2, 3, 4, 6\}$

30. Let R be a relation on a set A and $B \subseteq A$. Describe how to create the matrix of the restriction of R to B from $\mathbf{M}_R$.

31. Let R be a relation on a set A and $B \subseteq A$. Describe how to create the digraph for the restriction of R to B from the digraph of R.

32. Let S be the product set $\{1, 2, 3\} \times \{a, b\}$. How many relations are there on S?

4.3 PATHS IN RELATIONS AND DIGRAPHS

Suppose that R is a relation on a set A. A **path of length n** in R from a to b is a finite sequence $\pi : a, x_1, x_2, \ldots, x_{n-1}, b$, beginning with a and ending with b, such that

$$a \ R \ x_1, \ x_1 \ R \ x_2, \ \ldots, \ x_{n-1} \ R \ b.$$

Note that a path of length n involves $n + 1$ elements of A, although they are not necessarily distinct.

A path is most easily visualized with the aid of the digraph of the relation. It appears as a geometric *path* or succession of edges in such a digraph, where the indicated directions of the edges are followed, and in fact a path derives its name from this representation. Thus the length of a path is the number of edges in the path, where the vertices need not all be distinct.

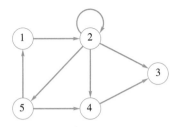

Figure 4.11

EXAMPLE 1

Consider the digraph in Figure 4.11. Then π_1: 1, 2, 5, 4, 3 is a path of length 4 from vertex 1 to vertex 3, π_2: 1, 2, 5, 1 is a path of length 3 from vertex 1 to itself, and π_3: 2, 2 is a path of length 1 from vertex 2 to itself. ∎

A path that begins and ends at the same vertex is called a **cycle**. In Example 1, π_2 and π_3 are cycles of lengths 3 and 1, respectively. It is clear that the paths of length 1 can be identified with the ordered pairs (x, y) that belong to R. Paths in a relation R can be used to define new relations that are quite useful. If n is a fixed positive integer, we define a relation R^n on A as follows: $x \, R^n \, y$ means that there is a path of length n from x to y in R. We may also define a relation R^∞ on A, by letting $x \, R^\infty \, y$ mean that there is some path in R from x to y. The length of such a path will depend, in general, on x and y. The relation R^∞ is sometimes called the **connectivity relation** for R.

Note that $R^n(x)$ consists of all vertices that can be reached from x by means of a path in R of length n. The set $R^\infty(x)$ consists of all vertices that can be reached from x by some path in R.

EXAMPLE 2

Let A be the set of all living human beings, and let R be the relation of mutual acquaintance. That is, $a \, R \, b$ means that a and b know one another. Then $a \, R^2 \, b$ if a and b have an acquaintance in common. In general, $a \, R^n \, b$ if a knows someone x_1, who knows $x_2, \ldots$, who knows x_{n-1}, who knows b. Finally, $a \, R^\infty \, b$ means that some chain of acquaintances exists that begins at a and ends at b. It is interesting (and unknown) whether every two Americans, say, are related by R^∞. ∎

EXAMPLE 3

Let A be a set of U.S. cities, and let $x \, R \, y$ if there is a direct flight from x to y on at least one airline. Then x and y are related by R^n if one can book a flight from x to y having exactly $n - 1$ intermediate stops, and $x \, R^\infty \, y$ if one can get from x to y by plane. ∎

EXAMPLE 4

Let $A = \{1, 2, 3, 4, 5, 6\}$. Let R be the relation whose digraph is shown in Figure 4.12. Figure 4.13 shows the digraph of the relation R^2 on A. A line connects two vertices in Figure 4.13 if and only if they are R^2-related, that is, if and only if there is a path of length two connecting those vertices in Figure 4.12. Thus

$$
\begin{array}{llll}
1 \, R^2 \, 2 & \text{since} & 1 \, R \, 2 & \text{and} & 2 \, R \, 2 \\
1 \, R^2 \, 4 & \text{since} & 1 \, R \, 2 & \text{and} & 2 \, R \, 4 \\
1 \, R^2 \, 5 & \text{since} & 1 \, R \, 2 & \text{and} & 2 \, R \, 5 \\
2 \, R^2 \, 2 & \text{since} & 2 \, R \, 2 & \text{and} & 2 \, R \, 2 \\
2 \, R^2 \, 4 & \text{since} & 2 \, R \, 2 & \text{and} & 2 \, R \, 4 \\
2 \, R^2 \, 5 & \text{since} & 2 \, R \, 2 & \text{and} & 2 \, R \, 5 \\
2 \, R^2 \, 6 & \text{since} & 2 \, R \, 5 & \text{and} & 5 \, R \, 6 \\
3 \, R^2 \, 5 & \text{since} & 3 \, R \, 4 & \text{and} & 4 \, R \, 5 \\
4 \, R^2 \, 6 & \text{since} & 4 \, R \, 5 & \text{and} & 5 \, R \, 6.
\end{array}
$$

In a similar way, we can construct the digraph of R^n for any n. ∎

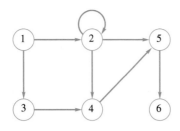

Figure 4.12

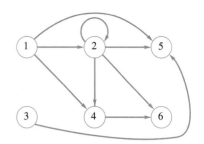

Figure 4.13

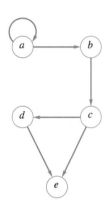

Figure 4.14

EXAMPLE 5   Let $A = \{a, b, c, d, e\}$ and

$$R = \{(a, a), (a, b), (b, c), (c, e), (c, d), (d, e)\}.$$

Compute (a) R^2; (b) R^∞.

Solution   (a) The digraph of R is shown in Figure 4.14.

$a\ R^2\ a$	since	$a\ R\ a$	and	$a\ R\ a.$
$a\ R^2\ b$	since	$a\ R\ a$	and	$a\ R\ b.$
$a\ R^2\ c$	since	$a\ R\ b$	and	$b\ R\ c.$
$b\ R^2\ e$	since	$b\ R\ c$	and	$c\ R\ e.$
$b\ R^2\ d$	since	$b\ R\ c$	and	$c\ R\ d.$
$c\ R^2\ e$	since	$c\ R\ d$	and	$d\ R\ e.$

Hence

$$R^2 = \{(a, a), (a, b), (a, c), (b, e), (b, d), (c, e)\}.$$

(b) To compute R^∞, we need all ordered pairs of vertices for which there is a path of any length from the first vertex to the second. From Figure 4.14 we see that

$$R^\infty = \{(a, a), (a, b), (a, c), (a, d), (a, e), (b, c),$$
$$(b, d), (b, e), (c, d), (c, e), (d, e)\}.$$

For example, $(a, d) \in R^\infty$, since there is a path of length 3 from a to d: a, b, c, d. Similarly, $(a, e) \in R^\infty$, since there is a path of length 3 from a to e: a, b, c, e as well as a path of length 4 from a to e: a, b, c, d, e.   ∎

If $|R|$ is large, it can be tedious and perhaps difficult to compute R^∞, or even R^2, from the set representation of R. However, $\mathbf{M}_R$ can be used to accomplish these tasks more efficiently.

Let R be a relation on a finite set $A = \{a_1, a_2, \ldots, a_n\}$, and let $\mathbf{M}_R$ be the $n \times n$ matrix representing R. We will show how the matrix $\mathbf{M}_{R^2}$, of R^2, can be computed from $\mathbf{M}_R$.

Theorem 1   If R is a relation on $A = \{a_1, a_2, \ldots, a_n\}$, then $\mathbf{M}_{R^2} = \mathbf{M}_R \odot \mathbf{M}_R$ (see Section 1.5).

Proof Let $\mathbf{M}_R = \begin{bmatrix} m_{ij} \end{bmatrix}$ and $\mathbf{M}_{R^2} = \begin{bmatrix} n_{ij} \end{bmatrix}$. By definition, the i, jth element of $\mathbf{M}_R \odot \mathbf{M}_R$ is equal to 1 if and only if row i of $\mathbf{M}_R$ and column j of $\mathbf{M}_R$ have a 1 in the same relative position, say position k. This means that $m_{ik} = 1$ and $m_{kj} = 1$ for some k, $1 \leq k \leq n$. By definition of the matrix $\mathbf{M}_R$, the preceding conditions mean that $a_i \ R \ a_k$ and $a_k \ R \ a_j$. Thus $a_i \ R^2 \ a_j$, and so $n_{ij} = 1$. We have therefore shown that position i, j of $\mathbf{M}_R \odot \mathbf{M}_R$ is equal to 1 if and only if $n_{ij} = 1$. This means that $\mathbf{M} \odot \mathbf{M}_R = \mathbf{M}_{R^2}$. ▼

For brevity, we will usually denote $\mathbf{M}_R \odot \mathbf{M}_R$ simply as $(\mathbf{M}_R)^2_\odot$ (the symbol $\odot$ reminds us that this is not the usual matrix product).

EXAMPLE 6 Let A and R be as in Example 5. Then

$$\mathbf{M}_R = \begin{bmatrix} 1 & 1 & 0 & 0 & 0 \\ 0 & 0 & 1 & 0 & 0 \\ 0 & 0 & 0 & 1 & 1 \\ 0 & 0 & 0 & 0 & 1 \\ 0 & 0 & 0 & 0 & 0 \end{bmatrix}.$$

From the preceding discussion, we see that

$$\mathbf{M}_{R^2} = \mathbf{M}_R \odot \mathbf{M}_R = \begin{bmatrix} 1 & 1 & 0 & 0 & 0 \\ 0 & 0 & 1 & 0 & 0 \\ 0 & 0 & 0 & 1 & 1 \\ 0 & 0 & 0 & 0 & 1 \\ 0 & 0 & 0 & 0 & 0 \end{bmatrix} \odot \begin{bmatrix} 1 & 1 & 0 & 0 & 0 \\ 0 & 0 & 1 & 0 & 0 \\ 0 & 0 & 0 & 1 & 1 \\ 0 & 0 & 0 & 0 & 1 \\ 0 & 0 & 0 & 0 & 0 \end{bmatrix}$$

$$= \begin{bmatrix} 1 & 1 & 1 & 0 & 0 \\ 0 & 0 & 0 & 1 & 1 \\ 0 & 0 & 0 & 0 & 1 \\ 0 & 0 & 0 & 0 & 0 \\ 0 & 0 & 0 & 0 & 0 \end{bmatrix}.$$

Computing $\mathbf{M}_{R^2}$ directly from R^2, we obtain the same result. ∎

We can see from Examples 5 and 6 that it is often easier to compute R^2 by computing $\mathbf{M}_R \odot \mathbf{M}_R$ instead of searching the digraph of R for all vertices that can be joined by a path of length 2. Similarly, we can show that $\mathbf{M}_{R^3} = \mathbf{M}_R \odot (\mathbf{M}_R \odot \mathbf{M}_R) = (\mathbf{M}_R)^3_\odot$. In fact, we now show by induction that these two results can be generalized.

Theorem 2 For $n \geq 2$ and R a relation on a finite set A, we have

$$\mathbf{M}_{R^n} = \mathbf{M}_R \odot \mathbf{M}_R \odot \cdots \odot \mathbf{M}_R \quad (n \text{ factors}).$$ ●

Proof Let P(n) be the assertion that the statement holds for an integer $n \geq 2$.

Basis Step: P(2) is true by Theorem 1.

Induction Step: We use P(k) to show P($k + 1$). Consider the matrix $\mathbf{M}_{R^{k+1}}$. Let $\mathbf{M}_{R^{k+1}} = [\, x_{ij} \,]$, $\mathbf{M}_{R^k} = [\, y_{ij} \,]$, and $\mathbf{M}_R = [\, m_{ij} \,]$. If $x_{ij} = 1$, we must have a path of length $k + 1$ from a_i to a_j. If we let a_s be the vertex that this path reaches just before the last vertex a_j, then there is a path of length k from a_i to a_s and a path of length 1 from a_s to a_j. Thus $y_{is} = 1$ and $m_{sj} = 1$, so $\mathbf{M}_{R^k} \odot \mathbf{M}_R$ has a 1 in position i, j. We can see, similarly, that if $\mathbf{M}_{R^k} \odot \mathbf{M}_R$ has a 1 in position i, j, then $x_{ij} = 1$. This means that $\mathbf{M}_{R^{k+1}} = \mathbf{M}_{R^k} \odot \mathbf{M}_R$.
Using

$$P(k): \mathbf{M}_{R^k} = \mathbf{M}_R \odot \cdots \odot \mathbf{M}_R \quad (k \text{ factors}),$$

we have

$$\mathbf{M}_{R^{k+1}} = \mathbf{M}_{R^k} \odot \mathbf{M}_R = (\mathbf{M}_R \odot \mathbf{M}_R \odot \cdots \odot \mathbf{M}_R) \odot \mathbf{M}_R$$

and hence

$$P(k + 1): \mathbf{M}_{R^{k+1}} = \mathbf{M}_R \odot \cdots \odot \mathbf{M}_R \odot \mathbf{M}_R \quad (k + 1 \text{ factors})$$

is true. Thus, by the principle of mathematical induction, P(n) is true for all $n \geq 2$. This proves the theorem. As before, we write $\mathbf{M}_R \odot \cdots \odot \mathbf{M}_R$ (n factors) as $(\mathbf{M}_R)_\odot^n$.

▼

Note that the key to an induction step is finding a useful connection between P(k) and P($k + 1$).

Now that we know how to compute the matrix of the relation R^n from the matrix of R, we would like to see how to compute the matrix of R^∞. We proceed as follows. Suppose that R is a relation on a finite set A, and $x \in A$, $y \in A$. We know that $x\ R^\infty\ y$ means that x and y are connected by a path in R of length n for some n. In general, n will depend on x and y, but, clearly, $x\ R^\infty\ y$ if and only if $x\ R\ y$ or $x\ R^2\ y$ or $x\ R^3\ y$ or $\ldots$. Thus the preceding statement tells us that $R^\infty = R \cup R^2 \cup R^3 \cup \cdots = \overset{\infty}{\underset{n=1}{\cup}} R^n$. If R and S are relations on A, the relation $R \cup S$ is defined by $x\ (R \cup S)\ y$ if and only if $x\ R\ y$ or $x\ S\ y$. (The relation $R \cup S$ will be discussed in more detail in Section 4.7.) The reader may verify that $\mathbf{M}_{R \cup S} = \mathbf{M}_R \vee \mathbf{M}_S$, and we will show this in Section 4.7. Thus

$$\mathbf{M}_{R^\infty} = \mathbf{M}_R \vee \mathbf{M}_{R^2} \vee \mathbf{M}_{R^3} \vee \cdots = \mathbf{M}_R \vee (\mathbf{M}_R)_\odot^2 \vee (\mathbf{M}_R)_\odot^3 \vee \cdots.$$

The **reachability** relation R^* of a relation R on a set A that has n elements is defined as follows: $x\ R^*\ y$ means that $x = y$ or $x\ R^\infty\ y$. The idea is that y is reachable from x if either y is x or there is some path from x to y. It is easily seen that $\mathbf{M}_{R^*} = \mathbf{M}_{R^\infty} \vee \mathbf{I}_n$, where $\mathbf{I}_n$ is the $n \times n$ identity matrix. Thus our discussion shows that

$$\mathbf{M}_{R^*} = \mathbf{I}_n \vee \mathbf{M}_R \vee (\mathbf{M}_R)_\odot^2 \vee (\mathbf{M}_R)_\odot^3 \vee \cdots.$$

Let $\pi_1\colon a, x_1, x_2, \ldots, x_{n-1}, b$ be a path in a relation R of length n from a to b, and let $\pi_2\colon b, y_1, y_2, \ldots, y_{m-1}, c$ be a path in R of length m from b to c. Then the **composition of π_1 and π_2** is the path $a, x_1, x_2, \ldots, b, y_1, y_2, \ldots, y_{m-1}, c$ of length $n + m$, which is denoted by $\pi_2 \circ \pi_1$. This is a path from a to c.

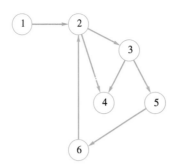

Figure 4.15

EXAMPLE 7   Consider the relation whose digraph is given in Figure 4.15 and the paths

$$\pi_1: 1, 2, 3 \quad \text{and} \quad \pi_2: 3, 5, 6, 2, 4.$$

Then the composition of π_1 and π_2 is the path $\pi_2 \circ \pi_1: 1, 2, 3, 5, 6, 2, 4$ from 1 to 4 of length 6.

4.3 Exercises

For Exercises 1 through 8, let R be the relation whose digraph is given in Figure 4.16.

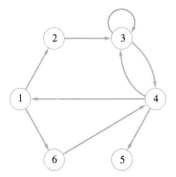

Figure 4.16

1. List all paths of length 1.

2. (a) List all paths of length 2 starting from vertex 2.
   (b) List all paths of length 2.

3. (a) List all paths of length 3 starting from vertex 3.
   (b) List all paths of length 3.

4. Find a cycle starting at vertex 2.

5. Find a cycle starting at vertex 6.

6. Draw the digraph of R^2.

7. Find $\mathbf{M}_{R^2}$.

8. (a) Find R^∞.
   (b) Find $\mathbf{M}_{R^\infty}$.

For Exercises 9 through 16, let R be the relation whose digraph is given in Figure 4.17.

9. List all paths of length 1.

10. (a) List all paths of length 2 starting from vertex c.
   (b) Find all paths of length 2.

11. (a) List all paths of length 3 starting from vertex a.
   (b) Find all paths of length 3.

12. Find a cycle starting at vertex c.

13. Find a cycle starting at vertex d.

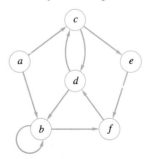

Figure 4.17

14. Find a cycle starting at vertex a.

15. Draw the digraph of R^2.

16. Find $\mathbf{M}_{R^2}$. Is this result consistent with the result of Exercise 15?

17. (a) Find $\mathbf{M}_{R^\infty}$.

   (b) Find R^∞.

18. Let R and S be relations on a set A. Show that

$$\mathbf{M}_{R \cup S} = \mathbf{M}_R \vee \mathbf{M}_S.$$

19. Let R be a relation on a set A that has n elements. Show that $\mathbf{M}_{R^*} = \mathbf{M}_{R^\infty} \vee \mathbf{I}_n$, where $\mathbf{I}_n$ is the $n \times n$ identity matrix.

In Exercises 20 through 23, let R be the relation whose digraph is given in Figure 4.18.

20. If π_1: 1, 2, 4, 3 and π_2: 3, 5, 6, 4, find the composition $\pi_2 \circ \pi_1$.

21. If π_1: 1, 7, 5 and π_2: 5, 6, 7, 4, 3, find the composition $\pi_2 \circ \pi_1$.

22. If π_1: 3, 4, 5, 6, and π_2: 6, 7, 4, 3, 5, find the composition $\pi_2 \circ \pi_1$.

23. If π_1: 2, 3, 5, 6, 7, and π_2: 7, 5, 6, 4, find the composition $\pi_2 \circ \pi_1$.

24. Let $A = \{1, 2, 3, 4, 5\}$ and R be the relation defined by $a \, R \, b$ if and only if $a < b$.

   (a) Compute R^2 and R^3.

   (b) Complete the following statement: $a \, R^2 \, b$ if and only if ____.

   (c) Complete the following statement: $a \, R^3 \, b$ if and only if ____.

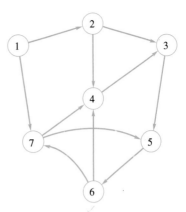

Figure 4.18

25. By Theorem 1, $\mathbf{M}_R \odot \mathbf{M}_R = \mathbf{M}_{R^2}$ so that $\mathbf{M}_R \odot \mathbf{M}_R$ shows where there are paths of length 2 in the digraph of R. Let

$$\mathbf{M}_R = \begin{bmatrix} 1 & 1 & 1 & 1 \\ 0 & 0 & 1 & 0 \\ 0 & 1 & 0 & 0 \\ 0 & 1 & 1 & 1 \end{bmatrix}.$$

What does $\mathbf{M}_R \cdot \mathbf{M}_R$ show? Justify your conclusion.

26. Is it possible to generalize the results of Exercise 25? For example, does $(\mathbf{M}_R)^3$ tell us anything useful about R?

27. Complete the following. The proof of Theorem 1 is ____ proof based on ____ of matrices.

28. (a) What about the statement of Theorem 2 indicates that an induction proof is appropriate?

   (b) What is the central idea of the induction step in the proof of Theorem 2?

4.4 PROPERTIES OF RELATIONS

In many applications to computer science and applied mathematics, we deal with relations on a set A rather than relations from A to B. Moreover, these relations often satisfy certain properties that will be discussed in this section.

Reflexive and Irreflexive Relations

A relation R on a set A is **reflexive** if $(a, a) \in R$ for all $a \in A$, that is, if $a \, R \, a$ for all $a \in A$. A relation R on a set A is **irreflexive** if $a \, \cancel{R} \, a$ for every $a \in A$.

Thus R is reflexive if every element $a \in A$ is related to itself and it is irreflexive if no element is related to itself.

EXAMPLE 1

(a) Let $\Delta = \{(a, a) \mid a \in A\}$, so that Δ is the relation of **equality** on the set A. Then Δ is reflexive, since $(a, a) \in \Delta$ for all $a \in A$.

(b) Let $R = \{(a, b) \in A \times A \mid a \neq b\}$, so that R is the relation of **inequality** on the set A. Then R is irreflexive, since $(a, a) \notin R$ for all $a \in A$.

(c) Let $A = \{1, 2, 3\}$, and let $R = \{(1, 1), (1, 2)\}$. Then R is not reflexive since $(2, 2) \notin R$ and $(3, 3) \notin R$. Also, R is not irreflexive, since $(1, 1) \in R$.

(d) Let A be a nonempty set. Let $R = \varnothing \subseteq A \times A$, the **empty relation**. Then R is not reflexive, since $(a, a) \notin R$ for all $a \in A$ (the empty set has no elements). However, R is irreflexive. ∎

We can identify a reflexive or irreflexive relation by its matrix as follows. The matrix of a reflexive relation must have all 1's on its main diagonal, while the matrix of an irreflexive relation must have all 0's on its main diagonal.

Similarly, we can characterize the digraph of a reflexive or irreflexive relation as follows. A reflexive relation has a cycle of length 1 at every vertex, while an irreflexive relation has no cycles of length 1. Another useful way of saying the same thing uses the equality relation Δ on a set A: R is reflexive if and only if $\Delta \subseteq R$, and R is irreflexive if and only if $\Delta \cap R = \varnothing$.

Finally, we may note that if R is reflexive on a set A, then $\text{Dom}(R) = \text{Ran}(R) = A$.

Symmetric, Asymmetric, and Antisymmetric Relations

A relation R on a set A is **symmetric** if whenever $a\,R\,b$, then $b\,R\,a$. It then follows that R is not symmetric if we have some a and $b \in A$ with $a\,R\,b$, but $b\,\not\!R\,a$. A relation R on a set A is **asymmetric** if whenever $a\,R\,b$, then $b\,\not\!R\,a$. It then follows that R is not asymmetric if we have some a and $b \in A$ with both $a\,R\,b$ and $b\,R\,a$.

A relation R on a set A is **antisymmetric** if whenever $a\,R\,b$ and $b\,R\,a$, then $a = b$. The contrapositive of this definition is that R is antisymmetric if whenever $a \neq b$, then $a\,\not\!R\,b$ or $b\,\not\!R\,a$. It follows that R is not antisymmetric if we have a and b in A, $a \neq b$, and both $a\,R\,b$ and $b\,R\,a$.

Given a relation R, we shall want to determine which properties hold for R. Keep in mind the following remark: A property fails to hold in general if we can find one situation where the property does not hold.

EXAMPLE 2   Let $A = Z$, the set of integers, and let

$$R = \{(a, b) \in A \times A \mid a < b\}$$

so that R is the relation **less than**. Is R symmetric, asymmetric, or antisymmetric?

Solution

Symmetry: If $a < b$, then it is not true that $b < a$, so R is not symmetric.

Asymmetry: If $a < b$, then $b \not< a$ (b is not less than a), so R is asymmetric.

Antisymmetry: If $a \neq b$, then either $a \not< b$ or $b \not< a$, so that R is antisymmetric. ∎

EXAMPLE 3   Let A be a set of people and let $R = \{(x, y) \in A \times A \mid x$ is a cousin of $y\}$. Then R is a symmetric relation (verify). ∎

EXAMPLE 4 Let $A = \{1, 2, 3, 4\}$ and let

$$R = \{(1, 2), (2, 2), (3, 4), (4, 1)\}.$$

Then R is not symmetric, since $(1, 2) \in R$, but $(2, 1) \notin R$. Also, R is not asymmetric, since $(2, 2) \in R$. Finally, R is antisymmetric, since if $a \neq b$, either $(a, b) \notin R$ or $(b, a) \notin R$. ∎

EXAMPLE 5 Let $A = Z^+$, the set of positive integers, and let

$$R = \{(a, b) \in A \times A \mid a \text{ divides } b\}.$$

Is R symmetric, asymmetric, or antisymmetric?

Solution If $a \mid b$, it does not follow that $b \mid a$, so R is not symmetric. For example, $2 \mid 4$, but $4 \nmid 2$.

If $a = b = 3$, say, then $a \, R \, b$ and $b \, R \, a$, so R is not asymmetric.

If $a \mid b$ and $b \mid a$, then $a = b$, so R is antisymmetric. (See Exercise 26 in Section 1.4.) ∎

We now relate symmetric, asymmetric, and antisymmetric properties of a relation to properties of its matrix. The matrix $\mathbf{M}_R = \left[\, m_{ij} \, \right]$ of a symmetric relation satisfies the property that

$$\text{if} \quad m_{ij} = 1, \quad \text{then} \quad m_{ji} = 1.$$

Moreover, if $m_{ji} = 0$, then $m_{ij} = 0$. Thus $\mathbf{M}_R$ is a matrix such that each pair of entries, symmetrically placed about the main diagonal, are either both 0 or both 1. It follows that $\mathbf{M}_R = \mathbf{M}_R^T$, so that $\mathbf{M}_R$ is a symmetric matrix (see Section 1.5).

The matrix $\mathbf{M}_R = \left[\, m_{ij} \, \right]$ of an asymmetric relation R satisfies the property that

$$\text{if} \quad m_{ij} = 1, \quad \text{then} \quad m_{ji} = 0.$$

If R is asymmetric, it follows that $m_{ii} = 0$ for all i; that is, the main diagonal of the matrix $\mathbf{M}_R$ consists entirely of 0's. This must be true since the asymmetric property implies that if $m_{ii} = 1$, then $m_{ii} = 0$, which is a contradiction.

Finally, the matrix $\mathbf{M}_R = \left[\, m_{ij} \, \right]$ of an antisymmetric relation R satisfies the property that if $i \neq j$, then $m_{ij} = 0$ or $m_{ji} = 0$.

EXAMPLE 6 Consider the matrices in Figure 4.19, each of which is the matrix of a relation, as indicated.

Relations R_1 and R_2 are symmetric since the matrices $\mathbf{M}_{R_1}$ and $\mathbf{M}_{R_2}$ are symmetric matrices. Relation R_3 is antisymmetric, since no symmetrically situated, off-diagonal positions of $\mathbf{M}_{R_3}$ both contain 1's. Such positions may both have 0's, however, and the diagonal elements are unrestricted. The relation R_3 is not asymmetric because $\mathbf{M}_{R_3}$ has 1's on the main diagonal.

Relation R_4 has none of the three properties: $\mathbf{M}_{R_4}$ is not symmetric. The presence of the 1's in positions 4, 1 and 1, 4 of $\mathbf{M}_{R_4}$ violates both asymmetry and antisymmetry.

Finally, R_5 is antisymmetric but not asymmetric, and R_6 is both asymmetric and antisymmetric. ∎

$$\begin{bmatrix} 1 & 0 & 1 \\ 0 & 0 & 1 \\ 1 & 1 & 1 \end{bmatrix} = \mathbf{M}_{R_1}$$

(a)

$$\begin{bmatrix} 0 & 1 & 1 & 0 \\ 1 & 1 & 0 & 0 \\ 1 & 0 & 1 & 1 \\ 0 & 0 & 1 & 1 \end{bmatrix} = \mathbf{M}_{R_2}$$

(b)

$$\begin{bmatrix} 1 & 1 & 1 \\ 0 & 1 & 0 \\ 0 & 0 & 0 \end{bmatrix} = \mathbf{M}_{R_3}$$

(c)

$$\begin{bmatrix} 0 & 0 & 1 & 1 \\ 0 & 0 & 1 & 0 \\ 0 & 0 & 0 & 1 \\ 1 & 0 & 0 & 0 \end{bmatrix} = \mathbf{M}_{R_4}$$

(d)

$$\begin{bmatrix} 1 & 0 & 0 & 1 \\ 0 & 1 & 1 & 1 \\ 0 & 0 & 1 & 0 \\ 0 & 0 & 0 & 1 \end{bmatrix} = \mathbf{M}_{R_5}$$

(e)

$$\begin{bmatrix} 0 & 1 & 1 & 1 \\ 0 & 0 & 1 & 0 \\ 0 & 0 & 0 & 1 \\ 0 & 0 & 0 & 0 \end{bmatrix} = \mathbf{M}_{R_6}$$

(f)

Figure 4.19

We now consider the digraphs of these three types of relations. If R is an asymmetric relation, then the digraph of R cannot simultaneously have an edge from vertex i to vertex j and an edge from vertex j to vertex i. This is true for any i and j, and in particular if i equals j. Thus there can be no cycles of length 1, and all edges are "one-way streets."

If R is an antisymmetric relation, then for different vertices i and j there cannot be an edge from vertex i to vertex j and an edge from vertex j to vertex i. When $i = j$, no condition is imposed. Thus there may be cycles of length 1, but again all edges are "one way."

We consider the digraphs of symmetric relations in more detail.

The digraph of a symmetric relation R has the property that if there is an edge from vertex i to vertex j, then there is an edge from vertex j to vertex i. Thus, if two vertices are connected by an edge, they must always be connected in both directions. Because of this, it is possible and quite useful to give a different representation of a symmetric relation. We keep the vertices as they appear in the digraph, but if two vertices a and b are connected by edges in each direction, we replace these two edges with one undirected edge, or a "two-way street." This undirected edge is just a single line without arrows and connects a and b. The resulting diagram will be called the **graph** of the symmetric relation. (Graph will be given a more general meaning in Chapter 8.)

EXAMPLE 7   Let $A = \{a, b, c, d, e\}$ and let R be the symmetric relation given by

$$R = \{(a, b), (b, a), (a, c), (c, a), (b, c), (c, b),$$
$$(b, e), (e, b), (e, d), (d, e), (c, d), (d, c)\}.$$

The usual digraph of R is shown in Figure 4.20(a), while Figure 4.20(b) shows the graph of R. Note that each undirected edge corresponds to two ordered pairs in the relation R. ■

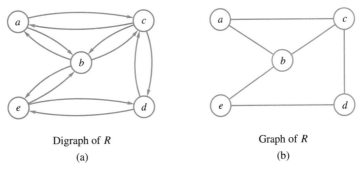

Digraph of R
(a)

Graph of R
(b)

Figure 4.20

An undirected edge between a and b, in the graph of a symmetric relation R, corresponds to a set $\{a, b\}$ such that $(a, b) \in R$ and $(b, a) \in R$. Sometimes we will also refer to such a set $\{a, b\}$ as an **undirected edge** of the relation R and call a and b **adjacent vertices**.

A symmetric relation R on a set A is called **connected** if there is a path from any element of A to any other element of A. This simply means that the graph of R is all in one piece. In Figure 4.21 we show the graphs of two symmetric relations. The graph in Figure 4.21(a) is connected, whereas that in Figure 4.21(b) is not connected.

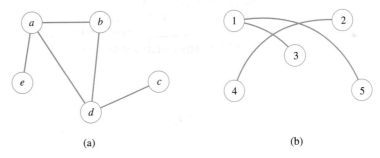

(a)

(b)

Figure 4.21

Transitive Relations

We say that a relation R on a set A is **transitive** if whenever $a\ R\ b$ and $b\ R\ c$, then $a\ R\ c$. It is often convenient to say what it means for a relation to be not transitive. A relation R on A is not transitive if there exist a, b, and c in A so that $a\ R\ b$ and $b\ R\ c$, but $a\ \not{R}\ c$. If such a, b, and c do not exist, then R is transitive.

EXAMPLE 8   Let $A = Z$, the set of integers, and let R be the relation less than. To see whether R is transitive, we assume that $a\ R\ b$ and $b\ R\ c$. Thus $a < b$ and $b < c$. It then follows that $a < c$, so $a\ R\ c$. Hence R is transitive. ■

EXAMPLE 9   Let $A = Z^+$ and let R be the relation considered in Example 5. Is R transitive?

> *Solution*   Suppose that $a \, R \, b$ and $b \, R \, c$, so that $a \mid b$ and $b \mid c$. It then does follow that $a \mid c$. [See Theorem 2(d) of Section 1.4.] Thus R is transitive.   ∎

EXAMPLE 10   Let $A = \{1, 2, 3, 4\}$ and let

$$R = \{(1, 2), (1, 3), (4, 2)\}.$$

Is R transitive?

> *Solution*   Since there are no elements a, b, and c in A such that $a \, R \, b$ and $b \, R \, c$, but $a \, \not{R} \, c$, we conclude that R is transitive.   ∎

A relation R is transitive if and only if its matrix $\mathbf{M}_R = \begin{bmatrix} m_{ij} \end{bmatrix}$ has the property

$$\text{if} \quad m_{ij} = 1 \quad \text{and} \; m_{jk} = 1, \quad \text{then} \quad m_{ik} = 1.$$

The left-hand side of this statement simply means that $(\mathbf{M}_R)^2_\odot$ has a 1 in position i, k. Thus the transitivity of R means that if $(\mathbf{M}_R)^2_\odot$ has a 1 in any position, then $\mathbf{M}_R$ must have a 1 in the same position. Thus, in particular, if $(\mathbf{M}_R)^2_\odot = \mathbf{M}_R$, then R is transitive. The converse is not true.

EXAMPLE 11   Let $A = \{1, 2, 3\}$ and let R be the relation on A whose matrix is

$$\mathbf{M}_R = \begin{bmatrix} 1 & 1 & 1 \\ 0 & 0 & 1 \\ 0 & 0 & 1 \end{bmatrix}.$$

Show that R is transitive.

> *Solution*   By direct computation, $(\mathbf{M}_R)^2_\odot = \mathbf{M}_R$; therefore, R is transitive.   ∎

To see what transitivity means for the digraph of a relation, we translate the definition of transitivity into geometric terms.

If we consider particular vertices a and c, the conditions $a \, R \, b$ and $b \, R \, c$ mean that there is a path of length 2 in R from a to c. In other words, $a \, R^2 \, c$. Therefore, we may rephrase the definition of transitivity as follows: If $a \, R^2 \, c$, then $a \, R \, c$; that is, $R^2 \subseteq R$ (as subsets of $A \times A$). In other words, if a and c are connected by a path of length 2 in R, then they must be connected by a path of length 1.

We can slightly generalize the foregoing geometric characterization of transitivity as follows.

Theorem 1   A relation R is transitive if and only if it satisfies the following property: If there is a path of length greater than 1 from vertex a to vertex b, there is a path of length 1 from a to b (that is, a is related to b). Algebraically stated, R is transitive if and only if $R^n \subseteq R$ for all $n \geq 1$.   ●

> **Proof**   The proof is left to the reader.   ▼

It will be convenient to have a restatement of some of these relational properties in terms of R-relative sets. We list these statements without proof.

Theorem 2   Let R be a relation on a set A. Then
(a) Reflexivity of R means that $a \in R(a)$ for all a in A.
(b) Symmetry of R means that $a \in R(b)$ if and only if $b \in R(a)$.
(c) Transitivity of R means that if $b \in R(a)$ and $c \in R(b)$, then $c \in R(a)$. ●

4.4 Exercises

In Exercises 1 through 8, let $A = \{1, 2, 3, 4\}$. *Determine whether the relation is reflexive, irreflexive, symmetric, asymmetric, antisymmetric, or transitive.*

1. $R = \{(1, 1), (1, 2), (2, 1), (2, 2), (3, 3), (3, 4),$
   $(4, 3), (4, 4)\}$

2. $R = \{(1, 2), (1, 3), (1, 4), (2, 3), (2, 4), (3, 4)\}$

3. $R = \{(1, 3), (1, 1), (3, 1), (1, 2), (3, 3), (4, 4)\}$

4. $R = \{(1, 1), (2, 2), (3, 3)\}$

5. $R = \varnothing$

6. $R = A \times A$

7. $R = \{(1, 2), (1, 3), (3, 1), (1, 1), (3, 3), (3, 2),$
   $(1, 4), (4, 2), (3, 4)\}$

8. $R = \{(1, 3), (4, 2), (2, 4), (3, 1), (2, 2)\}$

In Exercises 9 and 10 (Figures 4.22 and 4.23), let $A = \{1, 2, 3, 4, 5\}$. *Determine whether the relation R whose digraph is given is reflexive, irreflexive, symmetric, asymmetric, antisymmetric, or transitive.*

9.

10.

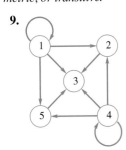

Figure 4.22

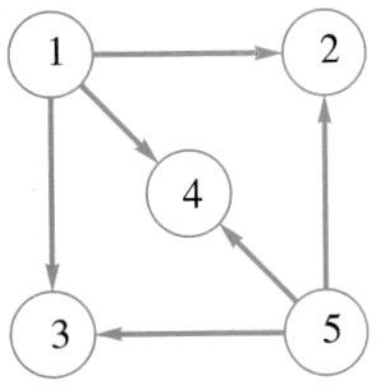

Figure 4.23

In Exercises 11 and 12, let $A = \{1, 2, 3, 4\}$. *Determine whether the relation R whose matrix M_R is given is reflexive, irreflexive, symmetric, asymmetric, antisymmetric, or transitive.*

11. $\begin{bmatrix} 0 & 1 & 0 & 1 \\ 1 & 0 & 1 & 1 \\ 0 & 1 & 0 & 0 \\ 1 & 1 & 0 & 0 \end{bmatrix}$ 12. $\begin{bmatrix} 1 & 1 & 0 & 0 \\ 1 & 1 & 0 & 0 \\ 0 & 0 & 1 & 0 \\ 0 & 0 & 0 & 1 \end{bmatrix}$

In Exercises 13 through 22 , determine whether the relation R on the set A is reflexive, irreflexive, symmetric, asymmetric, antisymmetric, or transitive.

13. $A = Z$; a R b if and only if $a \leq b + 1$.

14. $A = Z^+$; a R b if and only if $|a - b| \leq 2$.

15. $A = Z^+$; a R b if and only if $a = b^k$ for some $k \in Z^+$.

16. $A = Z$; a R b if and only if $a + b$ is even.

17. $A = Z$; a R b if and only if $|a - b| = 2$.

18. $A = $ the set of real numbers; a R b if and only if $a^2 + b^2 = 4$.

19. $A = Z^+$; a R b if and only if $GCD(a, b) = 1$. In this case, we say that a and b are **relatively prime**. (See Section 1.4 for GCD.)

20. $A = $ the set of all ordered pairs of real numbers; (a, b) R (c, d) if and only if $a = c$.

21. $S = \{1, 2, 3, 4\}$, $A = S \times S$; (a, b) R (c, d) if and only if $ad = bc$.

22. A is the set of all lines in the plane. l_1 R l_2 if and only if l_1 is parallel to l_2.

23. Let R be the following symmetric relation on the set $A = \{1, 2, 3, 4, 5\}$:

$$R = \{(1, 2), (2, 1), (3, 4), (4, 3), (3, 5), (5, 3),$$
$$(4, 5), (5, 4), (5, 5)\}.$$

Draw the graph of R.

24. Let $A = \{a, b, c, d\}$ and let R be the symmetric relation

$$R = \{(a, b), (b, a), (a, c), (c, a), (a, d), (d, a)\}.$$

Draw the graph of R.

25. Consider the graph of a symmetric relation R on $A = \{1, 2, 3, 4, 5, 6, 7\}$ shown in Figure 4.24. Determine R (list all pairs).

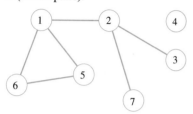

Figure 4.24

26. Consider the graph of a symmetric relation R on $A = \{a, b, c, d, e\}$ shown in Figure 4.25. Determine R (list all pairs).

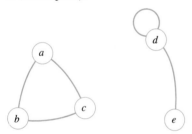

Figure 4.25

27. Let R be a symmetric relation given by its matrix $\mathbf{M}_R$. Describe a procedure for using $\mathbf{M}_R$ to determine if the graph of R is connected.

28. Let R be a relation on A and $B \subseteq A$. Which relational properties of R would be inherited by the restriction of R to B?

29. Prove or disprove that if a relation on a set A is transitive and irreflexive, then it is asymmetric.

30. Prove or disprove that if a relation R on A is transitive, then R^2 is also transitive.

31. Let R be a nonempty relation on a set A. Suppose that R is symmetric and transitive. Show that R is not irreflexive.

32. Prove that if a relation R on a set A is symmetric, then the relation R^2 is also symmetric.

33. Prove by induction that if a relation R on a set A is symmetric, then R^n is symmetric for $n \geq 1$.

34. Define a relation on Z^+ that is reflexive, symmetric, and transitive and has not been defined previously.

35. Give a direct proof of Theorem 1 of this section.

4.5 EQUIVALENCE RELATIONS

A relation R on a set A is called an **equivalence relation** if it is reflexive, symmetric, and transitive.

EXAMPLE 1   Let A be the set of all triangles in the plane and let R be the relation on A defined as follows:

$$R = \{(a, b) \in A \times A \mid a \text{ is congruent to } b\}.$$

It is easy to see that R is an equivalence relation.   ∎

EXAMPLE 2   Let $A = \{1, 2, 3, 4\}$ and let

$$R = \{(1, 1), (1, 2), (2, 1), (2, 2), (3, 4), (4, 3), (3, 3), (4, 4)\}.$$

It is easy to verify that R is an equivalence relation.   ∎

EXAMPLE 3   Let $A = Z$, the set of integers, and let R be defined by $a\,R\,b$ if and only if $a \leq b$. Is R an equivalence relation?

Solution   Since $a \leq a$, R is reflexive. If $a \leq b$, it need not follow that $b \leq a$, so R is not symmetric. Incidentally, R is transitive, since $a \leq b$ and $b \leq c$ imply that $a \leq c$. We see that R is not an equivalence relation.   ∎

EXAMPLE 4   Let $A = Z$ and let

$R = \{(a, b) \in A \times A \mid a \text{ and } b \text{ yield the same remainder when divided by 2}\}$.

In this case, we call 2 the **modulus** and write $a \equiv b \pmod 2$, read "a is **congruent to** b mod 2."

Show that congruence mod 2 is an equivalence relation.

Solution   First, clearly $a \equiv a \pmod 2$. Thus R is reflexive.

Second, if $a \equiv b \pmod 2$, then a and b yield the same remainder when divided by 2, so $b \equiv a \pmod 2$. R is symmetric.

Finally, suppose that $a \equiv b \pmod 2$ and $b \equiv c \pmod 2$. Then a, b, and c yield the same remainder when divided by 2. Thus, $a \equiv c \pmod 2$. Hence congruence mod 2 is an equivalence relation.   ∎

EXAMPLE 5   Let $A = Z$ and let $n \in Z^{+}$. We generalize the relation defined in Example 4 as follows. Let

$$R = \{(a, b) \in A \times A \mid a \equiv b \pmod n\}.$$

That is, $a \equiv b \pmod n$ if and only if a and b yield the same remainder when divided by n. Proceeding exactly as in Example 4, we can show that congruence mod n is an equivalence relation.   ∎

We note that if $a \equiv b \pmod n$, then $a = qn + r$ and $b = tn + r$ and $a - b$ is a multiple of n. Thus, $a \equiv b \pmod n$ if and only if $n \mid (a - b)$.

Equivalence Relations and Partitions

The following result shows that if $\mathcal{P}$ is a partition of a set A (see Section 4.1), then $\mathcal{P}$ can be used to construct an equivalence relation on A.

Theorem 1   Let $\mathcal{P}$ be a partition of a set A. Recall that the sets in $\mathcal{P}$ are called the blocks of $\mathcal{P}$. Define the relation R on A as follows:

$a R b$ if and only if a and b are members of the same block.

Then R is an equivalence relation on A.   ●

Proof
(1) If $a \in A$, then clearly a is in the same block as itself; so $a R a$.
(2) If $a R b$, then a and b are in the same block; so $b R a$.
(3) If $a R b$ and $b R c$, then a, b, and c must all lie in the same block of $\mathcal{P}$. Thus $a R c$.

Since R is reflexive, symmetric, and transitive, R is an equivalence relation. R will be called the **equivalence relation determined by** $\mathcal{P}$.   ▼

EXAMPLE 6   Let $A = \{1, 2, 3, 4\}$ and consider the partition $\mathscr{P} = \{\{1, 2, 3\}, \{4\}\}$ of A. Find the equivalence relation R on A determined by $\mathscr{P}$.

Solution   The blocks of $\mathscr{P}$ are $\{1, 2, 3\}$ and $\{4\}$. Each element in a block is related to every other element in the same block and only to those elements. Thus, in this case,

$$R = \{(1, 1), (1, 2), (1, 3), (2, 1), (2, 2), (2, 3), (3, 1), (3, 2), (3, 3), (4, 4)\}.$$

■

If $\mathscr{P}$ is a partition of A and R is the equivalence relation determined by $\mathscr{P}$, then the blocks of $\mathscr{P}$ can easily be described in terms of R. If A_1 is a block of $\mathscr{P}$ and $a \in A_1$, we see by definition that A_1 consists of all elements x of A with $a \, R \, x$. That is, $A_1 = R(a)$. Thus the partition $\mathscr{P}$ is $\{R(a) \mid a \in A\}$. In words, $\mathscr{P}$ consists of all distinct R-relative sets that arise from elements of A. For instance, in Example 6 the blocks $\{1, 2, 3\}$ and $\{4\}$ can be described, respectively, as $R(1)$ and $R(4)$. Of course, $\{1, 2, 3\}$ could also be described as $R(2)$ or $R(3)$, so this way of representing the blocks is not unique.

The foregoing construction of equivalence relations from partitions is very simple. We might be tempted to believe that few equivalence relations could be produced in this way. The fact is, as we will now show, that all equivalence relations on A can be produced from partitions.

We begin with the following result. Since its proof uses Theorem 2 of Section 4.4, the reader might first want to review that theorem.

___

Lemma 1[†]   Let R be an equivalence relation on a set A, and let $a \in A$ and $b \in A$. Then

$$a \, R \, b \quad \text{if and only if} \quad R(a) = R(b).$$

●

Proof   First suppose that $R(a) = R(b)$. Since R is reflexive, $b \in R(b)$; therefore, $b \in R(a)$, so $a \, R \, b$.

Conversely, suppose that $a \, R \, b$. Then note that

1. $b \in R(a)$ by definition. Therefore, since R is symmetric,
2. $a \in R(b)$, by Theorem 2(b) of Section 4.4.

We must show that $R(a) = R(b)$. First, choose an element $x \in R(b)$. Since R is transitive, the fact that $x \in R(b)$, together with (1), implies by Theorem 2(c) of Section 4.4 that $x \in R(a)$. Thus $R(b) \subseteq R(a)$. Now choose $y \in R(a)$. This fact and (2) imply, as before, that $y \in R(b)$. Thus $R(a) \subseteq R(b)$, so we must have $R(a) = R(b)$.

▼

Note the two-part structure of the lemma's proof. Because we want to prove a biconditional, $p \Leftrightarrow q$, we must show $q \Rightarrow p$ as well as $p \Rightarrow q$.

We now prove our main result.

___

Theorem 2   Let R be an equivalence relation on A, and let $\mathscr{P}$ be the collection of all distinct relative sets $R(a)$ for a in A. Then $\mathscr{P}$ is a partition of A, and R is the equivalence relation determined by $\mathscr{P}$.

●

___

[†]A lemma is a theorem whose main purpose is to aid in proving some other theorem.

Proof   According to the definition of a partition, we must show the following two properties:

> (a) Every element of A belongs to some relative set.
>
> (b) If $R(a)$ and $R(b)$ are not identical, then $R(a) \cap R(b) = \varnothing$.

Now property (a) is true, since $a \in R(a)$ by reflexivity of R. To show property (b) we prove the following equivalent statement:

$$\text{If} \quad R(a) \cap R(b) \neq \varnothing, \quad \text{then} \quad R(a) = R(b).$$

To prove this, we assume that $c \in R(a) \cap R(b)$. Then $a \; R \; c$ and $b \; R \; c$.

Since R is symmetric, we have $c \; R \; b$. Then $a \; R \; c$ and $c \; R \; b$, so, by transitivity of R, $a \; R \; b$. Lemma 1 then tells us that $R(a) = R(b)$. We have now proved that $\mathcal{P}$ is a partition. By Lemma 1 we see that $a \; R \; b$ if and only if a and b belong to the same block of $\mathcal{P}$. Thus $\mathcal{P}$ determines R, and the theorem is proved.   ▼

Note the use of the contrapositive in this proof.

If R is an equivalence relation on A, then the sets $R(a)$ are traditionally called **equivalence classes** of R. Some authors denote the class $R(a)$ by $[a]$ (see Section 9.3). The partition $\mathcal{P}$ constructed in Theorem 2 therefore consists of all equivalence classes of R, and this partition will be denoted by A/R. Recall that partitions of A are also called quotient sets of A, and the notation A/R reminds us that $\mathcal{P}$ is the quotient set of A that is constructed from and determines R.

EXAMPLE 7   Let R be the relation defined in Example 2. Determine A/R.

Solution   From Example 2 we have $R(1) = \{1, 2\} = R(2)$. Also, $R(3) = \{3, 4\} = R(4)$. Hence $A/R = \{\{1, 2\}, \{3, 4\}\}$.   ∎

EXAMPLE 8   Let R be the equivalence relation defined in Example 4. Determine A/R.

Solution   First, $R(0) = \{\dots, -6, -4, -2, 0, 2, 4, 6, 8, \dots\}$, the set of even integers, since the remainder is zero when each of these numbers is divided by 2. $R(1) = \{\dots, -5, -3, -1, 1, 3, 5, 7, \dots\}$, the set of odd integers, since each gives a remainder of 1 when divided by 2. Hence A/R consists of the set of even integers and the set of odd integers.   ∎

From Examples 7 and 8 we can extract a general procedure for determining partitions A/R for A finite or countable. The procedure is as follows:

***STEP 1*:**   Choose any element of A and compute the equivalence class $R(a)$.

***STEP 2*:**   If $R(a) \neq A$, choose an element b, not included in $R(a)$, and compute the equivalence class $R(b)$.

***STEP 3*:**   If A is not the union of previously computed equivalence classes, then choose an element x of A that is not in any of those equivalence classes and compute $R(x)$.

***STEP 4*:**   Repeat step 3 until all elements of A are included in the computed equivalence classes. If A is countable, this process could continue indefinitely. In that case, continue until a pattern emerges that allows you to describe or give a formula for all equivalence classes.

4.5 Exercises

In Exercises 1 and 2, let $A = \{a, b, c\}$. Determine whether the relation R whose matrix $\mathbf{M}_R$ is given is an equivalence relation.

1. $\mathbf{M}_R = \begin{bmatrix} 1 & 0 & 0 \\ 0 & 1 & 1 \\ 0 & 1 & 1 \end{bmatrix}$.   **2.** $\mathbf{M}_R = \begin{bmatrix} 1 & 0 & 1 \\ 0 & 1 & 0 \\ 0 & 0 & 1 \end{bmatrix}$.

In Exercises 3 and 4 (Figures 4.26 and 4.27), determine whether the relation R whose digraph is given is an equivalence relation.

3.

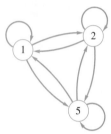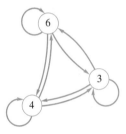

Figure 4.26

4.

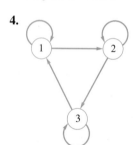

Figure 4.27

In Exercises 5 through 12, determine whether the relation R on the set A is an equivalence relation.

5. $A = \{a, b, c, d\}$,
$R = \{(a, a), (b, a), (b, b), (c, c), (d, d), (d, c)\}$

6. $A = \{1, 2, 3, 4, 5\}$, $R = \{(1, 1), (1, 2), (1, 3), (2, 1),$
$(2, 2), (3, 1), (2, 3), (3, 3), (4, 4), (3, 2), (5, 5)\}$

7. $A = \{1, 2, 3, 4\}$, $R = \{(1, 1), (1, 2), (2, 1), (2, 2), (3, 1),$
$(3, 3), (1, 3), (4, 1), (4, 4)\}$

8. A = the set of all members of the Software-of-the-Month Club; $a\ R\ b$ if and only if a and b buy the same number of programs.

9. A = the set of all members of the Software-of-the-Month Club; $a\ R\ b$ if and only if a and b buy the same programs.

10. A = the set of all people in the Social Security database; $a\ R\ b$ if and only if a and b have the same last name.

11. A = the set of all triangles in the plane; $a\ R\ b$ if and only if a is similar to b.

12. $A = Z^+ \times Z^+$; $(a, b)\ R\ (c, d)$ if and only if $b = d$.

13. If $\{\{a, c, e\}, \{b, d, f\}\}$ is a partition of the set $A = \{a, b, c, d, e, f\}$, determine the corresponding equivalence relation R.

14. If $\{\{1, 3, 5\}, \{2, 4\}\}$ is a partition of the set $A = \{1, 2, 3, 4, 5\}$, determine the corresponding equivalence relation R.

15. Let A and R be the set and relation defined in Example 5. Compute A/R.

16. Let $A = \{a, b, c, d, e\}$ and R be the relation on A defined by

$$\mathbf{M}_R = \begin{bmatrix} 1 & 1 & 1 & 0 & 1 \\ 1 & 1 & 1 & 0 & 1 \\ 1 & 1 & 1 & 0 & 1 \\ 0 & 0 & 0 & 1 & 0 \\ 1 & 1 & 1 & 0 & 1 \end{bmatrix}.$$

Compute A/R.

17. Let $S = \{1, 2, 3, 4, 5\}$ and $A = S \times S$. Define the following relation R on A: $(a, b)\ R\ (a', b')$ if and only if $ab' = a'b$.

(a) Show that R is an equivalence relation.

(b) Compute A/R.

18. Let $S = \{1, 2, 3, 4\}$ and $A = S \times S$. Define the following relation R on A: $(a, b)\ R\ (a', b')$ if and only if $a + b = a' + b'$.

(a) Show that R is an equivalence relation.

(b) Compute A/R.

19. A relation R on a set A is called **circular** if $a\ R\ b$ and $b\ R\ c$ imply $c\ R\ a$. Show that R is reflexive and circular if and only if it is an equivalence relation.

20. Show that if R_1 and R_2 are equivalence relations on A, then $R_1 \cap R_2$ is an equivalence relation on A.

21. Define an equivalence relation R on Z, the set of integers, different from that used in Examples 4 and 8 and whose corresponding partition contains exactly two infinite sets.

22. Define an equivalence relation R on Z, the set of integers, whose corresponding partition contains exactly three infinite sets.

In Exercises 23 and 24, *use the following definition. Given an equivalence relation R on a set where + is defined, the sum of R-relative sets, R(a) + R(b), is* $\{x \mid x = s + t, s \in R(a), t \in R(b)\}$.

23. Let R be the equivalence relation in Example 4. Show that $R(a) + R(b) = R(a + b)$ for all a, b.

24. Let R be the equivalence relation in Exercise 12. Show that $R(a) + R(b) = R(a + b)$ for all a, b.

4.6 COMPUTER REPRESENTATION OF RELATIONS AND DIGRAPHS

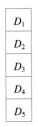

Figure 4.28

The most straightforward method of storing data items is to place them in a linear list or array. This generally corresponds to putting consecutive data items in consecutively numbered storage locations in a computer memory. Figure 4.28 illustrates this method for five data items $D_1, \ldots, D_5$. The method is an efficient use of space and provides, at least at the level of most programming languages, random access to the data. Thus the linear array might be A and the data would be in locations $A[1]$, $A[2]$, $A[3]$, $A[4]$, $A[5]$, and we would have access to any data item D_i by simply supplying its index i.

The main problem with this storage method is that we cannot insert new data between existing data without moving a possibly large number of items. Thus, to add another item E to the list in Figure 4.28 and place E between D_2 and D_3, we would have to move D_3 to $A[4]$, D_4 to $A[5]$, and D_5 to $A[6]$, if room exists, and then assign E to $A[3]$.

An alternative method of representing this sequence is by a **linked list**, shown in schematic fashion in Figure 4.29. The basic unit of information storage is the **storage cell**. We imagine such cells to have room for two information items. The first can be data (numbers or symbols), and the second item is a **pointer**, that is, a number that tells us (points to) the location of the next cell to be considered. Thus cells may be arranged sequentially, but the data items that they represent are not assumed to be in the same sequence. Instead, we discover the proper data sequence by following the pointers from each item to the next.

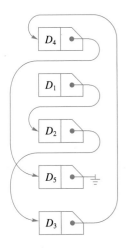

Figure 4.29

As shown in Figure 4.29, we represent the storage cell as a partitioned box $\boxed{\text{DATA} \mid \bullet}$, with a dot in the right-hand side representing a pointer. A line is drawn from each such dot to the cell that the corresponding pointer designates as next. The symbol $\bullet\!\!-\!\!\equiv$ means that data have ended and that no further pointers need be followed.

In practice, the concept of a linked list may be implemented using two linear arrays, a data array A and a pointer array P, as shown in Figure 4.30. Note that once we have accessed the data in location $A[i]$, then the number in location $P[i]$ gives, or points to, the index of A containing the next data item.

Thus, if we were at location $A[3]$, accessing data item D_2, then location $P[3]$ would contain 5, since the next data item, D_3, is located in $A[5]$. A zero in some location of P signifies that no more data items exist. In Figure 4.30, $P[4]$ is zero because $A[4]$ contains D_5, the last data item. In this scheme, we need two arrays for the data that we previously represented in a single array, and we have only sequential access. Thus we cannot locate D_2 directly, but must go through the links until we come to it. The big advantage of this method, however, is that the actual physical order of the data does not have to be the same as the logical, or natural, order. In the preceding example, the natural order is $D_1 D_2 D_3 D_4 D_5$, but the data are not stored this way. The links allow us to pass naturally through the data, no matter how they are stored. Thus it is easy to add new items anywhere. If we want to insert item

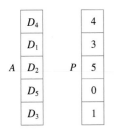

Figure 4.30

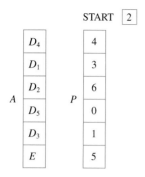

START 2

A		P
D_4		4
D_1		3
D_2		6
D_5		0
D_3		1
E		5

Figure 4.31

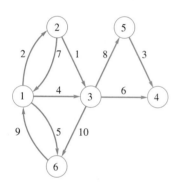

Figure 4.32

START	TAIL	HEAD	NEXT
2	1	3	9
	2	3	10
	2	1	4
	3	5	8
	5	4	1
	3	4	3
	3	6	0
	6	1	7
	1	6	6
	1	2	5

Figure 4.33

E between D_2 and D_3, we adjoin E to the end of the array A, change one pointer, and adjoin another pointer, as shown in Figure 4.31. This approach can be used no matter how long the list is. We should have one additional variable START holding the index of the first data item. In Figures 4.30 and 4.31, START would contain 2 since D_1 is in $A[2]$.

It does not matter how large the data item is, within computer constraints, so A might actually be a two-dimensional array or matrix. The first row would hold several numbers describing the first data item, the second row would describe the next item, and so on. The data can even be a pointer to the location of the actual data. Some programming languages implement pointers directly, but there are situations where it is advantageous to control linked lists directly as shown here.

The problem of storing information to represent a relation or its digraph also has two solutions similar to those presented previously for simple data. In the first place, we know from Section 4.2 that a relation R on A can be represented by an $n \times n$ matrix $\mathbf{M}_R$ if A has n elements. The matrix $\mathbf{M}_R$ has entries that are 0 or 1. Then a straightforward way of representing R in a computer would be by an $n \times n$ array having 0's and 1's stored in each location. Thus, if $A = \{1, 2\}$ and $R = \{(1, 1), (1, 2), (2, 2)\}$, then

$$\mathbf{M}_R = \begin{bmatrix} 1 & 1 \\ 0 & 1 \end{bmatrix}$$

and these data could be represented by a two-dimensional array MAT, where MAT[1, 1] = 1, MAT[1, 2] = 1, MAT[2, 1] = 0, and MAT[2, 2] = 1.

A second method of storing data for relations and digraphs uses the linked list idea described previously. For clarity, we use a graphical language. A linked list will be constructed that contains all the edges of the digraph, that is, the ordered pairs of numbers that determine those edges. The data can be represented by two arrays, TAIL and HEAD, giving the beginning vertex and end vertex, respectively, for all arrows. If we wish to make these edge data into a linked list, we will also need an array NEXT of pointers from each edge to the next edge.

Consider the relation whose digraph is shown in Figure 4.32. The vertices are the integers 1 through 6 and we arbitrarily number the edges as shown. If we wish to store the digraph in linked-list form so that the logical order coincides with the numbering of edges, we can use a scheme such as that illustrated in Figure 4.33. START contains 2, the index of the first data item, the edge (2, 3) (this edge is labeled with a 1 in Figure 4.32). This edge is stored in the second entries of TAIL and HEAD, respectively. Since NEXT[2] contains 10, the next edge is the one located in position 10 of TAIL and HEAD, that is, (1, 2) (labeled edge 2 in Figure 4.32).

NEXT[10] contains 5, so we go next to data position 5, which contains the edge (5, 4). This process continues until we reach edge (3, 6) in data position 7. This is the last edge, and this fact is indicated by having NEXT[7] contain 0. We use 0 as a pointer, indicating the absence of any more data.

If we trace through this process, we will see that we encounter the edges in exactly the order corresponding to their numbering. We can arrange, in a similar way, to pass through the edges in any desired order.

This scheme and the numerous equivalent variations of it have important disadvantages. In many algorithms, it is efficient to locate a vertex and then immedi-

VERT	TAIL	HEAD	NEXT
10	1	2	0
2	2	3	3
4	2	1	0
0	3	5	6
5	5	4	0
8	3	4	7
	3	6	0
	6	1	0
	1	6	1
	1	3	9

Figure 4.34

VERT	TAIL	HEAD	NEXT
9	1	2	0
3	2	3	0
6	2	1	2
0	3	5	7
5	5	4	0
8	3	4	4
	3	6	0
	6	1	0
	1	6	10
	1	3	1

Figure 4.35

ately begin to investigate the edges that begin or end with this vertex. This is not possible in general with the storage mechanism shown in Figure 4.33, so we now give a modification of it. We use an additional linear array VERT having one position for each vertex in the digraph. For each vertex I, VERT[I] is the index, in TAIL and HEAD, of the first edge we wish to consider leaving vertex I. In the digraph of Figure 4.32, the first edge could be taken to be the edge with the smallest number labeling it. Thus VERT, like NEXT, contains pointers to edges. For each vertex I, we must arrange the pointers in NEXT so that they link together all edges leaving I, starting with the edge pointed to by VERT[I]. The last of these edges is made to point to zero in each case. In a sense, the data arrays TAIL and HEAD really contain several linked lists of edges, one list for each vertex.

This method is shown in Figure 4.34 for the digraph of Figure 4.32. Here VERT[1] contains 10, so the first edge leaving vertex 1 must be stored in the tenth data position. This is edge (1, 3). Since NEXT[10] = 9, the next edge leaving vertex 1 is (1, 6) located in data position 9. Again NEXT[9] = 1, which points us to the edge (1, 2) in data position 1. Since NEXT[1] = 0, we have come to the end of those edges that begin at vertex 1. The order of the edges chosen here differs from the numbering in Figure 4.32.

We then proceed to VERT[2] and get a pointer to position 2 in the data. This contains the first edge leaving vertex 2, that is, (2, 3), and we can follow the pointers to visit all edges coming from vertex 2. In a similar way, we can trace through the edges (if any) coming from each vertex. Note that VERT[4] = 0, signifying that there are no edges beginning at vertex 4.

Figure 4.35 shows an alternative to Figure 4.34 for describing the digraph. The reader should check the accuracy of the method described in Figure 4.35. We remind the reader again that the ordering of the edges leaving each vertex can be chosen arbitrarily.

We see then that we have (at least) two methods for storing the data for a relation or digraph, one using the matrix of the relation and one using linked lists. A number of factors determines the choice of method to be used for storage. The total number of elements n in the set A, the number of ordered pairs in R or the ratio of this number to n^2 (the maximum possible number of ordered pairs), and the possible information that is to be extracted from R are all considerations. An analysis of such factors will determine which of the storage methods is superior. We will consider two cases.

Suppose that $A = \{1, 2, \ldots, N\}$, and let R be a relation on A, whose matrix $\mathbf{M}_R$ is represented by the array MAT. Suppose that R contains P ordered pairs so that MAT contains exactly P ones. First, we will consider the problem of adding a pair (I, J) to R and, second, the problem of testing R for transitivity.

Adding (I, J) to R is accomplished by the statement

$$\text{MAT}[I, J] \leftarrow 1.$$

This is extremely simple with the matrix storage method.

Now, consider the following algorithm, which assigns RESULT the value T (true) or F (false), depending on whether R is or is not transitive. We note that TRANS itself does not report whether R is transitive or not.

ALGORITHM TRANS
1.   RESULT ← T
2.   **FOR** $I = 1$ **THRU** N
    a.   **FOR** $J = 1$ **THRU** N
        1.   **IF** $(MAT[I, J] = 1)$ **THEN**
            a.   **FOR** $K = 1$ **THRU** N
                1.   **IF** $(MAT[J, K] = 1$ and $MAT[I, K] = 0)$ **THEN**
                    a.   RESULT ← F
END OF ALGORITHM TRANS

Here RESULT is originally set to T, and it is changed only if a situation is found where $(I, J) \in R$ and $(J, K) \in R$, but $(I, K) \notin R$ (a situation that violates transitivity).

We now provide a count of the number of steps required by algorithm TRANS. Observe that I and J each range from 1 to N. If (I, J) is not in R, we only perform the one test "**IF** $MAT[I, J] = 1$," which will be false, and the rest of the algorithm will not be executed. Since $N^2 - P$ ordered pairs do not belong to R, we have $N^2 - P$ steps that must be executed for such elements. If $(I, J) \in R$, then the test "**IF** $MAT[I, J] = 1$" will be true and an additional loop

  a.   **FOR** $K = 1$ **THRU** N
      1.   **IF** $(MAT[J, K] = 1$ and $MAT[I, K] = 0)$ **THEN**
          a.   RESULT ← F

of N steps will be executed. Since R contains P ordered pairs, we have PN steps for such elements. Thus the total number of steps required by algorithm TRANS is

$$T_A = PN + (N^2 - P).$$

Suppose that $P = kN^2$, where $0 \leq k \leq 1$, since P must be between 0 and N^2. Then algorithm TRANS tests for transitivity in

$$T_A = kN^3 + (1 - k)N^2$$

steps.

Now consider the same digraph represented by our linked-list scheme using VERT, TAIL, HEAD, and NEXT. First we deal with the problem of adding an edge (I, J). We assume that TAIL, HEAD, and NEXT have additional unused position available and that the total number of edges is counted by a variable P. Then the following algorithm adds an edge (I, J) to the relation R.

ALGORITHM ADDEDGE
1. $P \leftarrow P + 1$
2. $TAIL[P] \leftarrow I$
3. $HEAD[P] \leftarrow J$
4. $NEXT[P] \leftarrow VERT[I]$
5. $VERT[I] \leftarrow P$
END OF ALGORITHM ADDEDGE

Figure 4.36 shows the situation diagrammatically in pointer form, both before and after the addition of edge (I, J). VERT$[I]$ now points to the new edge, and the

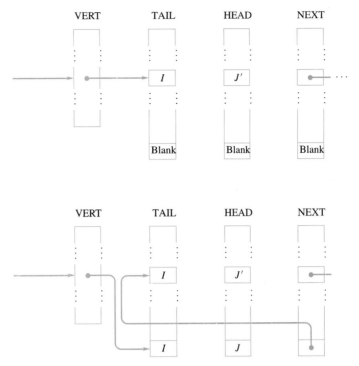

Figure 4.36

pointer from that edge goes to the edge previously pointed to by VERT[I], that is, (I, J'). This method is not too involved, but clearly the matrix storage method has the advantage for the task of adding an edge.

ALGORITHM NEWTRANS
1. RESULT ← T
2. **FOR** $I = 1$ **THRU** N
    a. X ← VERT[I]
    b. **WHILE** ($X \neq 0$)
        1. J ← HEAD[X]
        2. Y ← VERT[J]
        3. **WHILE** ($Y \neq 0$)
            a. K ← HEAD[Y]
            b. TEST ← EDGE[I, K]
            c. **IF** (TEST) **THEN**
                1. Y ← NEXT[Y]
            d. **ELSE**
                1. RESULT ← F
                2. Y ← NEXT[Y]
        4. X ← NEXT[X]
END OF ALGORITHM NEWTRANS

The reader should follow the steps of this algorithm with several simple examples. For each vertex I, it searches through all paths of length 2 beginning at I

and checks these for transitivity. Thus it eventually checks each path of length 2 to see if there is an equivalent direct path. Algorithm NEWTRANS is somewhat longer than algorithm TRANS, which corresponds to the matrix method of storage, and NEWTRANS also uses the function EDGE; but it is much more like the human method of determining the transitivity of R. Moreover, NEWTRANS may be more efficient.

Let us analyze the average number of steps that algorithm NEWTRANS takes to test for transitivity. Each of the P edges begins at a unique vertex, so, on the average, $P/N = D$ edges begin at a vertex. It is not hard to see that a function EDGE, such as needed in NEWTRANS, can be made to take an average of about D steps, since it must check all edges beginning at a particular vertex. The main **FOR** loop of NEWTRANS will be executed N times, and each subordinate **WHILE** statement will average about D executions. Since the last **WHILE** calls EDGE each time, we see that the entire algorithm will average about ND^3 execution steps. As before, we suppose that $P = kN^2$ with $0 \leq k \leq 1$. Then NEWTRANS averages about

$$T_L = N \left(\frac{kN^2}{N} \right)^3 = k^3 N^4 \quad \text{steps.}$$

Recall that algorithm TRANS, using matrix storage, required about $T_A = kN^3 + (1 - k)N^2$ steps.

Consider now the ratio T_L/T_A of the average number of steps needed with linked storage versus the number of steps needed with matrix storage to test R for transitivity. Thus

$$\frac{T_L}{T_A} = \frac{k^3 N^4}{kN^3 + (1 - k)N^2} = \frac{k^2 N}{1 + \left(\frac{1}{k} - 1 \right) \frac{1}{N}}.$$

When k is close to 1, that is, when there are many edges, then T_L/T_A is nearly N, so $T_L \approx T_A N$, and the linked-list method averages N times as many steps as the matrix-storage method. Thus the matrix-storage method is N times faster than the linked-list method in most cases.

On the other hand, if k is very small, then T_L/T_A may be nearly zero. This means that if the number of edges is small compared with N^2, it is, on average, considerably more efficient to test for transitivity in a linked-list storage method than with adjacency matrix storage.

We have, of course, made some oversimplifications. All steps do not take the same time to execute, and each algorithm to test for transitivity may be shortened by halting the search when the first counterexample to transitivity is discovered. In spite of this, the conclusions remain true and illustrate the important point that the choice of a data structure to represent objects such as sets, relations, and digraphs has an important effect on the efficiency with which information about the objects may be extracted.

Virtually all relations and digraphs of practical importance are too large to be explored by hand. Thus the computer storage of relations and the algorithmic implementation of methods for exploring them are of great importance.

4.6 Exercises

1. Verify that the linked-list arrangement of Figure 4.35 correctly describes the digraph of Figure 4.32.

2. Construct a function EDGE(I, J) (in pseudocode) that returns the value T (true) if the pair (i, j) is in R and F (false) otherwise. Assume that the relation R is given by arrays VERT, TAIL, HEAD, and NEXT, as described in this section.

3. Show that the function EDGE of Exercise 2 runs in an average of D steps, where $D = P/N$, P is the number of edges of R, and N is the number of vertices of R. (*Hint*: Let P_{ij} be the number of edges running from vertex i to vertex j. Express the total number of steps executed by EDGE for each pair of vertices and then average. Use the fact that $\sum_{i=1}^{} \sum_{j=1}^{} P_{ij} = P$.)

4. Let NUM be a linear array holding N positive integers, and let NEXT be a linear array of the same length. Suppose that START is a pointer to a "first" integer in NUM, and for each I, NEXT[I] points to the "next" integer in NUM to be considered. If NEXT[I] = 0, the list ends. Write a function LOOK(NUM, NEXT, START, N, K) in pseudocode to search NUM using the pointers in NEXT for an integer K. If K is found, the position of K in NUM is returned. If not, LOOK prints "NOT FOUND."

5. Let $A = \{1, 2, 3, 4\}$ and let $R = \{(1, 1), (1, 2), (1, 3), (2, 3), (2, 4), (3, 1), (3, 4), (4, 2)\}$ be a relation on A. Compute both the matrix $\mathbf{M}_R$ and the values of arrays VERT, TAIL, HEAD, and NEXT desribing R as a linked list. You may link in any reasonable way.

6. Let $A = \{1, 2, 3, 4\}$ and let R be the relation whose digraph is shown in Figure 4.37. Describe arrays VERT, TAIL, HEAD, and NEXT, setting up a linked-list representation of R, so that the edges out of each vertex are reached in the list in increasing order (relative to their numbering in Figure 4.37).

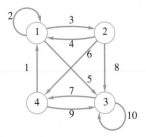

Figure 4.37

7. Consider the following arrays.

$$\text{VERT} = [1, 2, 6, 4]$$
$$\text{TAIL} = [1, 2, 2, 4, 4, 3, 4, 1]$$
$$\text{HEAD} = [2, 2, 3, 3, 4, 4, 1, 3]$$
$$\text{NEXT} = [8, 3, 0, 5, 7, 0, 0, 0]$$

These describe a relation R on the set $A = \{1, 2, 3, 4\}$. Compute both the digraph of R and the matrix $\mathbf{M}_R$.

8. The following arrays describe a relation R on the set $A = \{1, 2, 3, 4, 5\}$. Compute both the digraph of R and the matrix $\mathbf{M}_R$.

$$\text{VERT} = [6, 2, 8, 7, 10]$$
$$\text{TAIL} = [2, 2, 2, 2, 1, 1, 4, 3, 4, 5]$$
$$\text{HEAD} = [4, 3, 5, 1, 2, 3, 5, 4, 2, 4]$$
$$\text{NEXT} = [3, 1, 4, 0, 0, 5, 9, 0, 0, 0]$$

9. Let $A = \{1, 2, 3, 4, 5\}$ and let R be a relation on A such that

$$\mathbf{M}_R = \begin{bmatrix} 1 & 0 & 0 & 1 & 0 \\ 0 & 1 & 1 & 0 & 0 \\ 0 & 0 & 0 & 1 & 0 \\ 1 & 0 & 1 & 0 & 1 \\ 0 & 1 & 0 & 0 & 1 \end{bmatrix}.$$

Construct a linked-list representation, VERT, TAIL, HEAD, NEXT, for the relation R.

10. Let $A = \{a, b, c, d, e\}$ and let R be a relation described by

$$\mathbf{M}_R = \begin{bmatrix} 1 & 0 & 0 & 1 & 0 \\ 0 & 0 & 1 & 1 & 0 \\ 1 & 1 & 0 & 0 & 1 \\ 0 & 1 & 0 & 1 & 0 \\ 1 & 0 & 0 & 0 & 1 \end{bmatrix}.$$

Construct a linked-list representation, VERT, TAIL, HEAD, NEXT, for R.

11. Let $A = \{a, b, c, d\}$ and let R be a relation on A such that

$$\mathbf{M}_R = \begin{bmatrix} 1 & 1 & 0 & 1 \\ 0 & 1 & 1 & 0 \\ 0 & 1 & 1 & 1 \\ 1 & 1 & 1 & 1 \end{bmatrix}.$$

Construct a linked-list representation, VERT, TAIL, HEAD, NEXT, for the relation R.

12. Let $A = \{F, M, R, W\}$ and let R be a relation on A such that

$$\mathbf{M}_R = \begin{bmatrix} 1 & 1 & 0 & 1 \\ 1 & 1 & 0 & 1 \\ 1 & 1 & 1 & 1 \\ 1 & 1 & 0 & 1 \end{bmatrix}.$$

Construct a linked-list representation, VERT, TAIL, HEAD, NEXT, for the relation R.

4.7 OPERATIONS ON RELATIONS

Now that we have investigated the classification of relations by properties they do or do not have, we next define some operations on relations. As described in Section 1.6, relations, together with these operations, and the accompanying properties form a mathematical structure.

Let R and S be relations from a set A to a set B. Then, if we remember that R and S are simply subsets of $A \times B$, we can use set operations on R and S. For example, the complement of R, $\overline{R}$, is referred to as the **complementary relation**. It is, of course, a relation from A to B that can be expressed simply in terms of R:

$$a \, \overline{R} \, b \quad \text{if and only if} \quad a \, \not\mathrel{R} \, b.$$

We can also form the intersection $R \cap S$ and the union $R \cup S$ of the relations R and S. In relational terms, we see that $a \, R \cap S \, b$ means that $a \, R \, b$ and $a \, S \, b$. All our set-theoretic operations can be used in this way to produce new relations. The reader should try to give a relational description of the relation $R \oplus S$ (see Section 1.2).

A different type of operation on a relation R from A to B is the formation of the **inverse**, usually written R^{-1}. The relation R^{-1} is a relation from B to A (reverse order from R) defined by

$$b \, R^{-1} \, a \quad \text{if and only if} \quad a \, R \, b.$$

It is clear from this that $(R^{-1})^{-1} = R$. It is not hard to see that $\text{Dom}(R^{-1}) = \text{Ran}(R)$ and $\text{Ran}(R^{-1}) = \text{Dom}(R)$. We leave these simple facts for the reader to check.

EXAMPLE 1   Let $A = \{1, 2, 3, 4\}$ and $B = \{a, b, c\}$. Let

$$R = \{(1, a), (1, b), (2, b), (2, c), (3, b), (4, a)\}$$

and

$$S = \{(1, b), (2, c), (3, b), (4, b)\}.$$

Compute (a) $\overline{R}$;   (b) $R \cap S$;   (c) $R \cup S$;   and (d) R^{-1}.

Solution

(a) We first find

$$A \times B = \{(1, a), (1, b), (1, c), (2, a), (2, b), (2, c), (3, a),$$
$$= (3, b), (3, c), (4, a), (4, b), (4, c)\}.$$

Then the complement of R in $A \times B$ is

$$\overline{R} = \{(1, c), (2, a), (3, a), (3, c), (4, b), (4, c)\}.$$

(b) We have $R \cap S = \{(1, b), (3, b), (2, c)\}$.

(c) We have

$$R \cup S = \{(1, a), (1, b), (2, b), (2, c), (3, b), (4, a), (4, b)\}.$$

(d) Since $(x, y) \in R^{-1}$ if and only if $(y, x) \in R$, we have

$$R^{-1} = \{(a, 1), (b, 1), (b, 2), (c, 2), (b, 3), (a, 4)\}. \qquad \blacksquare$$

EXAMPLE 2   Let $A = \mathbb{R}$. Let R be the relation $\leq$ on A and let S be $\geq$. Then the complement of R is the relation $>$, since $a \not\leq b$ means that $a > b$. Similarly, the complement of S is $<$. On the other hand, $R^{-1} = S$, since for any numbers a and b,

$a \ R^{-1} \ b$ if and only if $b \ R \ a$ if and only if $b \leq a$ if and only if $a \geq b$.

Similarly, we have $S^{-1} = R$. Also, we note that $R \cap S$ is the relation of equality, since $a \ (R \cap S) \ b$ if and only if $a \leq b$ and $a \geq b$ if and only if $a = b$. Since, for any a and b, $a \leq b$ or $a \geq b$ must hold, we see that $R \cup S = A \times A$; that is, $R \cup S$ is the *universal* relation in which any a is related to any b. $\qquad \blacksquare$

EXAMPLE 3   Let $A = \{a, b, c, d, e\}$ and let R and S be two relations on A whose corresponding digraphs are shown in Figure 4.38. Then the reader can verify the following facts:

$$\overline{R} = \{(a, a), (b, b), (a, c), (b, a), (c, b), (c, d), (c, e), (c, a), (d, b),$$
$$(d, a), (d, e), (e, b), (e, a), (e, d), (e, c)\}$$

$$R^{-1} = \{(b, a), (e, b), (c, c), (c, d), (d, d), (d, b), (c, b), (d, a), (e, e), (e, a)\}$$

$$R \cap S = \{(a, b), (b, e), (c, c)\}. \qquad \blacksquare$$

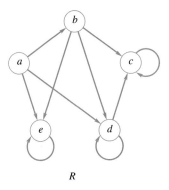

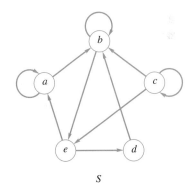

R   S

Figure 4.38

EXAMPLE 4   Let $A = \{1, 2, 3\}$ and let R and S be relations on A. Suppose that the matrices of R and S are

$$\mathbf{M}_R = \begin{bmatrix} 1 & 0 & 1 \\ 0 & 1 & 1 \\ 0 & 0 & 0 \end{bmatrix} \quad \text{and} \quad \mathbf{M}_S = \begin{bmatrix} 0 & 1 & 1 \\ 1 & 1 & 0 \\ 0 & 1 & 0 \end{bmatrix}.$$

Then we can verify that

$$\mathbf{M}_{\overline{R}} = \begin{bmatrix} 0 & 1 & 0 \\ 1 & 0 & 0 \\ 1 & 1 & 1 \end{bmatrix}, \qquad \mathbf{M}_{R^{-1}} = \begin{bmatrix} 1 & 0 & 0 \\ 0 & 1 & 0 \\ 1 & 1 & 0 \end{bmatrix},$$

$$\mathbf{M}_{R \cap S} = \begin{bmatrix} 0 & 0 & 1 \\ 0 & 1 & 0 \\ 0 & 0 & 0 \end{bmatrix}, \qquad \mathbf{M}_{R \cup S} = \begin{bmatrix} 1 & 1 & 1 \\ 1 & 1 & 1 \\ 0 & 1 & 0 \end{bmatrix}.$$

■

Example 4 illustrates some general facts. Recalling the operations on Boolean matrices from Section 1.5, we can show (Exercise 27) that if R and S are relations on set A, then

$$\mathbf{M}_{R \cap S} = \mathbf{M}_R \wedge \mathbf{M}_S$$
$$\mathbf{M}_{R \cup S} = \mathbf{M}_R \vee \mathbf{M}_S$$
$$\mathbf{M}_{R^{-1}} = (\mathbf{M}_R)^T.$$

Moreover, if $\mathbf{M}$ is a Boolean matrix, we define the **complement** $\overline{\mathbf{M}}$ of $\mathbf{M}$ as the matrix obtained from $\mathbf{M}$ by replacing every 1 in $\mathbf{M}$ by a 0 and every 0 by a 1. Thus, if

$$\mathbf{M} = \begin{bmatrix} 1 & 0 & 0 \\ 0 & 1 & 1 \\ 1 & 0 & 0 \end{bmatrix},$$

then

$$\overline{\mathbf{M}} = \begin{bmatrix} 0 & 1 & 1 \\ 1 & 0 & 0 \\ 0 & 1 & 1 \end{bmatrix}.$$

We can also show (Exercise 27) that if R is a relation on a set A, then

$$\mathbf{M}_{\overline{R}} = \overline{\mathbf{M}}_R.$$

We know that a symmetric relation is a relation R such that $\mathbf{M}_R = (\mathbf{M}_R)^T$, and since $(\mathbf{M}_R)^T = \mathbf{M}_{R^{-1}}$, we see that R is symmetric if and only if $R = R^{-1}$.

We now prove a few useful properties about combinations of relations.

Theorem 1   Suppose that R and S are relations from A to B.
   (a) If $R \subseteq S$, then $R^{-1} \subseteq S^{-1}$.
   (b) If $R \subseteq S$, then $\overline{S} \subseteq \overline{R}$.
   (c) $(R \cap S)^{-1} = R^{-1} \cap S^{-1}$ and $(R \cup S)^{-1} = R^{-1} \cup S^{-1}$.
   (d) $\overline{R \cap S} = \overline{R} \cup \overline{S}$ and $\overline{R \cup S} = \overline{R} \cap \overline{S}$.

●

Proof   Parts (b) and (d) are special cases of general set properties proved in Section 1.2.

We now prove part (a). Suppose that $R \subseteq S$ and let $(a, b) \in R^{-1}$. Then $(b, a) \in R$, so $(b, a) \in S$. This, in turn, implies that $(a, b) \in S^{-1}$. Since each element of R^{-1} is in S^{-1}, we are done.

We next prove part (c). For the first part, suppose that $(a, b) \in (R \cap S)^{-1}$. Then $(b, a) \in R \cap S$, so $(b, a) \in R$ and $(b, a) \in S$. This means that $(a, b) \in R^{-1}$ and $(a, b) \in S^{-1}$, so $(a, b) \in R^{-1} \cap S^{-1}$. The converse containment can be proved by reversing the steps. A similar argument works to show that $(R \cup S)^{-1} = R^{-1} \cup S^{-1}$.

▼

The relations $\overline{R}$ and R^{-1} can be used to check if R has the properties of relations that we presented in Section 4.4. For instance, we saw earlier that R is symmetric if and only if $R = R^{-1}$. Here are some other connections between operations on relations and properties of relations.

Theorem 2   Let R and S be relations on a set A.
    (a) If R is reflexive, so is R^{-1}.
    (b) If R and S are reflexive, then so are $R \cap S$ and $R \cup S$.
    (c) R is reflexive if and only if $\overline{R}$ is irreflexive.   ●

Proof   Let Δ be the equality relation on A. We know that R is reflexive if and only if $\Delta \subseteq R$. Clearly, $\Delta = \Delta^{-1}$, so if $\Delta \subseteq R$, then $\Delta = \Delta^{-1} \subseteq R^{-1}$ by Theorem 1, so R^{-1} is also reflexive. This proves part (a). To prove part (b), we note that if $\Delta \subseteq R$ and $\Delta \subseteq S$, then $\Delta \subseteq R \cap S$ and $\Delta \subseteq R \cup S$. To show part (c), we note that a relation S is irreflexive if and only if $S \cap \Delta = \varnothing$. Then R is reflexive if and only if $\Delta \subseteq R$ if and only if $\Delta \cap \overline{R} = \varnothing$ if and only if $\overline{R}$ is irreflexive.   ▼

EXAMPLE 5   Let $A = \{1, 2, 3\}$ and consider the two reflexive relations

$$R = \{(1, 1), (1, 2), (1, 3), (2, 2), (3, 3)\}$$

and

$$S = \{(1, 1), (1, 2), (2, 2), (3, 2), (3, 3)\}.$$

Then

    (a) $R^{-1} = \{(1, 1), (2, 1), (3, 1), (2, 2), (3, 3)\}$; R and R^{-1} are both reflexive.
    (b) $\overline{R} = \{(2, 1), (2, 3), (3, 1), (3, 2)\}$ is irreflexive while R is reflexive.
    (c) $R \cap S = \{(1, 1), (1, 2), (2, 2), (3, 3)\}$ and $R \cup S = \{(1, 1), (1, 2), (1, 3), (2, 2), (3, 2), (3, 3)\}$ are both reflexive.   ■

Theorem 3   Let R be a relation on a set A. Then
    (a) R is symmetric if and only if $R = R^{-1}$.
    (b) R is antisymmetric if and only if $R \cap R^{-1} \subseteq \Delta$.
    (c) R is asymmetric if and only if $R \cap R^{-1} = \varnothing$.   ●

Proof   The proof is straightforward and is left as an exercise.   ▼

Theorem 4   Let R and S be relations on A.

(a) If R is symmetric, so are R^{-1} and $\overline{R}$.

(b) If R and S are symmetric, so are $R \cap S$ and $R \cup S$.

Proof   If R is symmetric, $R = R^{-1}$ and thus $(R^{-1})^{-1} = R = R^{-1}$, which means that R^{-1} is also symmetric. Also, $(a, b) \in (\overline{R})^{-1}$ if and only if $(b, a) \in \overline{R}$ if and only if $(b, a) \notin R$ if and only if $(a, b) \notin R^{-1} = R$ if and only if $(a, b) \in \overline{R}$, so $\overline{R}$ is symmetric and part (a) is proved. The proof of part (b) follows immediately from Theorem 1(c).   ▽

EXAMPLE 6   Let $A = \{1, 2, 3\}$ and consider the symmetric relations

$$R = \{(1, 1), (1, 2), (2, 1), (1, 3), (3, 1)\}$$

and

$$S = \{(1, 1), (1, 2), (2, 1), (2, 2), (3, 3)\}.$$

Then

(a) $R^{-1} = \{(1, 1), (2, 1), (1, 2), (3, 1), (1, 3)\}$ and $\overline{R} = \{(2, 2), (2, 3), (3, 2), (3, 3)\}$; R^{-1} and $\overline{R}$ are symmetric.

(b) $R \cap S = \{(1, 1), (1, 2), (2, 1)\}$ and $R \cup S = \{(1, 1), (1, 2), (1, 3), (2, 1), (2, 2), (3, 1), (3, 3)\}$, which are both symmetric.   ■

Theorem 5   Let R and S be relations on A.

(a) $(R \cap S)^2 \subseteq R^2 \cap S^2$.

(b) If R and S are transitive, so is $R \cap S$.

(c) If R and S are equivalence relations, so is $R \cap S$.

Proof   We prove part (a) geometrically. We have $a \ (R \cap S)^2 \ b$ if and only if there is a path of length 2 from a to b in $R \cap S$. Both edges of this path lie in R and in S, so $a \ R^2 \ b$ and $a \ S^2 \ b$, which implies that $a \ (R^2 \cap S^2) \ b$. To show part (b), recall from Section 4.4 that a relation T is transitive if and only if $T^2 \subseteq T$. If R and S are transitive, then $R^2 \subseteq R$, $S^2 \subseteq S$, so $(R \cap S)^2 \subseteq R^2 \cap S^2$ [by part (a)] $\subseteq R \cap S$, so $R \cap S$ is transitive. We next prove part (c). Relations R and S are each reflexive, symmetric, and transitive. The same properties hold for $R \cap S$ from Theorems 2(b), 4(b), and 5(b), respectively. Hence $R \cap S$ is an equivalence relation.   ▽

EXAMPLE 7   Let R and S be equivalence relations on a finite set A, and let A/R and A/S be the corresponding partitions (see Section 4.5). Since $R \cap S$ is an equivalence relation, it corresponds to a partition $A/(R \cap S)$. We now describe $A/(R \cap S)$ in terms of A/R and A/S. Let W be a block of $A/(R \cap S)$ and suppose that a and b belong to W. Then $a \ (R \cap S) \ b$, so $a \ R \ b$ and $a \ S \ b$. Thus a and b belong to the same block, say X, of A/R and to the same block, say Y, of A/S. This means that $W \subseteq X \cap Y$. The steps in this argument are reversible; therefore, $W = X \cap Y$. Thus we can directly compute the partition $A/(R \cap S)$ by forming all possible intersections of blocks in A/R with blocks in A/S.   ■

Closures

If R is a relation on a set A, it may well happen that R lacks some of the important relational properties discussed in Section 4.4, especially reflexivity, symmetry, and transitivity. If R does not possess a particular property, we may wish to add pairs to R until we get a relation that *does* have the required property. Naturally, we want to add as few new pairs as possible, so what we need to find is the *smallest* relation R_1 on A that contains R and possesses the property we desire. Sometimes R_1 does not exist. If a relation such as R_1 does exist, we call it the **closure** of R with respect to the property in question.

EXAMPLE 8   Suppose that R is a relation on a set A, and R is not reflexive. This can only occur because some pairs of the diagonal relation Δ are not in R. Thus $R_1 = R \cup \Delta$ is the smallest reflexive relation on A containing R; that is, the **reflexive closure** of R is $R \cup \Delta$.   ■

EXAMPLE 9   Suppose now that R is a relation on A that is not symmetric. Then there must exist pairs (x, y) in R such that (y, x) is not in R. Of course, $(y, x) \in R^{-1}$, so if R is to be symmetric we must add all pairs from R^{-1}; that is, we must enlarge R to $R \cup R^{-1}$. Clearly, $(R \cup R^{-1})^{-1} = R \cup R^{-1}$, so $R \cup R^{-1}$ is the smallest symmetric relation containing R; that is, $R \cup R^{-1}$ is the **symmetric closure** of R.

If $A = \{a, b, c, d\}$ and $R = \{(a, b), (b, c), (a, c), (c, d)\}$, then $R^{-1} = \{(b, a), (c, b), (c, a), (d, c)\}$, so the symmetric closure of R is

$$R \cup R^{-1} = \{(a, b), (b, a), (b, c), (c, b), (a, c), (c, a), (c, d), (d, c)\}.$$ ■

The symmetric closure of a relation R is very easy to visualize geometrically. All edges in the digraph of R become "two-way streets" in $R \cup R^{-1}$. Thus the graph of the symmetric closure of R is simply the digraph of R with all edges made bidirectional. We show in Figure 4.39(a) the digraph of the relation R of Example 9. Figure 4.39(b) shows the graph of the symmetric closure $R \cup R^{-1}$.

The **transitive closure** of a relation R is the smallest transitive relation containing R. We will discuss the transitive closure in the next section.

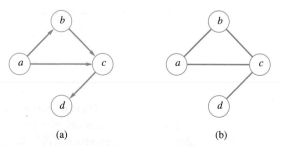

(a)   (b)

Figure 4.39

Composition

Now suppose that A, B, and C are sets, R is a relation from A to B, and S is a relation from B to C. We can then define a new relation, the **composition** of R and S, written $S \circ R$. The relation $S \circ R$ is a relation from A to C and is defined as follows. If a is in A and c is in C, then a $(S \circ R)$ c if and only if for some b in B, we have a R b and b S c. In other words, a is related to c by $S \circ R$ if we can get from a to c in two stages: first to an intermediate vertex b by relation R and then from b to c by relation S. The relation $S \circ R$ might be thought of as "S following R" since it represents the combined effect of two relations, first R, then S.

EXAMPLE 10 Let $A = \{1, 2, 3, 4\}$, $R = \{(1, 2), (1, 1), (1, 3), (2, 4), (3, 2)\}$, and $S = \{(1, 4), (1, 3), (2, 3), (3, 1), (4, 1)\}$. Since $(1, 2) \in R$ and $(2, 3) \in S$, we must have $(1, 3) \in S \circ R$. Similarly, since $(1, 1) \in R$ and $(1, 4) \in S$, we see that $(1, 4) \in S \circ R$. Proceeding in this way, we find that $S \circ R = \{(1, 4), (1, 3), (1, 1), (2, 1), (3, 3)\}$. ∎

The following result shows how to compute relative sets for the composition of two relations.

Theorem 6 Let R be a relation from A to B and let S be a relation from B to C. Then, if A_1 is any subset of A, we have

$$(S \circ R)(A_1) = S(R(A_1)).$$ ●

Proof If an element $z \in C$ is in $(S \circ R)(A_1)$, then x $(S \circ R)$ z for some x in A_1. By the definition of composition, this means that x R y and y S z for some y in B. Thus $y \in R(x)$, so $z \in S(R(x))$. Since $\{x\} \subseteq A_1$, Theorem 1(a) of Section 4.2 tells us that $S(R(x)) \subseteq S(R(A_1))$. Hence $z \in S(R(A_1))$, so $(S \circ R)(A_1) \subseteq S(R(A_1))$.

Conversely, suppose that $z \in S(R(A_1))$. Then $z \in S(y)$ for some y in $R(A_1)$ and, similarly, $y \in R(x)$ for some x in A_1. This means that x R y and y S z, so x $(S \circ R)$ z. Thus $z \in (S \circ R)(A_1)$, so $S(R(A_1)) \subseteq (S \circ R)(A_1)$. This proves the theorem. ▼

EXAMPLE 11 Let $A = \{a, b, c\}$ and let R and S be relations on A whose matrices are

$$\mathbf{M}_R = \begin{bmatrix} 1 & 0 & 1 \\ 1 & 1 & 1 \\ 0 & 1 & 0 \end{bmatrix}, \qquad \mathbf{M}_S = \begin{bmatrix} 1 & 0 & 0 \\ 0 & 1 & 1 \\ 1 & 0 & 1 \end{bmatrix}.$$

We see from the matrices that

$$(a, a) \in R \quad \text{and} \quad (a, a) \in S, \quad \text{so} \quad (a, a) \in S \circ R$$
$$(a, c) \in R \quad \text{and} \quad (c, a) \in S, \quad \text{so} \quad (a, a) \in S \circ R$$
$$(a, c) \in R \quad \text{and} \quad (c, c) \in S, \quad \text{so} \quad (a, c) \in S \circ R.$$

It is easily seen that $(a, b) \notin S \circ R$ since, if we had $(a, x) \in R$ and $(x, b) \in S$, then matrix $\mathbf{M}_R$ tells us that x would have to be a or c; but matrix $\mathbf{M}_S$ tells us that neither (a, b) nor (c, b) is an element of S.

We see that the first row of $\mathbf{M}_{S \circ R}$ is 1   0   1. The reader may show by similar analysis that

$$\mathbf{M}_{S \circ R} = \begin{bmatrix} 1 & 0 & 1 \\ 1 & 1 & 1 \\ 0 & 1 & 1 \end{bmatrix}.$$

We note that $\mathbf{M}_{S \circ R} = \mathbf{M}_R \odot \mathbf{M}_S$ (Verify this).   ∎

Example 11 illustrates a general and useful fact. Let A, B, and C be finite sets with n, p, and m elements, respectively, let R be a relation from A to B, and let S be a relation from B to C. Then R and S have Boolean matrices $\mathbf{M}_R$ and $\mathbf{M}_S$ with respective sizes $n \times p$ and $p \times m$. Thus $\mathbf{M}_R \odot \mathbf{M}_S$ can be computed, and it equals $\mathbf{M}_{S \circ R}$.

To see this let $A = \{a_1, \dots, a_n\}$, $B = \{b_1, \dots, b_p\}$, and $C = \{c_1, \dots, c_m\}$. Also, suppose that $\mathbf{M}_R = \begin{bmatrix} r_{ij} \end{bmatrix}$, $\mathbf{M}_S = \begin{bmatrix} s_{ij} \end{bmatrix}$, and $\mathbf{M}_{S \circ R} = \begin{bmatrix} t_{ij} \end{bmatrix}$. Then $t_{ij} = 1$ if and only if $(a_i, c_j) \in S \circ R$, which means that for some k, $(a_i, b_k) \in R$ and $(b_k, c_j) \in S$. In other words, $r_{ik} = 1$ and $s_{kj} = 1$ for some k between 1 and p. This condition is identical to the condition needed for $\mathbf{M}_R \odot \mathbf{M}_S$ to have a 1 in position i, j, and thus $\mathbf{M}_{S \circ R}$ and $\mathbf{M}_R \odot \mathbf{M}_S$ are equal.

In the special case where R and S are equal, we have $S \circ R = R^2$ and $\mathbf{M}_{S \circ R} = \mathbf{M}_{R^2} = \mathbf{M}_R \odot \mathbf{M}_R$, as was shown in Section 4.3.

EXAMPLE 12   Let us redo Example 10 using matrices. We see that

$$\mathbf{M}_R = \begin{bmatrix} 1 & 1 & 1 & 0 \\ 0 & 0 & 0 & 1 \\ 0 & 1 & 0 & 0 \\ 0 & 0 & 0 & 0 \end{bmatrix} \quad \text{and} \quad \mathbf{M}_S = \begin{bmatrix} 0 & 0 & 1 & 1 \\ 0 & 0 & 1 & 0 \\ 1 & 0 & 0 & 0 \\ 1 & 0 & 0 & 0 \end{bmatrix}.$$

Then

$$\mathbf{M}_R \odot \mathbf{M}_S = \begin{bmatrix} 1 & 0 & 1 & 1 \\ 1 & 0 & 0 & 0 \\ 0 & 0 & 1 & 0 \\ 0 & 0 & 0 & 0 \end{bmatrix},$$

so

$$S \circ R = \{(1, 1), (1, 3), (1, 4), (2, 1), (3, 3)\}$$

as we found before. In cases where the number of pairs in R and S is large, the matrix method is much more reliable.   ∎

Theorem 7   Let A, B, C, and D be sets, R a relation from A to B, S a relation from B to C, and T a relation from C to D. Then

$$T \circ (S \circ R) = (T \circ S) \circ R.$$ ●

Proof   The relations R, S, and T are determined by their Boolean matrices $\mathbf{M}_R$, $\mathbf{M}_S$, and $\mathbf{M}_T$, respectively. As we showed after Example 11, the matrix of the composition is the Boolean matrix product; that is, $\mathbf{M}_{S \circ R} = \mathbf{M}_R \odot \mathbf{M}_S$. Thus

$$\mathbf{M}_{T \circ (S \circ R)} = \mathbf{M}_{S \circ R} \odot \mathbf{M}_T = (\mathbf{M}_R \odot \mathbf{M}_S) \odot \mathbf{M}_T.$$

Similarly,

$$\mathbf{M}_{(T \circ S) \circ R} = \mathbf{M}_R \odot (\mathbf{M}_S \odot \mathbf{M}_T).$$

Since Boolean matrix multiplication is associative [see Exercise 31 of Section 1.5], we must have

$$(\mathbf{M}_R \odot \mathbf{M}_S) \odot \mathbf{M}_T = \mathbf{M}_R \odot (\mathbf{M}_S \odot \mathbf{M}_T),$$

and therefore

$$\mathbf{M}_{T \circ (S \circ R)} = \mathbf{M}_{(T \circ S) \circ R}.$$

Then

$$T \circ (S \circ R) = (T \circ S) \circ R$$

since these relations have the same matrices.   ▼

The proof illustrates the advantage of having several ways to represent a relation. Here using the matrix of the relation produces a simple proof.

In general, $R \circ S \neq S \circ R$, as shown in the following example.

EXAMPLE 13   Let $A = \{a, b\}$, $R = \{(a, a), (b, a), (b, b)\}$, and $S = \{(a, b), (b, a), (b, b)\}$. Then $S \circ R = \{(a, b), (b, a), (b, b)\}$, while $R \circ S = \{(a, a), (a, b), (b, a), (b, b)\}$.   ■

Theorem 8   Let A, B, and C be sets, R a relation from A to B, and S a relation from B to C. Then $(S \circ R)^{-1} = R^{-1} \circ S^{-1}$.   ●

Proof   Let $c \in C$ and $a \in A$. Then $(c, a) \in (S \circ R)^{-1}$ if and only if $(a, c) \in S \circ R$, that is, if and only if there is a $b \in B$ with $(a, b) \in R$ and $(b, c) \in S$. Finally, this is equivalent to the statement that $(c, b) \in S^{-1}$ and $(b, a) \in R^{-1}$; that is, $(c, a) \in R^{-1} \circ S^{-1}$.   ▼

4.7 Exercises

In Exercises 1 and 2, let R and S be the given relations from A to B. Compute (a) $\overline{R}$; (b) $R \cap S$; (c) $R \cup S$; (d) S^{-1}.

1. $A = B = \{1, 2, 3\}$;
    $R = \{(1, 1), (1, 2), (2, 3), (3, 1)\}$;
    $S = \{(2, 1), (3, 1), (3, 2), (3, 3)\}$.

2. $A = \{a, b, c\}$; $B = \{1, 2, 3\}$;
    $R = \{(a, 1), (b, 1), (c, 2), (c, 3)\}$;
    $S = \{(a, 1), (a, 2), (b, 1), (b, 2)\}$.

3. Let $A = $ a set of people. Let a R b if and only if a and b are brothers; let a S b if and only if a and b are sisters. Describe $R \cup S$.

4. Let $A = $ a set of people. Let a R b if and only if a is older than b; let a S b if and only if a is a brother of b. Describe $R \cap S$.

5. Let $A = $ a set of people. Let a R b if and only if a is the father of b; let a S b if and only if a is the mother of b. Describe $R \cup S$.

6. Let $A = \{2, 3, 6, 12\}$ and let R and S be the following relations on A: x R y if and only if $2 \mid (x - y)$; x S y if and only if $3 \mid (x - y)$. Compute
    (a) $\overline{R}$;   (b) $R \cap S$;   (c) $R \cup S$;   (d) S^{-1}.

In Exercises 7 and 8, let R and S be two relations whose corresponding digraphs are shown in Figures 4.40 and 4.41. Compute (a) $\overline{R}$; (b) $R \cap S$; (c) $R \cup S$; (d) S^{-1}.

7.

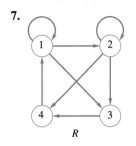 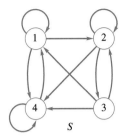

R S

Figure 4.40

8.

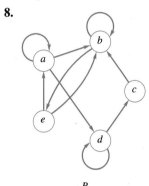 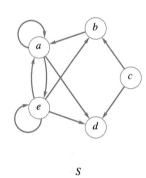

R S

Figure 4.41

In Exercises 9 and 10, let $A = \{1, 2, 3\}$ and $B = \{1, 2, 3, 4\}$. Let R and S be the relations from A to B whose matrices are given. Compute (a) $\overline{S}$; (b) $R \cap S$; (c) $R \cup S$; (d) R^{-1}.

9. $\mathbf{M}_R = \begin{bmatrix} 1 & 1 & 0 & 1 \\ 0 & 0 & 0 & 1 \\ 1 & 1 & 1 & 0 \end{bmatrix}$, $\mathbf{M}_S = \begin{bmatrix} 0 & 1 & 1 & 0 \\ 1 & 0 & 0 & 1 \\ 1 & 1 & 0 & 0 \end{bmatrix}$.

10. $\mathbf{M}_R = \begin{bmatrix} 1 & 0 & 1 & 0 \\ 0 & 0 & 0 & 1 \\ 1 & 1 & 1 & 0 \end{bmatrix}$, $\mathbf{M}_S = \begin{bmatrix} 1 & 1 & 1 & 1 \\ 0 & 0 & 0 & 1 \\ 0 & 1 & 0 & 1 \end{bmatrix}$.

In Exercises 11 and 12, let $A = \{1, 2, 3, 4\}$ and $B = \{1, 2, 3\}$. Given the matrices $\mathbf{M}_R$ and $\mathbf{M}_S$ of the relations R and S from A to B, compute (a) $\mathbf{M}_{R \cap S}$; (b) $\mathbf{M}_{R \cup S}$; (c) $\mathbf{M}_{R^{-1}}$; (d) $\mathbf{M}_{\overline{S}}$.

11. $\mathbf{M}_R = \begin{bmatrix} 1 & 0 & 1 \\ 0 & 1 & 1 \\ 0 & 1 & 0 \\ 1 & 0 & 1 \end{bmatrix}$, $\mathbf{M}_S = \begin{bmatrix} 0 & 1 & 0 \\ 1 & 0 & 1 \\ 1 & 0 & 1 \\ 1 & 1 & 1 \end{bmatrix}$.

12. $\mathbf{M}_R = \begin{bmatrix} 0 & 1 & 0 \\ 0 & 1 & 1 \\ 0 & 0 & 1 \\ 1 & 1 & 1 \end{bmatrix}$, $\mathbf{M}_S = \begin{bmatrix} 1 & 0 & 1 \\ 1 & 0 & 1 \\ 0 & 1 & 0 \\ 0 & 1 & 0 \end{bmatrix}$.

13. Let $A = B = \{1, 2, 3, 4\}$, $R = \{(1, 1), (1, 3), (2, 3), (3, 1), (4, 2), (4, 4)\}$, and $S = \{(1, 2), (2, 3), (3, 1), (3, 2), (4, 3)\}$. Compute (a) $\mathbf{M}_{R \cap S}$; (b) $\mathbf{M}_{R \cup S}$; (c) $\mathbf{M}_{R^{-1}}$; (d) $\mathbf{M}_{\overline{S}}$.

14. Let

$$A = \{1, 2, 3, 4, 5, 6\},$$
$$R = \{(1, 2), (1, 1), (2, 1), (2, 2), (3, 3), (4, 4), (5, 5), (5, 6), (6, 5), (6, 6)\}, \quad \text{and}$$
$$S = \{(1, 1), (1, 2), (1, 3), (2, 1), (2, 2), (2, 3), (3, 1), (3, 2), (3, 3), (4, 6), (4, 4), (6, 4), (6, 6), (5, 5)\}$$

be equivalence relations on A. Compute the partition corresponding to $R \cap S$.

15. Let $A = \{a, b, c, d, e\}$ and let the equivalence relations R and S on A be given by

$$\mathbf{M}_R = \begin{bmatrix} 1 & 1 & 1 & 1 & 0 \\ 1 & 1 & 1 & 1 & 0 \\ 1 & 1 & 1 & 1 & 0 \\ 1 & 1 & 1 & 1 & 0 \\ 0 & 0 & 0 & 0 & 1 \end{bmatrix}$$

$$\mathbf{M}_S = \begin{bmatrix} 1 & 0 & 0 & 0 & 0 \\ 0 & 1 & 1 & 0 & 0 \\ 0 & 1 & 1 & 0 & 0 \\ 0 & 0 & 0 & 1 & 1 \\ 0 & 0 & 0 & 1 & 1 \end{bmatrix}.$$

Compute the partition of A corresponding to $R \cap S$.

16. Let $A = \{1, 2, 3, 4\}$ and $R = \{(2, 1), (2, 3), (3, 2), (3, 3), (2, 2), (4, 2)\}$.
(a) Find the reflexive closure of R.
(b) Find the symmetric closure of R.

17. Let R be the relation whose matrix is

$$\begin{bmatrix} 1 & 0 & 0 & 1 & 1 \\ 0 & 0 & 1 & 0 & 1 \\ 1 & 1 & 1 & 0 & 0 \\ 0 & 1 & 1 & 0 & 0 \\ 0 & 0 & 1 & 0 & 1 \end{bmatrix}.$$

(a) Find the reflexive closure of R.
(b) Find the symmetric closure of R.

18. Explain why the concept of closure is not applicable for irreflexivity, asymmetry, or antisymmetry.

19. Let $A = B = C =$ the set of real numbers. Let R and S be the following relations from A to B and from B to C, respectively:

$$R = \{(a, b) \mid a \le 2b\}$$
$$S = \{(b, c) \mid b \le 3c\}.$$

(a) Is $(1, 5) \in S \circ R$?

(b) Is $(2, 3) \in S \circ R$?

(c) Describe $S \circ R$.

20. Let $A = \{1, 2, 3, 4\}$. Let

$$R = \{(1, 1), (1, 2), (2, 3), (2, 4), (3, 4), (4, 1), (4, 2)\}$$
$$S = \{(3, 1), (4, 4), (2, 3), (2, 4), (1, 1), (1, 4)\}.$$

(a) Is $(1, 3) \in R \circ R$?

(b) Is $(4, 3) \in S \circ R$?

(c) Is $(1, 1) \in R \circ S$?

(d) Compute $R \circ R$.

(e) Compute $S \circ R$.

(f) Compute $R \circ S$.

(g) Compute $S \circ S$.

21. (a) Which properties of relations on a set A are preserved by composition? Prove your conclusion.

(b) If R and S are equivalence relations on a set A, is $S \circ R$ an equivalence relation on A? Prove your conclusion.

In Exercises 22 and 23, let $A = \{1, 2, 3, 4, 5\}$ and let $\mathbf{M}_R$ and $\mathbf{M}_S$ be the matrices of the relations R and S on A. Compute (a) $\mathbf{M}_{R \circ R}$; (b) $\mathbf{M}_{S \circ R}$; (c) $\mathbf{M}_{R \circ S}$; (d) $\mathbf{M}_{S \circ S}$.

22.

$$\mathbf{M}_R = \begin{bmatrix} 1 & 0 & 1 & 1 & 1 \\ 0 & 1 & 1 & 0 & 0 \\ 1 & 0 & 0 & 1 & 0 \\ 1 & 0 & 1 & 0 & 0 \\ 0 & 1 & 1 & 1 & 1 \end{bmatrix},$$

$$\mathbf{M}_S = \begin{bmatrix} 1 & 0 & 0 & 1 & 0 \\ 1 & 0 & 1 & 0 & 0 \\ 1 & 0 & 1 & 0 & 0 \\ 0 & 1 & 1 & 1 & 1 \\ 1 & 0 & 0 & 0 & 1 \end{bmatrix}.$$

23.

$$\mathbf{M}_R = \begin{bmatrix} 1 & 1 & 0 & 0 & 1 \\ 0 & 0 & 0 & 1 & 0 \\ 1 & 1 & 0 & 0 & 1 \\ 0 & 1 & 0 & 1 & 1 \\ 1 & 0 & 0 & 0 & 0 \end{bmatrix},$$

$$\mathbf{M}_S = \begin{bmatrix} 0 & 0 & 0 & 1 & 1 \\ 1 & 0 & 0 & 0 & 1 \\ 0 & 1 & 0 & 1 & 0 \\ 1 & 1 & 0 & 1 & 1 \\ 1 & 0 & 1 & 0 & 0 \end{bmatrix}.$$

24. (a) Let R and S be relations on a set A. If R and S are asymmetric, prove or disprove that $R \cap S$ and $R \cup S$ are asymmetric.

(b) Let R and S be relations on a set A. If R and S are antisymmetric, prove or disprove that $R \cap S$ and $R \cup S$ are antisymmetric.

25. Let R be a relation from A to B and let S and T be relations from B to C. Prove or disprove.

(a) $(S \cup T) \circ R = (S \circ R) \cup (T \circ R)$

(b) $(S \cap T) \circ R = (S \circ R) \cap (T \circ R)$

26. Let R and S be relations from A to B and let T be a relation from B to C. Show that if $R \subseteq S$, then $T \circ R \subseteq T \circ S$.

27. Show that if R and S are relations on a set A, then

(a) $\mathbf{M}_{R \cap S} = \mathbf{M}_R \wedge \mathbf{M}_S$

(b) $\mathbf{M}_{R \cup S} = \mathbf{M}_R \vee \mathbf{M}_S$

(c) $\mathbf{M}_{R^{-1}} = (\mathbf{M}_R)^T$

(d) $\mathbf{M}_{\overline{R}} = \overline{\mathbf{M}_R}$

28. Let R and S be relations on a set A. Prove that $(R \cap S)^n \subseteq R^n \cap S^n$, for $n \ge 1$.

In Exercises 29 through 31, let R and S be relations on a finite set A. Describe how to form the digraph of the specified relation directly from the digraphs of R and S.

29. R^{-1} **30.** $R \cap S$ **31.** $R \cup S$

32. Prove Theorem 3.

4.8   TRANSITIVE CLOSURE AND WARSHALL'S ALGORITHM

Transitive Closure

In this section we consider a construction that has several interpretations and many important applications. Suppose that R is a relation on a set A and that R is not transitive. We will show that the transitive closure of R (see Section 4.7) is just the connectivity relation R^{∞}, defined in Section 4.3.

Theorem 1   Let R be a relation on a set A. Then R^{∞} is the transitive closure of R. ●

Proof   We recall that if a and b are in the set A, then $a \ R^{\infty} \ b$ if and only if there is a path in R from a to b. Now R^{∞} is certainly transitive since, if $a \ R^{\infty} \ b$ and $b \ R^{\infty} \ c$, the composition of the paths from a to b and from b to c forms a path from a to c in R, and so $a \ R^{\infty} \ c$. To show that R^{∞} is the smallest transitive relation containing R, we must show that if S is any transitive relation on A and $R \subseteq S$, then $R^{\infty} \subseteq S$. Theorem 1 of Section 4.4 tells us that if S is transitive, then $S^{n} \subseteq S$ for all n; that is, if a and b are connected by a path of length n, then $a \ S \ b$. It follows that $S^{\infty} = \bigcup_{n=1}^{\infty} S^{n} \subseteq S$. It is also true that if $R \subseteq S$, then $R^{\infty} \subseteq S^{\infty}$, since any path in R is also a path in S. Putting these facts together, we see that if $R \subseteq S$ and S is transitive on A, then $R^{\infty} \subseteq S^{\infty} \subseteq S$. This means that R^{∞} is the smallest of all transitive relations on A that contain R. ▼

We see that R^{∞} has several interpretations. From a geometric point of view, it is called the connectivity relation, since it specifies which vertices are connected (by paths) to other vertices. If we include the relation Δ (see Section 4.4), then $R^{\infty} \cup \Delta$ is the reachability relation R^{*} (see Section 4.3), which is frequently more useful. On the other hand, from the algebraic point of view, R^{∞} is the transitive closure of R, as we have shown in Theorem 1. In this form, it plays important roles in the theory of equivalence relations and in the theory of certain languages (see Section 10.1).

EXAMPLE 1   Let $A = \{1, 2, 3, 4\}$, and let $R = \{(1, 2), (2, 3), (3, 4), (2, 1)\}$. Find the transitive closure of R.

Solution

METHOD 1:   The digraph of R is shown in Figure 4.42. Since R^{∞} is the transitive closure, we can proceed geometrically by computing all paths. We see that from vertex 1 we have paths to vertices 2, 3, 4, and 1. Note that the path from 1 to 1 proceeds from 1 to 2 to 1. Thus we see that the ordered pairs $(1, 1)$, $(1, 2)$, $(1, 3)$, and $(1, 4)$ are in R^{∞}. Starting from vertex 2, we have paths to vertices 2, 1, 3, and 4, so the ordered pairs $(2, 1)$, $(2, 2)$, $(2, 3)$, and $(2, 4)$ are in R^{∞}. The only other path is from vertex 3 to vertex 4, so we have

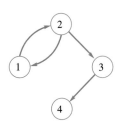

Figure 4.42

$$R^{\infty} = \{(1, 1), (1, 2), (1, 3), (1, 4), (2, 1), (2, 2), (2, 3), (2, 4), (3, 4)\}.$$

METHOD 2:   The matrix of R is

$$\mathbf{M}_{R} = \begin{bmatrix} 0 & 1 & 0 & 0 \\ 1 & 0 & 1 & 0 \\ 0 & 0 & 0 & 1 \\ 0 & 0 & 0 & 0 \end{bmatrix}.$$

We may proceed algebraically and compute the powers of $\mathbf{M}_R$. Thus

$$(\mathbf{M}_R)_{\odot}^2 = \begin{bmatrix} 1 & 0 & 1 & 0 \\ 0 & 1 & 0 & 1 \\ 0 & 0 & 0 & 0 \\ 0 & 0 & 0 & 0 \end{bmatrix}, \qquad (\mathbf{M}_R)_{\odot}^3 = \begin{bmatrix} 0 & 1 & 0 & 1 \\ 1 & 0 & 1 & 0 \\ 0 & 0 & 0 & 0 \\ 0 & 0 & 0 & 0 \end{bmatrix},$$

$$(\mathbf{M}_R)_{\odot}^4 = \begin{bmatrix} 1 & 0 & 1 & 0 \\ 0 & 1 & 0 & 1 \\ 0 & 0 & 0 & 0 \\ 0 & 0 & 0 & 0 \end{bmatrix}.$$

Continuing in this way, we can see that $(\mathbf{M}_R)_{\odot}^n$ equals $(\mathbf{M}_R)_{\odot}^2$ if n is even and equals $(\mathbf{M}_R)_{\odot}^3$ if n is odd and greater than 1. Thus

$$\mathbf{M}_{R^\infty} = \mathbf{M}_R \vee (\mathbf{M}_R)_{\odot}^2 \vee (\mathbf{M}_R)_{\odot}^3 = \begin{bmatrix} 1 & 1 & 1 & 1 \\ 1 & 1 & 1 & 1 \\ 0 & 0 & 0 & 1 \\ 0 & 0 & 0 & 0 \end{bmatrix}$$

and this gives the same relation as Method 1.   ■

In Example 1 we did not need to consider all powers R^n to obtain R^∞. This observation is true whenever the set A is finite, as we will now prove.

Theorem 2   Let A be a set with $|A| = n$, and let R be a relation on A. Then

$$R^\infty = R \cup R^2 \cup \cdots \cup R^n.$$

In other words, powers of R greater than n are not needed to compute R^∞.   ●

Proof   Let a and b be in A, and suppose that $a, x_1, x_2, \ldots, x_m, b$ is a path from a to b in R; that is, $(a, x_1), (x_1, x_2), \ldots, (x_m, b)$ are all in R. If x_i and x_j are the same vertex, say $i < j$, then the path can be divided into three sections. First, a path from a to x_i, then a path from x_i to x_j, and finally a path from x_j to b. The middle path is a cycle, since $x_i = x_j$, so we simply leave it out and put the remaining two paths together. This gives us a shorter path from a to b (see Figure 4.43).

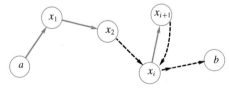

Figure 4.43

Now let $a, x_1, x_2, \ldots, x_k, b$ be the shortest path from a to b. If $a \neq b$, then all vertices $a, x_1, x_2, \ldots, x_k, b$ are distinct. Otherwise, the preceding discussion shows that we could find a shorter path. Thus the length of the path is at most $n - 1$ (since $|A| = n$). If $a = b$, then for similar reasons, the vertices $a, x_1, x_2, \ldots, x_k$ are distinct, so the length of the path is at most n. In other words, if $a \, R^\infty \, b$, then $a \, R^k \, b$, for some k, $1 \leq k \leq n$. Thus $R^\infty = R \cup R^2 \cup \cdots \cup R^n$.   ▼

The methods used to solve Example 1 each have certain difficulties. The graphical method is impractical for large sets and relations and is not systematic. The matrix method can be used in general and is systematic enough to be programmed for a computer, but it is inefficient and, for large matrices, can be prohibitively costly. Fortunately, a more efficient algorithm for computing transitive closure is available. It is known as Warshall's algorithm, after its creator, and we describe it next.

Warshall's Algorithm

Let R be a relation on a set $A = \{a_1, a_2, \ldots, a_n\}$. If $x_1, x_2, \ldots, x_m$ is a path in R, then any vertices other than x_1 and x_m are called **interior vertices** of the path. Now, for $1 \le k \le n$, we define a Boolean matrix $\mathbf{W}_k$ as follows. $\mathbf{W}_k$ has a 1 in position i, j if and only if there is a path from a_i to a_j in R whose interior vertices, if any, come from the set $\{a_1, a_2, \ldots, a_k\}$.

Since any vertex must come from the set $\{a_1, a_2, \ldots, a_n\}$, it follows that the matrix $\mathbf{W}_n$ has a 1 in position i, j if and only if some path in R connects a_i with a_j. In other words, $\mathbf{W}_n = \mathbf{M}_{R^\infty}$. If we define $\mathbf{W}_0$ to be $\mathbf{M}_R$, then we will have a sequence $\mathbf{W}_0, \mathbf{W}_1, \ldots, \mathbf{W}_n$ whose first term is $\mathbf{M}_R$ and whose last term is $\mathbf{M}_{R^\infty}$. We will show how to compute each matrix $\mathbf{W}_k$ from the previous matrix $\mathbf{W}_{k-1}$. Then we can begin with the matrix of R and proceed one step at a time until, in n steps, we reach the matrix of R^∞. This procedure is called **Warshall's algorithm**. The matrices $\mathbf{W}_k$ are different from the powers of the matrix $\mathbf{M}_R$, and this difference results in a considerable savings of steps in the computation of the transitive closure of R.

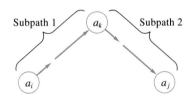

Subpath 1 a_k Subpath 2

a_i a_j

Figure 4.44

Suppose that $\mathbf{W}_k = \left[\, t_{ij} \,\right]$ and $\mathbf{W}_{k-1} = \left[\, s_{ij} \,\right]$. If $t_{ij} = 1$, then there must be a path from a_i to a_j whose interior vertices come from the set $\{a_1, a_2, \ldots, a_k\}$. If the vertex a_k is not an interior vertex of this path, then all interior vertices must actually come from the set $\{a_1, a_2, \ldots, a_{k-1}\}$, so $s_{ij} = 1$. If a_k is an interior vertex of the path, then the situation is as shown in Figure 4.44. As in the proof of Theorem 2, we may assume that all interior vertices are distinct. Thus a_k appears only once in the path, so all interior vertices of subpaths 1 and 2 must come from the set $\{a_1, a_2, \ldots, a_{k-1}\}$. This means that $s_{ik} = 1$ and $s_{kj} = 1$.

Thus $t_{ij} = 1$ if and only if either

$$(1) \quad s_{ij} = 1 \quad \text{or}$$
$$(2) \quad s_{ik} = 1 \quad \text{and} \quad s_{kj} = 1.$$

This is the basis for Warshall's algorithm. If $\mathbf{W}_{k-1}$ has a 1 in position i, j then, by (1), so will $\mathbf{W}_k$. By (2), a new 1 can be added in position i, j of $\mathbf{W}_k$ if and only if column k of $\mathbf{W}_{k-1}$ has a 1 in position i and row k of $\mathbf{W}_{k-1}$ has a 1 in position j. Thus we have the following procedure for computing $\mathbf{W}_k$ from $\mathbf{W}_{k-1}$.

STEP 1: First transfer to $\mathbf{W}_k$ all 1's in $\mathbf{W}_{k-1}$.

STEP 2: List the locations $p_1, p_2, \ldots$, in column k of $\mathbf{W}_{k-1}$, where the entry is 1, and the locations $q_1, q_2, \ldots$, in row k of $\mathbf{W}_{k-1}$, where the entry in 1.

STEP 3: Put 1's in all the positions p_i, q_j of $\mathbf{W}_k$ (if they are not already there).

EXAMPLE 2   Consider the relation R defined in Example 1. Then

$$\mathbf{W}_0 = \mathbf{M}_R = \begin{bmatrix} 0 & 1 & 0 & 0 \\ 1 & 0 & 1 & 0 \\ 0 & 0 & 0 & 1 \\ 0 & 0 & 0 & 0 \end{bmatrix}$$

and $n = 4$.

First we find $\mathbf{W}_1$ so that $k = 1$. $\mathbf{W}_0$ has 1's in location 2 of column 1 and location 2 of row 1. Thus $\mathbf{W}_1$ is just $\mathbf{W}_0$ with a new 1 in position 2, 2.

$$\mathbf{W}_1 = \begin{bmatrix} 0 & 1 & 0 & 0 \\ 1 & 1 & 1 & 0 \\ 0 & 0 & 0 & 1 \\ 0 & 0 & 0 & 0 \end{bmatrix}$$

Now we compute $\mathbf{W}_2$ so that $k = 2$. We must consult column 2 and row 2 of $\mathbf{W}_1$. Matrix $\mathbf{W}_1$ has 1's in locations 1 and 2 of column 2 and locations 1, 2, and 3 of row 2.

Thus, to obtain $\mathbf{W}_2$, we must put 1's in positions 1, 1; 1, 2; 1, 3; 2, 1; 2, 2; and 2, 3 of matrix $\mathbf{W}_1$ (if 1's are not already there). We see that

$$\mathbf{W}_2 = \begin{bmatrix} 1 & 1 & 1 & 0 \\ 1 & 1 & 1 & 0 \\ 0 & 0 & 0 & 1 \\ 0 & 0 & 0 & 0 \end{bmatrix}.$$

Proceeding, we see that column 3 of $\mathbf{W}_2$ has 1's in locations 1 and 2, and row 3 of $\mathbf{W}_2$ has a 1 in location 4. To obtain $\mathbf{W}_3$, we must put 1's in positions 1, 4 and 2, 4 of $\mathbf{W}_2$, so

$$\mathbf{W}_3 = \begin{bmatrix} 1 & 1 & 1 & 1 \\ 1 & 1 & 1 & 1 \\ 0 & 0 & 0 & 1 \\ 0 & 0 & 0 & 0 \end{bmatrix}.$$

Finally, $\mathbf{W}_3$ has 1's in locations 1, 2, 3 of column 4 and no 1's in row 4, so no new 1's are added and $\mathbf{M}_{R^\infty} = \mathbf{W}_4 = \mathbf{W}_3$. Thus we have obtained the same result as in Example 1.   ■

The procedure illustrated in Example 2 yields the following algorithm for computing the matrix, CLOSURE, of the transitive closure of a relation R represented by the $N \times N$ matrix MAT.

ALGORITHM WARSHALL
1.   CLOSURE ← MAT
2.   **FOR** $K = 1$ **THRU** N
    a.   **FOR** $I = 1$ **THRU** N
        1.   **FOR** $J = 1$ **THRU** N
            a.   CLOSURE $[I, J]$ ← CLOSURE $[I, J]$
                  $\vee$ (CLOSURE $[I, K] \wedge$ CLOSURE $[K, J]$)
END OF ALGORITHM WARSHALL

This algorithm was set up to proceed exactly as we have outlined previously. With some slight rearrangement of the steps, it can be made a little more efficient. If we think of the testing and assignment line as one step, then algorithm WARSHALL requires n^3 steps in all. The Boolean product of two $n \times n$ Boolean matrices $\mathbf{A}$ and $\mathbf{B}$ also requires n^3 steps, since we must compute n^2 entries, and each of these requires n comparisons. To compute all products $(\mathbf{M}_R)^2_{\odot}$, $(\mathbf{M}_R)^3_{\odot}, \ldots$, $(\mathbf{M}_R)^n_{\odot}$, we require $n^3(n-1)$ steps, since we will need $n-1$ matrix multiplications. The formula

$$\mathbf{M}_{R^\infty} = \mathbf{M}_R \vee (\mathbf{M}_R)^2_{\odot} \vee \cdots \vee (\mathbf{M}_R)^n_{\odot}, \tag{1}$$

if implemented directly, would require about n^4 steps without the final joins. Thus Warshall's algorithm is a significant improvement over direct computation of $\mathbf{M}_{R^\infty}$ using formula (1).

An interesting application of the transitive closure is to equivalence relations. We showed in Section 4.7 that if R and S are equivalence relations on a set A, then $R \cap S$ is also an equivalence relation on A. The relation $R \cap S$ is the largest equivalence relation contained in both R and S, since it is the largest subset of $A \times A$ contained in both R and S. We would like to know the smallest equivalence relation that contains both R and S. The natural candidate is $R \cup S$, but this relation is not necessarily transitive. The solution is given in the next theorem.

Theorem 3   If R and S are equivalence relations on a set A, then the smallest equivalence relation containing both R and S is $(R \cup S)^\infty$.   ●

Proof   Recall that Δ is the relation of equality on A and that a relation is reflexive if and only if it contains Δ. Then $\Delta \subseteq R$, $\Delta \subseteq S$ since both are reflexive, so $\Delta \subseteq R \cup S \subseteq (R \cup S)^\infty$, and $(R \cup S)^\infty$ is also reflexive.

Since R and S are symmetric, $R = R^{-1}$ and $S = S^{-1}$, so $(R \cup S)^{-1} = R^{-1} \cup S^{-1} = R \cup S$, and $R \cup S$ is also symmetric. Because of this, all paths in $R \cup S$ are "two-way streets," and it follows from the definitions that $(R \cup S)^\infty$ must also be symmetric. Since we already know that $(R \cup S)^\infty$ is transitive, it is an equivalence relation containing $R \cup S$. It is the smallest one, because no smaller set containing $R \cup S$ can be transitive, by definition of the transitive closure.   ▼

EXAMPLE 3   Let $A = \{1, 2, 3, 4, 5\}$, $R = \{(1, 1), (1, 2), (2, 1), (2, 2), (3, 3), (3, 4), (4, 3), (4, 4), (5, 5)\}$, and $S = \{(1, 1), (2, 2), (3, 3), (4, 4), (4, 5), (5, 4), (5, 5)\}$. The reader may verify that both R and S are equivalence relations. The partition A/R of A corresponding to R is $\{\{1, 2\}, \{3, 4\}, \{5\}\}$, and the partition A/S of A corresponding to S is $\{\{1\}, \{2\}, \{3\}, \{4, 5\}\}$. Find the smallest equivalence relation containing R and S, and compute the partition of A that it produces.

Solution   We have

$$\mathbf{M}_R = \begin{bmatrix} 1 & 1 & 0 & 0 & 0 \\ 1 & 1 & 0 & 0 & 0 \\ 0 & 0 & 1 & 1 & 0 \\ 0 & 0 & 1 & 1 & 0 \\ 0 & 0 & 0 & 0 & 1 \end{bmatrix} \quad \text{and} \quad \mathbf{M}_S = \begin{bmatrix} 1 & 0 & 0 & 0 & 0 \\ 0 & 1 & 0 & 0 & 0 \\ 0 & 0 & 1 & 0 & 0 \\ 0 & 0 & 0 & 1 & 1 \\ 0 & 0 & 0 & 1 & 1 \end{bmatrix},$$

so

$$\mathbf{M}_{R \cup S} = \mathbf{M}_R \vee \mathbf{M}_S = \begin{bmatrix} 1 & 1 & 0 & 0 & 0 \\ 1 & 1 & 0 & 0 & 0 \\ 0 & 0 & 1 & 1 & 0 \\ 0 & 0 & 1 & 1 & 1 \\ 0 & 0 & 0 & 1 & 1 \end{bmatrix}.$$

We now compute $\mathbf{M}_{(R \cup S)^\infty}$ by Warshall's algorithm. First, $\mathbf{W}_0 = \mathbf{M}_{R \cup S}$. We next compute $\mathbf{W}_1$, so $k = 1$. Since $\mathbf{W}_0$ has 1's in locations 1 and 2 of column 1 and in locations 1 and 2 of row 1, we find that no new 1's must be adjoined to $\mathbf{W}_1$. Thus

$$\mathbf{W}_1 = \mathbf{W}_0.$$

We now compute $\mathbf{W}_2$, so $k = 2$. Since $\mathbf{W}_1$ has 1's in locations 1 and 2 of column 2 and in locations 1 and 2 of row 2, we find that no new 1's must be added to $\mathbf{W}_1$. Thus

$$\mathbf{W}_2 = \mathbf{W}_1.$$

We next compute $\mathbf{W}_3$, so $k = 3$. Since $\mathbf{W}_2$ has 1's in locations 3 and 4 of column 3 and in locations 3 and 4 of row 3, we find that no new 1's must be added to $\mathbf{W}_2$. Thus

$$\mathbf{W}_3 = \mathbf{W}_2.$$

Things change when we now compute $\mathbf{W}_4$. Since $\mathbf{W}_3$ has 1's in locations 3, 4, and 5 of column 4 and in locations 3, 4, and 5 of row 4, we must add new 1's to $\mathbf{W}_3$ in positions 3, 5, and 5, 3. Thus

$$\mathbf{W}_4 = \begin{bmatrix} 1 & 1 & 0 & 0 & 0 \\ 1 & 1 & 0 & 0 & 0 \\ 0 & 0 & 1 & 1 & 1 \\ 0 & 0 & 1 & 1 & 1 \\ 0 & 0 & 1 & 1 & 1 \end{bmatrix}.$$

The reader may verify that $\mathbf{W}_5 = \mathbf{W}_4$ and thus

$$(R \cup S)^\infty = \{(1, 1), (1, 2), (2, 1), (2, 2), (3, 3), (3, 4), (3, 5), (4, 3),$$
$$(4, 4), (4, 5), (5, 3), (5, 4), (5, 5)\}.$$

The corresponding partition of A is then (verify) $\{\{1, 2\}, \{3, 4, 5\}\}$.   ■

4.8 Exercises

1. (a) Let $A = \{1, 2, 3\}$ and let $R = \{(1, 1), (1, 2), (2, 3),$ $(1, 3), (3, 1), (3, 2)\}$. Compute the matrix $\mathbf{M}_{R^\infty}$ of the transitive closure R by using the formula

$$\mathbf{M}_{R^\infty} = \mathbf{M}_R \vee (\mathbf{M}_R)^2_\odot \vee (\mathbf{M}_R)^3_\odot.$$

  (b) List the relation R^∞ whose matrix was computed in part (a).

2. For the relation R of Exercise 1, compute the transitive closure R^∞ by using Warshall's algorithm.

3. Let $A = \{a_1, a_2, a_3, a_4, a_5\}$ and let R be a relation on A whose matrix is

$$\mathbf{M}_R = \begin{bmatrix} 1 & 0 & 0 & 1 & 0 \\ 0 & 1 & 0 & 0 & 0 \\ 0 & 0 & 0 & 1 & 1 \\ 1 & 0 & 0 & 0 & 0 \\ 0 & 1 & 0 & 0 & 1 \end{bmatrix} = \mathbf{W}_0.$$

Compute $\mathbf{W}_1$, $\mathbf{W}_2$, and $\mathbf{W}_3$ as in Warshall's algorithm.

4. Find R^∞ for the relation in Exercise 3.

5. Prove that if R is reflexive and transitive, then $R^n = R$ for all n.

6. Let R be a relation on a set A, and let $S = R^2$. Prove that if $a, b \in A$, then $a\ S^\infty\ b$ if and only if there is a path in R from A to b having an even number of edges.

In Exercises 7 through 10, let $A = \{1, 2, 3, 4\}$. For the relation R whose matrix is given, find the matrix of the transitive closure by using Warshall's algorithm.

7. $\mathbf{M}_R = \begin{bmatrix} 1 & 0 & 0 & 1 \\ 1 & 1 & 0 & 0 \\ 0 & 0 & 1 & 0 \\ 0 & 0 & 0 & 1 \end{bmatrix}$ 8. $\mathbf{M}_R = \begin{bmatrix} 1 & 1 & 0 & 0 \\ 1 & 0 & 0 & 0 \\ 0 & 0 & 0 & 0 \\ 0 & 0 & 1 & 0 \end{bmatrix}$

9. $\mathbf{M}_R = \begin{bmatrix} 1 & 0 & 0 & 1 \\ 0 & 1 & 1 & 0 \\ 0 & 1 & 1 & 0 \\ 1 & 0 & 0 & 1 \end{bmatrix}$ 10. $\mathbf{M}_R = \begin{bmatrix} 0 & 0 & 0 & 1 \\ 1 & 0 & 0 & 1 \\ 0 & 1 & 0 & 1 \\ 0 & 0 & 1 & 0 \end{bmatrix}$

In Exercises 11 and 12, let $A = \{1, 2, 3, 4, 5\}$ and let R and S be the equivalence relations on A whose matrices are given. Compute the matrix of the smallest equivalence relation containing R and S, and list the elements of this relation.

11. $\mathbf{M}_R = \begin{bmatrix} 1 & 1 & 1 & 0 & 0 \\ 1 & 1 & 1 & 0 & 0 \\ 1 & 1 & 1 & 0 & 0 \\ 0 & 0 & 0 & 1 & 1 \\ 0 & 0 & 0 & 1 & 1 \end{bmatrix}$

$\mathbf{M}_S = \begin{bmatrix} 1 & 0 & 0 & 0 & 0 \\ 0 & 1 & 1 & 1 & 0 \\ 0 & 1 & 1 & 1 & 0 \\ 0 & 1 & 1 & 1 & 0 \\ 0 & 0 & 0 & 0 & 1 \end{bmatrix}$

12. $\mathbf{M}_R = \begin{bmatrix} 1 & 0 & 0 & 0 & 0 \\ 0 & 1 & 1 & 0 & 0 \\ 0 & 1 & 1 & 0 & 0 \\ 0 & 0 & 0 & 1 & 1 \\ 0 & 0 & 0 & 1 & 1 \end{bmatrix}$

$\mathbf{M}_S = \begin{bmatrix} 1 & 1 & 0 & 0 & 0 \\ 1 & 1 & 0 & 0 & 0 \\ 0 & 0 & 1 & 0 & 0 \\ 0 & 0 & 0 & 1 & 0 \\ 0 & 0 & 0 & 0 & 1 \end{bmatrix}$

13. Compute A/R, A/S, and the partition of A that corresponds to the equivalence relation found in Exercise 11.

14. Compute A/R, A/S, and the partition of A that corresponds to the equivalence relation found in Exercise 12.

15. Examine the results of Example 3 and Exercises 13 and 14. Based on these, give a procedure for producing $A/(R \cup S)^\infty$ from A/R and A/S. Explain why the procedure works.

16. Why is the procedure developed in Exercise 15 not a replacement for Warshall's algorithm?

17. Let $A = \{1, 2, 3, 4\}$ and let R and S be the relations on A described by

$$\mathbf{M}_R = \begin{bmatrix} 0 & 0 & 0 & 1 \\ 0 & 0 & 0 & 0 \\ 0 & 1 & 0 & 0 \\ 0 & 0 & 1 & 0 \end{bmatrix}$$

and

$$\mathbf{M}_S = \begin{bmatrix} 1 & 1 & 0 & 0 \\ 0 & 1 & 0 & 0 \\ 0 & 0 & 1 & 0 \\ 0 & 1 & 0 & 1 \end{bmatrix}.$$

Use Warshall's algorithm to compute the transitive closure of $R \cup S$.

18. Let $A = \{a, b, c, d, e\}$ and let R and S be the relations on A described by

$$\mathbf{M}_R = \begin{bmatrix} 1 & 0 & 1 & 0 & 1 \\ 0 & 0 & 0 & 1 & 0 \\ 1 & 0 & 0 & 0 & 0 \\ 0 & 0 & 1 & 1 & 0 \\ 1 & 0 & 1 & 0 & 0 \end{bmatrix}$$

and

$$\mathbf{M}_S = \begin{bmatrix} 0 & 1 & 0 & 1 & 0 \\ 1 & 1 & 0 & 0 & 1 \\ 1 & 1 & 1 & 0 & 0 \\ 0 & 1 & 0 & 0 & 0 \\ 0 & 1 & 0 & 1 & 0 \end{bmatrix}.$$

Use Warshall's algorithm to compute the transitive closure of $R \cup S$.

19. Outline the strategy of the proof of Theorem 1. What type of proof is it?

20. Outline the strategy of the proof of Theorem 2. What type of proof is it?

TIPS FOR PROOFS

Before beginning a proof, you should be able to restate the statement in your own words. Consider carefully what the statement says. For example, Theorem 1, Section 4.1, tells how to count the elements of $|A \times B|$. Thus to prove it you should try to apply one of the counting methods from Chapter 3.

Many statements about relations are statements about them as sets, and the techniques of Chapter 1 can be used to prove them. As an example, consider Section 4.2, Theorem 1. Remember that a very common way to show that two sets are equal is show that each is a subset of the other. We do have other representations for relations, too, and in some cases a proof based on matrix or digraph ideas may be clearer. Also, this chapter contains many facts about relational properties, operations on relations, and their interactions. These facts can form the basis of a proof at the name level rather than at the element level; see, for example, Section 4.7, Theorem 2.

Many definitions in this chapter are biconditional statements. A biconditional, p if and only if q, is generally proved in two parts: If p then q and if q then p. This is done for Lemma 1, Section 4.5. A frequently used grammatical structure is to introduce the second part of the proof with the word *conversely*. See Theorem 6, Section 4.7. Occasionally the proof of a biconditional is of the form $p \Leftrightarrow q \Leftrightarrow r \Leftrightarrow \cdots \Leftrightarrow t$ [Theorem 2(c), Section 4.7].

Checking whether a relation is an equivalence relation or has a certain relational property is the same as proving or disproving the statement R has the property $P(x)$. For this reason, you must work with generic elements or if R is small, check all cases.

Some exercises in this chapter ask you to analyze a proof and outline its strategy or identify its key points. These exercises should help you develop the habit of looking at all proofs for these features. Understanding how proofs you read are carried out will help you create proofs yourself. At this point you should be able to read a simple proof with understanding and recognize its structure.

KEY IDEAS FOR REVIEW

- $A \times B$ (product set or Cartesian product): $\{(a, b) \mid a \in A \text{ and } b \in B\}$
- $|A \times B| = |A| \, |B|$
- Partition or quotient set; see page 105
- Relation from A to B: subset of $A \times B$
- Domain and range of a relation: see page 109
- Relative sets $R(a)$, a in A, and $R(B)$, B a subset of A: see page 110
- Matrix of a relation: see page 111
- Digraph of a relation: pictorial representation of a relation: see page 112
- Path of length n from a to b in a relation R: finite sequence $a, x_1, x_2, \ldots, x_{n-1}, b$ such that $a \, R \, x_1, \, x_1 \, R \, x_2, \ldots, x_{n-1} \, R \, b$
- $x \, R^n \, y$ (R a relation on A): There is a path of length n from x to y in R
- $x \, R^\infty \, y$ (connectivity relation for R): Some path exists in R from x to y.
- Theorem: $\mathbf{M}_{R^n} = \mathbf{M}_R \odot \mathbf{M}_R \odot \cdots \odot \mathbf{M}_R$ (n factors)
- Properties of relations on a set A:

Reflexive	$(a, a) \in R$ for all $a \in A$
Irreflexive	$(a, a) \notin R$ for all $a \in A$
Symmetric	$(a, b) \in R$ implies that $(b, a) \in R$
Asymmetric	$(a, b) \in R$ implies that $(b, a) \notin R$
Antisymmetric	$(a, b) \in R$ and $(b, a) \in R$ imply that $a = b$
Transitive	$(a, b) \in R$ and $(b, c) \in R$ imply that $(a, c) \in R$

- Graph of a symmetric relation: see page 124
- Adjacent vertices: see page 125
- Equivalence relation: reflexive, symmetric, and transitive relation
- Equivalence relation determined by a partition: see page 129
- Linked-list computer representation of a relation: see pages 134–135
- $a \, \overline{R} \, b$ (complement of R): $a \, \overline{R} \, b$ if and only if $a \, \cancel{R} \, b$
- R^{-1}: $(x, y) \in R^{-1}$ if and only if $(y, x) \in R$
- $R \cup S$, $R \cap S$: see page 140
- $\mathbf{M}_{R \cap S} = \mathbf{M}_R \wedge \mathbf{M}_S$
- $\mathbf{M}_{R \cup S} = \mathbf{M}_R \vee \mathbf{M}_S$
- $\mathbf{M}_{R^{-1}} = (\mathbf{M}_R)^T$
- $\mathbf{M}_{\overline{R}} = \overline{\mathbf{M}_R}$
- If R and S are equivalence relations, so is $R \cap S$: see page 144
- $R \circ S$: see page 146
- $\mathbf{M}_{S \circ R} = \mathbf{M}_R \odot \mathbf{M}_S$: see page 147

- Theorem: R^∞ is the smallest transitive relation on A that contains R: see page 151
- Theorem: If $|A| = n$, $R^\infty = R \cup R^2 \cup \cdots \cup R^n$
- Warshall's algorithm: computes $\mathbf{M}_{R^\infty}$ efficiently;

see page 153
- Theorem: If R and S are equivalence relations on A, $(R \cup S)^\infty$ is the smallest equivalence relation on A containing both A and B.

CODING EXERCISES

For each of the following, write the requested program or subroutine in pseudocode (as described in Appendix A) or in a programming language that you know. Test your code either with a paper-and-pencil trace or with a computer run.

1. Write a program CROSS with input positive integers m and n and output the set $A \times B$ where $A = \{1, 2, 3, \ldots, m\}$ and $B = \{1, 2, 3, \ldots, n\}$.

2. (a) Write a subroutine that has as input the matrix of a relation and determines whether the relation is reflexive.

   (b) Write a subroutine that has as input the matrix of a relation and determines whether the relation is symmetric.

3. Write a program that has as input the matrix of a relation and determines whether the relation is an equivalence relation.

4. Let R and S be relations represented by matrices $\mathbf{M}_R$ and $\mathbf{M}_S$, respectively. Write a subroutine to produce the matrix of

   (a) $R \cup S$ (b) $R \cap S$ (c) $R \circ S$

5. Let R be a relation represented by the matrix $\mathbf{M}_R$. Write a subroutine to produce the matrix of

   (a) R^{-1} (b) $\overline{R}$

CHAPTER 4 SELF-TEST

1. Let $A = \{2, 5, 7\}$ and $B = \{x \mid x \in Z^+ \text{ and } x^3 < 100\}$.

   (a) What is $|A \times B|$?   (b) List $A \times B$.

2. Let A and B be subsets of the universal set U. Then $A \times B \subseteq U \times U$. Is $\overline{A \times B} = \overline{A} \times \overline{B}$? Justify your answer.

3. Give all two-element partitions of $\{a, b, c, d, e\}$.

4. Let $C = \{2, 8, 14, 18\}$. Define a relation on C by $x \, R \, y$ if and only if $x - y > 5$.

   (a) Draw the digraph of R.

   (b) Give $\mathbf{M}_R$.

5. Let $B = \{a, b, c, d\}$ and $R = \{(a, a), (a, b), (b, c), (c, d), (d, b)\}$.

   (a) Draw the digraphs of R and R^2.

   (b) Give $\mathbf{M}_R$ and $\mathbf{M}_{R^2}$.

   (c) Give $\mathbf{M}_{R^\infty}$.

6. Determine whether the relation R on the set A is reflexive, irreflexive, symmetric, asymmetric, antisymmetric, or transitive, if $A = Z^+$; $x \, R \, y$ if and only if $x \le 3y$.

7. Let $D = \{1, 2, 3, 4, 5, 6\}$ and R be the relation on D whose matrix is

$$\mathbf{M}_R = \begin{bmatrix} 1 & 0 & 0 & 0 & 0 & 1 \\ 0 & 1 & 1 & 0 & 1 & 0 \\ 0 & 0 & 0 & 1 & 0 & 1 \\ 1 & 0 & 0 & 1 & 0 & 1 \\ 0 & 0 & 1 & 0 & 1 & 0 \\ 0 & 1 & 0 & 0 & 1 & 1 \end{bmatrix}.$$

Determine whether R is reflexive, irreflexive, symmetric, asymmetric, antisymmetric, or transitive.

8. Suppose R is a relation on a set A and that R is asymmetric. Can R also be antisymmetric? Must R be antisymmetric? Explain your answers.

9. Let $B = \{1, 2, 3, 4, 5\}$, $A = B \times B$, and define R on A as follows: $(u, v) \, R \, (x, y)$ if and only if $u - v = x - y$.

   (a) Prove that R is an equivalence relation.

   (b) Find $[(2, 3)]$.

   (c) Compute A/R.

10. The following arrays describe a relation R on the set $A = \{1, 2, 3, 4\}$. Give the matrix of R.

$$\text{VERT} = [5, 3, 1, 8]$$
$$\text{TAIL} = [3, 3, 2, 2, 1, 1, 4, 4]$$
$$\text{HEAD} = [1, 4, 1, 3, 2, 3, 4, 2]$$
$$\text{NEXT} = [2, 0, 4, 0, 6, 0, 0, 7]$$

11. Let R and S be relations on $\{a, b, c, d, e\}$ where

$R = \{(a, b), (a, c), (b, c), (c, e), (e, a), (a, a), (d, c)\}$
and $S = \{(a, a), (a, b), (b, a), (c, c), (c, d), (d, e), (b, e), (e, d)\}$.

(a) Give R^{-1}.   (b) Compute $R \circ S$.

12. Let $R = \{(1, 4), (2, 1), (2, 5), (2, 4), (4, 3), (5, 3), (3, 2)\}$. Use Warshall's algorithm to find the matrix of the connectivity relation based on R.

5

FUNCTIONS

Prerequisites: *Chapter 4*

> In this chapter we focus our attention on a special type of relation, a function, that plays an important role in mathematics, computer science, and many applications. We also define some functions used in computer science and examine the growth of functions.

5.1 FUNCTIONS

In this section we define the notion of a function, a special type of relation. We study its basic properties and then discuss several special types of functions. A number of important applications of functions will occur in later sections of the book, so it is essential to get a good grasp of the material in this section.

Let A and B be nonempty sets. A **function** f from A to B, which is denoted $f: A \rightarrow B$, is a relation from A to B such that for all $a \in \text{Dom}(f)$, $f(a)$, the f-relative set of a, contains just one element of B. Naturally, if a is not in $\text{Dom}(f)$, then $f(a) = \varnothing$. If $f(a) = \{b\}$, it is traditional to identify the set $\{b\}$ with the element b and write $f(a) = b$. We will follow this custom, since no confusion results. The relation f can then be described as the set of pairs $\{(a, f(a)) \mid a \in \text{Dom}(f)\}$. Functions are also called **mappings** or **transformations**, since they can be geometrically viewed as rules that assign to each element $a \in A$ the unique element $f(a) \in B$ (see Figure 5.1). The element a is called an **argument** of the function f, and $f(a)$ is called the **value** of the function for the argument a and is also referred to as the **image** of a under f. Figure 5.1 is a schematic or pictorial display of our definition of a function, and we will use several other similar diagrams. They should not be confused with the digraph of the relation f, which we will not generally display.

Figure 5.1

EXAMPLE 1

Let $A = \{1, 2, 3, 4\}$ and $B = \{a, b, c, d\}$, and let

$$f = \{(1, a), (2, a), (3, d), (4, c)\}.$$

Here we have

$$f(1) = a$$
$$f(2) = a$$
$$f(3) = d$$
$$f(4) = c.$$

Since each set $f(n)$ is a single value, f is a function.

Note that the element $a \in B$ appears as the second element of two different ordered pairs in f. This does not conflict with the definition of a function. Thus a function may take the same value at two different elements of A.   ∎

EXAMPLE 2

Let $A = \{1, 2, 3\}$ and $B = \{x, y, z\}$. Consider the relations

$$R = \{(1, x), (2, x)\} \quad \text{and} \quad S = \{(1, x), (1, y), (2, z), (3, y)\}.$$

The relation S is not a function since $S(1) = \{x, y\}$. The relation R is a function with $\text{Dom}(R) = \{1, 2\}$ and $\text{Ran}(R) = \{x\}$.   ∎

EXAMPLE 3

Let P be a computer program that accepts an integer as input and produces an integer as output. Let $A = B = Z$. Then P determines a relation f_P defined as follows: $(m, n) \in f_P$ means that n is the output produced by program P when the input is m.

It is clear that f_P is a function, since any particular input corresponds to a unique output. (We assume that computer results are reproducible; that is, they are the same each time the program is run.)   ∎

Example 3 can be generalized to a program with any set A of possible inputs and set B of corresponding outputs. In general, therefore, we may think of functions as **input-output** relations.

EXAMPLE 4

Let $A = \mathbb{R}$ be the set of real numbers, and let $p(x) = a_0 + a_1x + \cdots + a_nx^n$ be a real polynomial. Then p may be viewed as a relation on $\mathbb{R}$. For each r in $\mathbb{R}$ we determine the relative set $p(r)$ by substituting r into the polynomial. Then, since all relative sets $p(r)$ are known, the relation p is determined. Since a unique value is produced by this substitution, the relation p is actually a function.   ∎

If the formula defining the function does not make sense for all elements of A, then the domain of the function is taken to be the set of elements for A for which the formula does make sense.

In elementary mathematics, the *formula* (in the case of Example 4, the polynomial) is usually confused with the *function* it produces. This is not harmful, unless the student comes to expect a formula for every type of function.

Suppose that, in the preceding construction, we used a formula that produced more than one element in $p(x)$, for example, $p(x) = \pm\sqrt{x}$. Then the resulting relation would not be a function. For this reason, in older texts, relations were sometimes called multiple-valued functions.

EXAMPLE 5

Figure 5.2

A **labeled digraph** is a digraph in which the vertices or the edges (or both) are labeled with information from a set. If V is the set of vertices and L is the set of labels of a labeled digraph, then the labeling of V can be specified to be a function $f: V \rightarrow L$, where, for each $v \in V$, $f(v)$ is the label we wish to attach to v. Similarly, we can define a labeling of the edges E as a function $g: E \rightarrow L$, where, for each $e \in E$, $g(e)$ is the label we wish to attach to e. An example of a labeled digraph is a map on which the vertices are labeled with the names of cities and the edges are labeled with the distances or travel times between the cities. Another example is a flow chart of a program in which the vertices are labeled with the steps that are to be performed at that point in the program; the edges indicate the flow from one part of the program to another part. Figure 5.2 shows an example of a labeled digraph. ∎

EXAMPLE 6

Let $A = B = Z$ and let $f: A \rightarrow B$ be defined by

$$f(a) = a + 1, \quad \text{for } a \in A.$$

Here, as in Example 4, f is defined by giving a formula for the values $f(a)$. ∎

EXAMPLE 7

Let $A = Z$ and let $B = \{0, 1\}$. Let $f: A \rightarrow B$ be found by

$$f(a) = \begin{cases} 0 & \text{if } a \text{ is even} \\ 1 & \text{if } a \text{ is odd.} \end{cases}$$

Then f is a function, since each set $f(a)$ consists of a single element. Unlike the situation in Examples 4 and 6, the elements $f(a)$ are not specified through an algebraic formula. Instead, a verbal description is given. ∎

EXAMPLE 8

Let A be an arbitrary nonempty set. The **identity function on A**, denoted by 1_A, is defined by $1_A(a) = a$. ∎

The reader may notice that 1_A is the relation we previously called Δ (see Section 4.4), which stands for the diagonal subset of $A \times A$. In the context of functions, the notation 1_A is preferred, since it emphasizes the input-output or functional nature of the relation. Clearly, if $A_1 \subseteq A$, then $1_A(A_1) = A_1$.

Suppose that $f: A \rightarrow B$ and $g: B \rightarrow C$ are functions. Then the composition of f and g, $g \circ f$ (see Section 4.7), is a relation. Let $a \in \text{Dom}(g \circ f)$. Then, by Theorem 6 of Section 4.7, $(g \circ f)(a) = g(f(a))$. Since f and g are functions, $f(a)$ consists of a single element $b \in B$, so $g(f(a)) = g(b)$. Since g is also a function, $g(b)$ contains just one element of C. Thus each set $(g \circ f)(a)$, for a in $\text{Dom}(g \circ f)$, contains just one element of C, so $g \circ f$ is a function. This is illustrated diagrammatically in Figure 5.3.

EXAMPLE 9

Let $A = B = Z$, and C be the set of even integers. Let $f: A \rightarrow B$ and $g: B \rightarrow C$ be defined by

$$f(a) = a + 1$$
$$g(b) = 2b.$$

Find $g \circ f$.

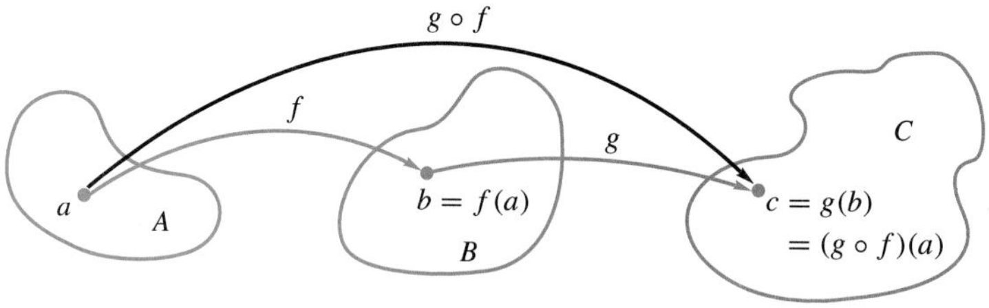

Figure 5.3

Solution   We have

$$(g \circ f)(a) = g(f(a)) = g(a + 1) = 2(a + 1).$$

Thus, if f and g are functions specified by giving formulas, then so is $g \circ f$ and the formula for $g \circ f$ is produced by substituting the formula for f into the formula for g.   ∎

Special Types of Functions

Let f be a function from A to B. Then we say that f is **everywhere defined** if $\text{Dom}(f) = A$. We say that f is **onto** if $\text{Ran}(f) = B$. Finally, we say that f is **one to one** if we cannot have $f(a) = f(a')$ for two distinct elements a and a' of A. The definition of one to one may be restated in the following equivalent form:

$$\text{If } f(a) = f(a'), \quad \text{then } a = a'.$$

The latter form is often easier to verify in particular examples.

EXAMPLE 10   Consider the function f defined in Example 1. Since $\text{Dom}(f) = A$, f is everywhere defined. On the other hand, $\text{Ran}(f) = \{a, c, d\} \neq B$; therefore, f is not onto. Since

$$f(1) = f(2) = a,$$

we can conclude that f is not one to one.   ∎

EXAMPLE 11   Consider the function f defined in Example 6. Which of the special properties, if any, does f possess?

Solution   Since the formula defining f makes sense for all integers, $\text{Dom}(f) = Z = A$, and so f is everywhere defined.
        Suppose that

$$f(a) = f(a')$$

for a and a' in A. Then

$$a + 1 = a' + 1$$

so

$$a = a'.$$

Hence f is one to one.

To see if f is onto, let b be an arbitrary element of B. Can we find an element $a \in A$ such that $f(a) = b$? Since

$$f(a) = a + 1,$$

we need an element a in A such that

$$a + 1 = b.$$

Of course,

$$a = b - 1$$

will satisfy the desired equation since $b - 1$ is in A. Hence $\text{Ran}(f) = B$; therefore, f is onto. ∎

EXAMPLE 12 Let $A = \{a_1, a_2, a_3\}$, $B = \{b_1, b_2, b_3\}$, $C = \{c_1, c_2\}$, and $D = \{d_1, d_2, d_3, d_4\}$. Consider the following four functions, from A to B, A to D, B to C, and D to B, respectively.

(a) $f_1 = \{(a_1, b_2), (a_2, b_3), (a_3, b_1)\}$
(b) $f_2 = \{(a_1, d_2), (a_2, d_1), (a_3, d_4)\}$
(c) $f_3 = \{(b_1, c_2), (b_2, c_2), (b_3, c_1)\}$
(d) $f_4 = \{(d_1, b_1), (d_2, b_2), (d_3, b_1)\}$

Determine whether each function is one to one, whether each function is onto, and whether each function is everywhere defined.

Solution

(a) f_1 is everywhere defined, one to one, and onto.
(b) f_2 is everywhere defined and one to one, but not onto.
(c) f_3 is everywhere defined and onto, but is not one to one.
(d) f_4 is not everywhere defined, not one to one, and not onto. ∎

If $f: A \to B$ is a one-to-one function, then f associates to each element a of $\text{Dom}(f)$ an element $b = f(a)$ of $\text{Ran}(f)$. Every b in $\text{Ran}(f)$ is matched, in this way, with one and only one element of $\text{Dom}(f)$. For this reason, such an f is often called a **bijection** between $\text{Dom}(f)$ and $\text{Ran}(f)$. If f is also everywhere defined and onto, then f is called a **one-to-one correspondence between A and B**.

EXAMPLE 13 Let $\mathcal{R}$ be the set of all equivalence relations on a given set A, and let Π be the set of all partitions on A. Then we can define a function $f: \mathcal{R} \to \Pi$ as follows. For each equivalence relation R on A, let $f(R) = A/R$, the partition of A that corresponds to R. The discussion in Section 4.5 shows that f is a one-to-one correspondence between $\mathcal{R}$ and Π. ∎

Invertible Functions

A function $f: A \to B$ is said to be **invertible** if its inverse relation, f^{-1}, is also a function. The next example shows that a function is not necessarily invertible.

EXAMPLE 14   Let f be the function of Example 1. Then

$$f^{-1} = \{(a, 1), (a, 2), (d, 3), (c, 4)\}.$$

We see that f^{-1} is not a function, since $f^{-1}(a) = \{1, 2\}$.   ∎

The following theorem is frequently used.

Theorem 1   Let $f: A \rightarrow B$ be a function.

(a) Then f^{-1} is a function from B to A if and only if f is one to one.

If f^{-1} is a function, then

(b) the function f^{-1} is also one to one.

(c) f^{-1} is everywhere defined if and only if f is onto.

(d) f^{-1} is onto if and only if f is everywhere defined.   ●

Proof

(a) We prove the following equivalent statement.

f^{-1} is not a function if and only if f is not one to one.

Suppose first that f^{-1} is not a function. Then, for some b in B, $f^{-1}(b)$ must contain at least two distinct elements, a_1 and a_2. Then $f(a_1) = b = f(a_2)$, so f is not one to one.

Conversely, suppose that f is not one to one. Then $f(a_1) = f(a_2) = b$ for two distinct elements a_1 and a_2 of A. Thus $f^{-1}(b)$ contains both a_1 and a_2, so f^{-1} cannot be a function.

(b) Since $(f^{-1})^{-1}$ is the function f, part (a) shows that f^{-1} is one to one.

(c) Recall that $\text{Dom}(f^{-1}) = \text{Ran}(f)$. Thus $B = \text{Dom}(f^{-1})$ if and only if $B = \text{Ran}(f)$. In other words, f^{-1} is everywhere defined if and only if f is onto.

(d) Since $\text{Ran}(f^{-1}) = \text{Dom}(f)$, $A = \text{Dom}(f)$ if and only if $A = \text{Ran}(f^{-1})$. That is, f is everywhere defined if and only if f^{-1} is onto.   ▼

As an immediate consequence of Theorem 1, we see that if f is a one-to-one correspondence between A and B, then f^{-1} is a one-to-one correspondence between B and A. Note also that if $f: A \rightarrow B$ is a one-to-one function, then the equation $b = f(a)$ is equivalent to $a = f^{-1}(b)$.

EXAMPLE 15   Consider the function f defined in Example 6. Since it is everywhere defined, one to one, and onto, f is a one-to-one correspondence between A and B. Thus f is invertible, and f^{-1} is a one-to-one correspondence between B and A.   ∎

EXAMPLE 16   Let $\mathbb{R}$ be the set of real numbers, and let $f: \mathbb{R} \rightarrow \mathbb{R}$ be defined by $f(x) = x^2$. Is f invertible?

Solution   We must determine whether f is one to one. Since

$$f(2) = f(-2) = 4,$$

we conclude that f is not one to one. Hence f is not invertible.   ∎

There are some useful results concerning the composition of functions. We summarize these in the following theorem.

Theorem 2   Let $f: A \to B$ be any function. Then

(a) $1_B \circ f = f$.

(b) $f \circ 1_A = f$.

If f is a one-to-one correspondence between A and B, then

(c) $f^{-1} \circ f = 1_A$.

(d) $f \circ f^{-1} = 1_B$.   ●

Proof

(a) $(1_B \circ f)(a) = 1_B(f(a)) = f(a)$, for all a in Dom(f). Thus, by Theorem 2 of Section 4.2, $1_B \circ f = f$.

(b) $(f \circ 1_A)(a) = f(1_A(a)) = f(a)$, for all a in Dom(f), so $f \circ 1_A = f$.

Suppose now that f is a one-to-one correspondence between A and B. As we pointed out, the equation $b = f(a)$ is equivalent to the equation $a = f^{-1}(b)$. Since f and f^{-1} are both everywhere defined and onto, this means that, for all a in A and b in B, $f(f^{-1}(b)) = b$ and $f^{-1}(f(a)) = a$. Then

(c) For all a in A, $1_A(a) = a = f^{-1}(f(a)) = (f^{-1} \circ f)(a)$. Thus $1_A = f^{-1} \circ f$.

(d) For all b in B, $1_B(b) = b = f(f^{-1}(b)) = (f \circ f^{-1})(b)$. Thus $1_B = f \circ f^{-1}$.   ▼

Theorem 3   (a) Let $f: A \to B$ and $g: B \to A$ be functions such that $g \circ f = 1_A$ and $f \circ g = 1_B$. Then f is a one-to-one correspondence between A and B, g is a one-to-one correspondence between B and A, and each is the inverse of the other.

(b) Let $f: A \to B$ and $g: B \to C$ be invertible. Then $g \circ f$ is invertible, and $(g \circ f)^{-1} = f^{-1} \circ g^{-1}$.   ●

Proof

(a) The assumptions mean that

$$g(f(a)) = a \quad \text{and} \quad f(g(b)) = b, \quad \text{for all } a \text{ in } A \text{ and } b \text{ in } B.$$

This shows in particular that Ran(f) $= B$ and Ran(g) $= A$, so each function is onto. If $f(a_1) = f(a_2)$, then $a_1 = g(f(a_1)) = g(f(a_2)) = a_2$. Thus f is one to one. In a similar way, we see that g is one to one, so both f and g are invertible. Note that f^{-1} is everywhere defined since Dom(f^{-1}) $=$ Ran(f) $= B$. Now, if b is any element in B,

$$f^{-1}(b) = f^{-1}(f(g(b))) = (f^{-1} \circ f)g(b)) = 1_A(g(b)) = g(b).$$

Thus $g = f^{-1}$, so also $f = (f^{-1})^{-1} = g^{-1}$. Then, since g and f are onto, f^{-1} and g^{-1} are onto, so f and g must be everywhere defined. This proves all parts of part (a).

(b) We know that $(g \circ f)^{-1} = f^{-1} \circ g^{-1}$, since this is true for any two relations. Since g^{-1} and f^{-1} are functions by assumption, so is their composition, and then $(g \circ f)^{-1}$ is a function. Thus $g \circ f$ is invertible.   ▼

EXAMPLE 17   Let $A = B = \mathbb{R}$, the set of real numbers. Let $f : A \to B$ be given by the formula $f(x) = 2x^3 - 1$ and let $g : B \to A$ be given by

$$g(y) = \sqrt[3]{\tfrac{1}{2}y + \tfrac{1}{2}}.$$

Show that f is a bijection between A and B and g is a bijection between B and A.

Solution   Let $x \in A$ and $y = f(x) = 2x^3 - 1$. Then $\tfrac{1}{2}(y+1) = x^3$; therefore,

$$x = \sqrt[3]{\tfrac{1}{2}y + \tfrac{1}{2}} = g(y) = g(f(x)) = (g \circ f)(x).$$

Thus $g \circ f = 1_A$. Similarly, $f \circ g = 1_B$, so by Theorem 3(a) both f and g are bijections.   ■

As Example 17 shows, it is often easier to show that a function, such as f, is one to one and onto by constructing an inverse instead of proceeding directly.

Finally, we discuss briefly some special results that hold when A and B are finite sets. Let $A = \{a_1, \ldots, a_n\}$ and $B = \{b_1, \ldots, b_n\}$, and let f be a function from A to B that is everywhere defined. If f is one to one, then $f(a_1), f(a_2), \ldots, f(a_n)$ are n distinct elements of B. Thus we must have all of B, so f is also onto. On the other hand, if f is onto, then $f(a_1), \ldots, f(a_n)$ form the entire set B, so they must all be different. Hence f is also one to one. We have therefore shown the following:

Theorem 4   Let A and B be two finite sets with the same number of elements, and let $f : A \to B$ be an everywhere defined function.

(a) If f is one to one, then f is onto.

(b) If f is onto, then f is one to one.   ●

Thus for finite sets A and B with the same number of elements, and particularly if $A = B$, we need only prove that a function is one to one *or* onto to show that it is a bijection. This is an application of the pigeonhole principle.

5.1 Exercises

1. Let $A = \{a, b, c, d\}$ and $B = \{1, 2, 3\}$. Determine whether the relation R from A to B is a function. If it is a function, give its range.

(a) $R = \{(a, 1), (b, 2), (c, 1), (d, 2)\}$

(b) $R = \{(a, 1), (b, 2), (a, 2), (c, 1), (d, 2)\}$

2. Let $A = \{a, b, c, d\}$ and $B = \{1, 2, 3\}$. Determine whether the relation R from A to B is a function. If it is a function, give its range.

(a) $R = \{(a, 3), (b, 2), (c, 1)\}$

(b) $R = \{(a, 1), (b, 1), (c, 1), (d, 1)\}$

3. Determine whether the relation R from A to B is a function.

   A = the set of all recipients of Medicare in the United States,

   $B = \{x \mid x \text{ is a nine-digit number}\}$,

   $a \mathrel{R} b$ if b is a's Social Security number.

4. Determine whether the relation R from A to B is a function.

   A = a set of people in the United States,

   $B = \{x \mid x \text{ is a nine-digit number}\}$,

   $a \mathrel{R} b$ if b is a's passport number.

In Exercises 5 through 8, verify that the formula yields a function from A to B.

5. $A = B = Z$; $f(a) = a^2$

6. $A = B = \mathbb{R}$; $f(a) = e^a$

7. $A = \mathbb{R}$, $B = \{0, 1\}$; let Z be the set of integers and note that $Z \subseteq \mathbb{R}$. Then for any real number a, let

$$f(a) = \begin{cases} 0 & \text{if } a \notin Z \\ 1 & \text{if } a \in Z. \end{cases}$$

8. $A = \mathbb{R}$, $B = Z$; $f(a) =$ the greatest integer less than or equal to a.

9. Let $A = B = C = \mathbb{R}$, and let $f: A \rightarrow B$, $g: B \rightarrow C$ be defined by $f(a) = a - 1$ and $g(b) = b^2$. Find

   (a) $(f \circ g)(2)$ (b) $(g \circ f)(2)$

   (c) $(g \circ f)(x)$ (d) $(f \circ g)(x)$

   (e) $(f \circ f)(y)$ (f) $(g \circ g)(y)$

10. Let $A = B = C = \mathbb{R}$, and let $f: A \rightarrow B$, $g: B \rightarrow C$ be defined by $f(a) = a + 1$ and $g(b) = b^2 + 2$. Find

   (a) $(g \circ f)(-2)$ (b) $(f \circ g)(-2)$

   (c) $(g \circ f)(x)$ (d) $(f \circ g)(x)$

   (e) $(f \circ f)(y)$ (f) $(g \circ g)(y)$

11. In each part, sets A and B and a function from A to B are given. Determine whether the function is one to one or onto (or both or neither).

   (a) $A = \{1, 2, 3, 4\} = B$;
      $f = \{(1, 1), (2, 3), (3, 4), (4, 2)\}$

   (b) $A = \{1, 2, 3\}$; $B = \{a, b, c, d\}$;
      $f = \{(1, a), (2, a), (3, c)\}$

12. In each part, sets A and B and a function from A to B are given. Determine whether the function is one to one or onto (or both or neither).

   (a) $A = \{\frac{1}{2}, \frac{1}{3}, \frac{1}{4}\}$; $B = \{x, y, z, w\}$;
      $f = \{(\frac{1}{2}, x), (\frac{1}{4}, y), (\frac{1}{3}, w)\}$

   (b) $A = \{1.1, 7, 0.06\}$; $B = \{p, q\}$;
      $f = \{(1.1, p), (7, q), (0.06, p)\}$

13. In each part, sets A and B and a function from A to B are given. Determine whether the function is one to one or onto (or both or neither).

   (a) $A = B = Z$; $f(a) = a - 1$

   (b) $A = \mathbb{R}$, $B = \{x \mid x \text{ is real and } x \geq 0\}$; $f(a) = |a|$

14. In each part, sets A and B and a function from A to B are given. Determine whether the function is one to one or onto (or both or neither).

   (a) $A = \mathbb{R} \times \mathbb{R}$, $B = \mathbb{R}$; $f((a, b)) = a$

   (b) Let $S = \{1, 2, 3\}$, $T = \{a, b\}$. Let $A = B = S \times T$ and let f be defined by $f(n, a) = (n, b), n = 1, 2, 3$, and $f(n, b) = (1, a), n = 1, 2, 3$.

15. In each part, sets A and B and a function from A to B are given. Determine whether the function is one to one or onto (or both or neither).

   (a) $A = B = \mathbb{R} \times \mathbb{R}$; $f((a, b)) = (a + b, a - b)$

   (b) $A = \mathbb{R}$, $B = \{x \mid x \text{ is real and } x \geq 0\}$; $f(a) = a^2$

16. Explain why Theorem 1(a) is equivalent to "f^{-1} is not a function if and only if f is not one to one".

17. Let $f: A \rightarrow B$ and $g: B \rightarrow A$. Verify that $g = f^{-1}$.

   (a) $A = B = \mathbb{R}$; $f(a) = \frac{a+1}{2}$, $g(b) = 2b - 1$

   (b) $A = \{x \mid x \text{ is real and } x \geq 0\}$; $B = \{y \mid y \text{ is real and } y \geq -1\}$; $f(a) = a^2 - 1$, $g(b) = \sqrt{b + 1}$

18. Let $f: A \rightarrow B$ and $g: B \rightarrow A$. Verify that $g = f^{-1}$.

   (a) $A = B = P(S)$, where S is a set. If $X \in P(S)$, let $f(X) = \overline{X} = g(X)$.

   (b) $A = B = \{1, 2, 3, 4\}$;
      $f = \{(1, 4), (2, 1), (3, 2), (4, 3)\}$;
      $g = \{(1, 2), (2, 3), (3, 4), (4, 1)\}$

19. Let f be a function from A to B. Find f^{-1}.

   (a) $A = \{x \mid x \text{ is real and } x \geq -1\}$; $B = \{x \mid x \text{ is real and } x \geq 0\}$; $f(a) = \sqrt{a + 1}$

   (b) $A = B = \mathbb{R}$; $f(a) = a^3 + 1$

20. Let f be a function from A to B. Find f^{-1}.

   (a) $A = B = \mathbb{R}$; $f(a) = \frac{2a-1}{3}$

   (b) $A = B = \{1, 2, 3, 4, 5\}$;
      $f = \{(1, 3), (2, 2), (3, 4), (4, 5), (5, 1)\}$

In Exercises 21 and 22, let f be a function from A = {1, 2, 3, 4} to B = {a, b, c, d}. Determine whether f^{-1} is a function.

21. $f = \{(1, a), (2, a), (3, c), (4, d)\}$

22. $f = \{(1, a), (2, c), (3, b), (4, d)\}$

23. Let $A = B = C = R$ and consider the functions $f: A \rightarrow B$ and $g: B \rightarrow C$ defined by $f(a) = 2a + 1$, $g(b) = b/3$. Verify Theorem 3(b): $(g \circ f)^{-1} = f^{-1} \circ g^{-1}$.

24. If a set A has n elements, how many functions are there from A to A?

25. If a set A has n elements, how many bijections are there from A to A?

26. If A has m elements and B has n elements, how many functions are there from A to B?

27. Complete the following proof.

If $f: A \rightarrow B$ and $g: B \rightarrow C$ are one-to-one functions, then $g \circ f$ is one to one.

Proof: Let $a_1, a_2 \in A$. Suppose $(g \circ f)(a_1) = (g \circ f)(a_2)$. Then $g(f(a_1)) = g(f(a_2))$ and $f(a_1) = f(a_2)$, because _____. Thus $a_1 = a_2$, because _____. Hence $g \circ f$ is one to one.

28. Complete the following proof.

If $f: A \rightarrow B$ and $g: B \rightarrow C$ are onto functions, then $g \circ f$ is onto.

Proof: Choose $x \in$ _____. Then there exists $y \in$ _____ such that $g(y) = x$. (Why?) Then there exists $z \in$ _____ such that $f(z) = y$ (why?) and $(g \circ f)(z) = x$. Hence, $g \circ f$ is onto.

29. Let $f: A \rightarrow B$ and $g: B \rightarrow C$ be functions. Show that if $g \circ f$ is one to one, then f is one to one.

30. Let $f: A \rightarrow B$ and $g: B \rightarrow C$ be functions. Show that if $g \circ f$ is onto, then g is onto.

31. Let A be a set, and let $f: A \rightarrow A$ be a bijection. For any integer $k \geq 1$, let $f^k = f \circ f \circ \cdots \circ f$ (k factors), and let $f^{-k} = f^{-1} \circ f^{-1} \circ \cdots \circ f^{-1}$ (k factors). Define f^0 to be 1_A. Then f^n is defined for all $n \in Z$. For any $a \in A$, let $O(a, f) = \{f^n(a) \mid n \in Z\}$. Prove that if $a_1, a_2 \in A$, and $O(a_1, f) \cap O(a_2, f) \neq \varnothing$, then $O(a_1, f) = O(a_2, f)$.

32. Let $f: A \rightarrow B$ be a function with finite domain and range. Suppose that $|\text{Dom}(f)| = n$ and $|\text{Ran}(f)| = m$. Prove that

(a) If f is one to one, then $m = n$.

(b) If f is not one to one, then $m < n$.

33. Let $|A| = |B| = n$ and let $f: A \rightarrow B$ be an everywhere defined function. Prove that the following three statements are equivalent.

(a) f is one to one.

(b) f is onto.

(c) f is a one-to-one correspondence (that is, f is one to one and onto).

34. Give a one-to-one correspondence between Z^+, the set of positive integers, and $A = \{x \mid x$ is a positive even integer$\}$.

35. Give a one-to-one correspondence between Z^+, the set of positive integers, and $A = \{x \mid x$ is a positive odd integer$\}$.

36. Based on Exercises 34 and 35, does $|Z^+| = |A| = |B|$? Justify your conclusion.

5.2 FUNCTIONS FOR COMPUTER SCIENCE

In previous chapters, we introduced on an informal basis some functions commonly used in computer science applications. In this section we review these and define some others.

EXAMPLE 1

Let A be a subset of the universal set $U = \{u_1, u_2, u_3, \ldots, u_n\}$. The **characteristic function of A** is defined as a function from U to $\{0, 1\}$ by the following:

$$f_A(u_i) = \begin{cases} 1 & \text{if } u_i \in A \\ 0 & \text{if } u_i \notin A. \end{cases}$$

If $A = \{4, 7, 9\}$ and $U = \{1, 2, 3, \ldots, 10\}$, then $f_A(2) = 0$, $f_A(4) = 1$, $f_A(7) = 1$, and $f_A(12)$ is undefined. It is easy to check that f_A is everywhere defined and onto, but is not one to one. ∎

EXAMPLE 2

In Section 1.4 we defined a family of mod-n functions, one for each positive integer n. Each f_n is a function from the nonnegative integers to the set $\{0, 1, 2, 3,$

..., $n - 1$}. For a fixed n, any nonnegative integer z can be written as $z = kn + r$ with $0 \leq r < n$. Then $f_n(z) = r$. We can also express this relation as $z \equiv r$ (mod n) (see Section 4.5). Each member of the mod function family is everywhere defined and onto, but not one to one. ∎

EXAMPLE 3   Let A be the set of nonnegative integers, $B = Z^+$, and let $f: A \to B$ be defined by $f(n) = n!$. ∎

EXAMPLE 4   The general version of the pigeonhole principle (Section 3.3) required the **floor function**, which is defined for rational numbers as $f(q)$ is the largest integer less than or equal to q. Here again is an example of a function that is not defined by a formula. The notation $\lfloor q \rfloor$ is often used for $f(q)$. Thus

$$f(1.5) = \lfloor 1.5 \rfloor = 1, \quad f(-3) = \lfloor -3 \rfloor = -3.$$ ∎

EXAMPLE 5   A function similar to that in Example 4 is the **ceiling function**, which is defined for rational numbers as $c(q)$ is the smallest integer greater than or equal to q. The notation $\lceil q \rceil$ is often used for $c(q)$. Thus

$$c(1.5) = \lceil 1.5 \rceil = 2, \quad c(-3) = \lceil -3 \rceil = -3.$$ ∎

Many common algebraic functions are used in computer science, often with domains restricted to subsets of the integers.

EXAMPLE 6
(a) Any polynomial with integer coefficients, p, can be used to define a function on Z as follows: If $p(x) = a_0 + a_1 x + a_2 x^2 + \cdots + a_n x^n$ and $z \in Z$, then $f(z)$ is the value of p evaluated at z.

(b) Let $A = B = Z^+$ and let $f: A \to B$ be defined by $f(z) = 2^z$. We call f the **base 2 exponential function**. Other bases may be used to define similar functions.

(c) Let $A = B = \mathbb{R}$ and let $f_n: A \to B$ be defined for each positive integer $n > 1$ as $f_n(x) = \log_n(x)$, the logarithm to the base n of x. In computer science applications, the bases 2 and 10 are particularly useful. ∎

In general, the unary operations discussed in previous sections can be used to create functions similar to the function in Example 3. The sets A and B in the definition of a function need not be sets of numbers, as seen in the following examples.

EXAMPLE 7
(a) Let A be a finite set and define $l: A^* \to Z$ as $l(w)$ is the length of the string w (see Section 1.3 for the definition of A^* and strings).

(b) Let B be a finite subset of the universal set U and define $pow(B)$ to be the power set of B. Then pow is a function from V, the power set of U, to the power set of V.

(c) Let $A = B =$ the set of all 2×2 matrices with real number entries and let $t(\mathbf{M}) = \mathbf{M}^T$, the transpose of $\mathbf{M}$. Then t is everywhere defined, onto, and one to one. ∎

EXAMPLE 8   (a) For elements of $Z^+ \times Z^+$, define $g(z_1, z_2)$ to be $\text{GCD}(z_1, z_2)$. Then g is a function from $Z^+ \times Z^+$ to Z^+. The GCD of two numbers is defined in Section 1.4.

(b) In a similar fashion we can define $m(z_1, z_2)$ to be $\text{LCM}(z_1, z_2)$.   ■

Another type of function, a Boolean function, plays a key role in nearly all computer programs. Let $B = \{\text{true}, \text{false}\}$. Then a function from a set A to B is called a **Boolean function**. The predicates in Section 2.1 are examples of Boolean functions.

EXAMPLE 9   Let $P(x)$: x is even and $Q(y)$: y is odd. Then P and Q are functions from Z to B. We see that $P(4)$ is true and $Q(4)$ is false. The predicate $R(x, y)$: x is even or y is odd is a Boolean function of two variables from $Z \times Z$ to B. Here $R(3, 4)$ is false and $R(6, 4)$ is true.   ■

Hashing Functions

In Section 4.6, two methods of storing the data for a relation or digraph in a computer were presented. Here we consider a more general problem of storing data. Suppose that we must store and later examine a large number of data records, customer accounts for example. In general we do not know how many records we may have to store at any given time. This suggests that linked-list storage is appropriate, because storage space is only used when we assign a record to it and we are not holding idle storage space. In order to examine a record we will have to be able to find it, so storing the data in a single linked list may not be practical because looking for an item may take a very long time (relatively speaking). One technique for handling such storage problems is to create a number of linked lists and to provide a method for deciding onto which list a new item should be linked. This method will also determine which list to search for a desired item. A key point is to attempt to assign an item to one of the lists at random. (Remember from Section 3.4 that this means each list has an equal chance of being selected.) This will have the effect of making the lists roughly the same size and thus keep the searching time about the same for any item.

Suppose we must maintain the customer records for a large company and will store the information as computer records. We begin by assigning each customer a unique seven-digit account number. A unique identifier for a record is called its **key**. For now we will not consider exactly how and what information will be stored for each customer account, but will describe only the storage of a location in the computer's memory where this information will be found. In order to determine to which list a particular record should be assigned, we create a **hashing function** from the set of keys to the set of list numbers. Hashing functions frequently use a mod-n function, as shown in the next example.

EXAMPLE 10   Suppose that (approximately) 10,000 customer account records must be stored and processed. The company's computer is capable of searching a list of 100 items in an acceptable amount of time. We decide to create 101 linked lists for storage, because if the hashing function works well in "randomly" assigning records to lists, we would expect to see roughly 100 records per list. We define a hashing function from the set of seven-digit account numbers to the set $\{0, 1, 2, 3, \ldots, 100\}$ as follows:

$$h(n) = n \ (\text{mod } 101).$$

That is, h is the mod-101 function. Thus,

$$h(2473871) = 2473871 \ (\text{mod } 101) = 78.$$

This means that the record with account number 2473871 will be assigned to list 78. Note that the range of h is the set $\{0, 1, 2, \ldots, 100\}$. ∎

Because the function h in Example 10 is not one to one, different account numbers may be assigned to the same list by the hashing function. If the first position on list 78 is already occupied when the record with key 2473871 is to be stored, we say a collision has occurred. There are many methods for resolving collisions. One very simple method that will be sufficient for our work is to insert the new record at the end of the existing list. Using this method, when we wish to find a record, its key will be hashed and the list $h(\text{key})$ will be searched sequentially.

Many other hashing functions are suitable for this situation. For example, we may break the seven-digit account number into a three-digit number and a four-digit number, add these, and then apply the mod 101 function. Many factors are considered in addition to the number of records to be stored; the speed with which an average-length list can be searched and the time needed to compute the list number for an account are two possible factors to be taken into account. For reasons that will not be discussed here, the modulus used in the mod function should be a prime. Determining a "good" hashing function for a particular application is a challenging task.

5.2 Exercises

1. Let f be the mod-10 function. Compute
   (a) $f(417)$ (b) $f(38)$ (c) $f(253)$

2. Let f be the mod-10 function. Compute
   (a) $f(81)$ (b) $f(316)$ (c) $f(1057)$

In Exercises 3 and 4, use the universal set $U = \{a, b, c, \ldots, y, z\}$ and the characteristic function for the specified subset to compute the requested function values.

3. $A = \{a, e, i, o, u\}$
   (a) $f_A(i)$ (b) $f_A(y)$ (c) $f_A(o)$

4. $A = \{m, n, o, p, q, r, z\}$
   (a) $f_B(a)$ (b) $f_B(m)$ (c) $f_B(s)$

5. Compute each of the following.
   (a) $\lfloor 2.78 \rfloor$ (b) $\lfloor -2.78 \rfloor$ (c) $\lfloor 14 \rfloor$
   (d) $\lfloor -17.3 \rfloor$ (e) $\lfloor 21.5 \rfloor$

6. Compute each of the following.
   (a) $\lceil 2.78 \rceil$ (b) $\lceil -2.78 \rceil$ (c) $\lceil 14 \rceil$
   (d) $\lceil -17.3 \rceil$ (e) $\lceil 21.5 \rceil$

In Exercises 7 and 8, compute the values indicated. Note that if the domain of these functions is Z^+, then each function is the explicit formula for an infinite sequence. Thus sequences can be viewed as a special type of function.

7. $f(n) = 3n^2 - 1$
   (a) $f(3)$ (b) $f(17)$
   (c) $f(5)$ (d) $f(12)$

8. $g(n) = 5 - 2n$
   (a) $g(4)$ (b) $g(14)$
   (c) $g(129)$ (d) $g(23)$

9. Let $f_2(n) = 2^n$. Compute each of the following.

(a) $f_2(1)$ (b) $f_2(3)$

(c) $f_2(5)$ (d) $f_2(10)$

10. Let $f_3(n) = 3^n$. Compute each of the following.

(a) $f_3(2)$ (b) $f_3(3)$

(c) $f_3(6)$ (d) $f_3(8)$

In Exercises 11 through 14, let $lg(x) = \log_2(x)$.

11. Compute each of the following.

(a) $lg(16)$ (b) $lg(128)$

(c) $lg(512)$ (d) $lg(1024)$

12. For each of the following find the largest integer less than or equal to the function value and the smallest integer greater than or equal to the function value.

(a) $lg(10)$ (b) $lg(25)$

13. For each of the following find the largest integer less than or equal to the function value and the smallest integer greater than or equal to the function value.

(a) $lg(50)$ (b) $lg(100)$

14. For each of the following find the largest integer less than or equal to the function value and the smallest integer greater than or equal to the function value.

(a) $lg(256)$ (b) $lg(500)$

15. Prove that the function in Example 7(c), t: {2×2 matrices with real entries} $\rightarrow$ { 2×2 matrices with real entries} is everywhere defined, onto, and one to one.

16. Let $A = \{a, b, c, d\}$. Let l be the function in Example 7(a).

(a) Prove that l is everywhere defined.

(b) Prove that l is not one to one.

(c) Prove or disprove that l is onto.

17. Let A be a set with n elements, S be the set of relations on A, and M the set of $n \times n$ Boolean matrices. Define $f: S \rightarrow M$ by $f(R) = \mathbf{M}_R$. Prove that f is a bijection between S and M.

18. Let p be a Boolean variable. How many different Boolean functions of p are there? How many different Boolean functions of two Boolean variables are there?

19. Build a table to represent the Boolean function $f(x, y, z) = (x' \wedge y) \vee z$ for all possible values of x, y, and z.

20. Let P be the propositional function defined by P$(x, y) = (x \vee y) \wedge \sim y$. Evaluate each of the following.

(a) P(true, true)

(b) P(false, true)

(c) P(true, false)

21. Let Q be the propositional function defined by Q(x): $\exists(y \in Z^+)(xy = 60)$. Evaluate each of the following.

(a) Q(3)   (b) Q(7)

(c) Q(-6)   (d) Q(15)

In Exercises 22 through 24, use the hashing function h, which takes the first three digits of the account number as one number and the last four digits as another number, adds them, and then applies the mod 59 function.

22. Assume that there are 7500 customer records to be stored using this hashing function.

(a) How many linked lists will be required for the storage of these records?

(b) If an approximately even distribution is achieved, roughly how many records will be stored by each linked list?

23. Determine to which list the given customer account should be attached.

(a) 3759273   (b) 7149021   (c) 5167249

24. Determine which list to search to find the given customer account.

(a) 2561384   (b) 6082376   (c) 4984620

25. Refer to Section 3.4, Exercise 33 for the average number of steps needed to search an array of length n for a key. Suppose a hashing function based on mod k is used to store m items. On average, how many steps will be required on average to search for a key?

26. Use the characteristic function of a set to prove that if $|A| = n$, then $|pow(A)| = 2^n$.

Exercises 27 through 33 use ideas from this section to complete a discussion begun in Section 3.5, Exercises 33 through 35. Pairs of parentheses are often used in mathematical expressions to indicate the order in which operations are to be done. A compiler (or interpreter) for a programming language must check that pairs of parentheses are properly placed. This may involve a number of things, but one simple check is that the number of left and right parentheses are equal and that in reading from left to right the number of left parentheses is always greater than or equal to the number of right parentheses read. An expression that passes this check is called well formed. The task here is to count the number of well-formed strings of n left and n right parentheses. This number is C_n, the nth Catalan number.

27. How many strings of n left and n right parentheses can be made (not just well-formed ones)?

28. List all well-formed strings of n left and n right parentheses for $n = 1, 2, 3$. What are the values of C_1, C_2, and C_3?

29. We will count the strings that are not well formed by making a one-to-one correspondence between them and a set of easier to count strings. Suppose $p_1 p_2 p_3 \ldots p_{2n}$ is not well formed; then there is a first p_i that is a right parenthesis and there are fewer left parentheses than right parentheses in $p_1 p_2 \ldots p_i$. How many fewer are there? So to the right of p_i the number of left parentheses is _____ than the number of right parentheses. Make a new string $q_1 q_2 \ldots q_{2n}$ as follows:

$$q_j = p_j, \quad j = 1, 2, \ldots, i$$

and

$$q_j = \begin{cases} (& \text{if } p_j =) \\ ) & \text{if } p_j = (\end{cases} \quad \text{for } j = i+1, i+2, \ldots, 2n.$$

This new string $q_1 q_2 \ldots q_{2n}$ has _____ left and _____ right parentheses. Explain your reasoning.

30. To complete the one-to-one correspondence between the p and the q strings of Exercise 29, we must show that any string with $n - 1$ left and $n + 1$ right parentheses can be paired with exactly one string with n left and n right parentheses that is not well formed. Let $r_1 r_2 r_3 \ldots r_{2n}$ consist of $n - 1$ left and $n + 1$ right parentheses. There must

be a first position j where the number of right parentheses is greater than the number of left parentheses. Why? So in $r_1 r_2 r_3 \ldots r_j$ there is one more right than left parenthesis. Hence in $r_{j+1} \ldots r_{2n}$, the number of left parentheses is _____ than the number of right parentheses. Make a new string $s_1 s_2 \ldots s_{2n}$ as follows:

$$s_k = r_k, \quad k = 1, 2, \ldots, j$$

and

$$s_k = \begin{cases} (& \text{if } r_k =) \\ ) & \text{if } r_k = (\end{cases} \quad \text{for } k = j+1, j+2, \ldots, 2n.$$

This new string $s_1 s_2 \ldots s_{2n}$ has _____ left and _____ right parentheses. Explain how you know $s_1 s_2 \ldots s_{2n}$ is not well formed.

31. Using the results of Exercises 29 and 30, the number of strings with n left and n right parentheses that are not well formed is equal to the number of strings with $n - 1$ left and $n + 1$ right parentheses. By Section 3.2, this number is _____.

32. Use the results of Exercises 27 and 31 to give a formula for C_n. Confirm this result by comparing its values with those found in Exercise 28.

33. Express C_n using the notation for combinations and without this notation.

5.3 GROWTH OF FUNCTIONS

In the earlier discussion of computer representations of relations (Section 4.6), we saw that one of the factors determining the choice of storage method is the efficiency of handling the data. In the example of testing to see if a relation is transitive, the average number of steps needed was computed for an algorithm with the relation stored as a matrix and for an algorithm with the relation stored using a linked list. The results were that it would take roughly $kn^3 + (1 - k)n^2$ steps using matrix storage and $k^3 n^4$ steps using a linked list, where the relation contains kn^2 ordered pairs. Although many details were ignored, these rough comparisons give enough information to make some decisions about appropriate data storage. In this section we apply some concepts from previous sections and lay the groundwork for more sophisticated analysis of algorithms.

The idea of one function growing more rapidly than another arises naturally when working with functions. In this section we formalize this notion.

EXAMPLE 1   Let R be a relation on a set A with $|A| = n$ and $|R| = \frac{1}{2}n^2$. If R is stored as a matrix, then $t(n) = \frac{1}{2}n^3 + \frac{1}{2}n^2$ is a function that describes (roughly) the average number of steps needed to determine if R is transitive using the algorithm TRANS

(Section 4.6). Storing R with a linked list and using NEWTRANS, the average number of steps needed is (roughly) given by $s(n) = \frac{1}{8}n^4$. Table 5.1 shows that s grows faster than t.

Table 5.1

n	$t(n)$	$s(n)$
10	550	1250
50	63,750	781,250
100	505,000	12,500,000

Let f and g be functions whose domains are subsets of Z^+, the positive integers. We say that f is $O(g)$, read "f is big-Oh of g," if there exist constants c and k such that $|f(n)| \leq c \cdot |g(n)|$ for all $n \geq k$. If f is $O(g)$, then f grows no faster than g does.

EXAMPLE 2   The function $f(n) = \frac{1}{2}n^3 + \frac{1}{2}n^2$ is $O(g)$ for $g(n) = n^3$. To see this, consider

$$\frac{1}{2}n^3 + \frac{1}{2}n^2 \leq \frac{1}{2}n^3 + \frac{1}{2}n^3, \qquad \text{if } n \geq 1.$$

Thus,

$$\frac{1}{2}n^3 + \frac{1}{2}n^2 \leq 1 \cdot n^3, \qquad \text{if } n \geq 1.$$

Choosing 1 for c and 1 for k, we have shown that $|f(n)| \leq c \cdot |g(n)|$ for all $n \geq 1$ and f is $O(g)$.

The reader can see from Example 2 that other choices of c, k, and even g are possible. If $|f(n)| \leq c|g(n)|$ for all $n \geq k$, then we have $|f(n)| \leq C \cdot |g(n)|$ for all $n \geq k$ for any $C \geq c$, and $|f(n)| \leq c \cdot |g(n)|$ for all $n \geq K$ for any $K \geq k$. For the function t in Example 2, t is $O(h)$ for $h(n) = dn^3$, if $d \geq 1$, since $|t(n)| \leq 1 \cdot |g(n)| \leq |h(n)|$. Observe also that t is $O(r(n))$ for $r(n) = n^4$, because $\frac{1}{2}n^3 + \frac{1}{2}n^2 \leq n^3 \leq n^4$ for all $n \geq 1$. When analyzing algorithms, we want to know the "slowest growing" simple function g for which f is $O(g)$.

It is common to replace g in $O(g)$ with the formula that defines g. Thus we write that t is $O(n^3)$. This is called big-O notation.

We say that f and g have the **same order** if f is $O(g)$ and g is $O(f)$.

EXAMPLE 3   Let $f(n) = 3n^4 - 5n^2$ and $g(n) = n^4$ be defined for positive integers n. Then f and g have the same order. First,

$$3n^4 - 5n^2 \leq 3n^4 + 5n^2$$
$$\leq 3n^4 + 5n^4, \qquad \text{if } n \geq 1$$
$$= 8n^4.$$

Let $c = 8$ and $k = 1$, then $|f(n)| \leq c \cdot |g(n)|$ for all $n \geq k$. Thus f is $O(g)$. Conversely, $n^4 = 3n^4 - 2n^4 \leq 3n^4 - 5n^2$ if $n \geq 2$. This is because if $n \geq 2$, then $n^2 > \frac{5}{2}$, $2n^2 > 5$, and $2n^4 > 5n^2$. Using 1 for c and 2 for k, we conclude that g is $O(f)$.

If f is $O(g)$ but g is not $O(f)$, we say that f is **lower order** than g or that f grows more slowly than g.

EXAMPLE 4

The function $f(n) = n^5$ is lower order than $g(n) = n^7$. Clearly, if $n \geq 1$, then $n^5 \leq n^7$. Suppose that there exist c and k such that $n^7 \leq cn^5$ for all $n \geq k$. Choose N so that $N > k$ and $N^2 > c$. Then $N^7 \leq cN^5 < N^2 \cdot N^5$, but this is a contradiction. Hence f is $O(g)$, but g is not $O(f)$, and f is lower order than g. This agrees with our experience that n^5 grows more slowly than n^7. ∎

We define a relation Θ, big-theta, on functions whose domains are subsets of Z^+ as $f \, \Theta \, g$ if and only if f and g have the same order.

Theorem 1   The relation Θ, big-theta, is an equivalence relation. ●

Proof   Clearly, Θ is reflexive since every function has the same order as itself. Because the definition of same order treats f and g in the same way, this definition is symmetric and the relation Θ is symmetric.

To see that Θ is transitive, suppose f and g have the same order. Then there exist c_1 and k_1 with $|f(n)| \leq c_1 \cdot |g(n)|$ for all $n \geq k_1$, and there exist c_2 and k_2 with $|g(n)| \leq c_2 \cdot |f(n)|$ for all $n \geq k_2$. Suppose that g and h have the same order; then there exist c_3, k_3 with $|g(n)| \leq c_3 \cdot |h(n)|$ for all $n \geq k_3$, and there exist c_4, k_4 with $|h(n)| \leq c_4 \cdot |g(n)|$ for all $n \geq k_4$.

Then $|f(n)| \leq c_1 \cdot |g(n)| \leq c_1(c_3 \cdot |h(n)|)$ if $n \geq k_1$ and $n \geq k_3$. Thus $|f(n)| \leq c_1c_3 \cdot |h(n)|$ for all $n \geq$ maximum of k_1 and k_3.

Similarly, $|h(n)| \leq c_2c_4 \cdot |f(n)|$ for all $n \geq$ maximum of k_2 and k_4. Thus f and h have the same order and Θ is transitive. ▼

The equivalence classes of Θ consist of functions that have the same order. We use any simple function in the equivalence class to represent the order of all functions in that class. One Θ-class is said to be **lower** than another Θ-class if a representative function from the first is of lower order than one from the second class. This means functions in the first class grow more slowly than those in the second. It is the Θ-class of a function that gives the information we need for algorithm analysis.

EXAMPLE 5

All functions that have the same order as $g(n) = n^3$ are said to have order $\Theta(n^3)$. The most common orders in computer science applications are $\Theta(1)$, $\Theta(n)$, $\Theta(n^2)$, $\Theta(n^3)$, $\Theta(lg(n))$, $\Theta(nlg(n))$, and $\Theta(2^n)$. Here $\Theta(1)$ represents the class of constant functions and lg is the base 2 log function. The continuous versions of some of these functions are shown in Figure 5.4. ∎

EXAMPLE 6

Every logarithmic function $f(n) = \log_b(n)$ has the same order as $g(n) = lg(n)$. There is a logarithmic change-of-base identity

$$\log_b(x) = \frac{\log_a(x)}{\log_a(b)}$$

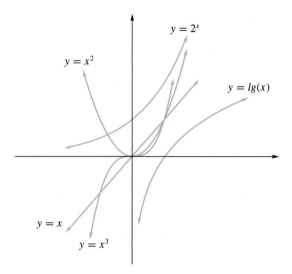

Figure 5.4

in which $\log_a(b)$ is a constant. Thus

$$\left|\log_b(n)\right| \le \frac{1}{lg(b)}|lg(n)|$$

and, conversely,

$$|lg(n)| \le lg(b) \cdot |\log_b(n)|.$$

Hence g is $O(f)$ and f is $O(g)$. ■

It is sometimes necessary to combine functions that give the number of steps required for pieces of an algorithm as is done in the analysis of TRANS (Section 4.6), where functions are added, and in the analysis of NEWTRANS, where functions are multiplied. There are some general rules regarding the ordering of the Θ-equivalence classes that can be used to determine the class of many functions and the class of the sum or product of previously classified functions.

Rules for Determining the Θ-Class of a Function

1. $\Theta(1)$ functions are constant and have zero growth, the slowest growth possible.
2. $\Theta(lg(n))$ is lower than $\Theta(n^k)$ if $k > 0$. This means that any logarithmic function grows more slowly than any power function with positive exponent.
3. $\Theta(n^a)$ is lower than $\Theta(n^b)$ if and only if $0 < a < b$.
4. $\Theta(a^n)$ is lower than $\Theta(b^n)$ if and only if $0 < a < b$.
5. $\Theta(n^k)$ is lower than $\Theta(a^n)$ for any power n^k and any $a > 1$. This means that any exponential function with base greater than 1 grows more rapidly than any power function.
6. If r is not zero, then $\Theta(rf) = \Theta(f)$ for any function f.

7. If h is a nonzero function and $\Theta(f)$ is lower than (or the same as) $\Theta(g)$, then $\Theta(fh)$ is lower than (or the same as) $\Theta(gh)$.

8. If $\Theta(f)$ is lower than $\Theta(g)$, then $\Theta(f + g) = \Theta(g)$.

EXAMPLE 7   Determine the Θ-class of each of the following.

(a) $f(n) = 4n^4 - 6n^7 + 25n^3$

(b) $g(n) = lg(n) - 3n$

(c) $h(n) = 1.1^n + n^{15}$

Solution

(a) By Rules 3, 6, and 8, the degree of the polynomial determines the Θ-class of a polynomial function. $\Theta(f) = \Theta(n^7)$.

(b) Using Rules 2, 6, and 8, we have that $\Theta(g) = \Theta(n)$.

(c) By Rules 5 and 8, $\Theta(h) = \Theta(1.1^n)$.   ■

EXAMPLE 8   Using the rules for ordering Θ-classes, arrange the following in order from lowest to highest.

$$\Theta(nlg(n)) \quad \Theta(1000n^2 - n) \quad \Theta(n^{0.2}) \quad \Theta(1,000,000) \quad \Theta(1.3^n) \quad \Theta(n + 10^7)$$

Solution $\Theta(1,000,000)$ is the class of constant functions, so it is the first on the list. By Rules 5 and 8, $\Theta(n + 10^7)$ is lower than $\Theta(1000n^2 - n)$, but higher than $\Theta(n^{0.2})$. To determine the position of $\Theta(nlg(n))$ on the list, we apply Rules 2 and 7. These give that $\Theta(nlg(n))$ is lower than $\Theta(n^2)$ and higher than $\Theta(n)$. Rule 5 says that $\Theta(1.3^n)$ is the highest class on this list. In order, the classes are

$$\Theta(1,000,000) \quad \Theta(n^{0.2}) \quad \Theta(n + 10^7)$$
$$\Theta(nlg(n)) \quad \Theta(1000n^2 - n) \quad \Theta(1.3^n).$$ ■

The Θ-class of a function that describes the number of steps performed by an algorithm is frequently referred to as the **running time** of the algorithm. For example, the algorithm TRANS has an average running time of n^3. In general, algorithms with exponential running times are impractical for all but very small values of n. In many cases the running time of an algorithm is estimated by examining best, worst, or average cases.

5.3 Exercises

In Exercises 1 and 2, let f be a function that describes the number of steps required to carry out a certain algorithm. The number of items to be processed is represented by n. For each function, describe what happens to the number of steps if the number of items is doubled.

1. (a) $f(n) = 1001$ (b) $f(n) = 3n$

(c) $f(n) = 5n^2$ (d) $f(n) = 2.5n^3$

2. (a) $f(n) = 1.4lg(n)$ (b) $f(n) = 2^n$

(c) $f(n) = nlg(n)$ (d) $f(n) = 100n^4$

3. Show that $g(n) = n!$ is $O(n^n)$.

4. Show that $h(n) = 1 + 2 + 3 + \cdots + n$ is $O(n^2)$.

5. Show that $f(n) = 8n + lg(n)$ is $O(n)$.

6. Show that $g(n) = n^2(7n - 2)$ is $O(n^3)$.

7. Show that $f(n) = nlg(n)$ is $O(g)$ for $g(n) = n^2$, but that g is not $O(f)$.

8. Show that $f(n) = n^{100}$ is $O(g)$ for $g(n) = 2^n$, but that g is not $O(f)$.

9. Show that f and g have the same order for $f(n) = 5n^2 + 4n + 3$ and $g(n) = n^2 + 100n$.

10. Show that f and g have the same order for $f(n) = lg(n^3)$ and $g(n) = \log_5(6n)$.

11. Determine which of the following are in the same Θ-class. A function may be in a class by itself.

$$f_1(n) = 5nlg(n), \quad f_2(n) = 6n^2 - 3n + 7,$$
$$f_3(n) = 1.5^n, \quad f_4(n) = lg(n^4),$$
$$f_5(n) = 13,463, \quad f_6(n) = -15n,$$
$$f_7(n) = lg(lg(n)), \quad f_8(n) = 9n^{0.7},$$
$$f_9(n) = n!, \quad f_{10}(n) = n + lg(n),$$
$$f_{11}(n) = \sqrt{n} + 12n, \quad f_{12}(n) = lg(n!)$$

12. Order the Θ-classes in Exercise 11 from lowest to highest.

In Exercises 13 through 18, analyze the operation performed by the given piece of pseudocode and write a function that describes the number of steps required. Give the Θ-class of the function.

13. 1. $A \leftarrow 1$
  2. $B \leftarrow 1$
  3. **UNTIL** $(B > 100)$
    a. $B \leftarrow 2A - 2$
    b. $A \leftarrow A + 3$

14. 1. $X \leftarrow 1$
  2. $Y \leftarrow 100$
  3. **WHILE** $(X < Y)$
    a. $X \leftarrow X + 2$
    b. $Y \leftarrow \frac{1}{2}Y$

15. 1. $I \leftarrow 1$
  2. $X \leftarrow 0$
  3. **WHILE** $(I \leq N)$
    a. $X \leftarrow X + 1$
    b. $I \leftarrow I + 1$

16. 1. $SUM \leftarrow 0$
  2. **FOR** $I = 0$ **THRU** $2(N - 1)$ **BY** 2
    a. $SUM \leftarrow SUM + I$

17. Assume that N is a power of 2.
  1. $X \leftarrow 1$
  2. $K \leftarrow N$
  3. **WHILE** $(K \geq 1)$
    a. $X \leftarrow 3X$
    b. $K \leftarrow \lfloor K/2 \rfloor$

18. SUBROUTINE MATMUL(A,B,N,M,P,Q;C)
  1. **IF** $(M = P)$ **THEN**
    a. **FOR** $I = 1$ **THRU** N
      1. **FOR** $J = 1$ **THRU** Q
        a. $C[I, J] \leftarrow 0$
        b. **FOR** $K = 1$ **THRU** M
          1. $C[I, J] \leftarrow C[I, J] + (A[I, K] \times B[K, J])$
  2. **ELSE**
    a. **CALL** PRINT ('INCOMPATIBLE')
  3. **RETURN**
  END OF SUBROUTINE MATMUL

19. Determine the Θ-class of the function defined in Section 1.3, Exercise 33. What is the running time for computing $F(N)$?

20. (a) Write a recurrence relation to count the number of ways a 3×3 square can be placed on an $n \times n$ square with the edges of the squares parallel.

  (b) What is the running time of an algorithm that uses the recurrence relation in (a) to count the number of placements?

21. Prove Rule 3.

22. Prove Rule 7.

23. Prove that if $\Theta(f) = \Theta(g) = \Theta(h)$, then $f + g$ is $O(h)$.

24. Prove that if $\Theta(f) = \Theta(g)$ and $c \neq 0$, then $\Theta(cf) = \Theta(g)$.

5.4 PERMUTATION FUNCTIONS

In this section we discuss bijections from a set A to itself. Of special importance is the case when A is finite. Bijections on a finite set occur in a wide variety of applications in mathematics, computer science, and physics.

A bijection from a set A to itself is called a **permutation** of A.

EXAMPLE 1   Let $A = \mathbb{R}$ and let $f : A \rightarrow A$ be defined by $f(a) = 2a + 1$. Since f is one to one and onto (verify), it follows that f is a permutation of A.   ∎

If $A = \{a_1, a_2, \ldots, a_n\}$ is a finite set and p is a bijection on A, we list the elements of A and the corresponding function values $p(a_1), p(a_2), \ldots, p(a_n)$ in the following form:

$$\begin{pmatrix} a_1 & a_2 & \cdots & a_n \\ p(a_1) & p(a_2) & \cdots & p(a_n) \end{pmatrix}. \tag{1}$$

Observe that (1) completely describes p since it gives the value of p for every element of A. We often write

$$p = \begin{pmatrix} a_1 & a_2 & \cdots & a_n \\ p(a_1) & p(a_2) & \cdots & p(a_n) \end{pmatrix}.$$

Thus, if p is a permutation of a finite set $A = \{a_1, a_2, \ldots, a_n\}$, then the sequence $p(a_1), p(a_2), \ldots, p(a_n)$ is just a rearrangement of the elements of A and so corresponds exactly to a permutation of A in the sense of Section 3.1.

EXAMPLE 2 Let $A = \{1, 2, 3\}$. Then all the permutations of A are

$$1_A = \begin{pmatrix} 1 & 2 & 3 \\ 1 & 2 & 3 \end{pmatrix}, \quad p_1 = \begin{pmatrix} 1 & 2 & 3 \\ 1 & 3 & 2 \end{pmatrix}, \quad p_2 = \begin{pmatrix} 1 & 2 & 3 \\ 2 & 1 & 3 \end{pmatrix},$$

$$p_3 = \begin{pmatrix} 1 & 2 & 3 \\ 2 & 3 & 1 \end{pmatrix}, \quad p_4 = \begin{pmatrix} 1 & 2 & 3 \\ 3 & 1 & 2 \end{pmatrix}, \quad p_5 = \begin{pmatrix} 1 & 2 & 3 \\ 3 & 2 & 1 \end{pmatrix}.$$

∎

EXAMPLE 3 Using the permutations of Example 2, compute (a) p_4^{-1}; (b) $p_3 \circ p_2$.

Solution

(a) Viewing p_4 as a function, we have

$$p_4 = \{(1, 3), (2, 1), (3, 2)\}.$$

Then

$$p_4^{-1} = \{(3, 1), (1, 2), (2, 3)\}$$

or, when written in increasing order of the first component of each ordered pair, we have

$$p_4^{-1} = \{(1, 2), (2, 3), (3, 1)\}.$$

Thus

$$p_4^{-1} = \begin{pmatrix} 1 & 2 & 3 \\ 2 & 3 & 1 \end{pmatrix} = p_3.$$

(b) The function p_2 takes 1 to 2 and p_3 takes 2 to 3, so $p_3 \circ p_2$ takes 1 to 3. Also, p_2 takes 2 to 1 and p_3 takes 1 to 2, so $p_3 \circ p_2$ takes 2 to 2. Finally, p_2 takes 3 to 3 and p_3 takes 3 to 1, so $p_3 \circ p_2$ takes 3 to 1. Thus

$$p_3 \circ p_2 = \begin{pmatrix} 1 & 2 & 3 \\ 3 & 2 & 1 \end{pmatrix}.$$

We may view the process of forming $p_3 \circ p_2$ as shown in Figure 5.5. Observe that $p_3 \circ p_2 = p_5$.

∎

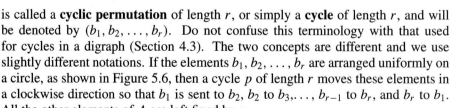

$$p_3 \circ p_2 = \begin{pmatrix} 1 & 2 & 3 \\ 2 & 3 & 1 \end{pmatrix} \circ \begin{pmatrix} 1 & 2 & 3 \\ 2 & 1 & 3 \end{pmatrix} = \begin{pmatrix} 1 & 2 & 3 \\ 3 & 2 & 1 \end{pmatrix}$$

Figure 5.5

The composition of two permutations is another permutation, usually referred to as the **product** of these permutations. In the remainder of this chapter, we will follow this convention.

Theorem 1　If $A = \{a_1, a_2, \ldots, a_n\}$ is a set containing n elements, then there are

$$n! = n \cdot (n - 1) \cdots 2 \cdot 1 \quad \text{permutations of } A.$$ ●

Proof　This result follows from Theorem 4 of Section 3.1 by letting $r = n$. ▼

Let $b_1, b_2, \ldots, b_r$ be r distinct elements of the set $A = \{a_1, a_2, \ldots, a_n\}$. The permutation $p: A \to A$ defined by

$$p(b_1) = b_2$$
$$p(b_2) = b_3$$
$$\vdots$$
$$p(b_{r-1}) = b_r$$
$$p(b_r) = b_1$$
$$p(x) = x, \quad \text{if } x \in A, x \notin \{b_1, b_2, \ldots, b_r\},$$

is called a **cyclic permutation** of length r, or simply a **cycle** of length r, and will be denoted by $(b_1, b_2, \ldots, b_r)$. Do not confuse this terminology with that used for cycles in a digraph (Section 4.3). The two concepts are different and we use slightly different notations. If the elements $b_1, b_2, \ldots, b_r$ are arranged uniformly on a circle, as shown in Figure 5.6, then a cycle p of length r moves these elements in a clockwise direction so that b_1 is sent to b_2, b_2 to $b_3, \ldots, b_{r-1}$ to b_r, and b_r to b_1. All the other elements of A are left fixed by p.

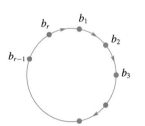

Figure 5.6

EXAMPLE 4　Let $A = \{1, 2, 3, 4, 5\}$. The cycle $(1, 3, 5)$ denotes the permutation

$$\begin{pmatrix} 1 & 2 & 3 & 4 & 5 \\ 3 & 2 & 5 & 4 & 1 \end{pmatrix}.$$ ■

Observe that if $p = (b_1, b_2, \ldots, b_r)$ is a cycle of length r, then we can also write p by starting with any b_i, $1 \le i \le r$, and moving in a clockwise direction, as shown in Figure 5.5. Thus, as cycles,

$$(3, 5, 8, 2) = (5, 8, 2, 3) = (8, 2, 3, 5) = (2, 3, 5, 8).$$

Note also that the notation for a cycle does not include the number of elements in the set A. Thus the cycle $(3, 2, 1, 4)$ could be a permutation of the set $\{1, 2, 3, 4\}$

or of $\{1, 2, 3, 4, 5, 6, 7, 8\}$. We need to be told explicitly the set on which a cycle is defined. It follows from the definition that a cycle on a set A is of length 1 if and only if it is the identity permutation, 1_A.

Since cycles are permutations, we can form their product. However, as we show in the following example, the product of two cycles need not be a cycle.

EXAMPLE 5 Let $A = \{1, 2, 3, 4, 5, 6\}$. Compute $(4, 1, 3, 5) \circ (5, 6, 3)$ and $(5, 6, 3) \circ (4, 1, 3, 5)$.

Solution We have

$$(4, 1, 3, 5) = \begin{pmatrix} 1 & 2 & 3 & 4 & 5 & 6 \\ 3 & 2 & 5 & 1 & 4 & 6 \end{pmatrix}$$

and

$$(5, 6, 3) = \begin{pmatrix} 1 & 2 & 3 & 4 & 5 & 6 \\ 1 & 2 & 5 & 4 & 6 & 3 \end{pmatrix}.$$

Then

$$(4, 1, 3, 5) \circ (5, 6, 3) = \begin{pmatrix} 1 & 2 & 3 & 4 & 5 & 6 \\ 3 & 2 & 5 & 1 & 4 & 6 \end{pmatrix} \circ \begin{pmatrix} 1 & 2 & 3 & 4 & 5 & 6 \\ 1 & 2 & 5 & 4 & 6 & 3 \end{pmatrix}$$

$$= \begin{pmatrix} 1 & 2 & 3 & 4 & 5 & 6 \\ 3 & 2 & 4 & 1 & 6 & 5 \end{pmatrix}$$

and

$$(5, 6, 3) \circ (4, 1, 3, 5) = \begin{pmatrix} 1 & 2 & 3 & 4 & 5 & 6 \\ 1 & 2 & 5 & 4 & 6 & 3 \end{pmatrix} \circ \begin{pmatrix} 1 & 2 & 3 & 4 & 5 & 6 \\ 3 & 2 & 5 & 1 & 4 & 6 \end{pmatrix}$$

$$= \begin{pmatrix} 1 & 2 & 3 & 4 & 5 & 6 \\ 5 & 2 & 6 & 1 & 4 & 3 \end{pmatrix}.$$

Observe that

$$(4, 1, 3, 5) \circ (5, 6, 3) \neq (5, 6, 3) \circ (4, 1, 3, 5)$$

and that neither product is a cycle. ∎

Two cycles of a set A are said to be **disjoint** if no element of A appears in both cycles.

EXAMPLE 6 Let $A = \{1, 2, 3, 4, 5, 6\}$. Then the cycles $(1, 2, 5)$ and $(3, 4, 6)$ are disjoint, whereas the cycles $(1, 2, 5)$ and $(2, 4, 6)$ are not. ∎

It is not difficult to show that if $p_1 = (a_1, a_2, \ldots, a_r)$ and $p_2 = (b_1, b_2, \ldots, b_s)$ are disjoint cycles of A, then $p_1 \circ p_2 = p_2 \circ p_1$. This can be seen by observing that p_1 affects only the a's, while p_2 affects only the b's.

We shall now present a fundamental theorem and, instead of giving its proof, we shall give an example that imitates the proof.

Theorem 2 A permutation of a finite set that is not the identity or a cycle can be written as a product of disjoint cycles of length ≥ 2. ●

EXAMPLE 7   Write the permutation

$$p = \begin{pmatrix} 1 & 2 & 3 & 4 & 5 & 6 & 7 & 8 \\ 3 & 4 & 6 & 5 & 2 & 1 & 8 & 7 \end{pmatrix}$$

of the set $A = \{1, 2, 3, 4, 5, 6, 7, 8\}$ as a product of disjoint cycles.

Solution   We start with 1 and find that $p(1) = 3$, $p(3) = 6$, and $p(6) = 1$, so we have the cycle $(1, 3, 6)$. Next we choose the first element of A that has not appeared in a previous cycle. We choose 2, and we have $p(2) = 4$, $p(4) = 5$, and $p(5) = 2$, so we obtain the cycle $(2, 4, 5)$. We now choose 7, the first element of A that has not appeared in a previous cycle. Since $p(7) = 8$ and $p(8) = 7$, we obtain the cycle $(7, 8)$. We can then write p as a product of disjoint cycles as

$$p = (7, 8) \circ (2, 4, 5) \circ (1, 3, 6).$$ ■

It is not difficult to show that in Theorem 2, when a permutation is written as a product of disjoint cycles, the product is unique except for the order of the cycles.

Even and Odd Permutations

A cycle of length 2 is called a **transposition**. That is, a transposition is a cycle $p = (a_i, a_j)$, where $p(a_i) = a_j$ and $p(a_j) = a_i$.

Observe that if $p = (a_i, a_j)$ is a transposition of A, then $p \circ p = 1_A$, the identity permutation of A.

Every cycle can be written as a product of transpositions. In fact,

$$(b_1, b_2, \ldots, b_r) = (b_1, b_r) \circ (b_1, b_{r-1}) \circ \cdots \circ (b_1, b_3) \circ (b_1, b_2).$$

This case be verified by induction on r, as follows:

Basis Step: If $r = 2$, then the cycle is just (b_1, b_2), which already has the proper form.

Induction Step: We use P(k) to show P($k + 1$). Let $(b_1, b_2, \ldots, b_k, b_{k+1})$ be a cycle of length $k + 1$. Then $(b_1, b_2, \ldots, b_k, b_{k+1}) = (b_1, b_{k+1}) \circ (b_1, b_2, \ldots, b_k)$, as may be verified by computing the composition. Using P(k), $(b_1, b_2, \ldots, b_k) = (b_1, b_k) \circ (b_1, b_{k-1}) \circ \cdots \circ (b_1, b_2)$. Thus, by substitution,

$$(b_1, b_2, \ldots, b_{k+1}) = (b_1, b_{k+1}) \circ (b_1, b_k) \circ \cdots \circ (b_1, b_3)(b_1, b_2).$$

This completes the induction step. Thus, by the principle of mathematical induction, the result holds for every cycle. For example,

$$(1, 2, 3, 4, 5) = (1, 5) \circ (1, 4) \circ (1, 3) \circ (1, 2).$$

We now obtain the following corollary of Theorem 2.

Corollary 1   Every permutation of a finite set with at least two elements can be written as a product of transpositions.   ●

Observe that the transpositions in Corollary 1 need not be disjoint.

EXAMPLE 8   Write the permutation p of Example 7 as a product of transpositions.

Solution   We have

$$p = (7, 8) \circ (2, 4, 5) \circ (1, 3, 6).$$

Since we can write

$$(1, 3, 6) = (1, 6) \circ (1, 3)$$
$$(2, 4, 5) = (2, 5) \circ (2, 4),$$

we have

$$p = (7, 8) \circ (2, 5) \circ (2, 4) \circ (1, 6) \circ (1, 3). \qquad \blacksquare$$

We have observed that every cycle can be written as a product of transpositions. However, this can be done in many different ways. For example,

$$(1, 2, 3) = (1, 3) \circ (1, 2)$$
$$= (2, 1) \circ (2, 3)$$
$$= (1, 3) \circ (3, 1) \circ (1, 3) \circ (1, 2) \circ (3, 2) \circ (2, 3).$$

It then follows that every permutation on a set of two or more elements can be written as a product of transpositions in many ways. However, the following theorem, whose proof we omit, brings some order to the situation.

Theorem 3   If a permutation of a finite set can be written as a product of an even number of transpositions, then it can never be written as a product of an odd number of transpositions, and conversely.   ●

A permutation of a finite set is called **even** if it can be written as a product of an even number of transpositions, and it is called **odd** if it can be written as a product of an odd number of transpositions.

EXAMPLE 9   Is the permutation

$$p = \begin{pmatrix} 1 & 2 & 3 & 4 & 5 & 6 & 7 \\ 2 & 4 & 5 & 7 & 6 & 3 & 1 \end{pmatrix}$$

even or odd?

Solution   We first write p as a product of disjoint cycles, obtaining

$$p = (3, 5, 6) \circ (1, 2, 4, 7). \qquad \text{(Verify this.)}$$

Next we write each of the cycles as a product of transpositions:

$$(1, 2, 4, 7) = (1, 7) \circ (1, 4) \circ (1, 2)$$
$$(3, 5, 6) = (3, 6) \circ (3, 5).$$

Then

$$p = (3, 6) \circ (3, 5) \circ (1, 7) \circ (1, 4) \circ (1, 2).$$

Since p is a product of an odd number of transpositions, it is an odd permutation.   ∎

From the definition of even and odd permutations, it follows (see Exercises 18 through 20) that

(a) The product of two even permutations is even.

(b) The product of two odd permutations is even.

(c) The product of an even and an odd permutation is odd.

Theorem 4   Let $A = \{a_1, a_2, \ldots, a_n\}$ be a finite set with n elements, $n \geq 2$. There are $n!/2$ even permutations and $n!/2$ odd permutations.   ●

Proof   Let A_n be the set of all even permutations of A, and let B_n be the set of all odd permutations. We shall define a function $f : A_n \to B_n$, which we show is one to one and onto, and this will show that A_n and B_n have the same number of elements.

Since $n \geq 2$, we can choose a particular transposition q_0 of A. Say that $q_0 = (a_{n-1}, a_n)$. We now define the function $f : A_n \to B_n$ by

$$f(p) = q_0 \circ p, \qquad p \in A_n.$$

Observe that if $p \in A_n$, then p is an even permutation, so $q_0 \circ p$ is an odd permutation and thus $f(p) \in B_n$. Suppose now that p_1 and p_2 are in A_n and

$$f(p_1) = f(p_2).$$

Then

$$q_0 \circ p_1 = q_0 \circ p_2. \tag{2}$$

We now compose each side of equation (2) with q_0:

$$q_0 \circ (q_0 \circ p_1) = q_0 \circ (q_0 \circ p_2);$$

so, by the associative property,

$$(q_0 \circ q_0) \circ p_1 = (q_0 \circ q_0) \circ p_2$$

or, since $q_0 \circ q_0 = 1_A$,

$$1_A \circ p_1 = 1_A \circ p_2$$
$$p_1 = p_2.$$

Thus f is one to one.

Now let $q \in B_n$. Then $q_0 \circ q \in A_n$, and

$$f(q_0 \circ q) = q_0 \circ (q_0 \circ q) = (q_0 \circ q_0) \circ q = 1_A \circ q = q,$$

which means that f is an onto function. Since $f : A_n \to B_n$ is one to one and onto, we conclude that A_n and B_n have the same number of elements. Note that $A_n \cap B_n = \varnothing$ since no permutation can be both even and odd. Also, by Theorem 1, $|A_n \cup B_n| = n!$. Thus, by Theorem 2 of Section 1.2,

$$n! = |A_n \cup B_n| = |A_n| + |B_n| - |A_n \cap B_n| = 2|A_n|.$$

We then have

$$|A_n| = |B_n| = \frac{n!}{2}.$$

▼

5.4 Exercises

1. Which of the following functions $f : \mathbb{R} \to \mathbb{R}$ are permutations of $\mathbb{R}$?

   (a) f is defined by $f(a) = a - 1$.

   (b) f is defined by $f(a) = a^2$.

2. Which of the following functions $f : \mathbb{R} \to \mathbb{R}$ are permutations of $\mathbb{R}$?

   (a) f is defined by $f(a) = a^3$.

   (b) f is defined by $f(a) = e^a$.

3. Which of the following functions $f : Z \to Z$ are permutations of Z?

   (a) f is defined by $f(a) = a + 1$.

   (b) f is defined by $f(a) = (a - 1)^2$.

4. Which of the following functions $f : Z \to Z$ are permutations of Z?

   (a) f is defined by $f(a) = a^2 + 1$.

   (b) f is defined by $f(a) = a^3 - 3$.

In Exercises 5 through 8, let $A = \{1, 2, 3, 4, 5, 6\}$ and

$$p_1 = \begin{pmatrix} 1 & 2 & 3 & 4 & 5 & 6 \\ 3 & 4 & 1 & 2 & 6 & 5 \end{pmatrix},$$

$$p_2 = \begin{pmatrix} 1 & 2 & 3 & 4 & 5 & 6 \\ 2 & 3 & 1 & 5 & 4 & 6 \end{pmatrix},$$

$$p_3 = \begin{pmatrix} 1 & 2 & 3 & 4 & 5 & 6 \\ 6 & 3 & 2 & 5 & 4 & 1 \end{pmatrix}.$$

5. Compute

   (a) p_1^{-1} (b) $p_3 \circ p_1$

6. Compute

   (a) $(p_2 \circ p_1) \circ p_2$ (b) $p_1 \circ (p_3 \circ p_2^{-1})$

7. Compute

   (a) p_3^{-1} (b) $p_1^{-1} \circ p_2^{-1}$

8. Compute

   (a) $(p_3 \circ p_2) \circ p_1$ (b) $p_3 \circ (p_2 \circ p_1)^{-1}$

In Exercises 9 and 10, let $A = \{1, 2, 3, 4, 5, 6, 7, 8\}$. Compute the products.

9. (a) $(3, 5, 7, 8) \circ (1, 3, 2)$

   (b) $(2, 6) \circ (3, 5, 7, 8) \circ (2, 5, 3, 4)$

10. (a) $(1, 4) \circ (2, 4, 5, 6) \circ (1, 4, 6, 7)$

    (b) $(5, 8) \circ (1, 2, 3, 4) \circ (3, 5, 6, 7)$

11. Let $A = \{a, b, c, d, e, f, g\}$. Compute the products.

    (a) $(a, f, g) \circ (b, c, d, e)$

    (b) $(f, g) \circ (b, c, f) \circ (a, b, c)$

In Exercises 12 and 13, let $A = \{1, 2, 3, 4, 5, 6, 7, 8\}$. Write each permutation as the product of disjoint cycles.

12. (a) $\begin{pmatrix} 1 & 2 & 3 & 4 & 5 & 6 & 7 & 8 \\ 4 & 3 & 2 & 5 & 1 & 8 & 7 & 6 \end{pmatrix}$

    (b) $\begin{pmatrix} 1 & 2 & 3 & 4 & 5 & 6 & 7 & 8 \\ 2 & 3 & 4 & 1 & 7 & 5 & 8 & 6 \end{pmatrix}$

13. (a) $\begin{pmatrix} 1 & 2 & 3 & 4 & 5 & 6 & 7 & 8 \\ 6 & 5 & 7 & 8 & 4 & 3 & 2 & 1 \end{pmatrix}$

    (b) $\begin{pmatrix} 1 & 2 & 3 & 4 & 5 & 6 & 7 & 8 \\ 2 & 3 & 1 & 4 & 6 & 7 & 8 & 5 \end{pmatrix}$

14. Let $A = \{a, b, c, d, e, f, g\}$. Write each permutation as the product of disjoint cycles.

    (a) $\begin{pmatrix} a & b & c & d & e & f & g \\ g & d & b & a & c & f & e \end{pmatrix}$

    (b) $\begin{pmatrix} a & b & c & d & e & f & g \\ d & e & a & b & g & f & c \end{pmatrix}$

15. Let $A = \{1, 2, 3, 4, 5, 6, 7, 8\}$. Write each permutation as a product of transpositions.

    (a) $(2, 1, 4, 5, 8, 6)$ (b) $(3, 1, 6) \circ (4, 8, 2, 5)$

In Exercises 16 and 17, let $A = \{1, 2, 3, 4, 5, 6, 7, 8\}$. Determine whether the permutation is even or odd.

16. (a) $\begin{pmatrix} 1 & 2 & 3 & 4 & 5 & 6 & 7 & 8 \\ 4 & 2 & 1 & 6 & 5 & 8 & 7 & 3 \end{pmatrix}$

    (b) $\begin{pmatrix} 1 & 2 & 3 & 4 & 5 & 6 & 7 & 8 \\ 7 & 3 & 4 & 2 & 1 & 8 & 6 & 5 \end{pmatrix}$

17. (a) $(6, 4, 2, 1, 5)$

    (b) $(4, 8) \circ (3, 5, 2, 1) \circ (2, 4, 7, 1)$

18. Prove that the product of two even permutations is even.

19. Prove that the product of two odd permutations is even.

20. Prove that the product of an even and an odd permutation is odd.

21. Let $A = \{1, 2, 3, 4, 5\}$. Let $f = (5, 2, 3)$ and $g = (3, 4, 1)$ be permutations of A. Compute each of the following and write the result as the product of disjoint cycles.

    (a) $f \circ g$ (b) $f^{-1} \circ g^{-1}$

22. Show that if p is a permutation of a finite set A, then $p^2 = p \circ p$ is a permutation of A.

23. Let $A = \{1, 2, 3, 4, 5, 6\}$ and

$$p = \begin{pmatrix} 1 & 2 & 3 & 4 & 5 & 6 \\ 2 & 4 & 3 & 1 & 5 & 6 \end{pmatrix}$$

be a permutation of A.

(a) Write p as a product of disjoint cycles.

(b) Compute p^{-1}.   (c) Compute p^2.

(d) Find the period of p, that is, the smallest positive integer k such that $p^k = 1_A$.

24. Let $A = \{1, 2, 3, 4, 5, 6\}$ and

$$p = \begin{pmatrix} 1 & 2 & 3 & 4 & 5 & 6 \\ 4 & 3 & 5 & 1 & 2 & 6 \end{pmatrix}$$

be a permutation of A.

(a) Write p as a product of disjoint cycles.

(b) Compute p^{-1}.   (c) Compute p^2.

(d) Find the period of p, that is, the smallest positive integer k such that $p^k = 1_A$.

25. (a) Use mathematical induction to show that if p is a permutation of a finite set A, then $p^n = p \circ p \circ \cdots \circ p$ is a permutation of A for $n \in Z^+$.

(b) If A is a finite set and p is a permutation of A, show that $p^m = 1_A$ for some $m \in Z^+$.

26. Let p be a permutation of a set A. Define the following relation R on A: $a \, R \, b$ if and only if $p^n(a) = b$ for some $n \in Z$. [p^0 is defined as the identity permutation and p^{-n} is defined as $(p^{-1})^n$.] Show that R is an equivalence relation and describe the equivalence classes.

27. Build a composition table for the permutations of $A = \{1, 2, 3\}$ given in Example 2.

28. Describe how to use the composition table in Exercise 27 to identify p^{-1} for any permutation p of A.

29. Find all subsets of $\{1_A, p_1, p_2, p_3, p_4, p_5\}$, the permutations in Example 2, that satisfy the closure property for composition.

30. For each permutation, p, of A in Example 2, determine its order. How does this relate to the subset in Exercise 29 to which p belongs?

TIPS FOR PROOFS

Before beginning a proof, you might find it helpful to consider what the statement does *not* say. This can help clarify your thinking about what facts and tools are available for the proof. Consider Theorem 4, Section 5.1. It does not say that if f is one to one, then f is onto. The additional facts that $|A| = |B|$ and that f is everywhere defined will need to be used in the proof.

To show that a function is one to one or onto, we need to use generic elements. See Example 11, Section 5.1. Either the definition of one-to-oneness or its contrapositive may be used to prove this property. We also have the fact that if $f : A \to B$ is everywhere defined and $|A| = |B| = n$, then f is one to one if and only if f is onto. In addition, if we wish to show f is one to one and onto, we may do this by constructing the inverse function f^{-1}. Establishing a one-to-one correspondence is a powerful counting strategy, because it allows us to count a different set than the original one. For example, Theorem 4, Section 5.4, and Exercises 29 through 31, Section 5.2.

To prove that f and g have the same order or one is of lower order than the other, the principal tools are the rules for Θ-classes or manipulation of inequalities (Section 5.3, Examples 2 and 3).

KEY IDEAS FOR REVIEW

- Function: see page 161
- Identity function, 1_A: $1_A(a) = a$
- One-to-one function f from A to B: $a \neq a'$ implies $f(a) \neq f(a')$
- Onto function f from A to B: $\text{Ran}(f) = B$
- Bijection: one-to-one and onto function
- One-to-one correspondence: onto, one-to-one, everywhere defined function

- If f is a function from A to B, $1_B \circ f = f$; $f \circ 1_A = f$
- If f is an invertible function from A to B, $f^{-1} \circ f = 1_A$; $f \circ f^{-1} = 1_B$
- $(g \circ f)^{-1} = f^{-1} \circ g^{-1}$
- Boolean function f: $\text{Ran}(f) \subseteq \{\text{true, false}\}$
- Hashing function: see page 172

- $O(g)$ (big Oh of g): see page 176.
- f and g of the same order: f is $O(g)$ and g is $O(f)$
- Theorem: The relation Θ, $f \Theta g$ if and only if f and g have the same order, is an equivalence relation.
- Lower Θ-class: see page 177
- Rules for determining Θ-class of a function: see page 178
- Running time of an algorithm: Θ-class of a function that describes the number of steps performed by the algorithm
- Permutation function: a bijection from a set A to itself
- Theorem: If A is a set that has n elements, then there are $n!$ permutations of A.
- Cycle of length r: $(b_1, b_2, \ldots, b_r)$; see page 182
- Theorem: A permutation of a finite set that is not the identity or a cycle can be written as a product of disjoint cycles.
- Transposition: a cycle of length 2

- Corollary: Every permutation of a finite set with at least two elements can be written as a product of transpositions.
- Even (odd) permutation: one that can be written as a product of an even (odd) number of transpositions
- Theorem: If a permutation of a finite set can be written as a product of an even number of transpositions, then it can never be written as a product of an odd number of transpositions, and conversely.
- The product of
  (a) Two even permutations is even.
  (b) Two odd permutations is even.
  (c) An even and an odd permutation is odd.
- Theorem: If A is a set that has n elements, then there are $n!/2$ even permutations and $n!/2$ odd permutations of A.

CODING EXERCISES

For each of the following, write the requested program or subroutine in pseudocode (as described in Appendix A) or in a programming language that you know. Test your code either with a paper-and-pencil trace or with a computer run.

1. Let $U = \{u_1, u_2, \ldots, u_n\}$ be the universal set for possible input sets. Write a function CHARFCN that given a set as input returns the characteristic function of the set as a sequence.

2. Write a function TRANSPOSE that, given an $n \times n$ matrix, returns its transpose.

3. Write a program that writes a given permutation as a product of disjoint cycles.

4. Write a program that writes a given permutation as a product of transpositions.

5. Use the program in Exercise 4 as a subroutine in a program that determines whether a given permutation is even or odd.

CHAPTER 5 SELF-TEST

1. Let $A = \{a, b, c, d\}$, $B = \{1, 2, 3\}$, and $R = \{(a, 2), (b, 1), (c, 2), (d, 1)\}$. Is R a function? Is R^{-1} a function? Explain your answers.

2. Let $A = B = \mathbb{R}$. Let $f: A \to B$ be the function defined by $f(x) = -5x + 8$. Show that f is one to one and onto.

3. Compute
   (a) $\lfloor 16.29 \rfloor$ (b) $\lfloor -1.6 \rfloor$

4. Compute
   (a) $\lceil 16.29 \rceil$ (b) $\lceil -1.6 \rceil$

5. Compute
   (a) $lg(1)$ (b) $lg(64)$

6. Let Q be the propositional function defined by
$$Q(x): \exists y \; xy = \begin{bmatrix} 1 & 0 \\ 0 & 1 \end{bmatrix}.$$
Evaluate $Q\left(\begin{bmatrix} 2 & 1 \\ 0 & 3 \end{bmatrix}\right)$ and $Q\left(\begin{bmatrix} 2 & 3 \\ 4 & 6 \end{bmatrix}\right)$.

7. Assume that 9,500 account records need to be stored using the hashing function h, which takes the first two digits of the account number as one number and the last four digits as another number, adds them, and then applies the mod-89 function.
   (a) How many linked lists will be needed?

(b) If an approximately even distribution of records is achieved, roughly how many records will be stored in each linked list?

(c) Compute $h(473810)$, $h(125332)$, and $h(308691)$.

8. Show that $f(n) = 2n^2 + 9n + 5$ is $O(n^2)$.

9. Determine the Θ-class of $f(n) = \lg(n) + n^2 + 2^n$.

10. Consider the following pseudocode.
   1. $X \leftarrow 10$
   2. $I \leftarrow 0$
   3. **UNTIL** $(I > N)$
       a. $X \leftarrow X + I$
       b. $I \leftarrow I + 2$

   Write a function of N that describes the number of steps required and give the Θ-class of the function.

11. Let $A = \{1, 2, 3, 4, 5, 6\}$ and let $p_1 = (3, 6, 2)$ and $p_2 = (5, 1, 4)$ be permutations of A.

(a) Compute $p_1 \circ p_2$ and write the result as a product of cycles and as the product of transpositions.

(b) Compute $p_1^{-1} \circ p_2^{-1}$.

12. Let $p_1 = \begin{pmatrix} 1 & 2 & 3 & 4 & 5 & 6 & 7 \\ 7 & 3 & 2 & 1 & 4 & 5 & 6 \end{pmatrix}$ and

$p_2 = \begin{pmatrix} 1 & 2 & 3 & 4 & 5 & 6 & 7 \\ 6 & 3 & 2 & 1 & 5 & 4 & 7 \end{pmatrix}$.

(a) Compute $p_1 \circ p_2$.

(b) Compute p_1^{-1}.

(c) Is p_1 an even or odd permutation? Explain.

6 ORDER RELATIONS AND STRUCTURES

Prerequisites: *Chapter 4*

In this chapter we study partially ordered sets, including lattices and Boolean algebras. These structures are useful in set theory, algebra, sorting and searching, and, especially in the case of Boolean algebras, in the construction of logical representations for computer circuits.

6.1 PARTIALLY ORDERED SETS

A relation R on a set A is called a **partial order** if R is reflexive, antisymmetric, and transitive. The set A together with the partial order R is called a **partially ordered set**, or simply a **poset**, and we will denote this poset by (A, R). If there is no possibility of confusion about the partial order, we may refer to the poset simply as A, rather than (A, R).

EXAMPLE 1

Let A be a collection of subsets of a set S. The relation $\subseteq$ of set inclusion is a partial order on A, so $(A, \subseteq)$ is a poset. ∎

EXAMPLE 2

Let Z^+ be the set of positive integers. The usual relation $\leq$ (less than or equal to) is a partial order on Z^+, as is $\geq$ (greater than or equal to). ∎

EXAMPLE 3

The relation of divisibility ($a \ R \ b$ if and only if $a \mid b$) is a partial order on Z^+. ∎

EXAMPLE 4

Let $\mathcal{R}$ be the set of all equivalence relations on a set A. Since $\mathcal{R}$ consists of subsets of $A \times A$, $\mathcal{R}$ is a partially ordered set under the partial order of set containment. If R and S are equivalence relations on A, the same property may be expressed in

191

relational notation as follows.

$$R \subseteq S \text{ if and only if } x \ R \ y \quad \text{implies} \quad x \ S \ y \text{ for all } x, y \text{ in } A.$$

Then $(\mathcal{R}, \subseteq)$ is a poset.   ■

EXAMPLE 5   The relation $<$ on Z^+ is not a partial order, since it is not reflexive.   ■

EXAMPLE 6   Let R be a partial order on a set A, and let R^{-1} be the inverse relation of R. Then R^{-1} is also a partial order. To see this, we recall the characterization of reflexive, antisymmetric, and transitive given in Section 4.4. If R has these three properties, then $\Delta \subseteq R$, $R \cap R^{-1} \subseteq \Delta$, and $R^2 \subseteq R$. By taking inverses, we have

$$\Delta = \Delta^{-1} \subseteq R^{-1}, \quad R^{-1} \cap (R^{-1})^{-1} = R^{-1} \cap R \subseteq \Delta, \quad \text{and} \quad (R^{-1})^2 \subseteq R^{-1},$$

so, by Section 4.4, R^{-1} is reflexive, antisymmetric, and transitive. Thus R^{-1} is also a partial order. The poset (A, R^{-1}) is called the **dual** of the poset (A, R), and the partial order R^{-1} is called the **dual** of the partial order R.   ■

The most familiar partial orders are the relations $\leq$ and $\geq$ on Z and $\mathbb{R}$. For this reason, when speaking in general of a partial order R on a set A, we shall often use the symbols $\leq$ or $\geq$ for R. This makes the properties of R more familiar and easier to remember. Thus the reader may see the symbol $\leq$ used for many different partial orders on different sets. Do not mistake this to mean that these relations are all the same or that they have anything to do with the familiar relation $\leq$ on Z or $\mathbb{R}$. If it becomes absolutely necessary to distinguish partial orders from one another, we may also use symbols such as $\leq_1, \leq', \geq_1, \geq'$, and so on, to denote partial orders.

We will observe the following convention. Whenever $(A, \leq)$ is a poset, we will always use the symbol $\geq$ for the partial order $\leq^{-1}$, and thus $(A, \geq)$ will be the dual poset. Similarly, the dual of poset $(A, \leq_1)$ will be denoted by $(A, \geq_1)$, and the dual of the poset $(B, \leq')$ will be denoted by $(B, \geq')$. Again, this convention is to remind us of the familiar dual posets $(Z, \leq)$ and $(Z, \geq)$, as well as the posets $(\mathbb{R}, \leq)$ and $(\mathbb{R}, \geq)$.

If $(A, \leq)$ is a poset, the elements a and b of A are said to be **comparable** if

$$a \leq b \quad \text{or} \quad b \leq a.$$

Observe that in a partially ordered set every pair of elements need not be comparable. For example, consider the poset in Example 3. The elements 2 and 7 are not comparable; since $2 \nmid 7$ and $7 \nmid 2$. Thus the word "partial" in partially ordered set means that some elements may not be comparable. If every pair of elements in a poset A is comparable, we say that A is a **linearly ordered** set, and the partial order is called a **linear order**. We also say that A is a **chain**.

EXAMPLE 7   The poset of Example 2 is linearly ordered.   ■

The following theorem is sometimes useful since it shows how to construct a new poset from given posets.

Theorem 1   If $(A, \leq)$ and $(B, \leq)$ are posets, then $(A \times B, \leq)$ is a poset, with partial order $\leq$ defined by

$$(a, b) \leq (a', b')   \text{if } a \leq a' \text{ in } A \text{ and } b \leq b' \text{ in } B.$$

Note that the symbol $\leq$ is being used to denote three distinct partial orders. The reader should find it easy to determine which of the three is meant at any time.   ●

Proof   If $(a, b) \in A \times B$, then $(a, b) \leq (a, b)$ since $a \leq a$ in A and $b \leq b$ in B, so $\leq$ satisfies the reflexive property in $A \times B$. Now suppose that $(a, b) \leq (a', b')$ and $(a', b') \leq (a, b)$, where a and $a' \in A$ and b and $b' \in B$. Then

$$a \leq a'   \text{and}   a' \leq a   \text{in } A$$

and

$$b \leq b'   \text{and}   b' \leq b   \text{in } B.$$

Since A and B are posets, the antisymmetry of the partial orders on A and B implies that

$$a = a'   \text{and}   b = b'.$$

Hence $\leq$ satisfies the antisymmetry property in $A \times B$.
    Finally, suppose that

$$(a, b) \leq (a', b')   \text{and}   (a', b') \leq (a'', b''),$$

where $a, a', a'' \in A$, and $b, b', b'' \in B$. Then

$$a \leq a'   \text{and}   a' \leq a'',$$

so $a \leq a''$, by the transitive property of the partial order on A. Similarly,

$$b \leq b'   \text{and}   b'' \leq b'',$$

so $b \leq b''$, by the transitive property of the partial order on B. Hence

$$(a, b) \leq (a'', b'').$$

Consequently, the transitive property holds for the partial order on $A \times B$, and we conclude that $A \times B$ is a poset.   ▼

    The partial order $\leq$ defined on the Cartesian product $A \times B$ is called the **product partial order**.
    If $(A, \leq)$ is a poset, we say that $a < b$ if $a \leq b$ but $a \neq b$. Suppose now that $(A, \leq)$ and $(B \leq)$ are posets. In Theorem 1 we have defined the product partial order on $A \times B$. Another useful partial order on $A \times B$, denoted by $\prec$, is defined as follows:

$$(a, b) \prec (a', b')   \text{if } a < a' \text{ or if } a = a' \text{ and } b \leq b'.$$

This ordering is called **lexicographic**, or "dictionary" order. The ordering of the elements in the first coordinate dominates, except in case of "ties," when attention passes to the second coordinate. If $(A, \leq)$ and $(B, \leq)$ are linearly ordered sets, then the lexicographic order $\prec$ on $A \times B$ is also a linear order.

EXAMPLE 8

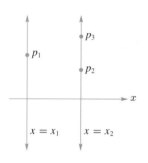

Figure 6.1

Let $A = \mathbb{R}$, with the usual ordering $\leq$. Then the plane $\mathbb{R}^2 = \mathbb{R} \times \mathbb{R}$ may be given lexicographic order. This is illustrated in Figure 6.1. We see that the plane is linearly ordered by lexicographic order. Each vertical line has the usual order, and points on one line are less than points on a line farther to the right. Thus, in Figure 6.1, $p_1 \prec p_2$, $p_1 \prec p_3$, and $p_2 \prec p_3$. ■

Lexicographic ordering is easily extended to Cartesian products $A_1 \times A_2 \times \cdots \times A_n$ as follows:

$$(a_1, a_2, \ldots, a_n) \prec (a_1', a_2', \ldots, a_n') \quad \text{if and only if}$$
$$a_1 < a_1' \quad \text{or}$$
$$a_1 = a_1' \quad \text{and } a_2 < a_2' \quad \text{or}$$
$$a_1 = a_1', \ a_2 = a_2', \quad \text{and } a_3 < a_3' \quad \text{or} \ldots$$
$$a_1 = a_1', \ a_2 = a_2', \quad \ldots, \quad a_{n-1} = a_{n-1}' \quad \text{and} \quad a_n \leq a_n'.$$

Thus the first coordinate dominates except for equality, in which case we consider the second coordinate. If equality holds again, we pass to the next coordinate, and so on.

EXAMPLE 9

Let $S = \{a, b, \ldots, z\}$ be the ordinary alphabet, linearly ordered in the usual way ($a \leq b, b \leq c, \ldots, y \leq z$). Then $S^n = S \times S \times \cdots \times S$ (n factors) can be identified with the set of all words having length n. Lexicographic order on S^n has the property that if $w_1 \prec w_2$ ($w_1, w_2 \in S^n$), then w_1 would precede w_2 in a dictionary listing. This fact accounts for the name of the ordering.

Thus *park* $\prec$ *part*, *help* $\prec$ *hind*, *jump* $\prec$ *mump*. The third is true since $j < m$; the second, since $h = h$, $e < i$; and the first is true since $p = p$, $a = a$, $r = r$, $k < t$. ■

If S is a poset, we can extend lexicographic order to S^* (see Section 1.3) in the following way.

If $x = a_1 a_2 \cdots a_n$ and $y = b_1 b_2 \cdots b_k$ are in S^* with $n \leq k$, we say that $x \prec y$ if $(a_1, \ldots, a_n) \prec (b_1, \ldots, b_n)$ in S^n under lexicographic ordering of S^n.

In the previous paragraph, we use the fact that the n-tuple $(a_1, a_2, \ldots, a_n) \in S^n$, and the string $a_1 a_2 \cdots a_n \in S^*$ are really the same sequence of length n, written in two different notations. The notations differ for historical reasons, and we will use them interchangeably depending on context.

EXAMPLE 10

Let S be $\{a, b, \ldots, z\}$, ordered as usual. Then S^* is the set of all possible "words" of any length, whether such words are meaningful or not.

Thus we have

$$help \prec helping$$

in S^* since

$$help \prec help$$

in S^4. Similarly, we have

$$helper \prec helping$$

since

$$helper \prec helpin$$

in S^6. As the example

$$help \prec helping$$

shows, this order includes *prefix order*; that is, any word is greater than all of its prefixes (beginning parts). This is also the way that words occur in the dictionary. Thus we have dictionary ordering again, but this time for words of any finite length.

■

Since a partial order is a relation, we can look at the digraph of any partial order on a finite set. We shall find that the digraphs of partial orders can be represented in a simpler manner than those of general relations. The following theorem provides the first result in this direction.

Theorem 2   The digraph of a partial order has no cycle of length greater than 1.   ●

Proof   Suppose that the digraph of the partial order $\leq$ on the set A contains a cycle of length $n \geq 2$. Then there exist distinct elements $a_1, a_2, \ldots, a_n \in A$ such that

$$a_1 \leq a_2, \quad a_2 \leq a_3, \quad \ldots, \quad a_{n-1} \leq a_n, \quad a_n \leq a_1.$$

By the transitivity of the partial order, used $n - 1$ times, $a_1 \leq a_n$. By antisymmetry, $a_n \leq a_1$ and $a_1 \leq a_n$ imply that $a_n = a_1$, a contradiction to the assumption that $a_1, a_2, \ldots, a_n$ are distinct.   ▼

Hasse Diagrams

Let A be a finite set. Theorem 2 has shown that the digraph of a partial order on A has only cycles of length 1. Indeed, since a partial order is reflexive, every vertex in the digraph of the partial order is contained in a cycle of length 1. To simplify matters, we shall delete all such cycles from the digraph. Thus the digraph shown in Figure 6.2(a) would be drawn as shown in Figure 6.2(b).

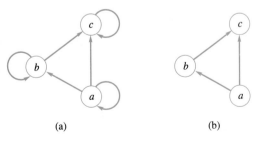

(a)   (b)

Figure 6.2

We shall also eliminate all edges that are implied by the transitive property. Thus, if $a \leq b$ and $b \leq c$, it follows that $a \leq c$. In this case, we omit the edge from a to c; however, we do draw the edges from a to b and from b to c. For example, the digraph shown in Figure 6.3(a) would be drawn as shown in Figure 6.3(b). We also agree to draw the digraph of a partial order with all edges pointing upward, so that

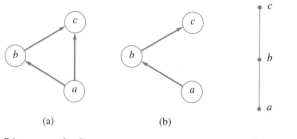

(a)   (b)

Figure 6.3   Figure 6.4

arrows may be omitted from the edges. Finally, we replace the circles representing the vertices by dots. Thus the diagram shown in Figure 6.4 gives the final form of the digraph shown in Figure 6.2(a). The resulting diagram of a partial order, much simpler than its digraph, is called the **Hasse diagram** of the partial order of the poset. Since the Hasse diagram completely describes the associated partial order, we shall find it to be a very useful tool.

EXAMPLE 11   Let $A = \{1, 2, 3, 4, 12\}$. Consider the partial order of divisibility on A. That is, if a and $b \in A$, $a \leq b$ if and only if $a \mid b$. Draw the Hasse diagram of the poset $(A, \leq)$.

Solution   The Hasse diagram is shown in Figure 6.5. To emphasize the simplicity of the Hasse diagram, we show in Figure 6.6 the digraph of the poset in Figure 6.5.   ■

EXAMPLE 12   Let $S = \{a, b, c\}$ and $A = P(S)$. Draw the Hasse diagram of the poset A with the partial order $\subseteq$ (set inclusion).

Solution   We first determine A, obtaining

$$A = \{\varnothing, \{a\}, \{b\}, \{c\}, \{a, b\}, \{a, c\}, \{b, c\}, \{a, b, c\}\}.$$

The Hasse diagram can then be drawn as shown in Figure 6.7.   ■

Observe that the Hasse diagram of a finite linearly ordered set is always of the form shown in Figure 6.8.

It is easily seen that if $(A, \leq)$ is a poset and $(A, \geq)$ is the dual poset, then the Hasse diagram of $(A, \geq)$ is just the Hasse diagram of $(A, \leq)$ turned upside down.

EXAMPLE 13   Figure 6.9(a) shows the Hasse diagram of a poset $(A, \leq)$, where

$$A = \{a, b, c, d, e, f\}.$$

Figure 6.9(b) shows the Hasse diagram of the dual poset $(A, \geq)$. Notice that, as stated, each of these diagrams can be constructed by turning the other upside down.   ■

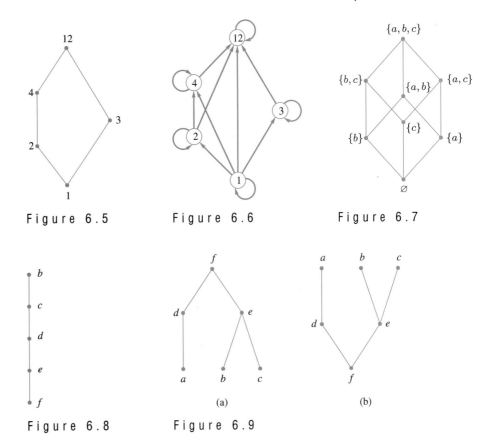

Figure 6.5   Figure 6.6   Figure 6.7

Figure 6.8   Figure 6.9

Topological Sorting

If A is a poset with partial order $\leq$, we sometimes need to find a linear order $\prec$ for the set A that will merely be an extension of the given partial order in the sense that if $a \leq b$, then $a \prec b$. The process of constructing a linear order such as $\prec$ is called **topological sorting**. This problem might arise when we have to enter a finite poset A into a computer. The elements of A must be entered in some order, and we might want them entered so that the partial order is preserved. That is, if $a \leq b$, then a is entered before b. A topological sorting $\prec$ will give an order of entry of the elements that meets this condition.

EXAMPLE 14   Give a topological sorting for the poset whose Hasse diagram is shown in Figure 6.10.

Solution   The partial order $\prec$ whose Hasse diagram is shown in Figure 6.11(a) is clearly a linear order. It is easy to see that every pair in $\leq$ is also in the order $\prec$, so $\prec$ is a topological sorting of the partial order $\leq$. Figures 6.11(b) and (c) show two other solutions to this problem.   ■

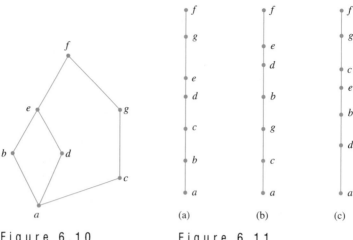

Figure 6.10   Figure 6.11

Isomorphism

Let $(A, \leq)$ and $(A', \leq')$ be posets and let $f : A \to A'$ be a one-to-one correspondence between A and A'. The function f is called an **isomorphism** from $(A, \leq)$ to $(A', \leq')$ if, for any a and b in A,

$$a \leq b \quad \text{if and only if} \quad f(a) \leq' f(b).$$

If $f : A \to A'$ is an isomorphism, we say that $(A, \leq)$ and $(A', \leq')$ are **isomorphic** posets.

EXAMPLE 15   Let A be the set Z^+ of positive integers, and let $\leq$ be the usual partial order on A (see Example 2). Let A' be the set of positive even integers, and let $\leq'$ be the usual partial order on A'. The function $f : A \to A'$ given by

$$f(a) = 2a$$

is an isomorphism from $(A, \leq)$ to $(A', \leq')$.

First, f is one to one since, if $f(a) = f(b)$, then $2a = 2b$, so $a = b$. Next, $\text{Dom}(f) = A$, so f is everywhere defined. Finally, if $c \in A'$, then $c = 2a$ for some $a \in Z^+$; therefore, $c = f(a)$. This shows that f is onto, so we see that f is a one-to-one correspondence. Finally, if a and b are elements of A, then it is clear that $a \leq b$ if and only if $2a \leq 2b$. Thus f is an isomorphism. ∎

Suppose that $f : A \to A'$ is an isomorphism from a poset $(A, \leq)$ to a poset $(A', \leq')$. Suppose also that B is a subset of A, and $B' = f(B)$ is the corresponding subset of A'. Then we see from the definition of isomorphism that the following principle must hold.

Theorem 3
Principle of Correspondence   If the elements of B have any property relating to one another or to other elements of A, and if this property can be defined entirely in terms of the relation $\leq$, then the elements of B' must possess exactly the same property, defined in terms of $\leq'$.   ●

EXAMPLE 16

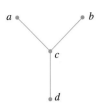

Figure 6.12

Let $(A, \le)$ be the poset whose Hasse diagram is shown in Figure 6.12, and suppose that f is an isomorphism from $(A, \le)$ to some other poset $(A', \le')$. Note first that $d \le x$ for any x in A (later we will call an element such as d a "least element" of A). Then the corresponding element $f(d)$ in A' must satisfy the property $f(d) \le' y$ for all y in A'. As another example, note that $a \not\le b$ and $b \not\le a$. Such a pair is called **incomparable** in A. It then follows from the principle of correspondence that $f(a)$ and $f(b)$ must be incomparable in A'. ∎

For a finite poset, one of the objects that is defined entirely in terms of the partial order is its Hasse diagram. It follows from the principle of correspondence that two finite isomorphic posets must have the same Hasse diagrams.

To be precise, let $(A, \le)$ and $(A', \le')$ be finite posets, let $f \colon A \to A'$ be a one-to-one correspondence, and let H be any Hasse diagram of $(A, \le)$. Then

1. If f is an isomorphism and each label a of H is replaced by $f(a)$, then H will become a Hasse diagram for $(A', \le')$.

Conversely,

2. If H becomes a Hasse diagram for $(A', \le')$, whenever each label a is replaced by $f(a)$, then f is an isomorphism.

This justifies the name "isomorphism," since isomorphic posets have the same (*iso-*) "shape" (*morph*) as described by their Hasse diagrams.

EXAMPLE 17

Let $A = \{1, 2, 3, 6\}$ and let $\le$ be the relation | (divides). Figure 6.13(a) shows the Hasse diagram for $(A, \le)$. Let

$$A' = P(\{a, b\}) = \{\varnothing, \{a\}, \{b\}, \{a, b\}\},$$

and let $\le'$ be set containment, $\subseteq$. If $f \colon A \to A'$ is defined by

$$f(1) = \varnothing, \quad f(2) = \{a\}, \quad f(3) = \{b\}, \quad f(6) = \{a, b\},$$

then it is easily seen that f is a one-to-one correspondence. If each label $a \in A$ of the Hasse diagram is replaced by $f(a)$, the result is as shown in Figure 6.13(b). Since this is clearly a Hasse diagram for $(A', \le')$, the function f is an isomorphism. ∎

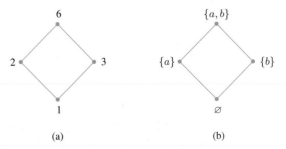

(a)   (b)

Figure 6.13

6.1 Exercises

1. Determine whether the relation R is a partial order on the set A.

   (a) $A = Z$, and $a\ R\ b$ if and only if $a = 2b$.

   (b) $A = Z$, and $a\ R\ b$ if and only if $b^2 \mid a$.

2. Determine whether the relation R is a partial order on the set A.

   (a) $A = Z$, and $a\ R\ b$ if and only if $a = b^k$ for some $k \in Z^+$. Note that k depends on a and b.

   (b) $A = \mathbb{R}$, and $a\ R\ b$ if and only if $a \le b$.

3. Determine whether the relation R is a linear order on the set A.

   (a) $A = \mathbb{R}$, and $a\ R\ b$ if and only if $a \le b$.

   (b) $A = \mathbb{R}$, and $a\ R\ b$ if and only if $a \ge b$.

4. Determine whether the relation R is a linear order on the set A.

   (a) $A = P(S)$, where S is a set. The relation R is set inclusion.

   (b) $A = \mathbb{R} \times \mathbb{R}$, and $(a, b)\ R\ (a', b')$ if and only if $a \le a'$ and $b \le b'$, where $\le$ is the usual partial order on $\mathbb{R}$.

5. On the set $A = \{a, b, c\}$, find all partial orders $\le$ in which $a \le b$.

6. What can you say about the relation R on a set A if R is a partial order and an equivalence relation?

7. Outline the structure of the proof given for Theorem 1.

8. Outline the structure of the proof given for Theorem 2.

In Exercises 9 and 10, determine the Hasse diagram of the relation R.

9. $A = \{1, 2, 3, 4\}$, $R = \{(1, 1), (1, 2), (2, 2), (2, 4), (1, 3), (3, 3), (3, 4), (1, 4), (4, 4)\}$.

10. $A = \{a, b, c, d, e\}$, $R = \{(a, a), (b, b), (c, c), (a, c), (c, d), (c, e), (a, d), (d, d), (a, e), (b, c), (b, d), (b, e), (e, e)\}$.

Describe the ordered pairs in the relation determined by the Hasse diagram on the set $A = \{1, 2, 3, 4\}$ in Figures 6.14 and 6.15.

11.

12.

Figure 6.14   Figure 6.15

In Exercises 13 and 14, determine the Hasse diagram of the partial order having the given digraph (Figures 6.16 and 6.17).

13.
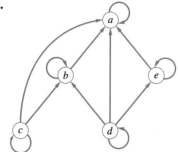

Figure 6.16

14.
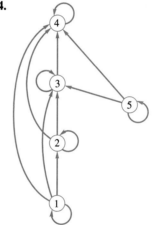

Figure 6.17

15. Determine the Hasse diagram of the relation on $A = \{1, 2, 3, 4, 5\}$ whose matrix is shown.

$$\begin{bmatrix} 1 & 1 & 1 & 1 & 1 \\ 0 & 1 & 1 & 1 & 1 \\ 0 & 0 & 1 & 1 & 1 \\ 0 & 0 & 0 & 1 & 1 \\ 0 & 0 & 0 & 0 & 1 \end{bmatrix}$$

16. Determine the Hasse diagram of the relation on $A = \{1, 2, 3, 4, 5\}$ whose matrix is shown.

$$\begin{bmatrix} 1 & 0 & 1 & 1 & 1 \\ 0 & 1 & 1 & 1 & 1 \\ 0 & 0 & 1 & 1 & 1 \\ 0 & 0 & 0 & 1 & 0 \\ 0 & 0 & 0 & 0 & 1 \end{bmatrix}$$

In Exercises 17 and 18, *determine the matrix of the partial order whose Hasse diagram is given (Figures 6.18 and 6.19).*

17.

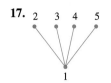

Figure 6.18

18.

Figure 6.19

19. Let $A = \{\Box, A, B, C, E, O, M, P, S\}$ have the usual alphabetical order, where $\Box$ represents a "blank" character and $\Box \leq x$ for all $x \in A$. Arrange the following in lexicographic order (as elements of $A \times A \times A \times A$).

(a) MOP□   (b) MOPE   (c) CAP□

(d) MAP□   (e) BASE   (f) ACE□

(g) MACE   (h) CAPE

20. Let $A = Z^+ \times Z^+$ have lexicographic order. Mark each of the following as true or false.

(a) $(2, 12) \prec (5, 3)$ (b) $(3, 6) \prec (3, 24)$

(c) $(4, 8) \prec (4, 6)$ (d) $(15, 92) \prec (12, 3)$

In Exercises 21 through 24, *consider the partial order of divisibility on the set A. Draw the Hasse diagram of the poset and determine which posets are linearly ordered.*

21. $A = \{1, 2, 3, 5, 6, 10, 15, 30\}$

22. $A = \{2, 4, 8, 16, 32\}$

23. $A = \{3, 6, 12, 36, 72\}$

24. $A = \{1, 2, 3, 4, 5, 6, 10, 12, 15, 30, 60\}$

25. Describe how to use $\mathbf{M}_R$ to determine if R is a partial order.

In Exercises 26 and 27, *draw the Hasse diagram of a topological sorting of the given poset (Figures 6.20 and 6.21).*

26.

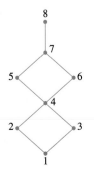

Figure 6.20

27.

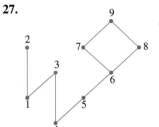

Figure 6.21

28. If $(A, \leq)$ is a poset and A' is a subset of A, show that $(A', \leq')$ is also a poset, where $\leq'$ is the restriction of $\leq$ to A'.

29. Show that if R is a linear order on the set A, then R^{-1} is also a linear order on A.

30. A relation R on a set A is called a **quasiorder** if it is transitive and irreflexive. Let $A = P(S)$ be the power set of a set S, and consider the following relation R on A: $U \ R \ T$ if and only if $U \subset T$ (proper containment). Show that R is a quasiorder.

31. Let $A = \{x \mid x$ is a real number and $-5 \leq x \leq 20\}$. Show that the usual relation $<$ is a quasiorder (see Exercise 30) on A.

32. If R is a quasiorder on A (see Exercise 30), prove that R^{-1} is also a quasiorder on A.

33. Modify the relation in Example 3 to produce a quasiorder on Z^+.

34. Let $B = \{2, 3, 6, 9, 12, 18, 24\}$ and let $A = B \times B$. Define the following relation on A: $(a, b) \prec (a', b')$ if and only if $a \mid a'$ and $b \leq b'$, where $\leq$ is the usual partial order. Show that $\prec$ is a partial order.

35. Let $A = \{1, 2, 3, 5, 6, 10, 15, 30\}$ and consider the partial order $\leq$ of divisibility on A. That is, define $a \leq b$ to mean that $a \mid b$. Let $A' = P(S)$, where $S = \{e, f, g\}$, be the poset with partial order $\subseteq$. Show that $(A, \leq)$ and $(A', \subseteq)$ are isomorphic.

36. Let $A = \{1, 2, 4, 8\}$ and let $\leq$ be the partial order of divisibility on A. Let $A' = \{0, 1, 2, 3\}$ and let $\leq'$ be the usual relation "less than or equal to" on integers. Show that $(A, \leq)$ and $(A', \leq')$ are isomorphic posets.

6.2 EXTREMAL ELEMENTS OF PARTIALLY ORDERED SETS

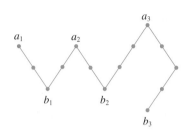

Figure 6.22

Certain elements in a poset are of special importance for many of the properties and applications of posets. In this section we discuss these elements, and in later sections we shall see the important role played by them. In this section we consider a poset $(A, \leq)$.

An element $a \in A$ is called a **maximal element** of A if there is no element c in A such that $a < c$ (see Section 6.1). An element $b \in A$ is called a **minimal element** of A if there is no element c in A such that $c < b$.

It follows immediately that, if $(A, \leq)$ is a poset and $(A, \geq)$ is its dual poset, an element $a \in A$ is a maximal element of $(A, \geq)$ if and only if a is a minimal element of $(A, \leq)$. Also, a is a minimal element of $(A, \geq)$ if and only if it is a maximal element of $(A, \leq)$.

EXAMPLE 1 Consider the poset A whose Hasse diagram is shown in Figure 6.22. The elements a_1, a_2, and a_3 are maximal elements of A, and the elements b_1, b_2, and b_3 are minimal elements. Observe that, since there is no line between b_2 and b_3, we can conclude neither that $b_3 \leq b_2$ nor that $b_2 \leq b_3$.   ∎

EXAMPLE 2 Let A be the poset of nonnegative real numbers with the usual partial order $\leq$. Then 0 is a minimal element of A. There are no maximal elements of A.   ∎

EXAMPLE 3 The poset Z with the usual partial order $\leq$ has no maximal elements and has no minimal elements.   ∎

Theorem 1 Let A be a finite nonempty poset with partial order $\leq$. Then A has at least one maximal element and at least one minimal element.   ●

Proof Let a be any element of A. If a is not maximal, we can find an element $a_1 \in A$ such that $a < a_1$. If a_1 is not maximal, we can find an element $a_2 \in A$ such that $a_1 < a_2$. This argument cannot be continued indefinitely, since A is a finite set. Thus we eventually obtain the finite chain

$$a < a_1 < a_2 < \cdots < a_{k-1} < a_k,$$

which cannot be extended. Hence we cannot have $a_k < b$ for any $b \in A$, so a_k is a maximal element of $(A, \leq)$.

This same argument says that the dual poset $(A, \geq)$ has a maximal element, so $(A, \leq)$ has a minimal element.   ▼

By using the concept of a minimal element, we can give an algorithm for finding a topological sorting of a given finite poset $(A, \leq)$. We remark first that if $a \in A$ and $B = A - \{a\}$, then B is also a poset under the restriction of $\leq$ to $B \times B$ (see Section 4.2). We then have the following algorithm, which produces a linear array named SORT. We assume that SORT is ordered by increasing index, that is, SORT[1] $\prec$ SORT[2] $\prec \cdots$. The relation $\prec$ on A defined in this way is a topological sorting of $(A, \leq)$.

◆ **ALGORITHM**   for finding a topological sorting of a finite poset $(A, \leq)$.

> **Step 1**   Choose a minimal element a of A.
>
> **Step 2**   Make a the next entry of SORT and replace A with $A - \{a\}$.
>
> **Step 3**   Repeat steps 1 and 2 until $A = \{\ \}$.
>
> End of Algorithm

EXAMPLE 4

Let $A = \{a, b, c, d, e\}$, and let the Hasse diagram of a partial order $\leq$ on A be as shown in Figure 6.23(a). A minimal element of this poset is the vertex labeled d (we could also have chosen e). We put d in SORT[1] and in Figure 6.23(b) we show the Hasse diagram of $A - \{d\}$. A minimal element of the new A is e, so e becomes SORT[2], and $A - \{e\}$ is shown in Figure 6.23(c). This process continues until we have exhausted A and filled SORT. Figure 6.23(f) shows the completed array SORT and the Hasse diagram of the poset corresponding to SORT. This is a topological sorting of $(A, \leq)$.   ■

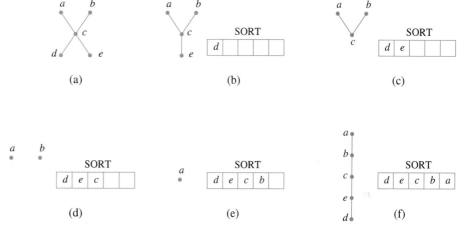

Figure 6.23

An element $a \in A$ is called a **greatest element** of A if $x \leq a$ for all $x \in A$. An element $a \in A$ is called a **least element** of A if $a \leq x$ for all $x \in A$.

As before, an element a of $(A, \leq)$ is a greatest (or least) element if and only if it is a least (or greatest) element of $(A, \geq)$.

EXAMPLE 5

Consider the poset defined in Example 2. Then 0 is a least element; there is no greatest element.   ■

EXAMPLE 6

Let $S = \{a, b, c\}$ and consider the poset $A = P(S)$ defined in Example 12 of Section 6.1. The empty set is a least element of A, and the set S is a greatest element of A.   ■

EXAMPLE 7

The poset Z with the usual partial order has neither a least nor a greatest element.   ■

Theorem 2   A poset has at most one greatest element and at most one least element.   ●

Proof   Suppose that a and b are greatest elements of a poset A. Then, since b is a greatest element, we have $a \leq b$. Similarly, since a is a greatest element, we have $b \leq a$. Hence $a = b$ by the antisymmetry property. Thus, if the poset has a greatest element, it only has one such element. Since this fact is true for all posets, the dual poset $(A, \geq)$ has at most one greatest element, so $(A, \leq)$ also has at most one least element.   ▼

The greatest element of a poset, if it exists, is denoted by I and is often called the **unit element**. Similarly, the least element of a poset, if it exists, is denoted by 0 and is often called the **zero element**.

Consider a poset and a subset B of A. An element $a \in A$ is called an **upper bound** of B if $b \leq a$ for all $b \in B$. An element $a \in A$ is called a **lower bound** of B if $a \leq b$ for all $b \in B$.

EXAMPLE 8   Consider the poset $A = \{a, b, c, d, e, f, g, h\}$, whose Hasse diagram is shown in Figure 6.24. Find all upper and lower bounds of the following subsets of A: (a) $B_1 = \{a, b\}$; (b) $B_2 = \{c, d, e\}$.

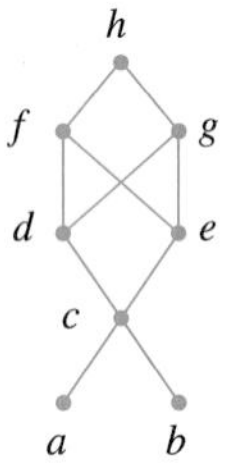

Solution

(a) B_1 has no lower bounds; its upper bounds are c, d, e, f, g, and h.

(b) The upper bounds of B_2 are f, g, and h; its lower bounds are c, a, and b.   ■

Figure 6.24

As Example 8 shows, a subset B of a poset may or may not have upper or lower bounds (in A). Moreover, an upper or lower bound of B may or may not belong to B itself.

Let A be a poset and B a subset of A. An element $a \in A$ is called a **least upper bound** (LUB) of B if a is an upper bound of B and $a \leq a'$, whenever a' is an upper bound of B. Thus $a = \mathrm{LUB}(B)$ if $b \leq a$ for all $b \in B$, and if whenever $a' \in A$ is also an upper bound of B, then $a \leq a'$.

Similarly, an element $a \in A$ is called a **greatest lower bound** (GLB) of B if a is a lower bound of B and $a' \leq a$, whenever a' is a lower bound of B. Thus $a = \mathrm{GLB}(B)$ if $a \leq b$ for all $b \in B$, and if whenever $a' \in A$ is also a lower bound of B, then $a' \leq a$.

As usual, upper bounds in $(A, \leq)$ correspond to lower bounds in $(A, \geq)$ (for the same set of elements), and lower bounds in $(A, \leq)$ correspond to upper bounds in $(A, \leq)$. Similar statements hold for greatest lower bounds and least upper bounds.

EXAMPLE 9   Let A be the poset considered in Example 8 with subsets B_1 and B_2 as defined in that example. Find all least upper bounds and all greatest lower bounds of (a) B_1; (b) B_2.

Solution

(a) Since B_1 has no lower bounds, it has no greatest lower bounds. However,

$$\mathrm{LUB}(B_1) = c.$$

(b) Since the lower bounds of B_2 are c, a, and b, we find that

$$GLB(B_2) = c.$$

The upper bounds of B_2 are f, g, and h. Since f and g are not comparable, we conclude that B_2 has no least upper bound. ∎

Theorem 3   Let $(A, \leq)$ be a poset. Then a subset B of A has at most one LUB and at most one GLB. ●

Proof   The proof is similar to the proof of Theorem 2. ▼

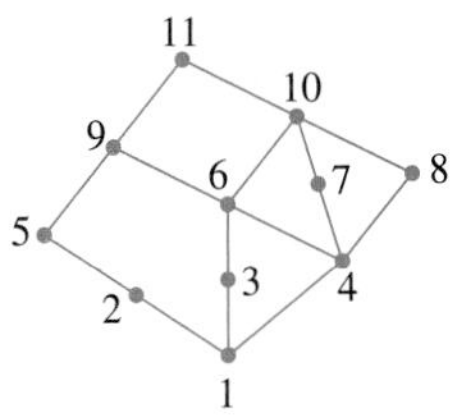

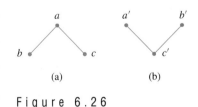

Figure 6.25

We conclude this section with some remarks about LUB and GLB in a finite poset A, as viewed from the Hasse diagram of A. Let $B = \{b_1, b_2, \ldots, b_r\}$. If $a = \text{LUB}(B)$, then a is the first vertex that can be reached from $b_1, b_2, \ldots, b_r$ by upward paths. Similarly, if $a = \text{GLB}(B)$, then a is the first vertex that can be reached from $b_1, b_2, \ldots, b_r$ by downward paths.

EXAMPLE 10   Let $A = \{1, 2, 3, 4, 5, \ldots, 11\}$ be the poset whose Hasse diagram is shown in Figure 6.25. Find the LUB and GLB of $B = \{6, 7, 10\}$, if they exist.

Solution   Exploring all upward paths from vertices 6, 7, and 10, we find that $\text{LUB}(B) = 10$. Similarly, by examining all downward paths from 6, 7, and 10, we find that $\text{GLB}(B) = 4$. ∎

The next result follows immediately from the principle of correspondence (see Section 6.1).

Theorem 4   Suppose that $(A, \leq)$ and $(A', \leq')$ are isomorphic posets under the isomorphism $f: A \rightarrow A'$.
  (a) If a is a maximal (minimal) element of $(A, \leq)$, then $f(a)$ is a maximal (minimal) element of $(A', \leq')$.
  (b) If a is the greatest (least) element of $(A, \leq)$, then $f(a)$ is the greatest (least) element of $(A', \leq')$.
  (c) If a is an upper bound (lower bound, least upper bound, greatest lower bound) of a subset B of A, then $f(a)$ is an upper bound (lower bound, least upper bound, greatest lower bound) for the subset $f(B)$ of A'.
  (d) If every subset of $(A, \leq)$ has a LUB (GLB), then every subset of $(A', \leq')$ has a LUB (GLB). ●

Figure 6.26

EXAMPLE 11   Show that the posets $(A, \leq)$ and $(A', \leq')$, whose Hasse diagrams are shown in Figures 6.26(a) and (b), respectively, are not isomorphic.

Solution   The two posets are not isomorphic because $(A, \leq)$ has a greatest element a, while $(A', \leq')$ does not have a greatest element. We could also argue that they are not isomorphic because $(A, \leq)$ does not have a least element, while $(A', \leq')$ does have a least element. ∎

6.2 Exercises

In Exercises 1 through 8, determine all maximal and minimal elements of the poset.

1.

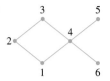

Figure 6.27

2.

Figure 6.28

3.

Figure 6.29

4.

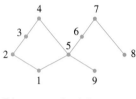

Figure 6.30

5. $A = \mathbb{R}$ with the usual partial order $\leq$.

6. $A = \{x \mid x$ is a real number and $0 \leq x < 1\}$ with the usual partial order $\leq$.

7. $A = \{x \mid x$ is a real number and $0 < x \leq 1\}$ with the usual partial order $\leq$.

8. $A = \{2, 3, 4, 6, 8, 24, 48\}$ with the partial order of divisibility.

In Exercises 9 through 16, determine the greatest and least elements, if they exist, of the poset.

9.

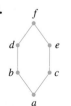

Figure 6.31

10.

Figure 6.32

11.

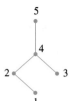

Figure 6.33

12.

Figure 6.34

13. $A = \{x \mid x$ is a real number and $0 < x < 1\}$ with the usual partial order $\leq$.

14. $A = \{x \mid x$ is a real number and $0 \leq x \leq 1\}$ with the usual partial order $\leq$.

15. $A = \{2, 4, 6, 8, 12, 18, 24, 36, 72\}$ with the partial order of divisibility.

16. $A = \{2, 3, 4, 6, 12, 18, 24, 36\}$ with the partial order of divisibility.

In Exercises 17 and 18, determine if the statements are equivalent. Justify your conclusion.

17. (a) If $a \in A$ is a maximal element, then there is no $c \in A$ such that $a < c$.

(b) If $a \in A$ is a maximal element, then for all $b \in A$, $b \leq a$.

18. (a) If $a \in A$ is a minimal element, then there is no $c \in A$ such that $c < a$.

(b) If $a \in A$ is a minimal element, then for all $b \in A$, $a \leq b$.

19. Determine if the given statement is true or false. Explain your reasoning.

(a) A non-empty finite poset has a maximal element.

(b) A non-empty finite poset has a greatest element.

(c) A non-empty finite poset has a minimal element.

(d) A non-empty finite poset has a least element.

20. Prove that if $(A, \leq)$ has a greatest element, then $(A, \leq)$ has a unique greatest element.

21. Prove that if $(A, \leq)$ has a least element, then $(A, \leq)$ has a unique least element.

22. Prove Theorem 3.

In Exercises 23 through 30 find, if they exist, (a) all upper bounds of B; (b) all lower bounds of B; (c) the least upper bound of B; (d) the greatest lower bound of B.

23.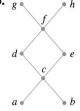

$B = \{c, d, e\}$

Figure 6.35

24.

$B = \{1, 2, 3, 4, 5\}$

Figure 6.36

25.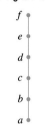

$B = \{b, c, d\}$

Figure 6.37

26.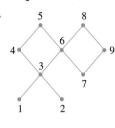

$B = \{3, 4, 6\}$

Figure 6.38

27. $(A, \leq)$ is the poset in Exercise 23; $B = \{b, g, h\}$.

28. (a) $(A, \leq)$ is the poset in Exercise 26; $B = \{4, 6, 9\}$.

   (b) $(A, \leq)$ is the poset in Exercise 26; $B = \{3, 4, 8\}$.

29. $A = \mathbb{R}$ and $\leq$ denotes the usual partial order; $B = \{x \mid x \text{ is a real number and } 1 < x < 2\}$.

30. $A = \mathbb{R}$ and $\leq$ denotes the usual partial order; $B = \{x \mid x \text{ is a real number and } 1 \leq x < 2\}$.

31. Construct the Hasse diagram of a topological sorting of the poset whose Hasse diagram is shown in Figure 6.35. Use the algorithm SORT.

32. Construct the Hasse diagram of a topological sorting of the poset whose Hasse diagram is shown in Figure 6.36. Use the algorithm SORT.

33. Let R be a partial order on a finite set A. Describe how to use $\mathbf{M}_R$ to find the least and greatest elements of A if they exist.

6.3 LATTICES

A **lattice** is a poset $(L, \leq)$ in which every subset $\{a, b\}$ consisting of two elements has a least upper bound and a greatest lower bound. We denote LUB($\{a, b\}$) by $a \vee b$ and call it the **join** of a and b. Similarly, we denote GLB($\{a, b\}$) by $a \wedge b$ and call it the **meet** of a and b. Lattice structures often appear in computing and mathematical applications. Observe that a lattice is a mathematical structure as described in Section 1.6, with two binary operations, join and meet.

EXAMPLE 1 Let S be a set and let $L = P(S)$. As we have seen, $\subseteq$, containment, is a partial order on L. Let A and B belong to the poset $(L, \subseteq)$. Then $A \vee B$ is the set $A \cup B$. To see this, note that $A \subseteq A \cup B$, $B \subseteq A \cup B$, and, if $A \subseteq C$ and $B \subseteq C$, then it follows that $A \cup B \subseteq C$. Similarly, we can show that the element $A \wedge B$ in $(L, \subseteq)$ is the set $A \cap B$. Thus, L is a lattice. ■

EXAMPLE 2 Consider the poset $(\mathbb{Z}^+, \leq)$, where for a and b in $\mathbb{Z}^+$, $a \leq b$ if and only if $a \mid b$. Then L is a lattice in which the join and meet of a and b are their least common multiple and greatest common divisor, respectively (see Section 1.4). That is,

$$a \vee b = \text{LCM}(a, b) \quad \text{and} \quad a \wedge b = \text{GCD}(a, b).$$ ■

EXAMPLE 3 Let n be a positive integer and let D_n be the set of all positive divisors of n. Then D_n is a lattice under the relation of divisibility as considered in Example 2. Thus, if $n = 20$, we have $D_{20} = \{1, 2, 4, 5, 10, 20\}$. The Hasse diagram of D_{20} is shown

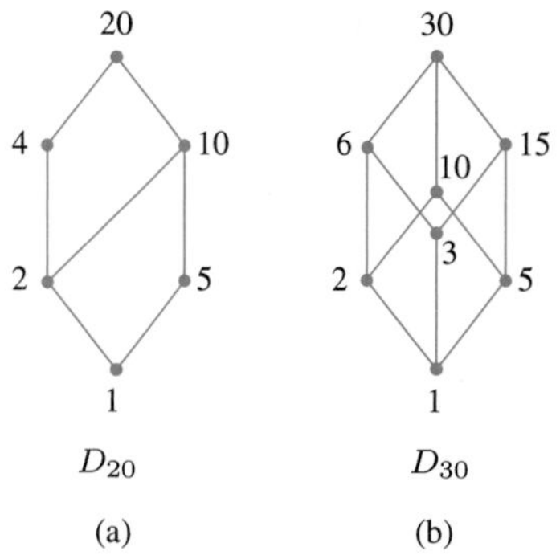

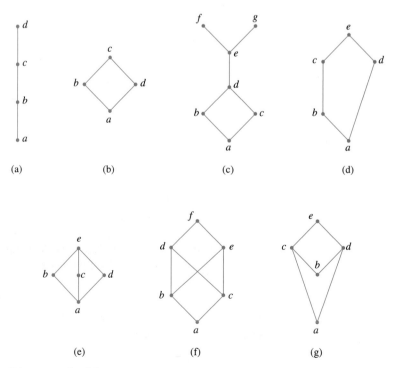

Figure 6.39

in Figure 6.39(a). If $n = 30$, we have $D_{30} = \{1, 2, 3, 5, 6, 10, 15, 30\}$. The Hasse diagram of D_{30} is shown in Figure 6.39(b).   ■

EXAMPLE 4   Which of the Hasse diagrams in Figure 6.40 represent lattices?

Solution   Hasse diagrams (a), (b), (d), and (e) represent lattices. Diagram (c) does not represent a lattice because $f \vee g$ does not exist. Diagram (f) does not represent a lattice because neither $d \wedge e$ nor $b \vee c$ exist. Diagram (g) does not represent a lattice because $c \wedge d$ does not exist.   ■

Figure 6.40

EXAMPLE 5

We have already observed in Example 4 of Section 6.1 that the set $\mathcal{R}$ of all equivalence relations on a set A is a poset under the partial order of set containment. We can now conclude that $\mathcal{R}$ is a lattice where the meet of the equivalence relations R and S is their intersection $R \cap S$ and their join is $(R \cup S)^\infty$, the transitive closure of their union (see Section 4.8). ∎

Let $(L, \leq)$ be a poset and let $(L, \geq)$ be the dual poset. If $(L, \leq)$ is a lattice, we can show that $(L, \geq)$ is also a lattice. In fact, for any a and b in L, the least upper bound of a and b in $(L, \leq)$ is equal to the greatest lower bound of a and b in $(L, \geq)$. Similarly, the greatest lower bound of a and b in $(L, \leq)$ is equal to the least upper bound of a and b in $(L, \geq)$. If L is a finite set, this property can easily be seen by examining the Hasse diagrams of the poset and its dual.

EXAMPLE 6

Let S be a set and $L = P(S)$. Then $(L, \subseteq)$ is a lattice, and its dual lattice is $(L, \supseteq)$, where $\subseteq$ is "contained in" and $\supseteq$ is "contains." The discussion preceding this example then shows that in the poset $(L, \supseteq)$ the join $A \vee B$ is the set $A \cap B$, and the meet $A \wedge B$ is the set $A \cup B$. ∎

Theorem 1

If $(L_1, \leq)$ and $(L_2, \leq)$ are lattices, then $(L, \leq)$ is a lattice, where $L = L_1 \times L_2$, and the partial order $\leq$ of L is the product partial order. ●

Proof   We denote the join and meet in L_1 by $\vee_1$ and $\wedge_1$, respectively, and the join and meet in L_2 by $\vee_2$ and $\wedge_2$, respectively. We already know from Theorem 1 of Section 6.1 that L is a poset. We now need to show that if (a_1, b_1) and $(a_2, b_2) \in L$, then $(a_1, b_1) \vee (a_2, b_2)$ and $(a_1, b_1) \wedge (a_2, b_2)$ exist in L. We leave it as an exercise to verify that

$$(a_1, b_1) \vee (a_2, b_2) = (a_1 \vee_1 a_2, b_1 \vee_2 b_2)$$
$$(a_1, b_1) \wedge (a_2, b_2) = (a_1 \wedge_1 a_2, b_1 \wedge_2 b_2).$$

Thus L is a lattice. ▼

EXAMPLE 7

Let L_1 and L_2 be the lattices shown in Figures 6.41(a) and (b), respectively. Then $L = L_1 \times L_2$ is the lattice shown in Figure 6.41(c). ∎

Let $(L, \leq)$ be a lattice. A nonempty subset S of L is called a **sublattice** of L if $a \vee b \in S$ and $a \wedge b \in S$ whenever $a \in S$ and $b \in S$.

EXAMPLE 8

The lattice D_n of all positive divisors of n (see Example 3) is a sublattice of the lattice Z^+ under the relation of divisibility (see Example 2). ∎

EXAMPLE 9

Consider the lattice L shown in Figure 6.42(a). The partially ordered subset S_b shown in Figure 6.42(b) is not a sublattice of L since $a \wedge b \notin S_b$. The partially ordered subset S_c in Figure 6.42(c) is not a sublattice of L since $a \vee b \notin S_c$. Observe, however, that S_c is a lattice when considered as a poset by itself. The partially ordered subset S_d in Figure 6.42(d) is a sublattice of L. ∎

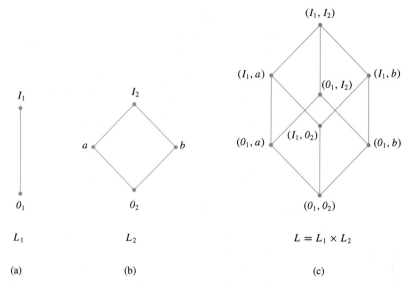

Figure 6.41

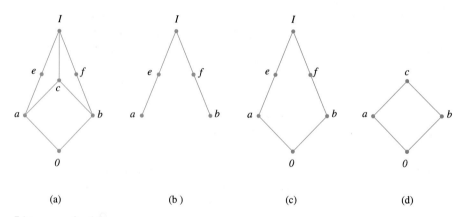

Figure 6.42

Isomorphic Lattices

If $f: L_1 \rightarrow L_2$ is an isomorphism from the poset $(L_1, \leq_1)$ to the poset $(L_2, \leq_2)$, then Theorem 4 of Section 6.2 tells us that L_1 is a lattice if and only if L_2 is a lattice. In fact, if a and b are elements of L_1, then $f(a \wedge b) = f(a) \wedge f(b)$ and $f(a \vee b) = f(a) \vee f(b)$. If two lattices are isomorphic, as posets, we say they are **isomorphic lattices**.

EXAMPLE 10   Let L be the lattice D_6, and let L' be the lattice $P(S)$ under the relation of containment, where $S = \{a, b\}$. These posets were discussed in Example 16 of Section 6.1, where they were shown to be isomorphic. Thus, since both are lattices, they are isomorphic lattices.   ■

If $f: A \to B$ is a one-to-one correspondence from a lattice $(A, \leq)$ to a set B, then we can use the function f to define a partial order $\leq'$ on B. If b_1 and b_2 are in B, then $b_1 = f(a_1)$ and $b_2 = f(a_2)$ for some unique elements a_1 and a_2 of A.

Define $b_1 \leq' b_2$ (in B) if and only if $a_1 \leq a_2$ (in A). If A and B are finite, then we can describe this process geometrically as follows. Construct the Hasse diagram for $(A, \leq)$. Then replace each label a by the corresponding element $f(a)$ of B. The result is the Hasse diagram of the partial order $\leq'$ on B.

When B is given the partial order $\leq'$, f will be an isomorphism from the poset $(A, \leq)$ to the poset $(B, \leq')$. To see this, note that f is already assumed to be a one-to-one correspondence. The definition of $\leq'$ states that, for any a_1 and a_2 in A, $a_1 \leq a_2$ if and only if $f(a_1) \leq' f(a_2)$. Thus f is an isomorphism. Since $(A, \leq)$ is a lattice, so is $(B, \leq')$, and they are isomorphic lattices.

EXAMPLE 11

If A is a set, let $\mathcal{R}$ be the set of all equivalence relations on A and let Π be the set of all partitions on A. In Example 13 of Section 5.1 we constructed a one-to-one correspondence f from $\mathcal{R}$ to Π. In Example 4 of Section 6.1, we considered the partial order $\subseteq$ on $\mathcal{R}$. From this partial order we can construct, using f as explained before, a partial order $\leq'$ on Π. By construction, if $\mathcal{P}_1$ and $\mathcal{P}_2$ are partitions of A, and R_1 and R_2, respectively, are the equivalence relations corresponding to these partitions, then $\mathcal{P}_1 \leq \mathcal{P}_2$ will mean that $R_1 \subseteq R_2$. Since we showed in Example 5 that $(\mathcal{R}, \subseteq)$ is a lattice, and we know that f is an isomorphism, it follows that $(\Pi, \leq')$ is also a lattice. In Exercise 33 we describe the partial order $\leq'$ directly in terms of the partitions themselves. ∎

Properties of Lattices

Before proving a number of the properties of lattices, we recall the meaning of $a \vee b$ and $a \wedge b$.

1. $a \leq a \vee b$ and $b \leq a \vee b$; $a \vee b$ is an upper bound of a and b.
2. If $a \leq c$ and $b \leq c$, then $a \vee b \leq c$; $a \vee b$ is the least upper bound of a and b.
1'. $a \wedge b \leq a$ and $a \wedge b \leq b$; $a \wedge b$ is a lower bound of a and b.
2'. If $c \leq a$ and $c \leq b$, then $c \leq a \wedge b$; $a \wedge b$ is the greatest lower bound of a and b.

Theorem 2 Let L be a lattice. Then for every a and b in L,

(a) $a \vee b = b$ if and only if $a \leq b$.
(b) $a \wedge b = a$ if and only if $a \leq b$.
(c) $a \wedge b = a$ if and only if $a \vee b = b$. ●

Proof

(a) Suppose that $a \vee b = b$. Since $a \leq a \vee b = b$, we have $a \leq b$. Conversely, if $a \leq b$, then, since $b \leq b$, b is an upper bound of a and b; so by definition of least upper bound we have $a \vee b \leq b$. Since $a \vee b$ is an upper bound, $b \leq a \vee b$, so $a \vee b = b$.

(b) The proof is analogous to the proof of part (a), and we leave it as an exercise for the reader.

(c) The proof follows from parts (a) and (b).   ▼

EXAMPLE 12   Let L be a linearly ordered set. If a and $b \in L$, then either $a \leq b$ or $b \leq a$. It follows from Theorem 2 that L is a lattice, since every pair of elements has a least upper bound and a greatest lower bound.   ■

Theorem 3   Let L be a lattice. Then

1. *Idempotent Properties*
   (a) $a \vee a = a$
   (b) $a \wedge a = a$

2. *Commutative Properties*
   (a) $a \vee b = b \vee a$
   (b) $a \wedge b = b \wedge a$

3. *Associative Properties*
   (a) $a \vee (b \vee c) = (a \vee b) \vee c$
   (b) $a \wedge (b \wedge c) = (a \wedge b) \wedge c$

4. *Absorption Properties*
   (a) $a \vee (a \wedge b) = a$
   (b) $a \wedge (a \vee b) = a$ ●

Proof

1. The statements follow from the definition of LUB and GLB.

2. The definition of LUB and GLB treat a and b symmetrically, so the results follow.

3. (a) From the definition of LUB, we have $a \leq a \vee (b \vee c)$ and $b \vee c \leq a \vee (b \vee c)$. Moreover, $b \leq b \vee c$ and $c \leq b \vee c$, so, by transitivity, $b \leq a \vee (b \vee c)$ and $c \leq a \vee (b \vee c)$. Thus $a \vee (b \vee c)$ is an upper bound of a and b, so by definition of least upper bound we have

$$a \vee b \leq a \vee (b \vee c).$$

Since $a \vee (b \vee c)$ is an upper bound of $a \vee b$ and c, we obtain

$$(a \vee b) \vee c \leq a \vee (b \vee c).$$

Similarly, $a \vee (b \vee c) \leq (a \vee b) \vee c$. By the antisymmetry of $\leq$, property 3(a) follows.

(b) The proof is analogous to the proof of part (a) and we omit it.

4. (a) Since $a \wedge b \leq a$ and $a \leq a$, we see that a is an upper bound of $a \wedge b$ and a; so $a \vee (a \wedge b) \leq a$. On the other hand, by the definition of LUB, we have $a \leq a \vee (a \wedge b)$, so $a \vee (a \wedge b) = a$.

(b) The proof is analogous to the proof of part (a) and we omit it.   ▼

It follows from property 3 that we can write $a \vee (b \vee c)$ and $(a \vee b) \vee c$ merely as $a \vee b \vee c$, and similarly for $a \wedge b \wedge c$. Moreover, we can write

$$\text{LUB}(\{a_1, a_2, \ldots, a_n\}) \quad \text{as} \quad a_1 \vee a_2 \vee \cdots \vee a_n$$
$$\text{GLB}(\{a_1, a_2, \ldots, a_n\}) \quad \text{as} \quad a_1 \wedge a_2 \wedge \cdots \wedge a_n,$$

since we can show by induction that these joins and meets are independent of the grouping of the terms.

Theorem 4 Let L be a lattice. Then, for every a, b, and c in L,

    1. If $a \leq b$, then
        (a) $a \vee c \leq b \vee c$.
        (b) $a \wedge c \leq b \wedge c$.
    2. $a \leq c$ and $b \leq c$ if and only if $a \vee b \leq c$.
    3. $c \leq a$ and $c \leq b$ if and only if $c \leq a \wedge b$.
    4. If $a \leq b$ and $c \leq d$, then
        (a) $a \vee c \leq b \vee d$.
        (b) $a \wedge c \leq b \wedge d$.

Proof The proof is left as an exercise. ▼

Special Types of Lattices

A lattice L is said to be **bounded** if it has a greatest element I and a least element 0 (see Section 6.2).

EXAMPLE 13 The lattice Z^+ under the partial order of divisibility, as defined in Example 2, is not a bounded lattice since it has a least element, the number 1, but no greatest element. ■

EXAMPLE 14 The lattice Z under the partial order $\leq$ is not bounded since it has neither a greatest nor a least element. ■

EXAMPLE 15 The lattice $P(S)$ of all subsets of a set S, as defined in Example 1, is bounded. Its greatest element is S and its least element is $\varnothing$. ■

If L is a bounded lattice, then for all $a \in A$,

$$0 \leq a \leq I$$
$$a \vee 0 = a, \quad a \wedge 0 = 0$$
$$a \vee I = I, \quad a \wedge I = a.$$

Theorem 5 Let $L = \{a_1, a_2, \ldots, a_n\}$ be a finite lattice. Then L is bounded.

Proof The greatest element of L is $a_1 \vee a_2 \vee \cdots \vee a_n$, and its least element is $a_1 \wedge a_2 \wedge \cdots \wedge a_n$. ■

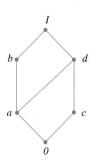

Figure 6.43

Note that the proof of Theorem 5 is a constructive proof. We show that L is bounded by constructing the greatest and the least elements.

A lattice L is called **distributive** if for any elements a, b, and c in L we have the following **distributive properties**:

1. $a \wedge (b \vee c) = (a \wedge b) \vee (a \wedge c)$
2. $a \vee (b \wedge c) = (a \vee b) \wedge (a \vee c)$

If L is not distributive, we say that L is **nondistributive**.

We leave it as an exercise to show that the distributive property holds when any two of the elements a, b, or c are equal or when any one of the elements is 0 or I. This observation reduces the number of cases that must be checked in verifying that a distributive property holds. However, verification of a distributive property is generally a tedious task.

EXAMPLE 16   For a set S, the lattice $P(S)$ is distributive, since union and intersection (the join and meet, respectively) each satisfy the distributive property shown in Section 1.2.   ■

EXAMPLE 17   The lattice shown in Figure 6.43 is distributive, as can be seen by verifying the distributive properties for all ordered triples chosen from the elements a, b, c, and d.   ■

EXAMPLE 18   Show that the lattices pictured in Figure 6.44 are nondistributive.

Solution

(a) We have
$$a \wedge (b \vee c) = a \wedge I = a$$
while
$$(a \wedge b) \vee (a \wedge c) = b \vee 0 = b.$$

(b) Observe that
$$a \wedge (b \vee c) = a \wedge I = a$$
while
$$(a \wedge b) \vee (a \wedge c) = 0 \vee 0 = 0.$$ ■

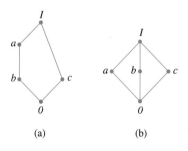

Figure 6.44

The nondistributive lattices discussed in Example 18 are useful for showing that a given lattice is nondistributive, as the following theorem, whose proof we omit, asserts.

Theorem 6   A lattice L is nondistributive if and only if it contains a sublattice that is isomorphic to one of the two lattices of Example 18.   ●

Theorem 6 can be used quite efficiently by inspecting the Hasse diagram of L.

Let L be a bounded lattice with greatest element I and least element 0, and let $a \in L$. An element $a' \in L$ is called a **complement** of a if
$$a \vee a' = I \quad \text{and} \quad a \wedge a' = 0.$$

Observe that
$$0' = I \quad \text{and} \quad I' = 0.$$

EXAMPLE 19 The lattice $L = P(S)$ is such that every element has a complement, since if $A \in L$, then its set complement $\overline{A}$ has the properties $A \vee \overline{A} = S$ and $A \wedge \overline{A} = \varnothing$. That is, the set complement is also the complement in the lattice L. ∎

EXAMPLE 20 The lattices in Figure 6.44 each have the property that every element has a complement. The element c in both cases has two complements, a and b. ∎

EXAMPLE 21 Consider the lattices D_{20} and D_{30} discussed in Example 3 and shown in Figure 6.39. Observe that every element in D_{30} has a complement. For example, if $a = 5$, then $a' = 6$. However, the elements 2 and 10 in D_{20} have no complements. ∎

Examples 20 and 21 show that an element a in a lattice need not have a complement, and it may have more than one complement. However, for a bounded distributive lattice, the situation is more restrictive, as shown by the following theorem.

Theorem 7 Let L be a bounded distributive lattice. If a complement exists, it is unique. ●

Proof Let a' and a'' be complements of the element $a \in L$. Then

$$a \vee a' = I, \quad a \vee a'' = I$$
$$a \wedge a' = 0, \quad a \wedge a'' = 0.$$

Using the distributive laws, we obtain

$$a' = a' \vee 0 = a' \vee (a \wedge a'')$$
$$= (a' \vee a) \wedge (a' \vee a'')$$
$$= I \wedge (a' \vee a'') = a' \vee a''.$$

Also,

$$a'' = a'' \vee 0 = a'' \vee (a \wedge a')$$
$$= (a'' \vee a) \wedge (a'' \vee a')$$
$$= I \wedge (a' \vee a'') = a' \vee a''.$$

Hence $a' = a''$. ▼

The proof of Theorem 7 is a direct proof, but it is not obvious how the representations of a' and a'' were chosen. There is some trial and error involved in this sort of proof, but we expect to use the hypothesis that L is bounded and that L is distributive. An alternative proof is outlined in Exercise 34.

A lattice L is called **complemented** if it is bounded and if every element in L has a complement.

EXAMPLE 22 The lattice $L = P(S)$ is complemented. Observe that in this case each element of L has a unique complement, which can be seen directly or is implied by Theorem 7.

∎

EXAMPLE 23 The lattices discussed in Example 20 and shown in Figure 6.44 are complemented. In this case, the complements are not unique. ∎

6.3 Exercises

In Exercises 1 through 6 (Figures 6.45 through 6.50), determine whether the Hasse diagram represents a lattice.

1.

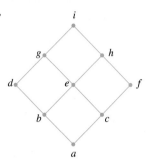

Figure 6.45

2.

Figure 6.46

3.

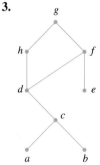

Figure 6.47

4.

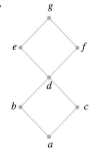

Figure 6.48

5.

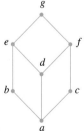

Figure 6.49

6.

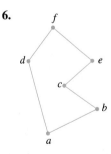

Figure 6.50

7. Is the poset $A = \{2, 3, 6, 12, 24, 36, 72\}$ under the relation of divisibility a lattice?

8. Amplify the explanations in the solution of Example 4 by explaining why the specified object does not exist.

9. If L_1 and L_2 are the lattices shown in Figure 6.51, draw the Hasse diagram of $L_1 \times L_2$ with the product partial order.

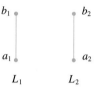

Figure 6.51

10. Complete the proof of Theorem 1 by verifying that $(a_1, b_1) \vee (a_2, b_2) = (a_1 \vee_1 a_2, b_1 \vee_2 b_2)$ and $(a_1, b_1) \wedge (a_2, b_2) = (a_1 \wedge_1 a_2, b_1 \wedge_2 b_2)$.

11. Let $L = P(S)$ be the lattice of all subsets of a set S under the relation of containment. Let T be a subset of S. Show that $P(T)$ is a sublattice of L.

12. Let L be a lattice and let a and b be elements of L such that $a \leq b$. The **interval** $[a, b]$ is defined as the set of all $x \in L$ such that $a \leq x \leq b$. Prove that $[a, b]$ is a sublattice of L.

13. Show that a subset of a linearly ordered poset is a sublattice.

14. Find all sublattices of D_{24} that contain at least five elements.

15. Give the Hasse diagrams of all nonisomorphic lattices that have one, two, three, four, or five elements.

16. Show that if a bounded lattice has two or more elements, then $0 \neq I$.

17. Prove Theorem 2(b).

18. Show that the lattice D_n is distributive for any n.

19. (a) In Example 17 (Figure 6.43), how many ordered triples must be checked to see if the lattice is distributive?

(b) In Example 18 (Figure 6.44), what is the maximum number of ordered triples that would need to be checked to show that the lattice is not distributive?

20. Show that a sublattice of a distributive lattice is distributive.

21. Show that if L_1 and L_2 are distributive lattices, then $L = L_1 \times L_2$ is also distributive, where the order of L is the product of the orders in L_1 and L_2.

22. Prove that if a and b are elements in a bounded, distributive lattice and if a has a complement a', then

$$a \vee (a' \wedge b) = a \vee b$$
$$a \wedge (a' \vee b) = a \wedge b.$$

23. Let L be a distributive lattice. Show that if there exists an a with $a \wedge x = a \wedge y$ and $a \vee x = a \vee y$, then $x = y$.

24. A lattice is said to be **modular** if, for all a, b, c, $a \le c$ implies that $a \vee (b \wedge c) = (a \vee b) \wedge c$.

  (a) Show that a distributive lattice is modular.

  (b) Show that the lattice shown in Figure 6.52 is a nondistributive lattice that is modular.

Figure 6.52

25. Find the complement of each element in D_{42}.

In Exercises 26 *through* 29 (*Figures* 6.53 *through* 6.56), *determine whether each lattice is distributive, complemented, or both.*

26.

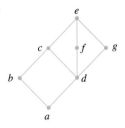

Figure 6.53

27.

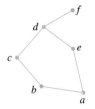

Figure 6.54

28.

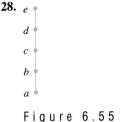

Figure 6.55

29.

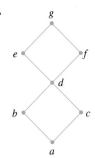

Figure 6.56

30. Prove Theorem 4, part (2).

31. Let L be a bounded lattice with at least two elements. Show that no element of L is its own complement.

32. Consider the complemented lattice shown in Figure 6.57. Give the complements of each element.

Figure 6.57

33. Let $\mathcal{P}_1 = \{A_1, A_2, \ldots\}$, $\mathcal{P}_2 = \{B_1, B_2, \ldots\}$ be two partitions of a set S. Show that $\mathcal{P}_1 \le \mathcal{P}_2$ (see the definition in Example 11) if and only if each A_i is contained in some B_j.

34. Complete the following proof of Theorem 7.
Proof: Let a' and a'' be complements of $a \in L$. Then $a' = a' \wedge I = $ _____. Also, $a'' = a'' \wedge I = $ _____. Hence $a' = a''$.

6.4 FINITE BOOLEAN ALGEBRAS

In this section we discuss a certain type of lattice that has a great many applications in computer science. We have seen in Example 6 of Section 6.3 that if S is a set, $L = P(S)$, and $\subseteq$ is the usual relation of containment, then the poset $(L, \subseteq)$ is a lattice. These lattices have many properties that are not shared by lattices in general. For this reason they are easier to work with, and they play a more important role in various applications.

    We will restrict our attention to the lattices $(P(S), \subseteq)$, where S is a finite set, and we begin by finding all essentially different examples.

Theorem 1   If $S_1 = \{x_1, x_2, \ldots, x_n\}$ and $S_2 = \{y_1, y_2, \ldots, y_n\}$ are any two finite sets with n elements, then the lattices $(P(S_1), \subseteq)$ and $(P(S_2), \subseteq)$ are isomorphic. Consequently, the Hasse diagrams of these lattices may be drawn identically.

Proof   Arrange the sets as shown in Figure 6.58 so that each element of S_1 is directly over the correspondingly numbered element in S_2. For each subset A of S_1, let $f(A)$ be the subset of S_2 consisting of all elements that correspond to the elements of A. Figure 6.59 shows a typical subset A of S_1 and the corresponding subset $f(A)$ of S_2. It is easily seen that the function f is a one-to-one correspondence from subsets of S_1 to subsets of S_2. Equally clear is the fact that if A and B are any subsets of S_1, then $A \subseteq B$ if and only if $f(A) \subseteq f(B)$. We omit the details. Thus the lattices $(P(S_1), \subseteq)$ and $(P(S_2), \subseteq)$ are isomorphic.   ∎

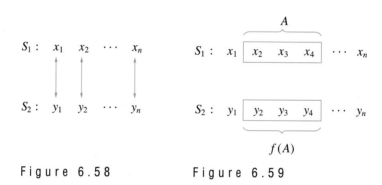

Figure 6.58   Figure 6.59

The essential point of this theorem is that the lattice $(P(S), \subseteq)$ is completely determined as a poset by the number $|S|$ and does not depend in any way on the nature of the elements in S.

EXAMPLE 1   Figures 6.60(a) and (b) show Hasse diagrams for the lattices $(P(S), \subseteq)$ and $(P(T), \subseteq)$, respectively, where $S = \{a, b, c\}$ and $T = \{2, 3, 5\}$. It is clear from this figure that the two lattices are isomorphic. In fact, we see that one possible isomorphism $f: S \to T$ is given by

$$f(\{a\}) = \{2\}, \qquad f(\{b\}) = \{3\}, \qquad f(\{c\}) = \{5\},$$
$$f(\{a, b\}) = \{2, 3\}, \qquad f(\{b, c\}) = \{3, 5\}, \qquad f(\{a, c\}) = \{2, 5\},$$
$$f(\{a, b, c\}) = \{2, 3, 5\}, \qquad f(\varnothing) = \varnothing.$$ ∎

Thus, for each $n = 0, 1, 2, \ldots$, there is only one type of lattice having the form $(P(S), \subseteq)$. This lattice depends only on n, not on S, and it has 2^n elements, as was shown in Example 2 of Section 3.1. Recall from Section 1.3 that if a set S has n elements, then all subsets of S can be represented by sequences of 0's and 1's of length n. We can therefore label the Hasse diagram of a lattice $(P(S), \subseteq)$ by such sequences. In doing so, we free the diagram from dependence on a particular set S and emphasize the fact that it depends only on n.

EXAMPLE 2   Figure 6.60(c) shows how the diagrams that appear in Figures 6.60(a) and (b) can be labeled by sequences of 0's and 1's. This labeling serves equally well to describe

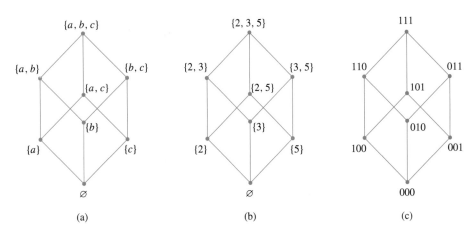

Figure 6.60

the lattice of Figure 6.60(a) or (b), or for that matter the lattice $(P(S), \subseteq)$ that arises from any set S having three elements. ∎

If the Hasse diagram of the lattice corresponding to a set with n elements is labeled by sequences of 0's and 1's of length n, as described previously, then the resulting lattice is named B_n. The properties of the partial order on B_n can be described directly as follows. If $x = a_1 a_2 \cdots a_n$ and $y = b_1 b_2 \cdots b_n$ are two elements of B_n, then

1. $x \le y$ if and only if $a_k \le b_k$ (as numbers 0 or 1) for $k = 1, 2, \ldots, n$.
2. $x \wedge y = c_1 c_2 \cdots c_n$, where $c_k = \min\{a_k, b_k\}$.
3. $x \vee y = d_1 d_2 \cdots d_n$, where $d_k = \max\{a_k, b_k\}$.
4. x has a complement $x' = z_1 z_2 \cdots z_n$, where $z_k = 1$ if $x_k = 0$, and $z_k = 0$ if $x_k = 1$.

The truth of these statements can be seen by noting that $(B_n, \le)$ is isomorphic with $(P(S), \subseteq)$, so each x and y in B_n correspond to subsets A and B of S. Then $x \le y$, $x \wedge y$, $x \vee y$, and x' correspond to $A \subseteq B$, $A \cap B$, $A \cup B$, and $\overline{A}$ (set complement), respectively (verify). Figure 6.61 shows the Hasse diagrams of the lattices B_n for $n = 0, 1, 2, 3$.

We have seen that each lattice $(P(S), \subseteq)$ is isomorphic with B_n, where $n = |S|$. Other lattices may also be isomorphic with one of the B_n and thus possess all the special properties that the B_n possess.

EXAMPLE 3   In Example 17 of Section 6.1, we considered the lattice D_6 consisting of all positive integer divisors of 6 under the partial order of divisibility. The Hasse diagram of D_6 is shown in that example, and we now see that D_6 is isomorphic with B_2. In fact, $f: D_6 \to B_2$ is an isomorphism, where

$$f(1) = 00, \quad f(2) = 10, \quad f(3) = 01, \quad f(6) = 11.$$

∎

We are therefore led to make the following definition. A finite lattice is called a **Boolean algebra** if it is isomorphic with B_n for some nonnegative integer n. Thus

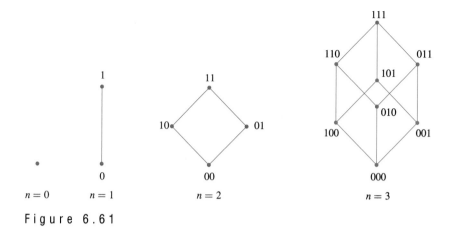

Figure 6.61

each B_n is a Boolean algebra and so is each lattice $(P(S), \subseteq)$, where S is a finite set. Example 3 shows that D_6 is also a Boolean algebra.

We will work only with finite posets in this section. For the curious, however, we note that there are infinite posets that share all the relevant properties of the lattices $(P(S), \subseteq)$ (for infinite sets S, of course), but that are not isomorphic with one these lattices. This necessitates the restriction of our definition of Boolean algebra to the finite case, which is sufficient for the applications that we present.

EXAMPLE 4   Consider the lattices D_{20} and D_{30} of all positive integer divisors of 20 and 30, respectively, under the partial order of divisibility. These posets were introduced in Example 3 of Section 6.3, and their Hasse diagrams were shown in Figure 6.39. Since D_{20} has six elements and $6 \neq 2^n$ for any integer $n \geq 0$, we conclude that D_{20} is not a Boolean algebra. The poset D_{30} has eight elements, and since $8 = 2^3$, it could be a Boolean algebra. By comparing Figure 6.39(b) and Figure 6.61, we see that D_{30} is isomorphic with B_3. In fact, we see that the one-to-one correspondence $f: D_{30} \rightarrow B_3$ defined by

$$f(1) = 000, \quad f(2) = 100, \quad f(3) = 010,$$
$$f(5) = 001, \quad f(6) = 110, \quad f(10) = 101,$$
$$f(15) = 011, \quad f(30) = 111$$

is an isomorphism. Thus D_{30} is a Boolean algebra.   ■

If a finite lattice L does not contain 2^n elements for some nonnegative integer n, we know that L cannot be a Boolean algebra. If $|L| = 2^n$, then L may or may not be a Boolean algebra. If L is relatively small, we may be able to compare its Hasse diagram with the Hasse diagram of B_n. In this way we saw in Example 4 that D_{30} is a Boolean algebra. However, this technique may not be practical if L is large. In that case, we may be able to show that L is a Boolean algebra by directly constructing an isomorphism with some B_n or, equivalently, with $(P(S), \subseteq)$ for some finite set S. Suppose, for example, that we want to know whether a lattice D_n is a Boolean algebra, and we want a method that works no matter how large n is. The following theorem gives a partial answer.

Theorem 2   Let

$$n = p_1 p_2 \cdots p_k,$$

where the p_i are distinct primes. Then D_n is a Boolean algebra.   ●

Proof   Let $S = \{p_1, p_2, \ldots, p_k\}$. If $T \subseteq S$ and a_T is the product of the primes in T, then $a_T \mid n$. Any divisor of n must be of the form a_T for some subset T of S (where we let $a_\varnothing = 1$). The reader may verify that if V and T are subsets of S, $V \subseteq T$ if and only if $a_V \mid a_T$. Also, it follows from the proof of Theorem 6 of Section 1.4 that $a_{V \cap T} = a_V \wedge a_T = \text{GCD}(a_V, a_T)$ and $a_{V \cup T} = a_V \vee a_T = \text{LCM}(a_V, a_T)$. Thus the function $f : P(S) \rightarrow D_n$ given by $f(T) = a_T$ is an isomorphism from $P(S)$ to D_n. Since $P(S)$ is a Boolean algebra, so is D_n.   ▼

EXAMPLE 5   Since $210 = 2 \cdot 3 \cdot 5 \cdot 7$, $66 = 2 \cdot 3 \cdot 11$, and $646 = 2 \cdot 17 \cdot 19$, we see from Theorem 2 that D_{210}, D_{66}, and D_{646} are all Boolean algebras.   ■

In other cases of large lattices L, we may be able to show that L is not a Boolean algebra by showing that the partial order of L does not have the necessary properties. A Boolean algebra is isomorphic with some B_n and therefore with some lattice $(P(S), \subseteq)$. Thus a Boolean algebra L must be a bounded lattice and a complemented lattice (see Section 6.3). In other words, it will have a greatest element I corresponding to the set S and a least element 0 corresponding to the subset $\varnothing$. Also, every element x of L will have a complement x'. According to Example 16, Section 6.3, L must also be distributive. The principle of correspondence (see Section 6.1) then tells us that the following rule holds.

Theorem 3
Substitution Rule for
Boolean Algebras   Any formula involving $\cup$ or $\cap$ that holds for arbitrary subsets of a set S will continue to hold for arbitrary elements of a Boolean algebra L if $\wedge$ is substituted for $\cap$ and $\vee$ for $\cup$.   ●

EXAMPLE 6   If L is any Boolean algebra and x, y, and z are in L, then the following three properties hold.

    1. $(x')' = x$ **Involution Property**
    2. $(x \wedge y)' = x' \vee y'$ $\Big\}$
    3. $(x \vee y)' = x' \wedge y'$ **De Morgan's Laws**

This is true by the substitution rule for Boolean algebras, since we know that the corresponding formulas

    $1'$. $\overline{(\overline{A})} = A$
    $2'$. $\overline{(A \cap B)} = \overline{A} \cup \overline{B}$
    $3'$. $\overline{(A \cup B)} = \overline{A} \cap \overline{B}$

hold for arbitrary subsets A and B of a set S.   ■

In a similar way, we can list other properties that must hold in any Boolean algebra by the substitution rule. Next we summarize all the basic properties of a

Boolean algebra $(L, \leq)$ and, next to each one, we list the corresponding property for subsets of a set S. We suppose that x, y, and z are arbitrary elements in L, and A, B, and C are arbitrary subsets of S. Also, we denote the greatest and least elements of L by I and 0, respectively.

1. $x \leq y$ if and only if $x \vee y = y$.
1'. $A \subseteq B$ if and only if $A \cup B = B$.

2. $x \leq y$ if and only if $x \wedge y = x$.
2'. $A \subseteq B$ if and only if $A \cap B = A$.

3. (a) $x \vee x = x$.
3'. (a) $A \cup A = A$.

    (b) $x \wedge x = x$.
    (b) $A \cap A = A$.

4. (a) $x \vee y = y \vee x$.
4'. (a) $A \cup B = B \cup A$.

    (b) $x \wedge y = y \wedge x$.
    (b) $A \cap B = B \cap A$.

5. (a) $x \vee (y \vee z) = (x \vee y) \vee z$.
5'. (a) $A \cup (B \cup C) = (A \cup B) \cup C$.

    (b) $x \wedge (y \wedge z) = (x \wedge y) \wedge z$.
    (b) $A \cap (B \cap C) = (A \cap B) \cap C$.

6. (a) $x \vee (x \wedge y) = x$.
6'. (a) $A \cup (A \cap B) = A$.

    (b) $x \wedge (x \vee y) = x$.
    (b) $A \cap (A \cup B) = A$.

7. $0 \leq x \leq I$ for all x in L.
7'. $\varnothing \subseteq A \subseteq S$ for all A in $P(S)$.

8. (a) $x \vee 0 = x$.
8'. (a) $A \cup \varnothing = A$.

    (b) $x \wedge 0 = 0$.
    (b) $A \cap \varnothing = \varnothing$.

9. (a) $x \vee I = I$.
9'. (a) $A \cup S = S$.

    (b) $x \wedge I = x$.
    (b) $A \cap S = A$.

10. (a) $x \wedge (y \vee z)$
     $= (x \wedge y) \vee (x \wedge z)$.
10'. (a) $A \cap (B \cup C)$
      $= (A \cap B) \cup (A \cap C)$.

    (b) $x \vee (y \wedge z)$
     $= (x \vee y) \wedge (x \vee z)$.
    (b) $A \cup (B \cap C)$
      $= (A \cup B) \cap (A \cup C)$.

11. Every element x has a *unique* complement x' satisfying
    (a) $x \vee x' = I$.
    (b) $x \wedge x' = 0$.
11'. Every element A has a *unique* complement $\overline{A}$ satisfying
    (a) $A \cup \overline{A} = S$.
    (b) $A \cap \overline{A} = \varnothing$.

12. (a) $0' = I$.
12'. (a) $\overline{\varnothing} = S$.

    (b) $I' = 0$.
    (b) $\overline{S} = \varnothing$.

13. $(x')' = x$.
13'. $\overline{(\overline{A})} = A$.

14. (a) $(x \wedge y)' = x' \vee y'$.
14'. (a) $\overline{(A \cap B)} = \overline{A} \cup \overline{B}$.

    (b) $(x \vee y)' = x' \wedge y'$.
    (b) $\overline{(A \cup B)} = \overline{A} \cap \overline{B}$.

Thus we may be able to show that a lattice L is not a Boolean algebra by showing that it does not possess one or more of these properties.

Figure 6.62

EXAMPLE 7   Show that the lattice whose Hasse diagram is shown in Figure 6.62 is not a Boolean algebra.

Solution   Elements a and e are both complements of c; that is, they both satisfy properties 11(a) and 11(b) with respect to the element c. But property 11 says that such an element is unique in any Boolean algebra. Thus the given lattice cannot be a Boolean algebra.   ∎

EXAMPLE 8   Show that if n is a positive integer and $p^2 \mid n$, where p is a prime number, then D_n is not a Boolean algebra.

Solution   Suppose that $p^2 \mid n$ so that $n = p^2 q$ for some positive integer q. Since p is also a divisor of n, p is an element of D_n. Thus, by the remarks given previously, if D_n is a Boolean algebra, then p must have a complement p'. Then GCD$(p, p') = 1$ and LCM$(p, p') = n$. By Theorem 6 of Section 1.4, $pp' = n$, so $p' = n/p = pq$. This shows that GCD$(p, pq) = 1$, which is impossible, since p and pq have p as a common divisor. Hence D_n cannot be a Boolean algebra. ∎

If we combine Example 8 and Theorem 2, we see that D_n is a Boolean algebra if and only if n is the product of distinct primes, that is, if and only if no prime divides n more than once.

EXAMPLE 9   If $n = 40$, then $n = 2^3 \cdot 5$, so 2 divides n three times. If $n = 75$, then $n = 3 \cdot 5^2$, so 5 divides n twice. Thus neither D_{40} nor D_{75} are Boolean algebras. ∎

Let us summarize what we have shown about Boolean algebras. We may attempt to show that a lattice L is a Boolean algebra by examining its Hasse diagram or constructing directly an isomorphism between L and B_n or $(P(S), \subseteq)$. We may attempt to show that L is not a Boolean algebra by checking the number of elements in L or the properties of its partial order. If L is a Boolean algebra, then we may use any of the properties 1 through 14 to manipulate or rewrite expressions involving elements of L. Simply proceed as if the elements were subsets and the manipulations were those that arise in set theory. We call such a lattice an algebra, because we use properties 1 through 14 just as the properties of real numbers are used in high school algebra.

From now on we will denote the Boolean algebra B_1 simply as B. Thus B contains only the two elements 0 and 1. It is sometimes useful to know that any of the Boolean algebras B_n can be described in terms of B. The following theorem gives this description.

Theorem 4   For any $n \geq 1$, B_n is the product $B \times B \times \cdots \times B$ of B, n factors, where $B \times B \times \cdots \times B$ is given the product partial order.   ●

Proof   By definition, B_n consists of all n-tuples of 0's and 1's, that is, all n-tuples of elements from B. Thus, as a set, B_n is equal to $B \times B \times \cdots \times B$ (n factors). Moreover, if $x = x_1 x_2 \cdots x_n$ and $y = y_1 y_2 \cdots y_n$ are two elements of B_n, then we know that

$$x \leq y \quad \text{if and only if} \quad x_k \leq y_k \qquad \text{for all } k.$$

Thus B_n, identified with $B \times B \times \cdots \times B$ (n factors), has the product partial order.   ▼

6.4 Exercises

In Exercises 1 through 10, determine whether the poset is a Boolean algebra. Explain.

1.

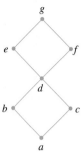

Figure 6.63

2.

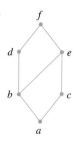

Figure 6.64

3.

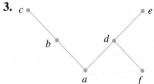

Figure 6.65

4.

Figure 6.66

5.

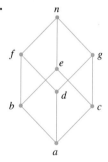

Figure 6.67

6.

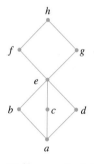

Figure 6.68

7.

Figure 6.69

8.

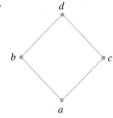

Figure 6.70

9. D_{385}

10. D_{60}

11. Are there any Boolean algebras having three elements? Why or why not?

12. Show that in a Boolean algebra, for any a and b, $a \leq b$ if and only if $b' \leq a'$.

13. Show that in a Boolean algebra, for any a and b, $a = b$ if and only if $(a \wedge b') \vee (a' \wedge b) = 0$.

14. Show that in a Boolean algebra, for any a, b, and c, if $a \leq b$, then $a \vee c \leq b \vee c$.

15. Show that in a Boolean algebra, for any a, b, and c, if $a \leq b$, then $a \wedge c \leq b \wedge c$.

16. Show that in a Boolean algebra the following statements are equivalent for any a and b.
(a) $a \vee b = b$
(b) $a \wedge b = a$
(c) $a' \vee b = I$
(d) $a \wedge b' = 0$
(e) $a \leq b$

17. Show that in a Boolean algebra, for any a and b,
$$(a \wedge b) \vee (a \wedge b') = a.$$

18. Show that in a Boolean algebra, for any a and b,
$$b \wedge (a \vee (a' \wedge (b \vee b'))) = b.$$

19. Show that in a Boolean algebra, for any a, b, and c,
$$(a \wedge b \wedge c) \vee (b \wedge c) = b \wedge c.$$

20. Show that in a Boolean algebra, for any a, b, and c,
$$((a \vee c) \wedge (b' \vee c))' = (a' \vee b) \wedge c'.$$

21. Show that in a Boolean algebra, for any a, b, and c, if $a \leq b$, then
$$a \vee (b \wedge c) = b \wedge (a \vee c).$$

22. Explain the connection between Examples 7 and 8.

For Exercises 23 through 26, let $A = \{a, b, c, d, e, f, g, h\}$ and R be the relation defined by

$$\mathbf{M}_R = \begin{bmatrix} 1 & 1 & 1 & 1 & 0 & 0 & 0 & 0 \\ 0 & 1 & 0 & 1 & 0 & 0 & 0 & 0 \\ 0 & 0 & 1 & 1 & 0 & 0 & 0 & 0 \\ 0 & 0 & 0 & 1 & 0 & 0 & 0 & 0 \\ 1 & 1 & 1 & 1 & 1 & 1 & 1 & 1 \\ 0 & 1 & 0 & 1 & 0 & 1 & 0 & 1 \\ 0 & 0 & 1 & 1 & 0 & 0 & 1 & 1 \\ 0 & 0 & 0 & 1 & 0 & 0 & 0 & 1 \end{bmatrix}.$$

23. Show that (A, R) is a poset.

24. Does the poset (A, R) have a least element? a greatest element? If so, identify them.

25. Show that the poset (A, R) is complemented and give all pairs of complements.

26. Prove or disprove that (A, R) is a Boolean algebra.

27. Let $A = \{a, b, c, d, e, f, g, h\}$ and R be the relation

defined by

$$\mathbf{M}_R = \begin{bmatrix} 1 & 1 & 1 & 1 & 1 & 1 & 1 & 1 \\ 0 & 1 & 0 & 0 & 1 & 1 & 1 & 1 \\ 0 & 0 & 1 & 0 & 1 & 1 & 1 & 1 \\ 0 & 0 & 0 & 1 & 0 & 1 & 1 & 1 \\ 0 & 0 & 0 & 0 & 1 & 1 & 1 & 1 \\ 0 & 0 & 0 & 0 & 0 & 1 & 0 & 1 \\ 0 & 0 & 0 & 0 & 0 & 0 & 1 & 1 \\ 0 & 0 & 0 & 0 & 0 & 0 & 0 & 1 \end{bmatrix}.$$

Prove or disprove that (A, R) is a Boolean algebra.

6.5 FUNCTIONS ON BOOLEAN ALGEBRAS

Tables listing the values of a function f for all elements of B_n, such as shown in Figure 6.71(a), are often called truth tables for f. This is because they are analogous with tables that arise in logic (see Section 2.1). Suppose that the x_k represent propositions, and $f(x_1, x_2, \ldots, x_n)$ represents a compound sentence constructed from the x_k's. If we think of the value 0 for a sentence as meaning that the sentence is false, and 1 as meaning that the sentence is true, then tables such as Figure 6.71(a) show us how truth or falsity of $f(x_1, x_2, \ldots, x_n)$ depends on the truth or falsity of its component sentences x_k. Thus such tables are called **truth tables**, even when they arise in areas other than logic, such as in Boolean algebras.

x_1	x_2	x_3	$f(x_1, x_2, x_3)$
0	0	0	0
0	0	1	1
0	1	0	1
0	1	1	0
1	0	0	1
1	0	1	0
1	1	0	1
1	1	1	0

(a)

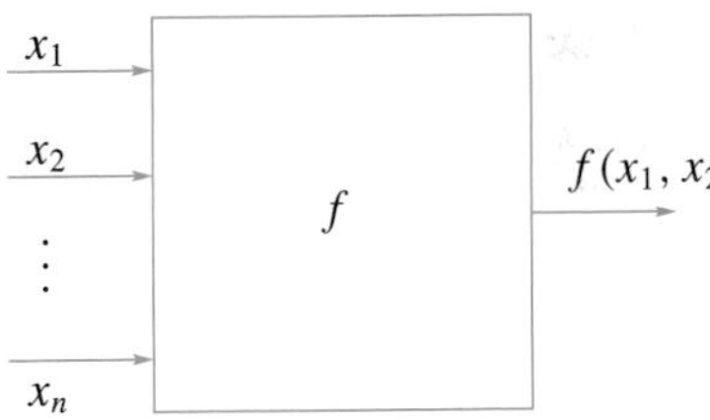

(b)

Figure 6.71

The reason that such functions are important is that, as shown schematically in Figure 6.71(b), they may be used to represent the output requirements of a circuit for any possible input values. Thus each x_i represents an input circuit capable of carrying two indicator voltages (one voltage for 0 and a different voltage for 1). The function f represents the desired output response in all cases. Such requirements occur at the design stage of all combinational and sequential computer circuitry.

Note carefully that the specification of a function $f: B_n \rightarrow B$ simply lists circuit requirements. It gives no indication of how these requirements can be met. One important way of producing functions from B_n to B is by using Boolean polynomials, which we now consider.

Boolean Polynomials

Let $x_1, x_2, \ldots, x_n$ be a set of n symbols or variables. A **Boolean polynomial** $p(x_1, x_2, \ldots, x_n)$ in the variables x_k, is defined recursively as follows:

1. $x_1, x_2, \ldots, x_n$ are all Boolean polynomials.
2. The symbols 0 and 1 are Boolean polynomials.
3. If $p(x_1, x_2, \ldots, x_n)$ and $q(x_1, x_2, \ldots, x_n)$ are two Boolean polynomials, then so are

$$p(x_1, x_2, \ldots, x_n) \vee q(x_1, x_2, \ldots, x_n)$$

and

$$p(x_1, x_2, \ldots, x_n) \wedge q(x_1, x_2, \ldots, x_n).$$

4. If $p(x_1, x_2, \ldots, x_n)$ is a Boolean polynomial, then so is

$$(p(x_1, x_2, \ldots, x_n))'.$$

By tradition, $(0)'$ is denoted $0'$, $(1)'$ is denoted $1'$, and $(x_k)'$ is denoted x_k'.

5. There are no Boolean polynomials in the variables x_k other than those that can be obtained by repeated use of rules 1, 2, 3, and 4.

Boolean polynomials are also called **Boolean expressions**.

EXAMPLE 1 The following are Boolean polynomials in the variables x, y, and z.

$$p_1(x, y, z) = (x \vee y) \wedge z$$
$$p_2(x, y, z) = (x \vee y') \vee (y \wedge 1)$$
$$p_3(x, y, z) = (x \vee (y' \wedge z)) \vee (x \wedge (y \wedge 1))$$
$$p_4(x, y, z) = (x \vee (y \vee z')) \wedge ((x' \wedge z)' \wedge (y' \vee 0))$$ ■

Ordinary polynomials in several variables, such as $x^2 y + z^4$, $xy + yz + x^2 y^2$, $x^3 y^3 + xz^4$, and so on, are generally interpreted as expressions representing algebraic computations with unspecified numbers. As such, they are subject to the usual rules of arithmetic. Thus the polynomials $x^2 + 2x + 1$ and $(x + 1)(x + 1)$ are considered equivalent, and so are $x(xy + yz)(x + z)$ and $x^3 y + 2x^2 yz + xyz^2$, since in each case we can turn one into the other with algebraic manipulation.

Similarly, Boolean polynomials may be interpreted as representing Boolean computations with unspecified elements of B, that is, with 0's and 1's. As such, these polynomials are subject to the rules of Boolean arithmetic, that is, to the rules obeyed by $\wedge$, $\vee$, and $'$ in Boolean algebras. As with ordinary polynomials, two Boolean polynomials are considered equivalent if we can turn one into the other with Boolean manipulations.

In Section 5.1 we showed how ordinary polynomials could produce functions by substitution. This process works whether the polynomials involve one or several variables. Thus the polynomial $xy + yz^3$ produces a function $f : \mathbb{R}^3 \to \mathbb{R}$ by letting $f(x, y, z) = xy + yz^3$. For example, $f(3, 4, 2) = (3)(4) + (4)(2^3)$ or 44. In a similar way, Boolean polynomials involving n variables produce functions from B_n to B. These Boolean functions are a natural generalization of those introduced in Section 5.2.

EXAMPLE 2   Consider the Boolean polynomial

$$p(x_1, x_2, x_3) = (x_1 \wedge x_2) \vee (x_1 \vee (x_2' \wedge x_3)).$$

Construct the truth table for the Boolean function $f: B_3 \rightarrow B$ determined by this Boolean polynomial.

Solution   The Boolean function $f: B_3 \rightarrow B$ is described by substituting all the 2^3 ordered triples of values from B for x_1, x_2, and x_3. The truth table for the resulting function is shown in Figure 6.72. ■

x_1	x_2	x_3	$f(x_1, x_2, x_3) = (x_1 \wedge x_2) \vee (x_1 \vee (x_2' \wedge x_3))$
0	0	0	0
0	0	1	1
0	1	0	0
0	1	1	0
1	0	0	1
1	0	1	1
1	1	0	1
1	1	1	1

Figure 6.72

Boolean polynomials can also be written in a graphical or schematic way. If x and y are variables, then the basic polynomials $x \vee y$, $x \wedge y$, and x' are shown schematically in Figure 6.73. Each symbol has lines for the variables on the left and a line on the right representing the polynomial as a whole. The symbol for $x \vee y$ is called an **or gate**, that for $x \wedge y$ is called an **and gate**, and the symbol for x' is called an **inverter**. The logical names arise because the truth tables showing the functions represented by $x \vee y$ and $x \wedge y$ are exact analogs of the truth table for the connectives "or" and "and," respectively.

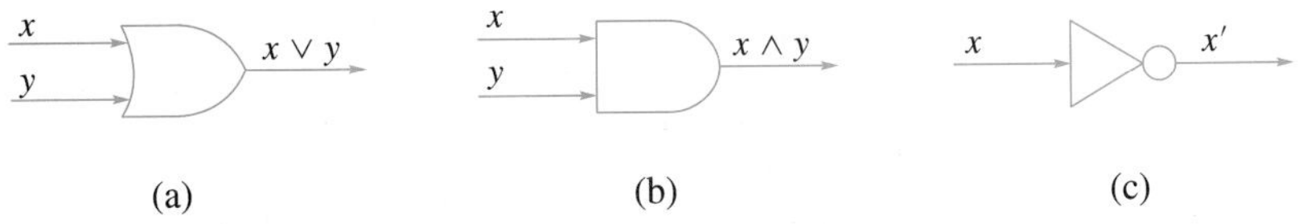

(a)   (b)   (c)

Figure 6.73

Recall that functions from B_n to B can be used to describe the desired behavior of circuits with n 0-or-1 inputs and one 0-or-1 output. In the case of the functions corresponding to the Boolean polynomials $x \vee y$, $x \wedge y$, and x', the desired circuits can be implemented, and the schematic forms of Figure 6.73 are also used to represent these circuits. By repeatedly substituting these schematic forms for $\vee$, $\wedge$, and $'$, we can make a schematic form to represent any Boolean polynomial. For the reasons given previously, such diagrams are called **logic diagrams** for the polynomial.

EXAMPLE 3   Let

$$p(x, y, z) = (x \wedge y) \vee (y \wedge z').$$

Figure 6.74(a) shows the truth table for the corresponding function $f : B_3 \to B$. Figure 6.74(b) shows the logic diagram for p. ∎

x	y	z	$f(x, y, z) = (x \wedge y) \vee (y \wedge z')$
0	0	0	0
0	0	1	0
0	1	0	1
0	1	1	0
1	0	0	0
1	0	1	0
1	1	0	1
1	1	1	1

(a)

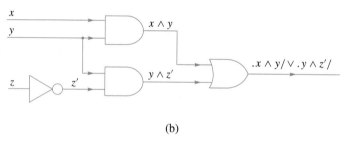

(b)

Figure 6.74

Suppose that p is a Boolean polynomial in n variables, and f is the corresponding function from B_n to B. We know that f may be viewed as a description of the behavior of a circuit having n inputs and one output. In the same way, the logic diagram of p can be viewed as a description of the construction of such a circuit, at least in terms of and gates, or gates, and inverters. Thus if the function f, describing the desired behavior of a circuit, can be produced by a Boolean polynomial p, then the logic diagram for p will give one way to construct a circuit having that behavior. In general, many different polynomials will produce the same function. The logic diagrams of these polynomials will represent alternative methods for constructing the desired circuit. It is almost impossible to overestimate the importance of these facts for the study of computer circuitry. Nearly all circuit design is done by software packages, but it is important to understand the basic principles of circuit design outlined in this section and the next.

6.5 Exercises

1. Consider the Boolean polynomial

$$p(x, y, z) = x \wedge (y \vee z').$$

If $B = \{0, 1\}$, compute the truth table of the function $f : B_3 \rightarrow B$ defined by p.

2. Consider the Boolean polynomial

$$p(x, y, z) = (x \vee y) \wedge (z \vee x').$$

If $B = \{0, 1\}$, compute the truth table of the function $f : B_3 \rightarrow B$ defined by p.

3. Consider the Boolean polynomial

$$p(x, y, z) = (x \wedge y') \vee (y \wedge (x' \vee y)).$$

If $B = \{0, 1\}$, compute the truth table of the function $f : B_3 \rightarrow B$ defined by p.

4. Consider the Boolean polynomial

$$p(x, y, z) = (x \wedge y) \vee (x' \wedge (y \wedge z')).$$

If $B = \{0, 1\}$, compute the truth table of the function $f : B_3 \rightarrow B$ defined by p.

In Exercises 5 through 8, apply the rules of Boolean arithmetic to show that the given Boolean polynomials are equivalent.

5. $(x \vee y) \wedge (x' \vee y); y$

6. $x \wedge (y \vee (y' \wedge (y \vee y'))); x$

7. $(z' \vee x) \wedge ((x \wedge y) \vee z) \wedge (z' \vee y); x \wedge y$

8. $[(x \wedge z) \vee (y' \vee z)'] \vee [(y \wedge z) \vee (x \wedge z')]; x \vee y$

In Exercises 9 through 12, rewrite the given Boolean polynomial to obtain the requested format.

9. $(x \wedge y' \wedge z) \vee (x \wedge y \wedge z)$; two variables and one operation

10. $(z \vee (y \wedge (x \vee x'))) \wedge (y \wedge z')'$; one variable

11. $(y \wedge z) \vee x' \vee (w \wedge w')' \vee (y \wedge z')$; two variables and two operations

12. $(x' \wedge y' \wedge z' \wedge w) \vee (x' \wedge z' \wedge w' \wedge y') \vee (w' \wedge x' \wedge y \wedge z') \vee (w \wedge x' \wedge y \wedge z')$; two variables and three operations

13. Construct a logic diagram implementing the function f of
(a) Exercise 1.   (b) Exercise 2.

14. Construct a logic diagram implementing the function f of
(a) Exercise 3.   (b) Exercise 4.

15. Give the Boolean function described by the logic diagram in Figure 6.75.

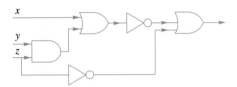

Figure 6.75

16. Give the Boolean function described by the logic diagram in Figure 6.76.

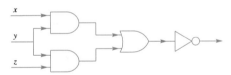

Figure 6.76

17. Use the properties of a Boolean algebra to refine the function in Exercise 15 to use the minimal number of variables and operations. Draw the logic diagram for the new function.

18. Use the properties of a Boolean algebra to refine the function in Exercise 16 to use the minimal number of variables and operations. Draw the logic diagram for the new function.

6.6 CIRCUIT DESIGNS

In Section 6.5 we considered functions from B_n to B, where B is the Boolean algebra $\{0, 1\}$. We noted that such functions can represent input-output requirements for models of many practical computer circuits. We also pointed out that if the function is given by some Boolean expression, then we can construct a logic diagram for it and thus model the implementation of the function. In this section we show that all functions from B_n to B are given by Boolean expressions, and thus logic diagrams can be constructed for any such function. Our discussion illustrates a method for finding a Boolean expression that produces a given function.

If $f: B_n \to B$, we will let $S(f) = \{b \in B_n \mid f(b) = 1\}$. We then have the following result.

Theorem 1   Let f, f_1, and f_2 be three functions from B_n to B.

(a) If $S(f) = S(f_1) \cup S(f_2)$, then $f(b) = f_1(b) \vee f_2(b)$ for all b in B.

(b) If $S(f) = S(f_1) \cap S(f_2)$, then $f(b) = f_1(b) \wedge f_2(b)$ for all b in B.

($\vee$ and $\wedge$ are LUB and GLB, respectively, in B.)

Proof

(a) Let $b \in B_n$. If $b \in S(f)$, then, by the definition of $S(f)$, $f(b) = 1$. Since $S(f) = S(f_1) \cup S(f_2)$, either $b \in S(f_1)$ or $b \in S(f_2)$, or both. In any case, $f_1(b) \vee f_2(b) = 1$. Now, if $b \notin S(f)$, then $f(b) = 0$. This means that $f_1(b) \vee f_2(b) = 0$. Thus, for all $b \in B_n$, $f(b) = f_1(b) \vee f_2(b)$.

(b) This part is proved in a manner completely analogous to that used in part (a). ∎

Recall that a function $f: B_n \to B$ can be viewed as a function $f(x_1, x_2, \ldots, x_n)$ of n variables, each of which may assume the values 0 or 1. If $E(x_1, x_2, \ldots, x_n)$ is a Boolean expression, then the function that it produces is generated by substituting all combinations of 0's and 1's for the x_i's in the expression.

EXAMPLE 1   Let $f_1: B_2 \to B$ be produced by the expression $E(x, y) = x'$, and let $f_2: B_2 \to B$ be produced by the expression $E(x, y) = y'$. Then the truth tables of f_1 and f_2 are shown in Figures 6.77(a) and (b), respectively. Let $f: B_2 \to B$ be the function whose truth table is shown in Figure 6.77(c). Clearly, $S(f) = S(f_1) \cup S(f_2)$, since f_1 is 1 at the elements $(0, 0)$ and $(0, 1)$ of B_2, f_2 is 1 at the elements $(0, 0)$ and $(1, 0)$ of B_2, and f is 1 at the elements $(0, 0)$, $(0, 1)$, and $(1, 0)$ of B_2. By Theorem 1, $f = f_1 \vee f_2$, so a Boolean expression that produces f is $x' \vee y'$. This is easily verified. ∎

x	y	$f_1(x, y)$		x	y	$f_2(x, y)$		x	y	$f(x, y)$
0	0	1		0	0	1		0	0	1
0	1	1		0	1	0		0	1	1
1	0	0		1	0	1		1	0	1
1	1	0		1	1	0		1	1	0
		(a)				(b)				(c)

Figure 6.77

It is not hard to show that any function $f: B_n \to B$ for which $S(f)$ has exactly one element is produced by a Boolean expression. Table 6.1 shows the correspondence between functions of two variables that are 1 at just one element and the Boolean expression that produces these functions.

Table 6.1

$S(f)$	Expression Producing f
$\{(0, 0)\}$	$x' \wedge y'$
$\{(0, 1)\}$	$x' \wedge y$
$\{(1, 0)\}$	$x \wedge y'$
$\{(1, 1)\}$	$x \wedge y$

EXAMPLE 2

Let $f : B_2 \to B$ be the function whose truth table is shown in Figure 6.78(a). This function is equal to 1 only at the element $(0, 1)$ of B_2; that is, $S(f) = \{(0, 1)\}$. Thus $f(x, y) = 1$ only when $x = 0$ and $y = 1$. This is also true for the expression $E(x, y) = x' \wedge y$, so f is produced by this expression.

x	y	$f(x, y)$
0	0	1
0	1	1
1	0	0
1	1	0

(a)

x	y	z	$f(x, y, z)$
0	0	0	0
0	0	1	0
0	1	0	1
0	1	1	0
1	0	0	0
1	0	1	0
1	1	0	0
1	1	1	0

(b)

Figure 6.78

The function $f : B_3 \to B$ whose truth table is shown in Figure 6.78(b) has $S(f) = \{(0, 1, 1)\}$; that is, f equals 1 only when $x = 0$, $y = 1$, and $z = 1$. This is also true for the Boolean expression $x' \wedge y \wedge z$, which must therefore produce f. ■

If $b \in B_n$, then b is a sequence $(c_1, c_2, \ldots, c_n)$ of length n, where each c_k is 0 or 1. Let E_b be the Boolean expression $\overline{x}_1 \wedge \overline{x}_2 \wedge \cdots \overline{x}_n$, where $\overline{x}_k = x_k$ when $c_k = 1$ and $\overline{x}_k = x'_k$ when $c_k = 0$. Such an expression is called a **minterm**. Example 2 illustrates the fact that any function $f : B_n \to B$ for which $S(f)$ is a single element of B_n is produced by a minterm expression. In fact, if $S(f) = \{b\}$, it is easily seen that the minterm expression E_b produces f. We then have the following result.

Theorem 2   Any function $f : B_n \to B$ is produced by a Boolean expression.   ●

Proof   Let $S(f) = \{b_1, b_2, \ldots, b_k\}$, and for each i, let $f_i : B_n \to B$ be the function defined by

$$f_i(b_i) = 1$$
$$f_i(b) = 0, \qquad \text{if } b \neq b_i.$$

Then $S(f_i) = \{b_i\}$, so $S(f) = S(f_1) \cup \cdots S(f_n)$ and by Theorem 1,

$$f = f_1 \vee f_2 \vee \cdots f_n.$$

By the preceding discussion, each f_i is produced by the minterm E_{b_i}. Thus f is produced by the Boolean expression

$$E_{b_1} \vee E_{b_2} \vee \cdots \vee E_{b_n}$$

and this completes the proof.   ▼

EXAMPLE 3   Consider the function $f: B_3 \to B$ whose truth table is shown in Figure 6.79. Since $S(f) = \{(0, 1, 1), (1, 1, 1)\}$, Theorem 2 shows that f is produced by the Boolean expression $E(x, y, z) = E_{(0,1,1)} \vee E_{(1,1,1)} = (x' \wedge y \wedge z) \vee (x \wedge y \wedge z)$. This expression, however, is not the simplest Boolean expression that produces f. Using properties of Boolean algebras, we have

$$(x' \wedge y \wedge z) \vee (x \wedge y \wedge z) = (x' \vee x) \wedge (y \wedge z) = 1 \wedge (y \wedge z) = y \wedge z.$$

Thus f is also produced by the simple expression $y \wedge z$.   ■

x	y	z	$f(x, y, z)$
0	0	0	0
0	0	1	0
0	1	0	0
0	1	1	1
1	0	0	0
1	0	1	0
1	1	0	0
1	1	1	1

Figure 6.79

The process of writing a function as an "or" combination of minterms and simplifying the resulting expression can be systematized in various ways. We will demonstrate a graphical procedure utilizing what is known as a Karnaugh map. This procedure is easy for humans beings to use with functions $f: B_n \to B$, if n is not too large. We will illustrate the method for $n = 2$, 3, and 4. If n is large or if a programmable algorithm is desired, other techniques may be preferable.

We consider first the case where $n = 2$ so that f is a function of two variables, say x and y. In Figure 6.80(a), we show a 2×2 matrix of squares with each square containing one possible input b from B_2. In Figure 6.80(b), we have replaced each input b with the corresponding minterm E_b. The labeling of the squares in Figure 6.80 is for reference only. In the future we will not exhibit these labels, but we will assume that the reader remembers their locations. In Figure 6.80(b), we note that x' appears everywhere in the first row and x appears everywhere in the second row. We label these rows accordingly, and we perform a similar labeling of the columns.

	y'	y
x'	$x' \wedge y'$	$x' \wedge y$
x	$x \wedge y'$	$x \wedge y$

00	01
10	11

(a)   (b)

Figure 6.80

EXAMPLE 4   Let $f: B_2 \to B$ be the function whose truth table is shown in Figure 6.81(a). In Figure 6.81(b), we have arranged the values of f in the appropriate squares, and we have kept the row and column labels. The resulting 2×2 array of 0's and 1's is called the **Karnaugh map of** f. Since $S(f) = \{(0, 0), (0, 1)\}$, the corresponding expression for f is $(x' \wedge y') \vee (x' \wedge y) = x' \wedge (y' \vee y) = x'$.   ■

x	y	$f(x, y)$
0	0	1
0	1	1
1	0	0
1	1	0

Truth table of f

$$\begin{array}{c c c} & y' & y \\ x' & 1 & 1 \\ x & 0 & 0 \end{array}$$

Karnaugh map of f

(a)   (b)

Figure 6.81

The outcome of Example 4 is typical. When the 1-values of a function $f: B_2 \to B$ exactly fill one row or one column, the label of that row or column gives the Boolean expression for f. Of course, we already know that if the 1-values of f fill just one square, then f is produced by the corresponding minterm. It can be shown that the larger the rectangle of 1-values of f, the smaller the expression for f will be. Finally, if the 1-values of f do not lie in a rectangle, we can decompose these values into the union of (possibly overlapping) rectangles. Then, by Theorem 1, the Boolean expression for f can be found by computing the expressions corresponding to each rectangle and combining them with $\vee$.

EXAMPLE 5 Consider the function $f: B_2 \to B$ whose truth table is shown in Figure 6.82(a). In Figure 6.82(b), we show the Karnaugh map of f and decompose the 1-values into the two indicated rectangles. The expression for the function having 1's in the horizontal rectangle is x' (verify). The function having all its 1's in the vertical rectangle corresponds to the expression y' (verify). Thus f corresponds to the expression $x' \vee y'$. In Figure 6.82(c), we show a different decomposition of the 1-values of f into rectangles. This decomposition is also correct, but it leads to the more complex expression $y' \vee (x' \wedge y)$. We see that the decomposition into rectangles is not unique and that we should try to use the largest possible rectangles. ∎

x	y	$f(x, y)$
0	0	1
0	1	1
1	0	1
1	1	0

(a)

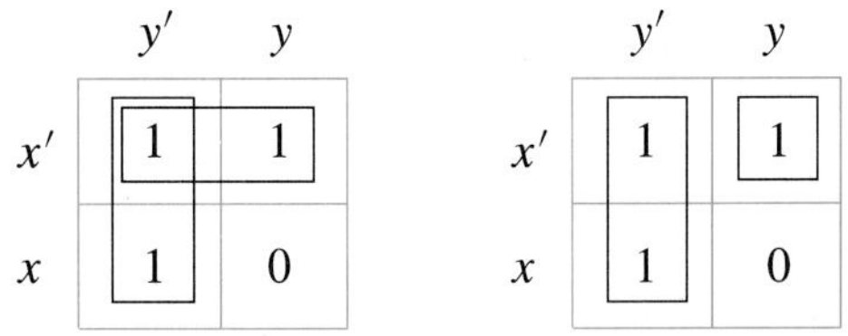

$$\begin{array}{c c c} & y' & y \\ x' & 1 & 1 \\ x & 1 & 0 \end{array}$$

(b)

$$\begin{array}{c c c} & y' & y \\ x' & 1 & 1 \\ x & 1 & 0 \end{array}$$

(c)

Figure 6.82

We now turn to the case of a function $f: B_3 \to B$, which we consider to be a function of x, y, and z. We could proceed as in the case of two variables and construct a cube of side 2 to contain the values of f. This would work, but three-dimensional figures are awkward to draw and use, and the idea would not generalize.

Instead, we use a rectangle of size 2×4. In Figures 6.83(a) and (b), respectively, we show the inputs (from B_3) and corresponding minterms for each square of such a rectangle.

	0 0	0 1	1 1	1 0
0	0 0 0	0 0 1	0 1 1	0 1 0
1	1 0 0	1 0 1	1 1 1	1 1 0

(a)

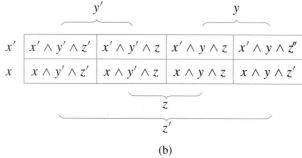

(b)

Figure 6.83

Consider the rectangular areas shown in Figure 6.84. If the 1-values for a function $f: B_3 \to B$ exactly fill one of the rectangles shown, then the Boolean expression for this function is one of the six expressions x, y, z, x', y', or z', as indicated in Figure 6.84.

Consider the situation shown in Figure 6.84(a). Theorem 1(a) shows that f can be computed by joining all the minterms corresponding to squares of the region with the symbol $\vee$. Thus f is produced by

$$(x' \wedge y' \wedge z') \vee (x' \wedge y' \wedge z) \vee (x \wedge y' \wedge z') \vee (x \wedge y' \wedge z)$$
$$= ((x' \vee x) \wedge (y' \wedge z')) \vee ((x' \vee x) \wedge (y' \wedge z))$$
$$= (1 \wedge (y' \wedge z')) \vee (1 \wedge (y' \wedge z))$$
$$= (y' \wedge z') \vee (y' \wedge z)$$
$$= y' \wedge (z' \vee z) = y' \wedge 1 = y'.$$

A similar computation shows that the other five regions are correctly labeled.

If we think of the left and right edges of our basic rectangle as glued together to make a cylinder, as we show in Figure 6.85, we can say that the six large regions shown in Figure 6.84 consist of any two adjacent columns of the cylinder, or of the top or bottom half-cylinder.

The six basic regions shown in Figure 6.84 are the only ones whose corresponding Boolean expressions need be considered. That is why we used them to label Figure 6.83(b), and we keep them as labels for all Karnaugh maps of functions from B_3 to B. Theorem 1(b) tells us that, if the 1-values of a function $f: B_3 \to B$ form exactly the intersection of two or three of the basic six regions, then a Boolean

Shaded region is y'

(a)

Shaded region is y

(b)

Shaded region is z

(c)

Shaded region is z'

(d)

Shaded region is x'

(e)

Shaded region is x

(f)

Figure 6.84

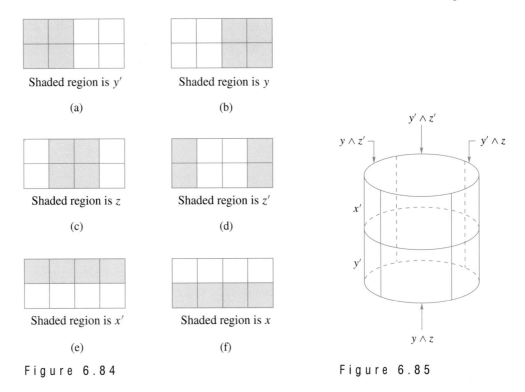

$y' \wedge z'$

$y \wedge z'$

$y' \wedge z$

x'

y'

$y \wedge z$

Figure 6.85

expression for f can be computed by combining the expressions for these basic regions with $\wedge$ symbols.

Thus, if the 1-values of the function f are as shown in Figure 6.86(a), then we get them by intersecting the regions shown in Figures 6.84(a) and (d). The Boolean expression for f is therefore $y' \wedge z'$. Similar derivations can be given for the other three columns. If the 1-values of f are as shown in Figure 6.86(b), we get them by intersecting the regions of Figures 6.84(c) and (e), so a Boolean expression for f is $z \wedge x'$. In a similar fashion, we can compute the expression for any function whose 1-values fill two horizontally adjacent squares. There are eight such functions if we again consider the rectangle to be formed into a cylinder. Thus we include the case where the 1-values of f are as shown in Figure 6.86(c). The resulting Boolean expression is $z' \wedge x'$.

If we intersect three of the basic regions and the intersection is not empty, the intersection must be a single square, and the resulting Boolean expression is a minterm. In Figure 6.86(d), the 1-values of f form the intersection of the three regions shown in Figures 6.84(a), (c), and (f). The corresponding minterm is $y' \wedge z \wedge x$. Thus we need not remember the placement of minterms in Figure 6.83(b), but instead may reconstruct it.

We have seen how to compute a Boolean expression for any function $f: B_3 \rightarrow B$ whose 1-values form a rectangle of adjacent squares (in the cylinder) of size $2^n \times 2^m$, $n = 0, 1$; $m = 0, 1, 2$. In general, if the set of 1-values of f do not form such a rectangle, we may write this set as the union of such rectangles. Then a Boolean expression for f is computed by combining the expression associated with each rectangle with $\vee$ symbols. This is true by Theorem 1(a). The preceding discussion shows that the larger the rectangles that are chosen, the simpler will be the resulting Boolean expression.

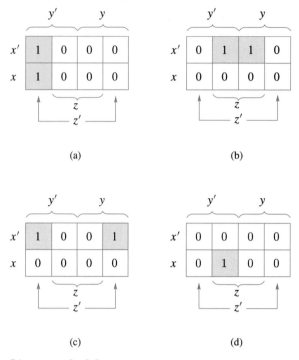

(a)   (b)

(c)   (d)

Figure 6.86

EXAMPLE 6   Consider the function f whose truth table and corresponding Karnaugh map are shown in Figure 6.87. The placement of the 1's can be derived by locating the corresponding inputs in Figure 6.83(a). One decomposition of the 1-values of f is shown in Figure 6.87(b). From this we see that a Boolean expression for f is $(y' \wedge z') \vee (x' \wedge y') \vee (y \wedge z)$.   ∎

x	y	z	$f(x, y, z)$
0	0	0	1
0	0	1	1
0	1	0	0
0	1	1	1
1	0	0	1
1	0	1	0
1	1	0	0
1	1	1	1

(a)   (b)

Figure 6.87

EXAMPLE 7 Figure 6.88 shows the truth table and corresponding Karnaugh map for a function f. The decomposition into rectangles shown in Figure 6.88(b) uses the idea that the first and last columns are considered adjacent (by wrapping around the cylinder). Thus the symbols are left open ended to signify that they join in one 2×2 rectangle corresponding to z'. The resulting Boolean expression is $z' \vee (x \wedge y)$ (verify). ■

x	y	z	$f(x, y, z)$
0	0	0	1
0	0	1	0
0	1	0	1
0	1	1	0
1	0	0	1
1	0	1	0
1	1	0	1
1	1	1	1

(a)

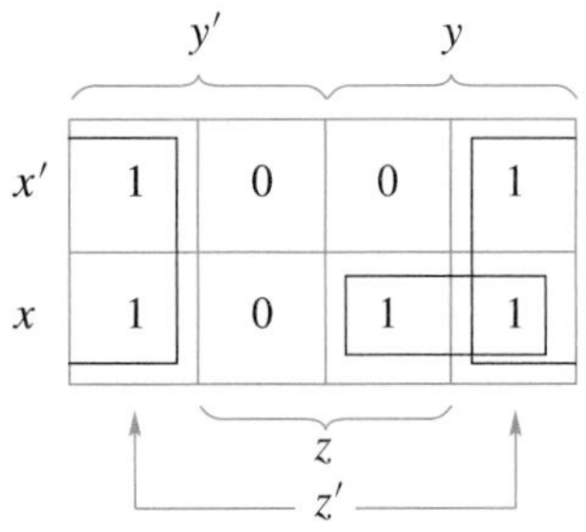

(b)

Figure 6.88

Finally, without additional comment, we present in Figure 6.89 the distribution of inputs and corresponding labeling of rectangles for the case of a function $f : B_4 \rightarrow B$, considered as a function of x, y, z, and w. Here again, we consider the first and last columns to be adjacent, and the first and last rows to be adjacent, both by wraparound, and we look for rectangles with sides of length some power of 2, so the length is 1, 2, or 4. The expression corresponding to such rectangles is given by intersecting the large labeled rectangles of Figure 6.90.

	00	01	11	10
00	0000	0001	0011	0010
01	0100	0101	0111	0110
11	1100	1101	1111	1110
10	1000	1001	1011	1010

(a)

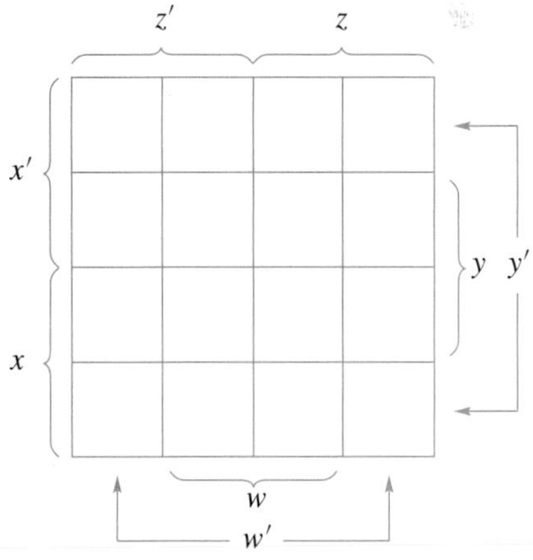

(b)

Figure 6.89

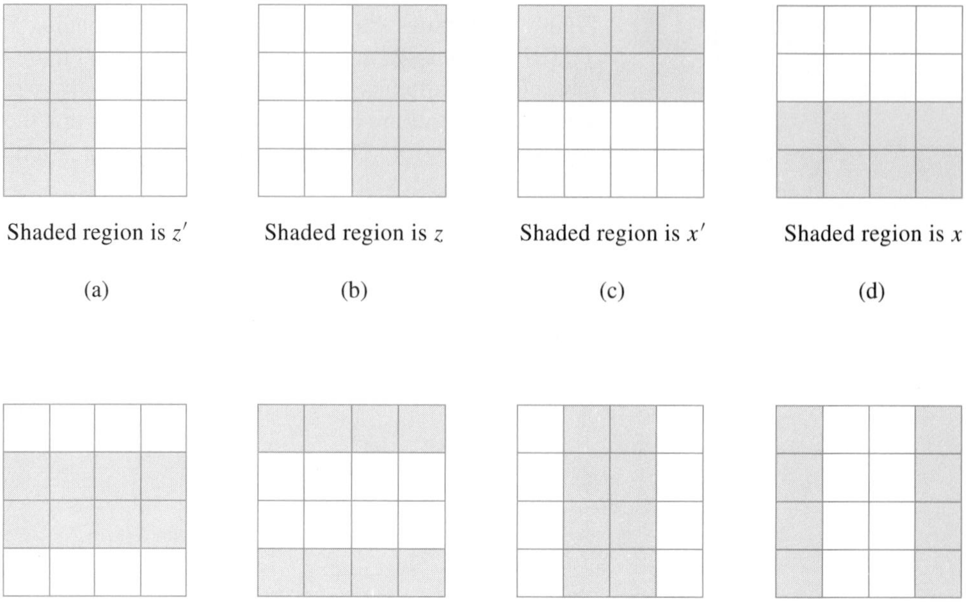

Shaded region is z' Shaded region is z Shaded region is x' Shaded region is x

(a)   (b)   (c)   (d)

Shaded region is y Shaded region is y' Shaded region is w Shaded region is w'

(e)   (f)   (g)   (h)

Figure 6.90

EXAMPLE 8

Figure 6.91 shows the Karnaugh map of a function $f: B_4 \to B$. The 1-values are placed by considering the location of inputs in Figure 6.89(a). Thus $f(0101) = 1$, $f(0001) = 0$, and so on.

The center 2×2 square represents the Boolean expression $w \wedge y$ (verify).

The four corners also form a 2×2 square, since the right and left edges and the top and bottom edges are considered adjacent. From a geometric point of view, we can see that if we wrap the rectangle around horizontally, getting a cylinder, then when we further wrap around vertically, we will get a torus or inner tube. On this inner tube, the four corners form a 2×2 square which represents the Boolean expression $w' \wedge y'$ (verify).

It then follows that the decomposition leads to the Boolean expression

$$(w \wedge y) \vee (w' \wedge y')$$

for f. ∎

EXAMPLE 9

In Figure 6.92, we show the Karnaugh map of a function $f: B_4 \to B$. The decomposition of 1-values into rectangles of sides 2^n, shown in this figure, again uses the wraparound property of top and bottom rows. The resulting expression for f is (verify)

$$(z' \wedge y') \vee (x' \wedge y' \wedge z) \vee (x \wedge y \wedge z \wedge w).$$

The first term comes from the 2×2 square formed by joining the 1×2 rectangle in the upper-left corner and the 1×2 rectangle in the lower-left corner. The second comes from the rectangle of size 1×2 in the upper-right corner, and the last is a minterm corresponding to the isolated square. ∎

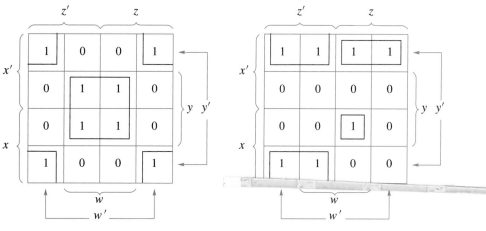

Figure 6.91   Figure 6.92

6.6 Exercises

In Exercises 1 through 6, construct Karnaugh maps for the functions whose truth tables are given.

1.

x	y	f(x, y)
0	0	1
0	1	0
1	0	0
1	1	1

2.

x	y	f(x, y)
0	0	1
0	1	0
1	0	1
1	1	0

3.

x	y	z	f(x, y, z)
0	0	0	1
0	0	1	1
0	1	0	0
0	1	1	0
1	0	0	1
1	0	1	0
1	1	0	1
1	1	1	0

4.

x	y	z	f(x, y, z)
0	0	0	0
0	0	1	1
0	1	0	1
0	1	1	1
1	0	0	0
1	0	1	0
1	1	0	0
1	1	1	1

5.

x	y	z	w	f(x, y, z, w)
0	0	0	0	1
0	0	0	1	0
0	0	1	0	1
0	0	1	1	0
0	1	0	0	0
0	1	0	1	1
0	1	1	0	1
0	1	1	1	0
1	0	0	0	0
1	0	0	1	0
1	0	1	0	0
1	0	1	1	0
1	1	0	0	1
1	1	0	1	0
1	1	1	0	1
1	1	1	1	0

6.

x	y	z	w	f(x, y, z, w)
0	0	0	0	0
0	0	0	1	0
0	0	1	0	1
0	0	1	1	0
0	1	0	0	0
0	1	0	1	0
0	1	1	0	1
0	1	1	1	0
1	0	0	0	0
1	0	0	1	0
1	0	1	0	0
1	0	1	1	1
1	1	0	0	0
1	1	0	1	0
1	1	1	0	1
1	1	1	1	1

7. Construct a Karnaugh map for the function f for which $S(f) = \{(0, 0, 1), (0, 1, 1), (1, 0, 1), (1, 1, 1)\}$.

8. Construct a Karnaugh map for the function f for which $S(f) = \{(0, 0, 0, 1), (0, 0, 1, 1), (1, 0, 1, 0), (1, 1, 0, 1), (0, 1, 0, 0), (1, 0, 0, 0)\}$.

In Exercises 9 through 16 (*Figures* 6.93 *through* 6.100), *Karnaugh maps of functions are given, and a decomposition of*

1-*values into rectangles is shown. Write the Boolean expression for these functions, which arise from the maps and rectangular decompositions.*

9.

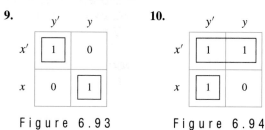

Figure 6.93

10.

Figure 6.94

11.

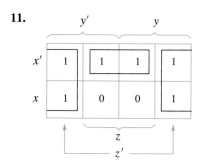

Figure 6.95

12.

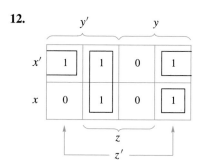

Figure 6.96

13.

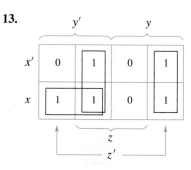

Figure 6.97

14.

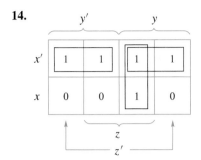

Figure 6.98

15.

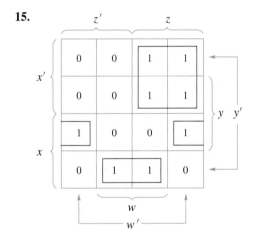

Figure 6.99

16.

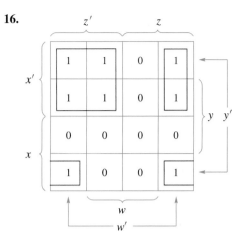

Figure 6.100

In Exercises 17 through 24, *use the Karnaugh map method to find a Boolean expression for the function* f.

17. Let f be the function of Exercise 1.

18. Let f be the function of Exercise 2.

19. Let f be the function of Exercise 3.

20. Let f be the function of Exercise 4.

21. Let f be the function of Exercise 5.

22. Let f be the function of Exercise 6.

23. Let f be the function of Exercise 7.

24. Let f be the function of Exercise 8.

TIPS FOR PROOFS

Statements of the form $\forall x, P(x)$ or $\sim(\exists x, Q(x))$ are candidates for proof by contradiction, since it is natural to explore what would happen if there were an object that did not have property P or if there were an object with property Q. This situation is seen in Theorem 2, Section 6.1. Statements about finite sets also often lend themselves to indirect proofs, as seen in Theorem 1, Section 6.2.

Many examples and exercises in this chapter ask that you check whether a specified lattice is a Boolean algebra or not. Remember that this is the same as proving that the lattice is a Boolean algebra. The fact that the number of elements is a power of 2 is a necessary condition, but not a sufficient condition. This means that if the number of elements is not a power

of 2, the lattice is not a Boolean algebra, but there are lattices with 2^n elements that are not Boolean algebras. The check on a divisibility lattice D_n is an easy one and worth memorizing. If the Hasse diagram is "small," then comparing it with those of the B_n is an efficient way to carry out the check. Attempting to construct an isomorphism between the given lattice and a known Boolean algebra is the next most efficient method. As a last resort, try to verify that the properties of a Boolean algebra are satisfied by the lattice. But be sensible about this. Try the "single" cases first—there is a unique least element 0; There is a unique greatest element I; $0' = I$; and so on—not those that require working through lots of cases, such as the associative property for $\wedge$.

KEY IDEAS FOR REVIEW

- Partial order on a set: relation that is reflexive, antisymmetric, and transitive

- Partially ordered set or poset: set together with a partial order

- Linearly ordered set: partially ordered set in which every pair of elements is comparable

- Theorem: If A and B are posets, then $A \times B$ is a poset with the product partial order.

- Dual of a poset $(A, \leq)$: the poset $(A, \geq)$, where $\geq$ denotes the inverse of $\leq$

- Hasse diagram: see page 196

- Topological sorting: see page 197

- Isomorphism of posets: see page 198

- Maximal (minimal) element of a poset: see page 202

- Theorem: A finite nonempty poset has at least one maximal element and at least one minimal element.

- Greatest (least) element of a poset A: see page 203

- Theorem: A poset has at most one greatest element and at most one least element.

- Upper (lower) bound of subset B of poset A: element $a \in A$ such that $b \leq a$ $(a \leq b)$ for all $b \in B$

- Least upper bound (greatest lower bound) of subset B of poset A: element $a \in A$ such that a is an upper (lower) bound of B and $a \leq a'$ $(a' \leq a)$, where a' is any upper (lower) bound of B

- Lattice: a poset in which every subset consisting of two elements has a LUB and a GLB

- Theorem: If L_1 and L_2 are lattices, then $L = L_1 \times L_2$ is a lattice.

- Theorem: Let L be a lattice, and $a, b \in L$. Then
  (a) $a \vee b = b$ if and only if $a \leq b$.
  (b) $a \wedge b = a$ if and only if $a \leq b$.
  (c) $a \wedge b = a$ if and only if $a \vee b = b$.

- Theorem: Let L be a lattice. Then
  1. (a) $a \vee a = a$
     (b) $a \wedge a = a$
  2. (a) $a \vee b = b \vee a$
     (b) $a \wedge b = b \wedge a$

  3. (a) $a \vee (b \vee c) = (a \vee b) \vee c$
     (b) $a \wedge (b \wedge c) = (a \wedge b) \wedge c$
  4. (a) $a \vee (a \wedge b) = a$
     (b) $a \wedge (a \vee b) = a$

- Theorem: Let L be a lattice, and $a, b, c \in L$.
  1. If $a \leq b$, then
     (a) $a \vee c \leq b \vee c$
     (b) $a \wedge c \leq b \wedge c$
  2. $a \leq c$ and $b \leq c$ if and only if $a \vee b \leq c$
  3. $c \leq a$ and $c \leq b$ if and only if $c \leq a \wedge b$
  4. If $a \leq b$ and $c \leq d$, then
     (a) $a \vee c \leq b \vee d$
     (b) $a \wedge c \leq b \wedge d$

- Isomorphic lattices: see page 210

- Bounded lattices: lattice that has a greatest element I and a least element 0

- Theorem: A finite lattice is bounded.

- Distributive lattice: lattice that satisfies the distributive laws:

$$a \wedge (b \vee c) = (a \wedge b) \vee (a \wedge c),$$
$$a \vee (b \wedge c) = (a \vee b) \wedge (a \vee c).$$

- Complement of a: element $a' \in L$ (bounded lattice) such that

$$a \vee a' = I \quad \text{and} \quad a \wedge a' = 0.$$

- Theorem: Let L be a bounded distributive lattice. If a complement exists, it is unique.

- Complemented lattice: bounded lattice in which every element has a complement

- Boolean algebra: a lattice isomorphic with $(P(S), \subseteq)$ for some finite set S

- Properties of a Boolean algebra: see page 222

- Truth tables: see page 225

- Boolean expression: see page 226

- Minterm: a Boolean expression of the form $\overline{x}_1 \wedge \overline{x}_2 \wedge \cdots \wedge \overline{x}_n$, where each $\overline{x}_k$ is either x_k or x'_k

- Theorem: Any function $f : B_n \to B$ is produced by a Boolean expression.

- Karnaugh map: see page 232

CODING EXERCISES

For each of the following, write the requested program or sub-routine in pseudocode (as described in Appendix A) or in a programming language that you know. Test your code either with a paper-and-pencil trace or with a computer run.

1. Write a subroutine that determines if a relation R represented by its matrix is a partial order.

For Exercises 2 through 4, let

$$B_n = \{(x_1, x_2, x_3, \ldots, x_n) \mid x_i \in \{0, 1\}\} \quad and \; x, \; y \in B_n.$$

2. Write a subroutine that determines if $x \leq y$.

3. (a) Write a function that computes $x \wedge y$.
   (b) Write a function that computes $x \vee y$.
   (c) Write a function that computes x'.

4. Write a subroutine that given x produces the corresponding minterm.

5. Let $B = \{0, 1\}$. Write a program that prints a truth table for the function $f : B_3 \rightarrow B$ defined by $p(x, y, z) = (x \wedge y') \vee (y \wedge (x' \vee y))$.

CHAPTER 6 SELF-TEST

1. Determine whether the given relation is a partial order. Explain your answer.
   (a) A is any set; $a \; R \; b$ if and only if $a = b$.
   (b) A is the set of parallel lines in the plane; $l_1 \, R \, l_2$ if and only if l_1 coincides with l_2 or l_1 is parallel to l_2.

2. Given the Hasse diagram in Figure 6.101,
   (a) draw the digraph of the partial order R defined;
   (b) draw the Hasse diagram of the partial order R^{-1}.

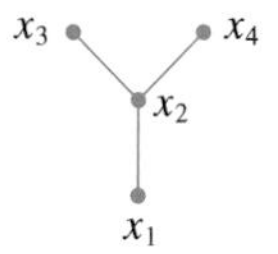

$x_3 \quad \bullet \quad \bullet \; x_4$
$\quad \quad \bullet \, x_2$
$\quad \quad x_1$

Figure 6.101

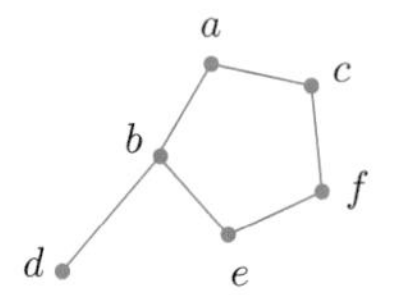

a
$\quad c$
b
$\quad \quad f$
$d \quad e$

Figure 6.102

3. Let $(A, \leq)$ be the poset whose Hasse diagram is given in Figure 6.102.
   (a) Find all minimal and maximal elements of A.
   (b) Find the least and greatest elements of A.

4. Let $A = \{2, 3, 4, 6, 8, 12, 24, 48\}$ and $\leq$ denote the partial order of divisibility; $B = \{4, 6, 12\}$. Find, if they exist,
   (a) all upper bounds of B;
   (b) all lower bounds of B;
   (c) the least upper bound of B;
   (d) the greatest lower bound of B.

5. Show that a linearly ordered poset is a distributive lattice.

6. Find the complement of each element in D_{105}.

7. Let $A = \{a, b, c, d\}$ and R be a relation on A whose matrix is

$$\mathbf{M}_R = \begin{bmatrix} 1 & 0 & 1 & 1 \\ 0 & 1 & 1 & 1 \\ 0 & 0 & 1 & 1 \\ 0 & 0 & 0 & 1 \end{bmatrix}.$$

   (a) Prove that R is a partial order.
   (b) Draw the Hasse diagram of R.

8. Let L be a lattice. Prove that for every a, b, and c in L, if $a \leq b$ and $c \leq d$, then $a \vee c \leq b \vee d$ and $a \wedge c \leq b \wedge d$.

9. Consider the Hasse diagrams given in Figure 6.103.
   (a) Which of these posets are not lattices? Explain.
   (b) Which of these posets are not Boolean algebras? Explain.

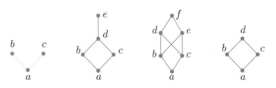

Figure 6.103

10. Let $(D_{63}, \leq)$ be the lattice of all positive divisors of 63 and $x \leq y$ means $x \mid y$.
    (a) Draw the Hasse diagram of the lattice.
    (b) Prove or disprove the statement: $(D_{63}, \leq)$ is a Boolean algebra.

11. (a) Write the Boolean expression represented by the logic diagram in Figure 6.104.

(b) Use the rules of Boolean arithmetic to find an expression using fewer operations that is equivalent to the expression found in part (a).

(c) Draw a logic diagram for the expression found in part (b).

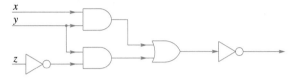

Figure 6.104

12. Use the Karnaugh map method to find a Boolean expression for the function f whose truth table is as follows.

x	y	z	$f(x, y, z)$
0	0	0	0
0	0	1	0
0	1	0	1
0	1	1	0
1	0	0	1
1	0	1	1
1	1	0	1
1	1	1	0

TREES

Prerequisites: Chapter 4

In this chapter we study a special type of relation that is exceptionally useful in a variety of computer science applications and that is usually represented by its digraph. These relations are essential for the construction of data bases and language compilers, to name just two important areas. They are called trees or sometimes rooted trees, because of the appearance of their digraphs.

7.1 TREES

Let A be a set, and let T be a relation on A. We say that T is a **tree** if there is a vertex v_0 in A with the property that there exists a unique path in T from v_0 to every other vertex in A, but no path from v_0 to v_0.

We show below that the vertex v_0, described in the definition of a tree, is unique. It is often called the **root** of the tree T, and T is then referred to as a **rooted tree**. We write (T, v_0) to denote a rooted tree T with root v_0.

If (T, v_0) is a rooted tree on the set A, an element v of A will often be referred to as a **vertex in T**. This terminology simplifies the discussion, since it often happens that the underlying set A of T is of no importance.

To help us see the nature of trees, we will prove some simple properties satisfied by trees.

Theorem 1

Let (T, v_0) be a rooted tree. Then

    (a) There are no cycles in T.

    (b) v_0 is the only root of T.

    (c) Each vertex in T, other than v_0, has in-degree one, and v_0 has in-degree zero.   ●

Proof

    (a) Suppose that there is a cycle q in T, beginning and ending at vertex v. By definition of a tree, we know that $v \neq v_0$, and there must be a path p from

v_0 to v. Then $q \circ p$ (see Section 4.3) is a path from v_0 to v that is different from p, and this contradicts the definition of a tree.

(b) If v_0' is another root of T, there is a path p from v_0 to v_0' and a path q from v_0' to v_0 (since v_0' is a root). Then $q \circ p$ is a cycle from v_0 to v_0, and this is impossible by definition. Hence the vertex v_0 is the unique root.

(c) Let w_1 be a vertex in T other than v_0. Then there is a unique path $v_0, \ldots, v_k, w_1$ from v_0 to w_1 in T. This means that $(v_k, w_1) \in T$, so w_1 has in-degree at least one. If the in-degree of w_1 is more than one, there must be distinct vertices w_2 and w_3 such that (w_2, w_1) and (w_3, w_1) are both in T. If $w_2 \neq v_0$ and $w_3 \neq v_0$, there are paths p_2 from v_0 to w_2 and p_3 from v_0 to w_3, by definition. Then $(w_2, w_1) \circ p_2$ and $(w_3, w_1) \circ p_3$ are two different paths from v_0 to w_1, and this contradicts the definition of a tree with root v_0. Hence, the in-degree of w_1 is one. We leave it as an exercise to complete the proof if $w_2 = v_0$ or $w_3 = v_0$ and to show that v_0 has in-degree zero. ▼

Theorem 1 summarizes the geometric properties of a tree. With these properties in mind, we can see how the digraph of a typical tree must look.

Let us first draw the root v_0. No edges enter v_0, but several may leave, and we draw these edges downward. The terminal vertices of the edges beginning at v_0 will be called the **level 1** vertices, while v_0 will be said to be at **level 0**. Also, v_0 is sometimes called the **parent** of these level 1 vertices, and the level 1 vertices are called the **offspring** of v_0. This is shown in Figure 7.1(a). Each vertex at level 1 has no other edges entering it, by part (c) of Theorem 1, but each of these vertices may have edges leaving the vertex. The edges leaving a vertex of level 1 are drawn downward and terminate at various vertices, which are said to be at **level 2**. Figure 7.1(b) shows the situation at this point. A parent-offspring relationship holds also for these levels (and at every consecutive pair of levels). For example, v_3 would be called the parent of the three offspring v_7, v_8, and v_9. The offspring of any one vertex are sometimes called **siblings**.

The process above continues for as many levels as are required to complete the digraph. If we view the digraph upside down, we will see why these relations are called trees. The largest level number of a tree is called the **height** of the tree.

We should note that a tree may have infinitely many levels and that any level

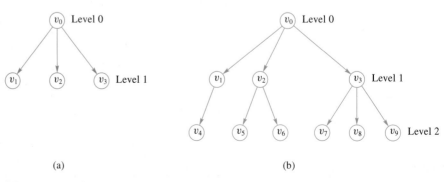

(a)   (b)

Figure 7.1

other than level 0 may contain an infinite number of vertices. In fact, any vertex could have infinitely many offspring. However, in all our future discussions, trees will be assumed to have a finite number of vertices. Thus the trees will always have a bottom (highest-numbered) level consisting of vertices with no offspring. The vertices of the tree that have no offspring are called the **leaves** of the tree.

The vertices of a tree that lie at any one level simply form a set of vertices in A. Often, however, it is useful to suppose that the offspring of each vertex of the tree are linearly ordered. Thus, if a vertex v has four offspring, we may assume that they are ordered, so we may refer to them as the first, second, third, or fourth offspring of v. Whenever we draw the digraph of a tree, we automatically assume some ordering at each level by arranging offspring from left to right. Such a tree will be called an **ordered tree**. Generally, ordering of offspring in a tree is not explicitly mentioned. If ordering is needed, it is usually introduced at the time when the need arises, and it often is specified by the way the digraph of the tree is drawn. The following relational properties of trees are easily verified.

Theorem 2   Let (T, v_0) be a rooted tree on a set A. Then

    (a) T is irreflexive.

    (b) T is asymmetric.

    (c) If $(a, b) \in T$ and $(b, c) \in T$, then $(a, c) \notin T$, for all a, b, and c in A.   ●

Proof   The proof is left as an exercise.   ▼

EXAMPLE 1   Let A be the set of all female descendants of a given woman v_0. We now define the following relation T on A: If v_1 and v_2 are elements of A, then $v_1 \; T \; v_2$ if and only if v_1 is the mother of v_2. The relation T on A is a rooted tree with root v_0.   ▫

EXAMPLE 2   Let $A = \{v_1, v_2, v_3, v_4, v_5, v_6, v_7, v_8, v_9, v_{10}\}$ and let $T = \{(v_2, v_3), (v_2, v_1), (v_4, v_5), (v_4, v_6), (v_5, v_8), (v_6, v_7), (v_4, v_2), (v_7, v_9), (v_7, v_{10})\}$. Show that T is a rooted tree and identify the root.

Solution   Since no paths begin at vertices v_1, v_3, v_8, v_9, and v_{10}, these vertices cannot be roots of a tree. There are no paths from vertices v_6, v_7, v_2, and v_5 to vertex v_4, so we must eliminate these vertices as possible roots. Thus, if T is a rooted tree, its root must be vertex v_4. It is easy to show that there is a path from v_4 to every other vertex. For example, the path v_4, v_6, v_7, v_9 leads from v_4 to v_9, since (v_4, v_6), (v_6, v_7), and (v_7, v_9) are all in T. We draw the digraph of T, beginning with vertex v_4, and with edges shown downward. The result is shown in Figure 7.2. A quick inspection of this digraph shows that paths from vertex v_4 to every other vertex are unique, and there are no paths from v_4 to v_4. Thus T is a tree with root v_4.   ■

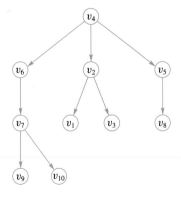

Figure 7.2

If n is a positive integer, we say that a tree T is an **n-tree** if every vertex has at most n offspring. If all vertices of T, other than the leaves, have exactly n offspring, we say that T is a **complete n-tree**. In particular, a 2-tree is often called a **binary tree**, and a complete 2-tree is often called a **complete binary tree**.

Binary trees are extremely important, since there are efficient methods of implementing them and searching through them on computers. We will see some of these methods in Section 7.3, and we will also see that any tree can be represented as a binary tree.

Let (T, v_0) be a rooted tree on the set A, and let v be a vertex of T. Let B be the set consisting of v and all its **descendants**, that is, all vertices of T that can be reached by a path beginning at v. Observe that $B \subseteq A$. Let $T(v)$ be the restriction of the relation T to B, that is, $T \cap (B \times B)$ (see Section 4.2). In other words, $T(v)$ is the tree that results from T in the following way. Delete all vertices that are not descendants of v and all edges that do not begin or end at any such vertex. Then we have the following result.

Theorem 3   If (T, v_0) is a rooted tree and $v \in T$, then $T(v)$ is also a rooted tree with root v. We will say that $T(v)$ is the **subtree** of T beginning at v.   ●

Proof   By definition of $T(v)$, we see that there is a path from v to every other vertex in $T(v)$. If there is a vertex w in $T(v)$ such that there are two distinct paths q and q' from v to w, and if p is the path in T from v_0 to v, then $q \circ p$ and $q' \circ p$ would be two distinct paths in T from v_0 to w. This is impossible, since T is a tree with root v_0. Thus each path from v to another vertex w in $T(v)$ must be unique. Also, if q is a cycle at v in $T(v)$, then q is also a cycle in T. This contradicts Theorem 1(a); therefore, q cannot exist. It follows that $T(v)$ is a tree with root v.   ▼

Subtrees and sublattices (Section 6.3) are examples of substructures of a mathematical structure. In general, if a set A and a collection of operations and their properties form a mathematical structure (Section 1.6), then a substructure of the same type is a subset of A with the same operations that satisfies all the properties that define this type of structure. This concept is used again in Chapter 9.

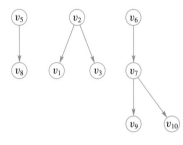

Figure 7.3

EXAMPLE 3   Consider the tree T of Example 2. This tree has root v_4 and is shown in Figure 7.2. In Figure 7.3 we have drawn the subtrees $T(v_5)$, $T(v_2)$, and $T(v_6)$ of T.   ■

7.1 Exercises

In Exercises 1 through 8, each relation R is defined on the set A. In each case determine if R is a tree and, if it is, find the root.

1. $A = \{a, b, c, d, e\}$
   $R = \{(a, d), (b, c), (c, a), (d, e)\}$

2. $A = \{a, b, c, d, e\}$
   $R = \{(a, b), (b, e), (c, d), (d, b), (c, a)\}$

3. $A = \{a, b, c, d, e, f\}$
   $R = \{(a, b), (c, e), (f, a), (f, c), (f, d)\}$

4. $A = \{1, 2, 3, 4, 5, 6\}$

   $R = \{(2, 1), (3, 4), (5, 2), (6, 5), (6, 3)\}$

5. $A = \{1, 2, 3, 4, 5, 6\}$
   $R = \{(1, 1), (2, 1), (2, 3), (3, 4), (4, 5), (4, 6)\}$

6. $A = \{1, 2, 3, 4, 5, 6\}$
   $R = \{(1, 2), (1, 3), (4, 5), (4, 6)\}$

7. $A = \{t, u, v, w, x, y, z\}$
   $R = \{(t, u), (u, w), (u, x), (u, v), (v, z), (v, y)\}$

8. $A = \{u, v, w, x, y, z\}$
   $R = \{(u, x), (u, v), (w, v), (x, z), (x, y)\}$

In Exercises 9 through 13, consider the rooted tree (T, v_0) shown in Figure 7.4.

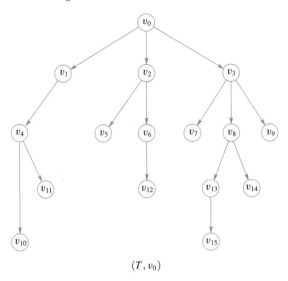

(T, v_0)

Figure 7.4

9. (a) List all level-3 vertices.

(b) List all leaves.

10. (a) What are the siblings of v_8?

(b) What are the descendants of v_8?

11. (a) Compute the tree $T(v_2)$.

(b) Compute the tree $T(v_3)$.

12. (a) What is the height of (T, v_0)?

(b) What is the height of $T(v_3)$?

13. Is (T, v_0) an n-tree? If so, for what integer n? Is (T, v_0) a complete n-tree? If so, for what integer n?

In Exercises 14 through 18, consider the rooted tree (T, v_0) shown in Figure 7.5.

14. (a) List all level-4 vertices.

(b) List all leaves.

15. (a) What are the siblings of v_2?

(b) What are the descendants of v_2?

16. (a) Compute the tree $T(v_4)$.

(b) Compute the tree $T(v_2)$.

17. (a) What is the height of (T, v_0)?

(b) What is the height of $T(v_4)$?

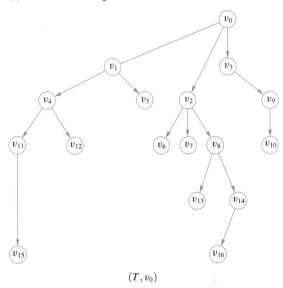

(T, v_0)

Figure 7.5

18. What is the minimal number of vertices that would need to be added to make (T, v_0) a complete 3-tree? Draw the new tree.

19. Show that the maximum number of vertices in a binary tree of height n is $2^{n+1} - 1$.

20. If T is a complete n-tree with exactly three levels, prove that the number of vertices of T must be $1 + kn$, where $2 \leq k \leq n + 1$.

21. Let T be a complete n-tree with m vertices of which k are nonleaves and l are leaves. (That is, $m = k + l$.) Prove that $m = nk + 1$ and $l = (n - 1)k + 1$.

22. Prove Theorem 2(a).

23. Prove Theorem 2(b).

24. Prove Theorem 2(c).

25. Let T be a tree. Suppose that T has r vertices and s edges. Find a formula relating r to s.

26. Draw all possible unordered trees on the set $S = \{a, b, c\}$.

27. What is the maximum height for a tree on $S = \{a, b, c, d, e\}$? Explain.

28. What is the maximum height for a complete binary tree on $S = \{a, b, c, d, e\}$?

29. Show that if (T, v_0) is a rooted tree, then v_0 has in-degree zero.

7.2 LABELED TREES

It is sometimes useful to label the vertices or edges of a digraph to indicate that the digraph is being used for a particular purpose. This is especially true for many uses of trees in computer science. We will now give a series of examples in which the sets of vertices of the trees are not important, but rather the utility of the tree is best emphasized by the labels on these vertices. Thus we will represent the vertices simply as dots and show the label of each vertex next to the dot representing that vertex.

Consider the fully parenthesized, algebraic expression

$$(3 - (2 \times x)) + ((x - 2) - (3 + x)).$$

We assume, in such an expression, that no operation such as $-, +, \times,$ or $\div$ can be performed until both of its arguments have been evaluated, that is, until all computations inside both the left and right arguments have been performed. Therefore, we cannot perform the central addition until we have evaluated $(3 - (2 \times x))$ and $((x - 2) - (3 + x))$. We cannot perform the central subtraction in $((x - 2) - (3 + x))$ until we evaluate $(x - 2)$ and $(3 + x)$, and so on. It is easy to see that each such expression has a **central operator**, corresponding to the last computation that can be performed. Thus $+$ is central to the main expression above, $-$ is central to $(3 - (2 \times x))$, and so on. An important graphical representation of such an expression is as a labeled binary tree. In this tree the root is labeled with the central operator of the main expression. The two offspring of the root are labeled with the central operator of the expressions for the left and right arguments, respectively. If either argument is a constant or variable, instead of an expression, this constant or variable is used to label the corresponding offspring vertex. This process continues until the expression is exhausted. Figure 7.6 shows the tree for the original expression of this example. To illustrate the technique further, we have shown in Figure 7.7 the tree corresponding to the full parenthesized expression

$$((3 \times (1 - x)) \div ((4 + (7 - (y + 2))) \times (7 + (x \div y)))).$$

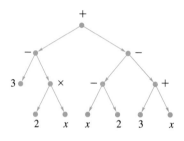

Figure 7.6

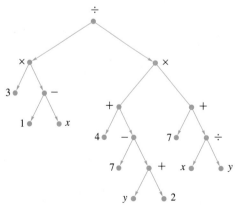

Figure 7.7

Our next example of a labeled tree is important for the computer implementation of a tree data structure. We start with an n-tree (T, v_0). Each vertex in T has at most n offspring. We imagine that each vertex *potentially* has exactly n offspring, which would be ordered from 1 to n, but that some of the offspring in the sequence may be missing. The remaining offspring are labeled with the position that they occupy in the hypothetical sequence. Thus the offspring of any vertex are labeled with distinct numbers from the set $\{1, 2, \ldots, n\}$.

Such a labeled digraph is sometimes called **positional**, and we will also use this term. Note that positional trees are also ordered trees. When drawing the digraphs of a positional tree, we will imagine that the n offspring positions for each vertex are arranged symmetrically below the vertex, and we place in its appropriate position each offspring that actually occurs.

Figure 7.8 shows the digraph of a positional 3-tree, with all actually occurring positions labeled. If offspring 1 of any vertex v actually exists, the edge from v to that offspring is drawn sloping to the left. Offspring 2 of any vertex v is drawn vertically downward from v, whenever it occurs. Similarly, offspring labeled 3 will be drawn to the right. Naturally, the root is not labeled, since it is not an offspring.

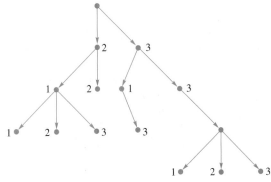

Figure 7.8

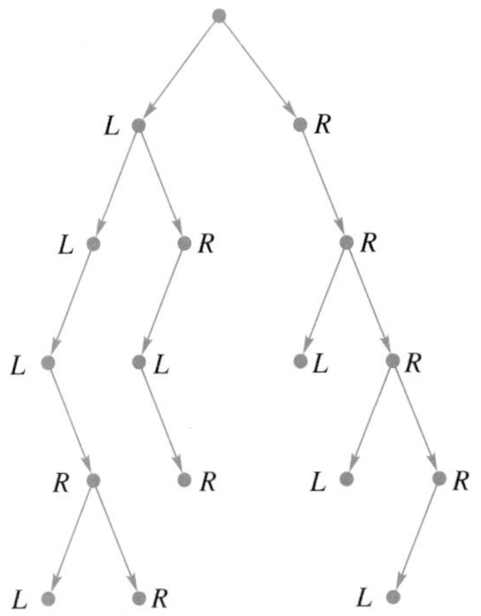

The **positional binary tree** is of special importance. In this case, for obvious reasons, the positions for potential offspring are often labeled *left* and *right*, instead of 1 and 2. Figure 7.9 shows the digraph of a positional binary tree, with offspring labeled L for left and R for right. Labeled trees may have several sets of labels, all in force simultaneously. We will usually omit the left-right labels on a positional binary tree in order to emphasize other useful labels. The positions of the offspring will then be indicated by the direction of the edges, as we have drawn them in Figure 7.9.

Figure 7.9

Computer Representation of Binary Positional Trees

In Section 4.6, we discussed an idealized information storage unit called a cell. A cell contains two items. One is data of some sort and the other is a pointer to the next cell, that is, an address where the next cell is located. A collection of such cells, linked together by their pointers, is called a linked list. The discussion in Section 4.6 included both a graphical representation of linked lists, and an implementation of them that used arrays.

We need here an extended version of this concept, called a **doubly linked list**, in which each cell contains two pointers and a data item. We use the pictorial symbol ⊸•│ │•⊸ to represent these new cells. The center space represents data storage and the two pointers, called the **left pointer** and the **right pointer**, are represented as before by dots and arrows. Once again we use the symbol •—⊥ for a pointer signifying no additional data. Sometimes a doubly linked list is arranged so that each cell points to both the next cell and the previous cell. This is useful if we want to search through a set of data items in either direction. Our use of doubly linked lists here is very different. We will use them to represent binary positional labeled trees. Each cell will correspond to a vertex, and the data part can contain a label for the vertex or a pointer to such a label. The left and right pointers will direct us to the left and right offspring vertices, if they exist. If either offspring fails to exist, the corresponding pointer will be •—⊥.

We implement this representation by using three arrays: LEFT holds pointers to the left offspring, RIGHT holds the pointers to the right offspring, and DATA holds information or labels related to each vertex, or pointers to such information. The value 0, used as a pointer, will signify that the corresponding offspring does not exist. To the linked list and the arrays we add a starting entry that points to the root of the tree.

EXAMPLE 1   We consider again the positional binary tree shown in Figure 7.6. In Figure 7.10(a), we represent this tree as a doubly linked list, in symbolic form. In Figure 7.10(b), we show the implementation of this list as a sequence of three arrays (see also Section 4.6). The first row of these arrays is just a starting point whose left pointer points to the root of the tree. As an example of how to interpret the three arrays, consider the fifth entry in the array DATA, which is ×. The fifth entry in LEFT is 6, which means that the left offspring of × is the sixth entry in DATA, or 2. Similarly, the fifth entry in RIGHT is 7, so the right offspring of × is the seventh entry in DATA, or x. ∎

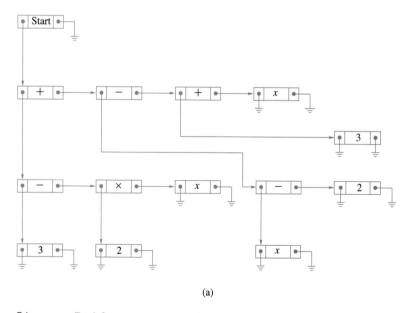

INDEX	LEFT	DATA	RIGHT
1	2	⊠	0
2	3	+	8
3	4	−	5
4	0	3	0
5	6	×	7
6	0	2	0
7	0	x	0
8	9	−	12
9	10	−	11
10	0	x	0
11	0	2	0
12	13	+	14
13	0	3	0
14	0	x	0

(a)

(b)

Figure 7.10

EXAMPLE 2   Now consider the tree of Figure 7.7. We represent this tree in Figure 7.11(a) as a doubly linked list. As before, Figure 7.11(b) shows the implementation of this linked list in three arrays. Again, the first entry is a starting point whose left pointer points to the root of the tree. We have listed the vertices in a somewhat unnatural order to show that, if the pointers are correctly determined, any ordering of vertices can be used.   ■

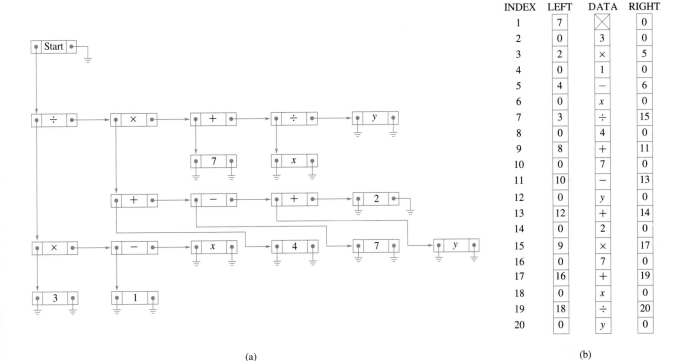

INDEX	LEFT	DATA	RIGHT
1	7	⊠	0
2	0	3	0
3	2	×	5
4	0	1	0
5	4	−	6
6	0	x	0
7	3	÷	15
8	0	4	0
9	8	+	11
10	0	7	0
11	10	−	13
12	0	y	0
13	12	+	14
14	0	2	0
15	9	×	17
16	0	7	0
17	16	+	19
18	0	x	0
19	18	÷	20
20	0	y	0

(a)   (b)

Figure 7.11

7.2 Exercises

In Exercises 1 through 10, construct the tree of the algebraic expression.

1. $(7 + (6 - 2)) - (x - (y - 4))$

2. $(x + (y - (x + y))) \times ((3 \div (2 \times 7)) \times 4)$

3. $3 - (x + (6 \times (4 \div (2 - 3))))$

4. $(((2 \times 7) + x) \div y) \div (3 - 11)$

5. $((2 + x) - (2 \times x)) - (x - 2)$

6. $(11 - (11 \times (11 + 11))) + (11 + (11 \times 11))$

7. $(3 - (2 - (11 - (9 - 4)))) \div (2 + (3 + (4 + 7)))$

8. $(x \div y) \div ((x \times 3) - (z \div 4))$

9. $((2 \times x) + (3 - (4 \times x))) + (x - (3 \times 11))$

10. $((1 + 1) + (1 - 2)) \div ((2 - x) + 1)$

11. Construct the digraphs of all distinct binary positional trees having three or fewer edges and height 2.

12. How many distinct binary positional trees are there with height 2?

13. How many distinct positional 3-trees are there with height 2?

14. Construct the digraphs of all distinct positional 3-trees having two or fewer edges.

15. The following is the doubly linked list representation of a binary positional labeled tree. Construct the digraph of this tree with each vertex labeled as indicated.

INDEX	LEFT	DATA	RIGHT
1	8	⊠	0
2	5	D	7
3	9	E	0
4	2	C	3
5	0	F	0
6	0	B	4
7	0	G	0
8	6	A	0
9	0	H	0

16. The following is the doubly linked list representation of a binary positional labeled tree. Construct the digraph of this tree with each vertex labeled as indicated.

INDEX	LEFT	DATA	RIGHT
1	9	⊠	0
2	10	M	7
3	0	Q	0
4	8	T	0
5	3	V	4
6	0	X	2
7	0	K	0
8	0	D	0
9	6	G	5
10	0	C	0

17. The following is the doubly linked list representation of a binary positional labeled tree. Construct the digraph of this tree with each vertex labeled as indicated.

INDEX	LEFT	DATA	RIGHT
1	7	⊠	0
2	0	c	9
3	2	a	6
4	0	t	0
5	0	s	0
6	10	a	11
7	3	n	8
8	4	d	13
9	0	f	0
10	0	r	0
11	0	o	0
12	0	g	0
13	12	s	5

18. Give arrays LEFT, DATA, and RIGHT describing the tree given in Figure 7.12 as a doubly linked list.

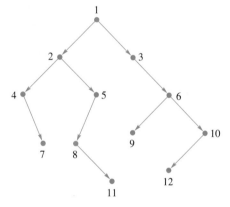

Figure 7.12

In Exercises 19 through 22, give arrays LEFT, DATA, and RIGHT describing the tree created in the indicated exercise.

19. Exercise 1

20. Exercise 4

21. Exercise 5

22. Exercise 8

7.3   TREE SEARCHING

There are many occasions when it is useful to consider each vertex of a tree T exactly once in some specific order. As each successive vertex is encountered, we may wish to take some action or perform some computation appropriate to the application being represented by the tree. For example, if the tree T is labeled, the label on each vertex may be displayed. If T is the tree of an algebraic expression, then at each vertex we may want to perform the computation indicated by the operator that labels that vertex. Performing appropriate tasks at a vertex will be called **visiting** the vertex. This is a convenient, nonspecific term that allows us to write algorithms without giving the details of what constitutes a "visit" in each particular case.

The process of visiting each vertex of a tree in some specific order will be called **searching** the tree or performing a **tree search**. In some texts, this process is called **walking** or **traversing** the tree.

Let us consider tree searches on binary positional trees. Recall that in a binary positional tree each vertex has two potential offspring. We denote these potential offspring by v_L (the left offspring) and v_R (the right offspring), and either or both may be missing. If a binary tree T is not positional, it may always be labeled so that it becomes positional.

Let T be a binary positional tree with root v. Then, if v_L exists, the subtree $T(v_L)$ (see Section 7.1) will be called the **left subtree** of T, and if v_R exists, the subtree $T(v_R)$ will be called the **right subtree** of T.

Note that $T(v_L)$, if it exists, is a positional binary tree with root v_L, and similarly $T(v_R)$ is a positional binary tree with root v_R. This notation allows us to specify searching algorithms in a natural and powerful recursive form. Recall that recursive algorithms are those that refer to themselves. We first describe a method of searching called a **preorder search**. For the moment, we leave the details of visiting a vertex of a tree unspecified. Consider the following algorithm for searching a positional binary tree T with root v.

◆ **ALGORITHM** PREORDER

> ***Step 1***   Visit v.
>
> ***Step 2***   If v_L exists, then apply this algorithm to $(T(v_L), v_L)$.
>
> ***Step 3***   If v_R exists, then apply this algorithm to $(T(v_R), v_R)$.
>
> End of Algorithm

Informally, we see that a preorder search of a tree consists of the following three steps:

> 1. Visit the root.
> 2. Search the left subtree if it exists.
> 3. Search the right subtree if it exists.

EXAMPLE 1

Let T be the labeled, positional binary tree whose digraph is shown in Figure 7.13(a). The root of this tree is the vertex labeled A. Suppose that, for any vertex v of T, visiting v prints out the label of v. Let us now apply the preorder search algorithm to this tree. Note first that if a tree consists only of one vertex, its root, then a search of this tree simply prints out the label of the root. In Figure 7.13(b), we have placed boxes around the subtrees of T and numbered these subtrees (in the corner of the boxes) for convenient reference.

According to PREORDER, applied to T, we will visit the root and print A, then search subtree 1, and then subtree 7. Applying PREORDER to subtree 1 results in visiting the root of subtree 1 and printing B, then searching subtree 2, and finally searching subtree 4. The search of subtree 2 first prints the symbol C and then searches subtree 3. Subtree 3 has just one vertex, and so, as previously mentioned, a

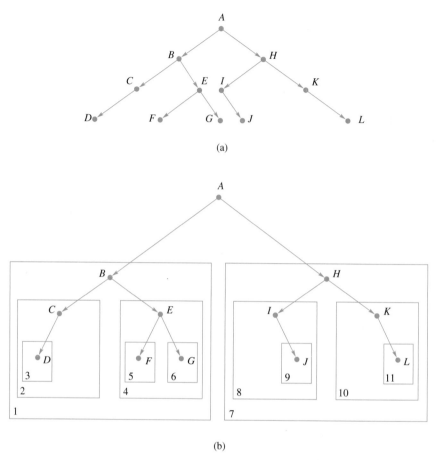

Figure 7.13

search of this tree yields just the symbol D. Up to this point, the search has yielded the string $ABCD$. Note that we have had to interrupt the search of each tree (except subtree 3, which is a leaf of T) in order to apply the search procedure to a subtree. Thus we cannot finish the search of T by searching subtree 7 until we apply the search procedure to subtrees 2 and 4. We could not complete the search of subtree 2 until we search subtree 3, and so on. The bookkeeping brought about by these interruptions produces the labels in the desired order, and recursion is a simple way to specify this bookkeeping.

Returning to the search, we have completed searching subtree 2, and we now must search subtree 4, since this is the right subtree of tree 1. Thus we print E and search 5 and 6 in order. These searches produce F and G. The search of subtree 1 is now complete, and we go to subtree 7. Applying the same procedure, we can see that the search of subtree 7 will ultimately produce the string $HIJKL$. The result, then, of the complete search of T is to print the string $ABCDEFGHIJKL$.   ■

EXAMPLE 2   Consider the completely parenthesized expression $(a - b) \times (c + (d \div e))$. Figure 7.14(a) shows the digraph of the labeled, positional binary tree representation of this expression. We apply the search procedure PREORDER to this tree, as we did

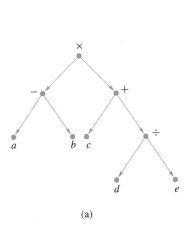

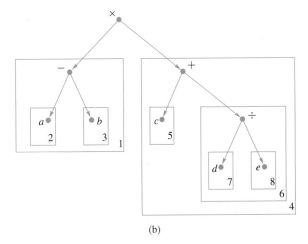

(a)   (b)

Figure 7.14

to the tree in Example 1. Figure 7.14(b) shows the various subtrees encountered in the search. Proceeding as in Example 1 and supposing again that visiting v simply prints out the label of v, we see that the string $\times - a\, b + c \div d\, e$ is the result of the search. This is the **prefix** or **Polish form** of the given algebraic expression. Once again, the numbering of the boxes in Figure 7.14(b) shows the order in which the algorithm PREORDER is applied to subtrees. ■

The Polish form of an algebraic expression is interesting because it represents the expression unambiguously, without the need for parentheses. To evaluate an expression in Polish form, proceed as follows. Move from left to right until we find a string of the form Fxy, where F is the symbol for a binary operation ($+$, $-$, $\times$, and so on) and x and y are numbers. Evaluate $x\, F\, y$ and substitute the answer for the string Fxy. Continue this procedure until only one number remains.

For example, in the preceding expression, suppose that $a = 6$, $b = 4$, $c = 5$, $d = 2$, and $e = 2$. Then we are to evaluate $\times - 6\, 4 + 5 \div 2\, 2$. This is done in the following sequence of steps.

1. $\times - 6\, 4 + 5 \div 2\, 2$
2. $\times\, 2 + 5 \div 2\, 2$ since the first string of the correct type is $-\, 6\, 4$ and $6 - 4 = 2$
3. $\times\, 2 + 5\, 1$ replacing $\div\, 2\, 2$ by $2 \div 2$ or 1
4. $\times\, 2\, 6$ replacing $+\, 5\, 1$ by $5 + 1$ or 6
5. 12 replacing $\times\, 2\, 6$ by 2×6

This example is one of the primary reasons for calling this type of search the preorder search, because here the operation symbol precedes the arguments.

Consider now the following informal descriptions of two other procedures for searching a positional binary tree T with root v.

◆ **ALGORITHM** INORDER

>   ***Step 1***   Search the left subtree $(T(v_L), v_L)$, if it exists.
>   ***Step 2***   Visit the root, v.
>   ***Step 3***   Search the right subtree $(T(v_R), v_R)$, if it exists.
>   End of Algorithm

◆ **ALGORITHM** POSTORDER

>   ***Step 1***   Search the left subtree $(T(v_L), v_L)$, if it exists.
>   ***Step 2***   Search the right subtree $(T(v_R), v_R)$, if it exists.
>   ***Step 3***   Visit the root, v.
>   End of Algorithm

As indicated by the naming of the algorithms, these searches are called, respectively, the **inorder** and **postorder** searches. The names indicate when the root of the (sub)tree is visited relative to when the left and right subtrees are searched. Informally, in a preorder search, the order is root, left, right; for an inorder search, it is left, root, right; and for a postorder search, it is left, right, root.

EXAMPLE 3   Consider the tree of Figure 7.13(b) and apply the algorithm INORDER to search it. First we must search subtree 1. This requires us to first search subtree 2, and this in turn requires us to search subtree 3. As before, a search of a tree with only one vertex simply prints the label of the vertex. Thus D is the first symbol printed. The search of subtree 2 continues by printing C and then stops, since there is no right subtree at C. We then visit the root of subtree 1 and print B, and then proceed to the search of subtree 4, which yields F, E, and G, in that order. We then visit the root of T and print A and proceed to search subtree 7. The reader may complete the analysis of the search of subtree 7 to show that the subtree yields the string *IJHKL*. Thus the complete search yields the string *DCBFEGAIJHKL*.

Suppose now that we apply algorithm POSTORDER to search the same tree. Again, the search of a tree with just one vertex will yield the label of that vertex. In general, we must search both the left and the right subtrees of a tree with root v before we print out the label at v.

Referring again to Figure 7.13(b), we see that both subtree 1 and subtree 7 must be searched before A is printed. Subtrees 2 and 4 must be searched before B is printed, and so on.

The search of subtree 2 requires us to search subtree 3, and D is the first symbol printed. The search of subtree 2 continues by printing C. We now search subtree 4, yielding F, G, and E. We next visit the root of subtree 1 and print B. Then we proceed with the search of subtree 7 and print the symbols J, I, L, K, and H. Finally, we visit the root of T and print A. Thus we print out the string *DCFGEBJILKHA*. ∎

EXAMPLE 4   Let us now apply the inorder and postorder searches to the algebraic expression tree of Example 2 (see Figure 7.14(a)). The use of INORDER produces the string

$a - b \times c + d \div e$. Notice that this is exactly the expression that we began with in Example 2, with all parentheses removed. Since the algebraic symbols lie between their arguments, this is often called the **infix notation**, and this explains the name INORDER. The preceding expression is ambiguous without parentheses. It could have come from the expression $a - (b \times ((c + d) \div e))$, which would have produced a different tree. Thus the tree cannot be recovered from the output of search procedure INORDER, while it can be shown that the tree is recoverable from the Polish form produced by PREORDER. For this reason, Polish notation is often better for computer applications, although infix form is more familiar to human beings.

The use of search procedure POSTORDER on this tree produces the string $ab - cde \div + \times$. This is the **postfix** or **reverse Polish** form of the expression. It is evaluated in a manner similar to that used for Polish form, except that the operator symbol is *after* its arguments rather than *before* them. If $a = 2$, $b = 1$, $c = 3$, $d = 4$, and $e = 2$, the preceding expression is evaluated in the following sequence of steps.

1. $2\,1 - 3\,4\,2 \div + \times$
2. $1\,3\,4\,2 \div + \times$ replacing $2\,1 -$ by $2 - 1$ or 1
3. $1\,3\,2 + \times$ replacing $4\,2 \div$ by $4 \div 2$ or 2
4. $1\,5 \times$ replacing $3\,2 +$ by $3 + 2$ or 5
5. 5 replacing $1\,5 \times$ by 1×5 or 5   ■

Reverse Polish form is also parentheses free, and from it we can recover the tree of the expression. It is used even more frequently than the Polish form and is the method of evaluating expressions in some calculators.

Searching General Trees

Until now, we have only shown how to search binary positional trees. We now show that any ordered tree T (see Section 7.1) may be represented as a binary positional tree that, although different from T, captures all the structure of T and can be used to re-create T. With the binary positional description of the tree, we may apply the computer representation and search methods previously developed. Since any tree may be ordered, we can use this technique on any (finite) tree.

Let T be any ordered tree and let A be the set of vertices of T. Define a binary positional tree $B(T)$ on the set of vertices A, as follows. If $v \in A$, then the left offspring v_L of v in $B(T)$ is the first offspring of v in T (in the given order of siblings in T), if it exists. The right offspring v_R of v in $B(T)$ is the next sibling of v in T (in the given order of siblings in T), if it exists.

EXAMPLE 5   Figure 7.15(a) shows the digraph of a labeled tree T. We assume that each set of siblings is ordered from left to right, as they are drawn. Thus the offspring of vertex 1, that is, vertices 2, 3, and 4, are ordered with vertex 2 first, 3 second, and 4 third. Similarly, the first offspring of vertex 5 is vertex 11, the second is vertex 12, and the third is vertex 13.

In Figure 7.15(b), we show the digraph of the corresponding binary positional tree, $B(T)$. To obtain Figure 7.15(b), we simply draw a left edge from each vertex v to its first offspring (if v has offspring). Then we draw a right edge from each vertex v to its next sibling (in the order given), if v has a next sibling. Thus the left edge from vertex 2, in Figure 7.15(b), goes to vertex 5, because vertex 5 is the first offspring of vertex 2 in the tree T. Also, the right edge from vertex 2, in Figure 7.15(b), goes to vertex 3, since vertex 3 is the next sibling in line (among all offspring of vertex 1). A doubly-linked-list representation of $B(T)$ is sometimes simply referred to as a **linked-list representation of T**. ■

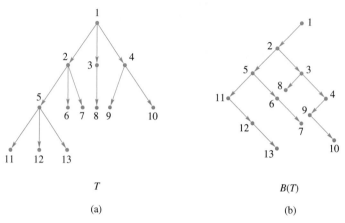

T

(a)

$B(T)$

(b)

Figure 7.15

EXAMPLE 6 Figure 7.16(a) shows the digraph of another labeled tree, with siblings ordered from left to right, as indicated. Figure 7.16(b) shows the digraph of the corresponding tree $B(T)$, and Figure 7.16(c) gives an array representation of $B(T)$. As mentioned, the data in Figure 7.16(c) would be called a linked-list representation of T. ■

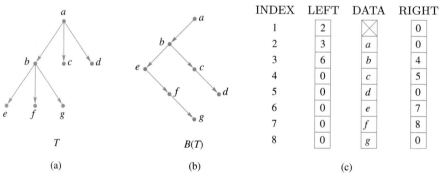

INDEX	LEFT	DATA	RIGHT
1	2	⊠	0
2	3	a	0
3	6	b	4
4	0	c	5
5	0	d	0
6	0	e	7
7	0	f	8
8	0	g	0

T

(a)

$B(T)$

(b)

(c)

Figure 7.16

Pseudocode Versions

The three search algorithms in this section have straightforward pseudocode versions, which we present here. In each, we assume that the subroutine VISIT has been previously defined.

SUBROUTINE PREORDER(T,v)
1. **CALL** VISIT (v)
2. **IF** (v_L exists) **THEN**
    a. **CALL** PREORDER($T(v_L)$, v_L)
3. **IF** (v_R exists) **THEN**
    a. **CALL** PREORDER($T(v_R)$, v_R)
4. **RETURN**
END OF SUBROUTINE PREORDER

SUBROUTINE INORDER(T,v)
1. **IF** (v_L exists) **THEN**
    a. **CALL** INORDER($T(v_L)$, v_L)
2. **CALL** VISIT(v)
3. **IF** (v_R exists) **THEN**
    a. **CALL** INORDER($T(v_R)$, v_R)
4. **RETURN**
END OF SUBROUTINE INORDER

SUBROUTINE POSTORDER(T,v)
1. **IF** (v_L exists) **THEN**
    a. **CALL** POSTORDER($T(v_L)$, v_L)
2. **IF** (v_R exists) **THEN**
    a. **CALL** POSTORDER($T(v_R)$, v_R)
3. **CALL** VISIT(v)
4. **RETURN**
END OF SUBROUTINE POSTORDER

7.3 Exercises

In Exercises 1 through 5 (Figures 7.17 through 7.21), the digraphs of labeled, positional binary trees are shown. In each case we suppose that visiting a node results in printing out the label of that node. For each exercise, show the result of performing a preorder search of the tree whose digraph is shown.

1.

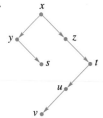

Figure 7.17

2.

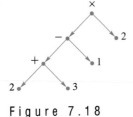

Figure 7.18

3.

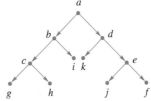

Figure 7.19

4.

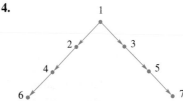

Figure 7.20

5.

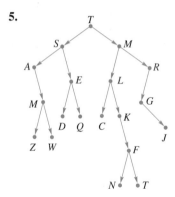

Figure 7.21

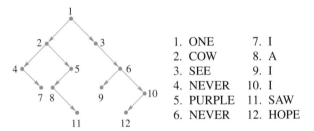

Figure 7.22

In Exercises 6 *through* 12, *visiting a node means printing out the label of the node.*

6. Show the result of performing an inorder search of the tree shown in Figure 7.17.

7. Show the result of performing an inorder search of the tree shown in Figure 7.18.

8. Show the result of performing an inorder search of the tree shown in Figure 7.19.

9. Show the result of performing an inorder search of the tree shown in Figure 7.20.

10. Show the result of performing an inorder search of the tree shown in Figure 7.21.

11. Show the result of performing a postorder search of the tree shown in Figure 7.17.

12. Show the result of performing a postorder search of the tree shown in Figure 7.18.

13. Show the result of performing a postorder search of the tree shown in Figure 7.19.

14. Show the result of performing a postorder search of the tree shown in Figure 7.20.

15. Show the result of performing a postorder search of the tree shown in Figure 7.21.

16. Consider the tree digraph shown in Figure 7.22 and the following list of words. Suppose that visiting a node of this tree means printing out the word corresponding to the number that labels that node. Print out the sentence that results from doing a postorder search of the tree.

In Exercises 17 *and* 18, *evaluate the expression, which is given in Polish, or prefix, notation.*

17. $\times - + 3\ 4 - 7\ 2 \div 12 \times 3 - 6\ 4$

18. $\div - \times 3\ x \times 4\ y + 15 \times 2 - 6\ y$, where x is 2 and y is 3.

In Exercises 19 *and* 20, *evaluate the expression, which is given in reverse Polish, or postfix, notation.*

19. $4\ 3\ 2 \div - 5 \times 4\ 2 \times 5 \times 3 \div \div$

20. $x\ 2 - 3 + 2\ 3\ y + - w\ 3 - \times \div$, where x is 7, y is 2, and w is 1.

21. Draw a binary tree whose preorder search produces the string *JBACDIHEGF*.

22. Draw a binary tree whose preorder search produces the string *CATSANDDOGS*.

23. Draw a binary tree whose postorder search produces the string *SEARCHING*.

24. Draw a binary tree whose postorder search produces the string *TREEHOUSE*.

25. (a) Every binary tree whose preorder search produces the string *JBACDIHEGF* must have 10 vertices. What else do the trees have in common?

    (b) Every binary tree whose postorder search produces the string *SEARCHING* must have 9 vertices. What else do the trees have in common?

26. Show that any element of the string *ABCDEF* may be the root of a binary tree whose inorder search produces this string.

In Exercises 27 *and* 28 (*Figures* 7.23 *and* 7.24), *draw the digraph of the binary positional tree* $B(T)$ *that corresponds to the tree shown. Label the vertices of* $B(T)$ *to show their correspondence to the vertices of* T.

27.

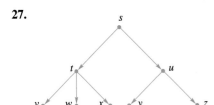

Figure 7.23

28.

A

L   G   O

R   I T H M   S

Figure 7.24

In Exercises 29 and 30, we give, in array form, the doubly-linked-list representation of a labeled tree T (not binary). Draw the digraph of both the labeled binary tree B(T) actually stored in the arrays and the labeled tree T of which B(T) is the binary representation.

29.

INDEX	LEFT	DATA	RIGHT
1	2	✕	0
2	3	a	0
3	4	b	5
4	6	c	7
5	8	d	0
6	0	e	10
7	0	f	0
8	0	g	11
9	0	h	0
10	0	i	9
11	0	j	12
12	0	k	0

30.

INDEX	LEFT	DATA	RIGHT
1	12	✕	0
2	0	T	0
3	0	W	0
4	2	O	3
5	0	B	0
6	0	R	0
7	0	A	0
8	6	N	7
9	5	C	8
10	4	H	9
11	0	E	10
12	11	S	0

In Exercises 31 and 32 (Figures 7.25 and 7.26), consider the digraph of the labeled binary positional tree shown. If this tree is the binary form B(T) of some tree T, draw the digraph of the labeled tree T.

31.

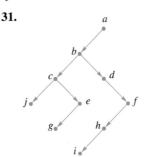

Figure 7.25

32.

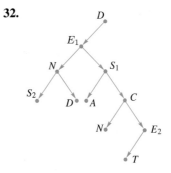

Figure 7.26

33. Finding information stored in a binary tree is a common task. We can improve the efficiency of such a search by having a "balanced" tree. An **AVL tree**, (T, v_0), is a binary tree where for each $v \in T$, the height of $T(v_L)$ and the height of $T(v_R)$ differ by at most one. For the given height, draw an AVL tree using the smallest possible number of vertices.

(a) $ht = 0$ (b) $ht = 1$

(c) $ht = 2$ (d) $ht = 3$

34. Write a recurrence relation for AVL_n, the minimum number of vertices needed for an AVL tree of height n. Justify your recurrence relation by giving a procedure for forming an AVL tree of height k with a minimum number of vertices from smaller AVL trees with minimal numbers of vertices.

7.4   UNDIRECTED TREES

An **undirected tree** is simply the symmetric closure of a tree (see Section 4.7); that is, it is a tree with all edges made bidirectional. As is the custom with symmetric relations, we represent an undirected tree by its graph, rather than by its digraph. The graph of an undirected tree T will have a single line without arrows connecting vertices a and b whenever (a, b) and (b, a) belong to T. The set $\{a, b\}$, where (a, b) and (b, a) are in T, is called an **undirected edge** of T (see Section 4.4). In this case, the vertices a and b are called **adjacent vertices**. Thus each undirected edge $\{a, b\}$ corresponds to two ordinary edges, (a, b) and (b, a). The lines in the graph of an undirected tree T correspond to the undirected edges in T.

EXAMPLE 1   Figure 7.27(a) shows the graph of an undirected tree T. In Figures 7.27(b) and (c), we show digraphs of ordinary trees T_1 and T_2, respectively, which have T as symmetric closure. This merely shows than an undirected tree will, in general, correspond to many directed trees. Labels are included to show the correspondence of underlying vertices in the three relations. Note that the graph of T in Figure 7.27(a) has six lines (undirected edges), although the relation T contains 12 pairs.■

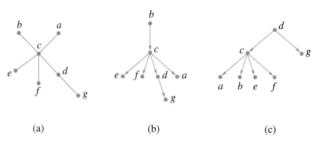

(a)   (b)   (c)

Figure 7.27

We want to present some useful alternative definitions of an undirected tree, and to do so we must make a few remarks about symmetric relations.

Let R be a symmetric relation, and let $p: v_1, v_2, \ldots, v_n$ be a path in R. We will say that p is **simple** if no two edges of p correspond to the same undirected edge. If, in addition, v_1 equals v_n (so that p is a cycle), we will call p a **simple cycle**.

EXAMPLE 2   Figure 7.28 shows the graph of a symmetric relation R. The path a, b, c, e, d is simple, but the path f, e, d, c, d, a is not simple, since d, c, and c, d correspond to the same undirected edge. Also, f, e, a, d, b, a, f and d, a, b, d are simple cycles, but f, e, d, c, e, f is not a simple cycle, since f, e and e, f correspond to the same undirected edge.   ■

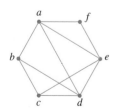

Figure 7.28

We will say that a symmetric relation R is **acyclic** if it contains no simple cycles. Recall (see Section 4.4) that a symmetric relation R is connected if there is a path in R from any vertex to any other vertex.

The following theorem provides a useful statement equivalent to the previous definition of an undirected tree.

Theorem 1   Let R be a symmetric relation on a set A. Then the following statements are equivalent.

(a) R is an undirected tree.

(b) R is connected and acyclic.   ●

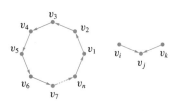

(a)   (b)

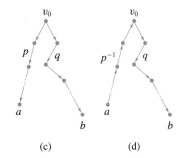

(c)   (d)

Figure 7.29

Proof   We will prove that part (a) implies part (b), and we will omit the proof that part (b) implies part (a). We suppose that R is an undirected tree, which means that R is the symmetric closure of some tree T on A. Note first that if $(a, b) \in R$, we must have either $(a, b) \in T$ or $(b, a) \in T$. In geometric terms, this means that every undirected edge in the graph of R appears in the digraph of T, directed one way or the other.

We will show by contradiction that R has no simple cycles. Suppose that R has a simple cycle $p : v_1, v_2, \ldots, v_n, v_1$. For each edge (v_i, v_j) in p, choose whichever pair (v_i, v_j) or (v_j, v_i) is in T. The result is a closed figure with edges in T, where each edge may be pointing in either direction. Now there are three possibilities. Either all arrows point clockwise, as in Figure 7.29(a), all point counterclockwise, or some pair must be as in Figure 7.29(b). Figure 7.29(b) is impossible, since in a tree T every vertex (except the root) has in-degree 1 (see Theorem 1 of Section 7.1). But either of the other two cases would mean that T contains a cycle, which is also impossible. Thus the existence of the cycle p in R leads to a contradiction and so is impossible.

We must also show that R is connected. Let v_0 be the root of the tree T. Then, if a and b are any vertices in A, there must be paths p from v_0 to a and q from v_0 to b, as shown in Figure 7.29(c). Now all paths in T are reversible in R, so the path $q \circ p^{-1}$, shown in Figure 7.29(d), connects a with b in R, where p^{-1} is the reverse path of p. Since a and b are arbitrary, R is connected, and part (b) is proved.   ▼

There are other useful characterizations of undirected trees. We state two of these without proof in the following theorem.

Theorem 2   Let R be a symmetric relation on a set A. Then R is an undirected tree if and only if either of the following statements is true.

(a) R is acyclic, and if any undirected edge is added to R, the new relation will not be acyclic.

(b) R is connected, and if any undirected edge is removed from R, the new relation will not be connected.   ●

Note that Theorems 1 and 2 tell us that an undirected tree must have exactly the "right" number of edges; one too many and a cycle will be created; one too few and the tree will become disconnected.

The following theorem will be useful in finding certain types of trees.

Theorem 3   A tree with n vertices has $n - 1$ edges.   ●

Proof   Because a tree is connected, there must be at least $n - 1$ edges to connect the n vertices. Suppose that there are more than $n - 1$ edges. Then either the root has in-degree 1 or some other vertex has in-degree at least 2. But by Theorem 1, Section 7.1, this is impossible. Thus there are exactly $n - 1$ edges.   ▼

Spanning Trees of Connected Relations

If R is a symmetric, connected relation on a set A, we say that a tree T on A is a **spanning tree** for R if T is a tree with exactly the same vertices as R and which can be obtained from R by deleting some edges of R.

EXAMPLE 3

The symmetric relation R whose graph is shown in Figure 7.30(a) has the tree T', whose digraph is shown in Figure 7.30(b), as a spanning tree. Also, the tree T'', whose digraph is shown in Figure 7.30(c), is a spanning tree for R. Since R, T', and T'' are all relations on the same set A, we have labeled the vertices to show the correspondence of elements. As this example illustrates, spanning trees are not unique.   ■

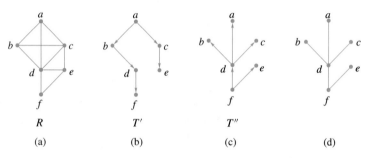

Figure 7.30

Sometimes there is interest in an **undirected spanning tree** for a symmetric connected relation R. This is just the symmetric closure of a spanning tree. Figure 7.30(d) shows an undirected spanning tree for R that is derived from the spanning tree of Figure 7.30(c). If R is a complicated relation that is symmetric and connected, it might be difficult to devise a scheme for searching R, that is, for visiting each of its vertices once in some systematic manner. If R is reduced to a spanning tree, the searching algorithms discussed in Section 7.3 can be used.

Theorem 2(b) suggests an algorithm for finding an undirected spanning tree for a relation R. Simply remove undirected edges from R until we reach a point where removal of one more undirected edge will result in a relation that is not connected. The result will be an undirected spanning tree.

EXAMPLE 4

In Figure 7.31(a), we repeat the graph of Figure 7.30(a). We then show the result of successive removal of undirected edges, culminating in Figure 7.31(f), the undirected spanning tree, which agrees with Figure 7.30(d).   ■

This algorithm is fine for small relations whose graphs are easily drawn. For large relations, perhaps stored in a computer, it is inefficient because at each stage we must check for connectedness, and this in itself requires a complicated algorithm. We now introduce a more efficient method, which also yields a spanning tree, rather than an undirected spanning tree.

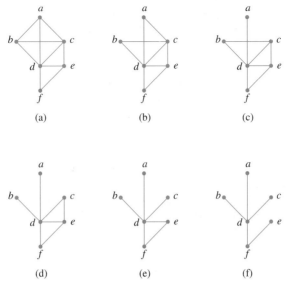

Figure 7.31

Let R be a relation on a set A, and let $a, b \in A$. Let $A_0 = A - \{a, b\}$, and $A' = A_0 \cup \{a'\}$, where a' is some new element not in A. Define a relation R' on A' as follows. Suppose $u, v \in A'$, $u \neq a'$, $v \neq a'$. Let $(a', u) \in R'$ if and only if $(a, u) \in R$ or $(b, u) \in R$. Let $(u, a') \in R'$ if and only if $(u, a) \in R$ or $(u, b) \in R$. Finally, let $(u, v) \in R'$ if and only if $(u, v) \in R$. We say that R' is a result of merging the vertices a and b.

Imagine, in the digraph of R, that the vertices are pins, and the edges are elastic bands that can be shrunk to zero length. Now physically move pins a and b together, shrinking the edge between them, if there is one, to zero length. The resulting digraph is the digraph of R'. If R is symmetric, we may perform the same operation on the graph of R.

EXAMPLE 5

Figure 7.32(a) shows the graph of a symmetric relation R. In Figure 7.32(b), we show the result of merging vertices v_0 and v_1 into a new vertex v_0'. In Figure 7.32(c), we show the result of merging vertices v_0' and v_2 of the relation whose graph is shown in Figure 7.32(b) into a new vertex v_0''. Notice in Figure 7.32(c) that the undirected edges that were previously present between v_0' and v_5 and between v_2 and v_5 have been combined into one undirected edge. ■

The algebraic form of this merging process is also very important. Let us restrict our attention to symmetric relations and their graphs. We know from Section 4.2 how to construct the matrix of a relation R.

If R is a relation on A, we will temporarily refer to elements of A as vertices of R. This will facilitate the discussion.

Suppose now that vertices a and b of a relation R are merged into a new vertex a' that replaces a and b to obtain the relation R'. To determine the matrix of R', we proceed as follows.

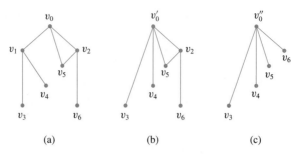

Figure 7.32

STEP 1:   Let row i represent vertex a and row j represent vertex b. Replace row i by the join of rows i and j. The join of two n-tuples of 0's and 1's has a 1 in some position exactly when either of those two n-tuples has a 1 in that position.

STEP 2:   Replace column i by the join of columns i and j.

STEP 3:   Restore the main diagonal to its original values in R.

STEP 4:   Delete row j and column j.

We make the following observation regarding step 3. If $e = (a, b) \in R$ and we merge a and b, then e would become a cycle of length 1 at a'. We do not want to create this situation, since it does not correspond to "shrinking (a, b) to zero." Step 3 corrects for this occurrence.

EXAMPLE 6   Figure 7.33 gives the matrices for the corresponding symmetric relations whose graphs are given in Figure 7.32. In Figure 7.33(b), we have merged vertices v_0 and v_1 into v_0'. Note that this is done by taking the join of the first two rows and entering the result in row 1, doing the same for the columns, then restoring the diagonal, and removing row 2 and column 2. If vertices v_0' and v_2 in the graph whose matrix is given by Figure 7.33(b) are merged, the resulting graph has the matrix given by Figure 7.33(c). ∎

$$
\begin{array}{c}
\begin{array}{cc}
 & \begin{array}{ccccccc} v_0 & v_1 & v_2 & v_3 & v_4 & v_5 & v_6 \end{array} \\
\begin{array}{c} v_0 \\ v_1 \\ v_2 \\ v_3 \\ v_4 \\ v_5 \\ v_6 \end{array} &
\left[\begin{array}{ccccccc}
0 & 1 & 1 & 0 & 0 & 1 & 0 \\
1 & 0 & 0 & 1 & 1 & 0 & 0 \\
1 & 0 & 0 & 0 & 0 & 1 & 1 \\
0 & 1 & 0 & 0 & 0 & 0 & 0 \\
0 & 1 & 0 & 0 & 0 & 0 & 0 \\
1 & 0 & 1 & 0 & 0 & 0 & 0 \\
0 & 0 & 1 & 0 & 0 & 0 & 0
\end{array}\right]
\end{array}
\end{array}
$$

(a)

$$
\begin{array}{cc}
 & \begin{array}{cccccc} v_0' & v_2 & v_3 & v_4 & v_5 & v_6 \end{array} \\
\begin{array}{c} v_0' \\ v_2 \\ v_3 \\ v_4 \\ v_5 \\ v_6 \end{array} &
\left[\begin{array}{cccccc}
0 & 1 & 1 & 1 & 1 & 0 \\
1 & 0 & 0 & 0 & 1 & 1 \\
1 & 0 & 0 & 0 & 0 & 0 \\
1 & 0 & 0 & 0 & 0 & 0 \\
1 & 1 & 0 & 0 & 0 & 0 \\
0 & 1 & 0 & 0 & 0 & 0
\end{array}\right]
\end{array}
$$

(b)

$$
\begin{array}{cc}
 & \begin{array}{ccccc} v_0'' & v_3 & v_4 & v_5 & v_6 \end{array} \\
\begin{array}{c} v_0'' \\ v_3 \\ v_4 \\ v_5 \\ v_6 \end{array} &
\left[\begin{array}{ccccc}
0 & 1 & 1 & 1 & 1 \\
1 & 0 & 0 & 0 & 0 \\
1 & 0 & 0 & 0 & 0 \\
1 & 0 & 0 & 0 & 0 \\
1 & 0 & 0 & 0 & 0
\end{array}\right]
\end{array}
$$

(c)

Figure 7.33

We can now give an algorithm for finding a spanning tree for a symmetric, connected relation R on the set $A = \{v_1, v_2, \ldots, v_n\}$. The method is a special case of an algorithm called **Prim's algorithm**. The steps are as follows:

STEP 1: Choose a vertex v_1 of R, and arrange the matrix of R so that the first row corresponds to v_1.

STEP 2: Choose a vertex v_2 of R such that $(v_1, v_2) \in R$, merge v_1 and v_2 into a new vertex v_1', representing $\{v_1, v_2\}$, and replace v_1 by v_1'. Compute the matrix of the resulting relation R'. Call the vertex v_1' a merged vertex.

STEP 3: Repeat steps 1 and 2 on R' and on all subsequent relations until a relation with a single vertex is obtained. At each step, keep a record of the set of original vertices that is represented by each merged vertex.

STEP 4: Construct the spanning tree as follows. At each stage, when merging vertices a and b, select an edge in R from one of the original vertices represented by a to one of the original vertices represented by b.

EXAMPLE 7

We apply Prim's algorithm to the symmetric relation whose graph is shown in Figure 7.34. In Table 7.1, we show the matrices that are obtained when the original set of vertices is reduced by merging until a single vertex is obtained, and at each stage we keep track of the set of original vertices represented by each merged vertex, as well as of the new vertex that is about to be merged.

Figure 7.34

Table 7.1

Matrix	Original Vertices Represented by Merged Vertices	New Vertex to Be Merged (with First Row)
$\begin{array}{c c c c c} & a & b & c & d \\ a & 0 & 0 & 1 & 1 \\ b & 0 & 0 & 1 & 1 \\ c & 1 & 1 & 0 & 0 \\ d & 1 & 1 & 0 & 0 \end{array}$	—	c
$\begin{array}{c c c c} & a' & b & d \\ a' & 0 & 1 & 1 \\ b & 1 & 0 & 1 \\ d & 1 & 1 & 0 \end{array}$	$a' \leftrightarrow \{a, c\}$	b
$\begin{array}{c c c} & a'' & d \\ a'' & 0 & 1 \\ d & 1 & 0 \end{array}$	$a'' \leftrightarrow \{a, c, b\}$	d
$\begin{array}{c c} & a''' \\ a''' & [\,0\,] \end{array}$	$a''' \leftrightarrow \{a, c, d, b\}$	—

The first vertex chosen is a, and we choose c as the vertex to be merged with a, since there is a 1 at vertex c in row 1. We also select the edge (a, c) from the original graph. At the second stage, there is a 1 at vertex b in row 1, so we merge b

Figure 7.35

with vertex a'. We select an edge in the original relation R from a vertex of $\{a, c\}$ to b, say (c, b). At the third stage, we have to merge d with vertex a''. Again, we need an edge in R from a vertex of $\{a, c, b\}$ to d, say (a, d). The selected edges (a, c), (c, b), and (a, d) form the spanning tree for R, which is shown in Figure 7.35. Note that the first vertex selected becomes the root of the spanning tree that is constructed.   ∎

7.4 Exercises

In Exercises 1 through 6, construct an undirected spanning tree for the connected graph G by removing edges in succession. Show the graph of the resulting undirected tree.

1. Let G be the graph shown in Figure 7.36.

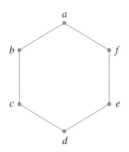

Figure 7.36

2. Let G be the graph shown in Figure 7.37.

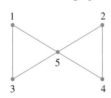

Figure 7.37

3. Let G be the graph shown in Figure 7.38.

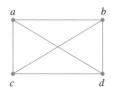

Figure 7.38

4. Let G be the graph shown in Figure 7.39.

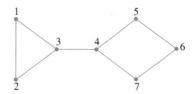

Figure 7.39

5. Let G be the graph shown in Figure 7.40.

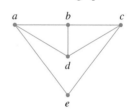

Figure 7.40

6. Let G be the graph shown in Figure 7.41.

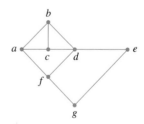

Figure 7.41

In Exercises 7 through 12 (Figures 7.36 through 7.41), use Prim's algorithm to construct a spanning tree for the connected graph shown. Use the indicated vertex as the root of the tree and draw the digraph of the spanning tree produced.

7. Figure 7.36; use e as the root.

8. Figure 7.37; use 5 as the root.

9. Figure 7.38; use c as the root.

10. Figure 7.39; use 4 as the root.

11. Figure 7.40; use e as the root.

12. Figure 7.41; use d as the root.

13. Consider the connected graph shown in Figure 7.42. Show the graphs of three different undirected spanning trees.

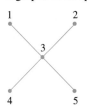

Figure 7.42

14. For the connected graph shown in Figure 7.43, show the graphs of all undirected spanning trees.

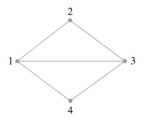

Figure 7.43

15. For the undirected tree shown in Figure 7.44, show the digraphs of all spanning trees. How many are there?

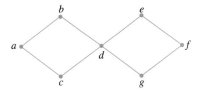

Figure 7.44

16. For the graph in Figure 7.45, give all spanning trees.

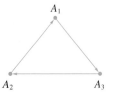

Figure 7.45   Figure 7.46

17. For the graph in Figure 7.46, give all spanning trees.

18. For the graph in Figure 7.47, how many different spanning trees are there?

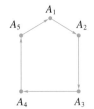

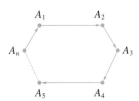

Figure 7.47   Figure 7.48

19. For the graph in Figure 7.48, how many different spanning trees are there?

20. State your conclusion for Figure 7.48 as a theorem and prove it.

21. Prove that a symmetric connected relation has a undirected spanning tree.

7.5 MINIMAL SPANNING TREES

In many applications of symmetric connected relations, the (undirected) graph of the relation models a situation in which the edges as well as the vertices carry information. A **weighted graph** is a graph for which each edge is labeled with a numerical value called its **weight**.

EXAMPLE 1   The small town of Social Circle maintains a system of walking trails between the recreational areas in town. The system is modeled by the weighted graph in Figure 7.49, where the weights represent the distances in kilometers between sites. ∎

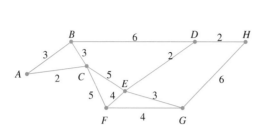

Figure 7.49

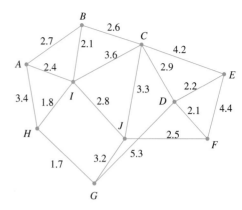

Figure 7.50

EXAMPLE 2   A communications company is investigating the costs of upgrading links between the relay stations it owns. The weighted graph in Figure 7.50 shows the stations and the cost in millions of dollars for upgrading each link.   ■

The weight of an edge (v_i, v_j) is sometimes referred to as the **distance between vertices** v_i and v_j. A vertex u is a **nearest neighbor of vertex v** if u and v are adjacent and no other vertex is joined to v by an edge of lesser weight than (u, v). Notice that, ungrammatically, v may have more than one nearest neighbor.

EXAMPLE 3   In the graph shown in Figure 7.49, vertex C is a nearest neighbor of vertex A. Vertices E and G are both nearest neighbors of vertex F.   ■

A vertex v is a **nearest neighbor of a set of vertices** $V = \{v_1, v_2, \ldots, v_k\}$ in a graph if v is adjacent to some member v_i of V and no other vertex adjacent to a member of V is joined by an edge of lesser weight than (v, v_i). This vertex v may belong to V.

EXAMPLE 4   Referring to the graph given in Figure 7.50, let $V = \{C, E, J\}$. Then vertex D is a nearest neighbor of V, because (D, E) has weight 2.2 and no other vertex adjacent to C, E, or J is linked by an edge of lesser weight to one of these vertices.   ■

With applications of weighted graphs, it is often necessary to find an undirected spanning tree for which the total weight of the edges in the tree is as small as possible. Such a spanning tree is called a **minimal spanning tree**. Prim's algorithm (Section 7.4) can easily be adapted to produce a minimal spanning tree for a weighted graph. We restate Prim's algorithm as it would be applied to a symmetric, connected relation given by its undirected weighted graph.

◆ **PRIM'S ALGORITHM**   Let R be a symmetric, connected relation with n vertices.

Step 1   Choose a vertex v_1 of R. Let $V = \{v_1\}$ and $E = \{\ \}$.

Step 2 Choose a nearest neighbor v_i of V that is adjacent to v_j, $v_j \in V$, and for which the edge (v_i, v_j) does not form a cycle with members of E. Add v_i to V and add (v_i, v_j) to E.

Step 3 Repeat step 2 until $|E| = n - 1$. Then V contains all n vertices of R, and E contains the edges of a minimal spanning tree for R.

End of Algorithm

In this version of Prim's algorithm, we begin at any vertex of R and construct a minimal spanning tree by adding an edge to a nearest neighbor of the set of vertices already linked, as long as adding this edge does not complete a cycle. This is an example of a **greedy algorithm**. At each stage we chose what is "best" based on local conditions, rather than looking at the global situation. Greedy algorithms do not always produce optimal solutions, but we can show that in this case the solution is optimal.

Theorem 1 Prim's algorithm, as given, produces a minimal spanning tree for the relation. ●

Proof Let R have n vertices. Let T be the spanning tree for R produced by Prim's algorithm. Suppose that the edges of T, in the order in which they were selected, are $t_1, t_2, \ldots, t_{n-1}$. For each i from 1 to $n-1$, we define T_i to be the tree with edges $t_1, t_2, \ldots, t_i$ and $T_0 = \{ \ \}$. Then $T_0 \subset T_1 \subset \cdots \subset T_{n-1} = T$. We now prove, by mathematical induction, that each T_i is contained in a minimal spanning tree for R.

Basis Step: Clearly P(0): $T_0 = \{ \ \}$ is contained in every minimal spanning tree for R.

Induction Step: We use P(k): T_k is contained in a minimal spanning tree T' for R to show P($k+1$): T_{k+1} is contained in a minimal spanning tree for R. By definition we have $\{t_1, t_2, \ldots, t_k\} \subseteq T'$. If t_{k+1} also belongs to T', then $T_{k+1} \subseteq T'$ and we have P($k+1$) is true. If t_{k+1} does not belong to T', then $T' \cup \{t_{k+1}\}$ must contain a cycle. (Why?) This cycle would be as shown in Figure 7.51 for some edges $s_1, s_2, \ldots, s_r$ in T'. Now the edges of this cycle cannot all be from T_k or T_{k+1} would contain this cycle. Let s_l be the edge with smallest index l that is not in T_k. Then s_l has one vertex in the tree T_k and one not in T_k. This means that when t_{k+1} was chosen by Prim's algorithm, s_l was also available. Thus the weight of s_l is at least as large as that of t_{k+1}. The spanning tree $(T' - \{s_l\}) \cup \{t_{k+1}\}$ contains T_{k+1}. The weight of this tree is less than or equal to the weight of T', so it is a minimal spanning tree for R. Thus, P($k+1$) is true. So $T_{n-1} = T$ is contained in a minimal spanning tree and must in fact be that minimal spanning tree. (Why?) ▼

F i g u r e 7 . 5 1

EXAMPLE 5 Social Circle, the town in Example 1, plans to pave some of the walking trails to make them bicycle paths as well. As a first stage, the town wants to link all the recreational areas with bicycle paths as cheaply as possible. Assuming that construction costs are the same on all parts of the system, use Prim's algorithm to find a plan for the town's paving.

Solution Referring to Figure 7.49, if we choose A as the first vertex the nearest neighbor is C, 2 kilometers away. So (A, C) is the first edge selected.

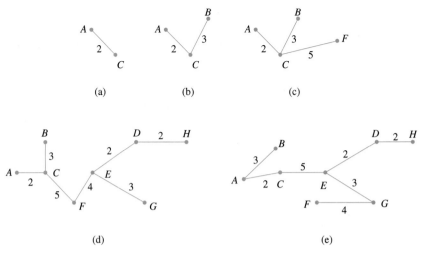

Figure 7.52

Considering the set of vertices $\{A, C\}$, B is the nearest neighbor, and we may choose either (A, B) or (B, C) as the next edge. Arbitrarily, we choose (B, C). B is a nearest neighbor for $\{A, B, C\}$, but the only edge available (A, B) would make a cycle, so we must move to a next nearest neighbor and choose (C, F) [or (C, E)]. Figures 7.52(a) through (c) show the beginning steps and Figure 7.52(d) shows a possible final result. Figure 7.52(e) shows a minimal spanning tree using Prim's algorithm beginning with vertex E. In either case, the bicycle paths would cover 21 kilometers. ■

EXAMPLE 6

A minimal spanning tree for the communication network in Example 2 may be found by using Prim's algorithm beginning at any vertex. Figure 7.53 shows a minimal spanning tree produced by beginning at I. The total cost of upgrading these links would be $20,200,000. ■

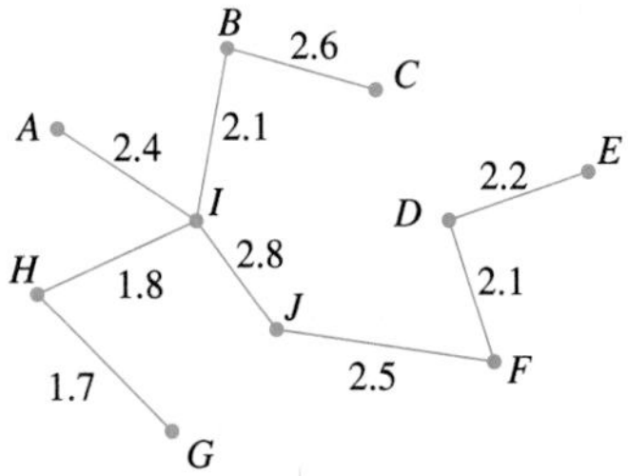

Figure 7.53

If a symmetric connected relation R has n vertices, then Prim's algorithm has running time $\Theta(n^2)$. (This can be improved somewhat.) If R has relatively few edges, a different algorithm may be more efficient. This is similar to the case for determining whether a relation is transitive, as seen in Section 4.6. Kruskal's algorithm is another example of a greedy algorithm that produces an optimal solution.

◆ **KRUSKAL'S ALGORITHM** Let R be a symmetric, connected relation with n vertices and let $S = \{e_1, e_2, \dots, e_k\}$ be the set of weighted edges of R.

Step 1 Choose an edge e_1 in S of least weight. Let $E = \{e_1\}$. Replace S with $S - \{e_1\}$.

Step 2 Select an edge e_i in S of least weight that will not make a cycle with members of E. Replace E with $E \cup \{e_i\}$ and S with $S - \{e_i\}$.

Step 3 Repeat step 2 until $|E| = n - 1$.

End of Algorithm

Since R has n vertices, the $n - 1$ edges in E will form a spanning tree. The selection process in step 2 guarantees that this is a minimal spanning tree. (We omit the proof.) Roughly speaking, the running time of Kruskal's algorithm is $\Theta(k \lg(k))$, where k is the number of edges in R.

EXAMPLE 7

A minimal spanning tree from Kruskal's algorithm for the walking trails in Example 1 is given in Figure 7.54. One sequence of edge selections is (D, E), (D, H), (A, C), (A, B), (E, G), (E, F), and (C, E) for a total weight of 21 kilometers. Naturally either of the algorithms for minimal spanning trees should produce trees of the same weight. ∎

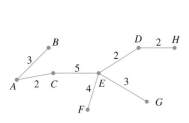

Figure 7.54

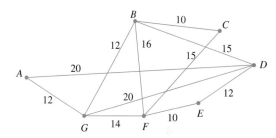

Figure 7.55

EXAMPLE 8

Use Kruskal's algorithm to find a minimal spanning tree for the relation given by the graph in Figure 7.55.

Solution Initially, there are two edges of least weight, (B, C) and (E, F). Both of these are selected. Next there are three edges, (A, G), (B, G), and (D, E), of weight 12. All of these may be added without creating any cycles. Edge (F, G) of weight 14 is the remaining edge of least weight. Adding (F, G) gives us six edges for a 7-vertex graph, so a minimal spanning tree has been found. ∎

7.5 Exercises

In Exercises 1 through 6, use Prim's algorithm as given in this section to find a minimal spanning tree for the connected graph indicated. Use the specified vertex as the initial vertex.

1. Let G be the graph shown in Figure 7.49. Begin at F.

2. Let G be the graph shown in Figure 7.50. Begin at A.

3. Let G be the graph shown in Figure 7.55. Begin at G.

4. Let G be the graph shown in Figure 7.56. Begin at E.

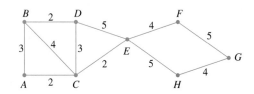

Figure 7.56

5. Let G be the graph shown in Figure 7.57. Begin at K.

6. Let G be the graph shown in Figure 7.57. Begin at M.

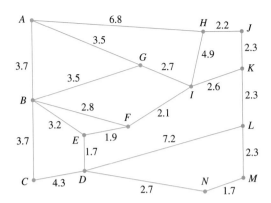

Figure 7.57

In Exercises 7 through 9, use Kruskal's algorithm to find a minimal spanning tree for the indicated graph.

7. Let G be the graph shown in Figure 7.50.

8. Let G be the graph shown in Figure 7.56.

9. Let G be the graph shown in Figure 7.57.

10. The distances between eight cities are given in the table below. Use Kruskal's algorithm to find a minimal spanning tree whose vertices are these cities. What is the total distance for the tree?

11. Suppose that in constructing a minimal spanning tree a certain edge must be included. Give a modified version of Kruskal's algorithm for this case.

12. Redo Exercise 10 with the requirement that the route from Atlanta to Augusta must be included. How much longer does this make the tree?

13. Modify Prim's algorithm to handle the case of finding a maximal spanning tree if a certain edge must be included in the tree.

14. Use the modification of Prim's algorithm developed in Exercise 13 on the graph shown in Figure 7.57 if the edge from D to L must be included in the tree.

15. Modify Kruskal's algorithm so that it will produce a maximal spanning tree, that is, one with the largest possible sum of the weights.

16. Suppose that the graph in Figure 7.57 represents possible flows through a system of pipes. Find a spanning tree that gives the maximum possible flow in this system.

17. Modify Prim's algorithm as given in this section to find a maximal spanning tree.

18. Use the modified Prim's algorithm from Exercise 17 to find a maximal spanning tree for the graph in Figure 7.57.

19. In Example 5, two different minimal spanning trees for the same graph were displayed. When will a weighted graph have a unique minimal spanning tree? Give reasons for your answer.

20. Give a simple condition on the weights of a graph that will guarantee that there is a unique maximal spanning tree for the graph.

21. Expand the proof of Theorem 1 by completing the following.
    (a) T must have $n - 1$ edges, because _____.
    (b) $T' \cup \{t_{k+1}\}$ must contain a cycle, because _____.
    (c) $(T' - \{s_l\}) \cup \{t_{k+1}\}$ is a spanning tree for R, because _____.
    (d) If T is contained in a minimal spanning tree for R, then T must be that tree. Why?

	Abbeville	Aiken	Allendale	Anderson	Asheville	Athens	Atlanta	Augusta
Abbeville		69	121	309	113	70	135	63
Aiken	69		52	97	170	117	163	16
Allendale	121	52		149	222	160	206	59
Anderson	30	97	149		92	63	122	93
Asheville	113	170	222	92		155	204	174
Athens	70	117	160	63	155		66	101
Atlanta	135	163	206	122	204	66		147
Augusta	63	16	59	93	174	101	147	

TIPS FOR PROOFS

The uniqueness of the root or of paths from the root to other vertices forms the basis of many indirect proofs for statements about trees. [See Theorem 1(c), Section 7.1.] Counting arguments are also common in proofs about trees; for example, Theorem 3, Section 7.4. Because a tree, like a lattice, is a relation with certain properties, the various facts and representa-

tions for relations are available to create proofs for tree theorems. We see this in Theorem 2, Section 7.1. The development of Prim's algorithm in Section 7.4 uses the matrix representation of a relation; this representation could be used in a proof that the running time of the algorithm is $\Theta(n^2)$.

KEY IDEAS FOR REVIEW

- Tree: relation on a finite set A such that there exists a vertex $v_0 \in A$ with the property that there is a unique path from v_0 to any other vertex in A and no path from v_0 to v_0.
- Root of tree: vertex v_0 in the preceding definition
- Rooted tree (T, v_0): tree T with root v_0
- Theorem. Let (T, v_0) be a rooted tree. Then
  (a) There are no cycles in T.
  (b) v_0 is the only root of T.
  (c) Each vertex in T, other than v_0, has in-degree one, and v_0 has in-degree zero.
- Level: see page 246
- Height of a tree: the largest level number of a tree
- Leaves: vertices having no offspring
- Theorem: Let T be a rooted tree on a set A. Then
  (a) T is irreflexive.
  (b) T is asymmetric.
  (c) If $(a, b) \in T$ and $b, c) \in T$, then $(a, c) \notin T$, for all a, b, and c in A.
- n-tree: tree in which every vertex has at most n offspring
- Complete n-tree: see page 247
- Binary tree: 2-tree
- Theorem. If (T, v_0) is a rooted tree and $v \in T$, then $T(v)$ is also a rooted tree with root v.
- $T(v)$: subtree of T beginning at v
- Positional binary tree: see page 251
- Computer representation of trees: see pages 251–252
- Preorder search: see page 255

- Inorder search: see page 258
- Postorder search: see page 258
- Reverse Polish notation: see page 259
- Searching general trees: see page 259
- Linked-list representation of a tree: see page 260
- Undirected tree: symmetric closure of a tree
- Simple path: No two edges correspond to the same undirected edge.
- Connected symmetric relation R: There is a path in R from any vertex to any other vertex.
- Theorem: A tree with n vertices has $n - 1$ edges.
- Spanning tree for symmetric connected relation R: tree reaching all the vertices of R and whose edges are edges of R
- Undirected spanning tree: symmetric closure of a spanning tree
- Prim's algorithm: see pages 269–270
- Weighted graph: a graph whose edges are each labeled with a numerical value
- Distance between vertices v_i and v_j: weight of (v_i, v_j)
- Nearest neighbor of v: see page 272
- Minimal spanning tree: undirected spanning tree for which the total weight of the edges is as small as possible
- Prim's algorithm (second version): see page 273
- Greedy algorithm: see page 273
- Kruskal's algorithm: see page 274

CODING EXERCISES

For each of the following, write the requested program or subroutine in pseudocode (as described in Appendix A) or in a programming language that you know. Test your code either with a paper-and-pencil trace or with a computer run.

1. Use the arrays LEFT, DATA, RIGHT (Section 7.2) in a program to store letters so that a postorder traversal of the tree created will print the letters out in alphabetical order.

2. Write a program that with input an ordered tree has as out-

put the corresponding binary positional tree (as described in Section 7.3).

3. Write a subroutine to carry out the merging of vertices as described in Prim's algorithm on page 268.

4. Write code for the second version of Prim's algorithm (Section 7.5).

5. Write code for Kruskal's algorithm.

CHAPTER 7 SELF-TEST

1. Determine if the relation $R = \{(1, 7), (2, 3), (4, 1), (2, 6), (4, 5), (5, 3), (4, 2)\}$ is a tree on the set $A = \{1, 2, 3, 4, 5, 6, 7\}$ If it is a tree, what is the root? If it is not a tree, make the least number of changes necessary to make it a tree and give the root.

2. Consider (T, v_0), the tree whose digraph is given in Figure 7.58.

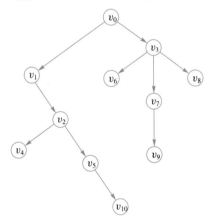

Figure 7.58

(a) What is the height of T?

(b) List the leaves of T.

(c) How many subtrees of T contain v_4?

(d) List the siblings of v_7.

3. Let (T, v_0) be a rooted tree. Prove that if any edge is removed from T, then the resulting relation cannot be a tree.

4. Construct the labeled tree representing the algebraic expression

$$(((x+3)(x+3) - (x-2)(x-2)) \div (6x - 5)) + (13 - \pi x)).$$

5. The arrays LEFT, DATA, RIGHT give a doubly-linked-list representation of a labeled binary, positional tree. Construct the digraph of this tree.

INDEX	LEFT	DATA	RIGHT
1	8	⌧	0
2	0	S	5
3	2	T	0
4	0	R	0
5	4	U	7
6	0	C	9
7	0	T	0
8	6	U	3
9	0	R	10
10	0	E	0

6. Here to visit a vertex means to print the contents of the vertex.

(a) Show the result of performing a preorder search on the tree constructed in Problem 5.

(b) Show the results of performing a postorder search on the tree constructed in Problem 5.

7. Create a binary tree for which the results of performing a preorder search are $S_1 TRE_1 S_2 S_3 E_2 D$ and for which the results of performing a postorder search are $DE_2 S_3 S_2 E_1 RTS_1$. Assume that to visit a vertex means to print the contents of the vertex.

8. Draw a complete 3-tree with height 3 using the smallest possible number of vertices.

9. The digraph of a labeled, binary positional tree is shown in Figure 7.59. Construct the digraph of the labeled ordered tree T' such that $T = B(T')$.

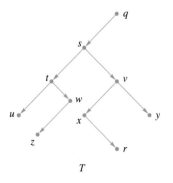

Figure 7.59

10. Give an undirected spanning tree for the relation whose graph is given in Figure 7.60.

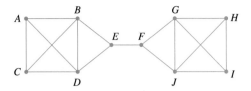

Figure 7.60

11. Use Prim's greedy algorithm to find a minimal spanning tree for the graph in Figure 7.61. Use vertex *E* as the initial vertex and list the edges in the order in which they are chosen.

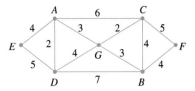

Figure 7.61

12. Use Kruskal's algorithm to find a minimal spanning tree for the graph in Figure 7.61. List the edges in the order in which they are chosen.

8

TOPICS IN GRAPH THEORY

Prerequisites: Chapters 3 and 5

> Graph theory begins with very simple geometric ideas and has many powerful applications. Some uses of graphs are discussed in Chapters 4, 6, and 7. In those chapters a graph is associated with the digraph of a symmetric relation. Here we give an alternate definition of graph that includes the more general multigraphs and is more appropriate for the applications developed in this chapter.

8.1 GRAPHS

A **graph** G consists of a finite set V of objects called **vertices**, a finite set E of objects called **edges**, and a function γ that assigns to each edge a subset $\{v, w\}$, where v and w are vertices (and may be the same). We will write $G = (V, E, \gamma)$ when we need to name the parts of G. If e is an edge, and $\gamma(e) = \{v, w\}$, we say that e is an edge between v and w and that e is determined by v and w. The vertices v and w are called the **end points** of e. If there is only one edge between v and w, we often identify e with the set $\{v, w\}$. This should cause no confusion. The restriction that there are only a finite number of vertices may be dropped, but for the discussion here all graphs have a finite number of vertices.

EXAMPLE 1

Let $V = \{1, 2, 3, 4\}$ and $E = \{e_1, e_2, e_3, e_4, e_5\}$. Let γ be defined by

$$\gamma(e_1) = \gamma(e_5) = \{1, 2\}, \quad \gamma(e_2) = \{4, 3\}, \quad \gamma(e_3) = \{1, 3\}, \quad \gamma(e_4) = \{2, 4\}.$$

Then $G = (V, E, \gamma)$ is a graph. ∎

Graphs are usually represented by pictures using a point for each vertex and a line for each edge. G in Example 1 is represented in Figure 8.1. We usually omit the names of the edges, since they have no intrinsic meaning. Also, we may want to put other more useful labels on the edges. We sometimes omit the labels on vertices as well if the graphical information is adequate for the discussion.

Graphs are often used to record information about relationships or connections. An edge between v_i and v_j indicates a connection between the objects v_i and v_j. In a pictorial representation of a graph, the connections are the most important information, and generally a number of different pictures may represent the same graph.

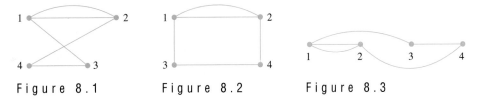

Figure 8.1   Figure 8.2   Figure 8.3

EXAMPLE 2   Figures 8.2 and 8.3 also represent the graph G given in Example 1.   ■

The **degree of a vertex** is the number of edges having that vertex as an end point. A graph may contain an edge from a vertex to itself; such an edge is referred to as a **loop**. A loop contributes 2 to the degree of a vertex, since that vertex serves as both endpoints of the loop.

EXAMPLE 3   (a) In the graph in Figure 8.4, the vertex A has degree 2, vertex B has degree 4, and vertex D has degree 3.

(b) In Figure 8.5, vertex a has degree 4, vertex e has degree 0, and vertex b has degree 2.

(c) Each vertex of the graph in Figure 8.6 has degree 2.   ■

A vertex with degree 0 will be called an **isolated** vertex. A pair of vertices that determine an edge are **adjacent** vertices.

EXAMPLE 4   In Figure 8.5, vertex e is an isolated vertex. In Figure 8.5, a and b are adjacent vertices; vertices a and d are not adjacent.   ◨

A **path** π **in a graph** G consists of a pair (V_π, E_π) of sequences; a vertex sequence $V_\pi : v_1, v_2, \ldots, v_k$ and an edge sequence $E_\pi : e_1, e_2, \ldots, e_{k-1}$ for which:

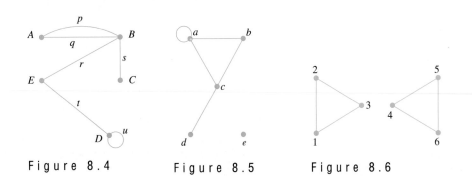

Figure 8.4   Figure 8.5   Figure 8.6

1. Each successive pair v_i, v_{i+1} of vertices is adjacent in G, and edge e_i has v_i and v_{i+1} as endpoints, for $i = 1, \ldots, k - 1$.
2. No edge occurs more than once in the edge sequence.

Thus we may begin at v_1 and travel through the edges $e_1, e_2, \ldots, e_{k-1}$ to v_k without using any edge twice.

A **circuit** is a path that begins and ends at the same vertex. In Chapter 4 we called such paths cycles; the word "circuit" is more common in general graph theory. A path is called **simple** if no vertex appears more than once in the vertex sequence, except possibly if $v_1 = v_k$. In this case the path is called a **simple circuit**. This expanded definition of path is needed to handle paths in multigraphs. For the graph in Figure 8.4 we define a path π_1 by sequences V_{π_1}: A, B, E, D, D and E_{π_1}: p, r, t, u, and the path π_2 by the sequences V_{π_2}: A, B, A and E_{π_2}: p, q. The vertices alone would not be sufficient to define these paths. For π_1 we would not know which edge to travel from A to B, p or q, and for π_2 we would not know which edge to use first and which to use second. The edges alone would not always be enough either, for if we only knew the edge sequence p, q for π_2, we would not know if the vertex sequence was A, B, A or B, A, B. If the vertices of a path have only one edge between each adjacent pair, then the edge sequence is completely determined by the vertex sequence. In this case we specify the path by the vertex sequence alone and write π: $v_1, v_2, \ldots, v_k$. Thus for the graph in Figure 8.2 we can write π_3: $1, 3, 4, 2$, since this path contains no adjacent vertices with two edges between them. In cases such as this, it is not necessary to label the edges.

EXAMPLE 5

(a) Paths π_1 and π_2 in the graph of Figure 8.4 were defined previously. Path π_1 is not simple, since vertex D appears twice, but π_2 is a simple circuit, since the only vertex appearing twice occurs at the beginning and at the end.

(b) The path π_4: D, E, B, C in the graph of Figure 8.4 is simple. Here no mention of edges is needed.

(c) Examples of paths in the graph of Figure 8.5 are π_5: a, b, c, a and π_6: d, c, a, a. Here π_5 is a simple circuit, but π_6 is not simple.

(d) In Figure 8.6 the vertex sequence 1, 2, 3, 2 does not specify a path, since the single edge between 2 and 3 would be traveled twice.

(e) The path π_7: c, a, b, c, d in Figure 8.5 is not simple.   ∎

A graph is called **connected** if there is a path from any vertex to any other vertex in the graph. Otherwise, the graph is **disconnected**. If the graph is disconnected, the various connected pieces are called the **components** of the graph.

EXAMPLE 6

The graphs in Figures 8.1 and 8.4 are connected. Those in Figures 8.5 and 8.6 are disconnected. The graph of Figure 8.6 has two components.   ∎

Some important special families of graphs will be useful in our discussions. We present them here.

U_2 U_5

Figure 8.7

1. For each integer $n \geq 1$, we let U_n denote the graph with n vertices and no edges. Figure 8.7 shows U_2 and U_5. We call U_n the **discrete graph** on n vertices.

2. For each integer $n \geq 1$, let K_n denote the graph with vertices $\{v_1, v_2, \ldots, v_n\}$ and with an edge $\{v_i, v_j\}$ for every i and j. In other words, every vertex in K_n is connected to every other vertex. In Figure 8.8 we show K_3, K_4, and K_5. The graph K_n is called the **complete graph** on n vertices. More generally, if each vertex of a graph has the same degree as every other vertex, the graph is called **regular**. The graphs U_n are also regular.

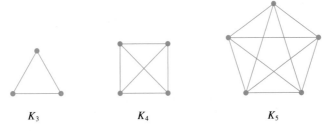

K_3 K_4 K_5

Figure 8.8

3. For each integer $n \geq 1$, we let L_n denote the graph with n vertices $\{v_1, v_2, \ldots, v_n\}$ and with edges $\{v_i, v_{i+1}\}$ for $1 \leq i < n$. We show L_2 and L_4 in Figure 8.9. We call L_n the **linear graph** on n vertices.

L_2 L_4

Figure 8.9

EXAMPLE 7

All the K_n and L_n are connected, while the U_n are disconnected for $n > 1$. In fact, the graph U_n has exactly n components. ∎

Subgraphs and Quotient Graphs

Suppose that $G = (V, E, \gamma)$ is a graph. Choose a subset E_1 of the edges in E and a subset V_1 of the vertices in V, so that V_1 contains (at least) all the end points of edges in E_1. Then $H = (V_1, E_1, \gamma_1)$ is also a graph where γ_1 is γ restricted to edges in E_1. Such a graph H is called a **subgraph** of G. Subgraphs play an important role in analyzing graph properties.

EXAMPLE 8

The graphs shown in Figure 8.11, 8.12, and 8.13 are each a subgraph of the graph shown in Figure 8.10. ∎

One of the most important subgraphs is the one that arises by deleting one edge and no vertices. If $G = (V, E, \gamma)$ is a graph and $e \in E$, then we denote by G_e the subgraph obtained by omitting the edge e from E and keeping all vertices. If G is the graph of Figure 8.10, and $e = \{a, b\}$, then G_e is the graph shown in Figure 8.13.

Our second important construction is defined for graphs without multiple edges between the same vertices. Suppose that $G = (V, E, \gamma)$ is such a graph and that R

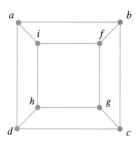

Figure 8.10

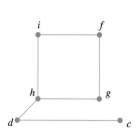

Figure 8.11

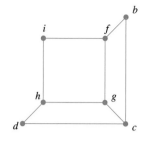

Figure 8.12

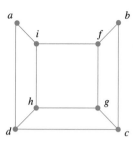

Figure 8.13

is an equivalence relation on the set V. Then we construct the **quotient graph** G^R in the following way. The vertices of G^R are the equivalence classes of V produced by R (see Section 4.5). If $[v]$ and $[w]$ are the equivalence classes of vertices v and w of G, then there is an edge in G^R from $[v]$ to $[w]$ if and only if some vertex in $[v]$ is connected to some vertex in $[w]$ in the graph G. Informally, this just says that we get G^R by merging all the vertices in each equivalence class into a single vertex and combining any edges that are superimposed by such a process.

 EXAMPLE 9   Let G be the graph of Figure 8.14 (which has no multiple edges), and let R be the equivalence relation on V defined by the partition

$$\{\{a, m, i\}, \{b, f, j\}, \{c, g, k\}, \{d, h, l\}\}.$$

Then G^R is shown in Figure 8.15.

If S is also an equivalence relation on V defined by the partition

$$\{\{i, j, k, l\}, \{a, m\}, \{f, b, c\}, \{d\}, \{g\}, \{h\}\},$$

then the quotient graph G^S is shown in Figure 8.16.   ∎

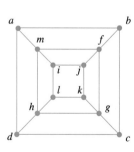

Figure 8.14

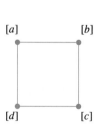

Figure 8.15

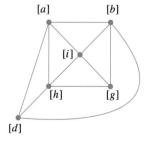

Figure 8.16

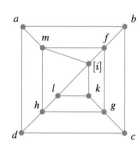

Figure 8.17

Again, one of the most important cases arises from using just one edge. If e is an edge between vertex v and vertex w in a graph $G = \{V, E, \gamma\}$, then we consider the equivalence relation whose partition consists of $\{v, w\}$ and $\{v_i\}$, for each $v_i \neq v$, $v_i \neq w$. That is, we merge v and w and leave everything else alone. The resulting quotient graph is denoted G^e. If G is the graph of Figure 8.14, and $e = \{i, j\}$, then G^e is the graph shown in Figure 8.17.

8.1 Exercises

In Exercises 1 through 4 (Figures 8.18 through 8.21), give V, the set of vertices, and E, the set of edges, for the graph $G = (V, E, \gamma)$ given in Figures 8.18 through 8.21.

1.
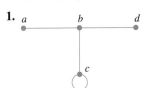
Figure 8.18

2.

Figure 8.19

3.

Figure 8.20

4.
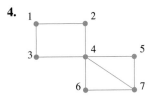
Figure 8.21

5. Draw a picture of the graph $G = (V, E, \gamma)$, where $V = \{a, b, c, d, e\}$, $E = \{e_1, e_2, e_3, e_4, e_5, e_6\}$, and $\gamma(e_1) = \gamma(e_5) = \{a, c\}$, $\gamma(e_2) = \{a, d\}$, $\gamma(e_3) = \{e, c\}$, $\gamma(e_4) = \{b, c\}$, and $\gamma(e_6) = \{e, d\}$.

6. Draw a picture of the graph $G = \{V, E, \gamma)$, where $V = \{a, b, c, d, e, f, g, h\}$, $E = \{e_1, e_2, \dots, e_9\}$, and $\gamma(e_1) = \{a, c\}$, $\gamma(e_2) = \{a, b\}$, $\gamma(e_3) = \{d, c\}$, $\gamma(e_4) = \{b, d\}$, $\gamma(e_5) = \{e, a\}$, $\gamma(e_6) = \{e, d\}$, $\gamma(e_7) = \{f, e\}$, $\gamma(e_8) = \{e, g\}$, and $\gamma(e_9) = \{f, g\}$.

7. Give the degree of each vertex in Figure 8.18.

8. Give the degree of each vertex in Figure 8.20.

9. List all paths that begin at a in Figure 8.19.

10. List three circuits that begin at 5 in Figure 8.21.

11. Draw the complete graph on seven vertices.

12. Consider K_n, the complete graph on n vertices. What is the degree of each vertex?

13. Which of the graphs in Exercises 1 through 4 are regular?

14. Give an example of a regular, connected graph on six vertices that is not complete.

15. Give an example of a graph on five vertices with exactly two components.

16. Give an example of a graph that is regular, but not complete, with each vertex having degree three.

For Exercises 17 through 20, use the graph G in Figure 8.22.

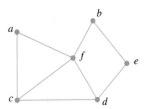
Figure 8.22

17. If R is the equivalence relation defined by the partition $\{\{a, f\}, \{e, b, d\}, \{c\}\}$, find the quotient graph, G^R.

18. If R is the equivalence relation defined by the partition $\{\{a, b\}, \{e\}, \{d\}, \{f, c\}\}$, find the quotient graph, G^R.

19. (a) Give the largest subgraph of G that does not contain f.

(b) Give the largest subgraph of G that does not contain a.

20. Let e_1 be the edge between c and f. Draw the graph of
(a) G_{e_1} (b) G^{e_1}

For Exercises 21 and 22, use the graph G in Figure 8.23.

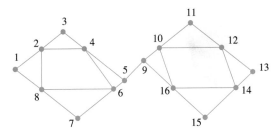
Figure 8.23

21. Let $R = \{(1, 1), (2, 2), (3, 3), (4, 4), (5, 5), (6, 6), (7, 7), (8, 8), (9, 9), (10, 10), (11, 11), (12, 12), (13, 13), (14, 14), (15, 15), (16, 16), (1, 10), (10, 1), (3, 12), (12, 3), (5, 14), (14, 5), (2, 11), (11, 2), (4, 13), (13, 4), (6, 15), (15, 6), (7, 16), (16, 7), (8, 9), (9, 8)\}$. Draw the quotient graph G^R.

22. Let $R = \{(1, 1), (2, 2), (3, 3), (4, 4), (5, 5), (6, 6), (7, 7), (8, 8), (9, 9), (10, 10), (11, 11), (12, 12), (13, 13), (14, 14), (15, 15), (16, 16), (1, 2), (2, 1), (3, 4), (4, 3), (5, 6), (6, 5), (7, 8), (8, 7), (9, 16), (16, 9), (10, 11), (11, 10), (12, 13), (13, 12), (14, 15), (15, 14)\}$. Draw the quotient graph G^R.

23. Complete the following statement. Every linear graph on n vertices must have _____ edges. Explain your answer.

24. What is the total number of edges in K_n, the complete graph on n vertices? Justify your answer.

25. Explain how the definition of digraph in Chapter 4 differs from the definition of graph in this section.

26. Prove that if a graph G has no loops or multiple edges, then twice the number of edges is equal to the sum of the degrees of all vertices.

27. Use the result of Exercise 26 to prove that if a graph G has no loops or multiple edges, then the number of vertices of odd degree is an even number.

8.2   EULER PATHS AND CIRCUITS

In this section and the next, we consider broad categories of problems for which graph theory is used. In the first type of problem, the task is to travel a path using each edge of the graph exactly once. It may or may not be necessary to begin and end at the same vertex. A simple example of this is the common puzzle problem that asks the solver to trace a geometric figure without lifting pencil from paper or tracing an edge more than once.

A path in a graph G is called an **Euler path** if it includes every edge exactly once. An **Euler circuit** is an Euler path that is a circuit. Leonhard Euler (1707–1783) worked in many areas of mathematics. The names "Euler path" and "Euler circuit" recognize his work with a problem called the Konigsberg Bridge problem.

EXAMPLE 1   Figure 8.24 shows the street map of a small neighborhood. A recycling ordinance has been passed, and those responsible for picking up the recyclables must start and end each trip by having the truck in the recycling terminal. They would like to plan the truck route so that the entire neighborhood can be covered and each street need be traveled only once. A graph can be constructed having one vertex for each intersection and an edge for each street between any two intersections. The problem then is to find an Euler circuit for this graph. ∎

EXAMPLE 2   (a) An Euler path in Figure 8.25 is $\pi: E, D, B, A, C, D$.
(b) One Euler circuit in the graph of Figure 8.26 is $\pi: 5, 3, 2, 1, 3, 4, 5$. ∎

Figure 8.24

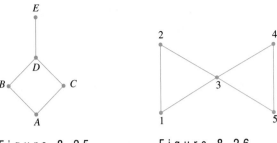

Figure 8.25   Figure 8.26

A little experimentation will show that no Euler circuit is possible for the graph in Figure 8.25. We also see that an Euler path is not possible for the graph in Figure 8.6 in Section 8.1. (Why?)

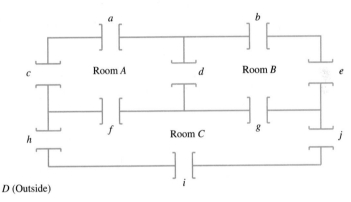

Figure 8.27

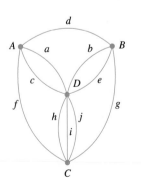

Figure 8.28

EXAMPLE 3

Consider the floor plan of a three-room structure that is shown in Figure 8.27. Each room is connected to every room that it shares a wall with and to the outside along each wall. The problem is this: Is it possible to begin in a room or outside and take a walk that goes through each door exactly once? This diagram can also be formulated as a graph where each room and the outside constitute a vertex and an edge corresponds to each door. A possible graph for this structure is shown in Figure 8.28. The translation of the problem is whether or not there exists an Euler path for this graph. We will solve this problem later. ■

Two questions arise naturally at this point. Is it possible to determine whether an Euler path or Euler circuit exists without actually finding it? If there must be an Euler path or circuit, is there an efficient way to find one?

Consider again the graphs in Example 2. In Figure 8.25 the edge $\{D, E\}$ must be either the first or the last traveled, because there is no other way to travel to or from vertex E. This means that if G has a vertex of degree 1, there cannot be an Euler circuit, and if there is an Euler path, it must begin or end at this vertex. A similar argument applies to any vertex v of odd degree, say $2n + 1$. We may travel in on one of these edges and out on another one n times, leaving one edge from v

untraveled. This last edge may be used for leaving v or arriving at v, but not both, so a circuit cannot be completed. We have just shown the first of the following results.

Theorem 1   (a) If a graph G has a vertex of odd degree, there can be no Euler circuit in G.

(b) If G is a connected graph and every vertex has even degree, then there is an Euler circuit in G. ●

Proof

(b) Suppose that there are connected graphs where every vertex has even degree, but there is no Euler circuit. Choose such a G with the smallest number of edges. G must have more than one vertex since, if there were only one vertex of even degree, there is clearly an Euler circuit.

We show first that G must have at least one circuit. If v is a fixed vertex of G, then since G is connected and has more than one vertex, there must be an edge between v and some other vertex v_1. This is a simple path (of length 1) and so simple paths exist. Let π_0 be a simple path in G having the longest possible length, and let its vertex sequence be $v_1, v_2, \ldots, v_s$. Since v_s has even degree and π_0 uses only one edge that has v_s as a vertex, there must be an edge e not in π_0 that also has v_s as a vertex. If the other vertex of e is not one of the v_i, then we could construct a simple path longer than π_0, which is a contradiction. Thus e has some v_i as its other vertex, and therefore we can construct a circuit $v_i, v_{i+1}, \ldots, v_s, v_i$ in G.

Since we now know that G has circuits, we may choose a circuit π in G that has the longest possible length. Since we assumed that G has no Euler circuits, π cannot contain all the edges of G. Let G_1 be the graph formed from G by deleting all edges in π (but no vertices). Since π is a circuit, deleting its edges will reduce the degree of every vertex by 0 or 2, so G_1 is also a graph with all vertices of even degree. The graph G_1 may not be connected, but we can choose a largest connected component (piece) and call this graph G_2 (G_2 may be G_1). Now G_2 has fewer edges than G, and so (because of the way G was chosen), G_2 must have an Euler path π'.

If π' passes through all the vertices on G, then π and π' clearly have vertices in common. If not, then there must be an edge in G between some vertex v' in π', and some vertex v not in π'. Otherwise we could not get from vertices in π' to the other vertices in G, and G would not be connected. Since e is not in π', it must have been deleted when G_1 was created from G, and so must be an edge in π. Then v' is also in the vertex sequence of π, and so in any case π and π' have at least one vertex v' in common. We can then construct a circuit in G that is longer than π by combining π and π' at v'. This is a contradiction, since π was chosen to be the longest possible circuit in G. Hence the existence of the graph G always produces a contradiction, and so no such graph is possible. ▼

The strategy of this proof is one we have used before: Suppose there is a largest (smallest) object and construct a larger (smaller) object of the same type thereby creating a contradiction. Here we have π, the longest possible circuit that begins and ends at v, in G, and we construct a longer circuit that begins and ends at v.

We have proved that if G has vertices of odd degree, it is not possible to construct an Euler circuit for G, but an Euler path may be possible. Our earlier discussion noted that a vertex of odd degree must be either the beginning or the end of any possible Euler path. We have the following theorem.

Theorem 2 (a) If a graph G has more than two vertices of odd degree, then there can be no Euler path in G.

(b) If G is connected and has exactly two vertices of odd degree, there is an Euler path in G. Any Euler path in G must begin at one vertex of odd degree and end at the other. ●

Proof

(a) Let v_1, v_2, v_3 be vertices of odd degree. Any possible Euler path must leave (or arrive at) each of v_1, v_2, v_3 with no way to return (or leave) since each of these vertices has odd degree. One vertex of these three vertices may be the beginning of the Euler path and another the end, but this leaves the third vertex at one end of an untraveled edge. Thus there is no Euler path.

(b) Let u and v be the two vertices of odd degree. Adding the edge $\{u, v\}$ to G produces a connected graph G' all of whose vertices have even degree. By Theorem 1(b), there is an Euler circuit π' in G'. Omitting $\{u, v\}$ from π' produces an Euler path that begins at u (or v) and ends at v (or u). ▼

EXAMPLE 4   Which of the graphs in Figures 8.29, 8.30, and 8.31 have an Euler circuit, an Euler path but not an Euler circuit, or neither?

Solution

(a) In Figure 8.29, each of the four vertices has degree 3; thus, by Theorems 1 and 2, there is neither an Euler circuit nor an Euler path.

(b) The graph in Figure 8.30 has exactly two vertices of odd degree. There is no Euler circuit, but there must be an Euler path.

(c) In Figure 8.31, every vertex has even degree; thus the graph must have an Euler circuit. ■

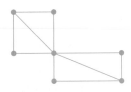

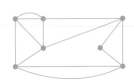

Figure 8.29   Figure 8.30   Figure 8.31

EXAMPLE 5

Let us return to Example 3. We see that the four vertices have degrees 4, 4, 5, and 7, respectively. Thus the problem can be solved by Theorem 2; that is, there is an Euler path. One is shown in Figure 8.32. Using the labels of Figure 8.28, this path π is specified by $V_\pi: C, D, C, A, D, A, B, D, B, C$ and $E_\pi: i, h, f, c, a, d, b, e, g, j$.

■

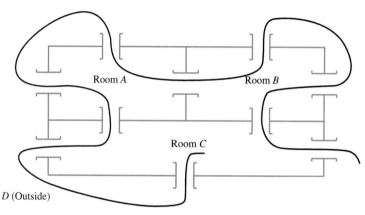

Figure 8.32

Figure 8.33

Theorems 1 and 2 are examples of existence theorems. They guarantee the existence of an object of a certain type, but they give no information on how to produce the object. There is one hint in Theorem 2(b) about how to proceed. In Figure 8.33, an Euler path must begin (or end) at B and end (or begin) at C. One possible Euler path for this graph is specified by $V_\pi: B, A, D, C, A, B, C$ and $E_\pi: a, c, d, f, b, e$.

We next give an algorithm that produces an Euler circuit for a connected graph with no vertices of odd degree. We require an additional definition before stating the algorithm. An edge is a **bridge** in a connected graph G if deleting it would create a disconnected graph. For example, in the graph of Figure 8.4, Section 8.1, $\{B, E\}$ is a bridge.

◆ **FLEURY'S ALGORITHM**

Let $G = (V, E, \gamma)$ be a connected graph with each vertex of even degree.

Step 1   Select an edge e_1 that is not a bridge in G. Let its vertices be v_1, v_2. Let π be specified by $V_\pi: v_1, v_2$ and $E_\pi: e_1$. Remove e_1 from E and v_1 and v_2 from V to create G_1.

Step 2   Suppose that $V_\pi: v_1, v_2, \ldots, v_k$ and $E_\pi: e_1, e_2, \ldots, e_{k-1}$ have been constructed so far, and that all of these edges and vertices have been removed from V and E to form G_{k-1}. Since v_k has even degree, and e_{k-1} ends there, there must be an edge e_k in G_{k-1} that also has v_k as a vertex. If there is more than one such edge, select one that is not a bridge for G_{k-1}. Denote the vertex of e_k other than v_k by v_{k+1}, and extend V_π and E_π to $V_\pi: v_1, v_2, \ldots, v_k, v_{k+1}$ and $E_\pi: e_1, e_2, \ldots, e_{k-1}, e_k$.

Step 3   Repeat Step 2 until no edges remain in E.

End of Algorithm

EXAMPLE 6   Use Fleury's algorithm to construct an Euler circuit for the graph in Figure 8.34.

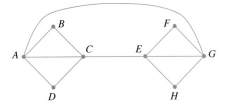

Figure 8.34

Solution   According to Step 1, we may begin anywhere. Arbitrarily choose vertex A. We summarize the results of applying Step 2 repeatedly in Table 8.1.

Table 8.1

Current Path	Next Edge	Reasoning
$\pi : A$	$\{A, B\}$	No edge from A is a bridge. Choose any one.
$\pi : A, B$	$\{B, C\}$	Only one edge from B remains.
$\pi : A, B, C$	$\{C, A\}$	No edge from C is a bridge. Choose any one.
$\pi : A, B, C, A$	$\{A, D\}$	No edge from A is a bridge. Choose any one.
$\pi : A, B, C, A, D$	$\{D, C\}$	Only one edge from D remains.
$\pi : A, B, C, A, D, C$	$\{C, E\}$	Only one edge from C remains.
$\pi : A, B, C, A, D, C, E$	$\{E, G\}$	No edge from E is a bridge. Choose any one.
$\pi : A, B, C, A, D, C, E, G$	$\{G, F\}$	$\{A, G\}$ is a bridge. Choose $\{G, F\}$ or $\{G, H\}$.
$\pi : A, B, C, A, D, C, E, G, F$	$\{F, E\}$	Only one edge from F remains.
$\pi : A, B, C, A, D, C, E, G, F, E$	$\{E, H\}$	Only one edge from E remains.
$\pi : A, B, C, A, D, C, E, G, F, E, H$	$\{H, G\}$	Only one edge from H remains.
$\pi : A, B, C, A, D, C, E, G, F, E, H, G$	$\{G, A\}$	Only one edge from G remains.
$\pi : A, B, C, A, D, C, E, G, F, E, H, G, A$		

The edges in Figure 8.35 have been numbered in the order of their choice in applying Step 2. In several places, other choices could have been made. In general, if a graph has an Euler circuit, it is likely to have several different Euler circuits.   ∎

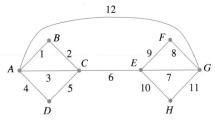

Figure 8.35

8.2 Exercises

In Exercises 1 *through* 8, *(Figures 8.36 through 8.43), tell whether the graph has an Euler circuit, an Euler path but no Euler circuit, or neither. Give reasons for your choice.*

1.

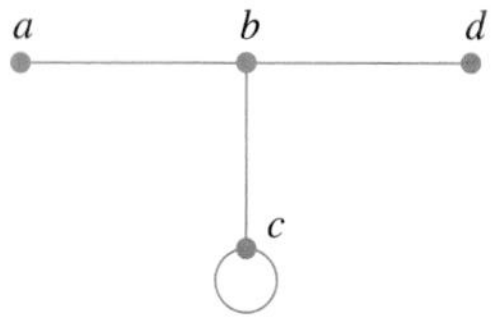

a   b   d

c

Figure 8.36

2.

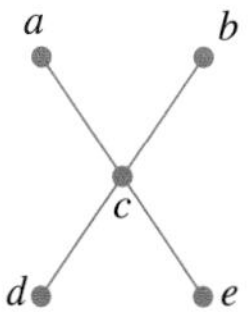

a   b

c

d   e

Figure 8.37

3.

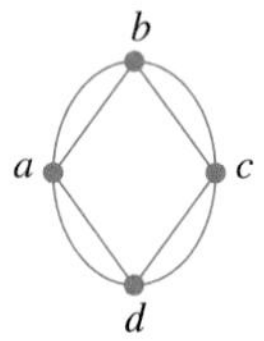

b

a   c

d

Figure 8.38

4.

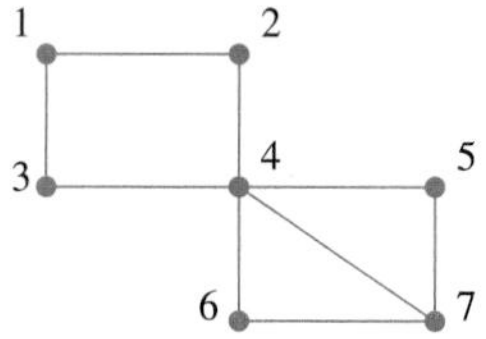

1   2

3   4   5

6   7

Figure 8.39

5.

Figure 8.40

6.

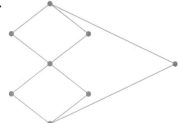

Figure 8.41

7.

Figure 8.42

8.

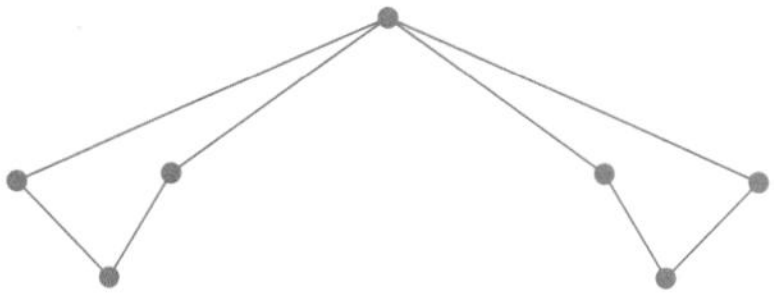

Figure 8.43

In Exercises 9 *and* 10, *(Figures 8.44 through 8.45), tell if it is possible to trace the figure without lifting the pencil. Explain your reasoning.*

9.

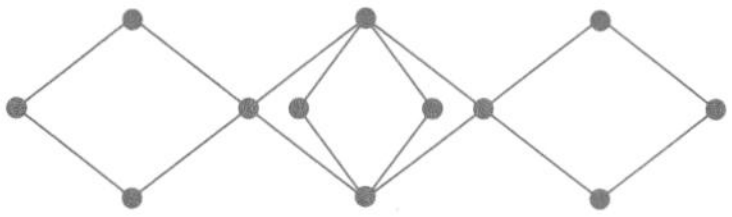

Figure 8.44

10.

Figure 8.45

11. Use Fleury's algorithm to produce an Euler circuit for the graph in Figure 8.46.

Figure 8.46

12. Use Fleury's algorithm to produce an Euler circuit for the graph in Figure 8.44.

13. An art museum arranged its current exhibit in the five rooms shown in Figure 8.47. Is there a way to tour the exhibit so that you pass through each door exactly once? If so, give a sketch of your tour.

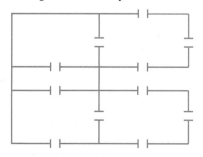

Figure 8.47

14. At the door of an historical mansion, you receive a copy of the floor plan for the house (Figure 8.48). Is it possible to visit every room in the house by passing through each door exactly once? Explain your reasoning.

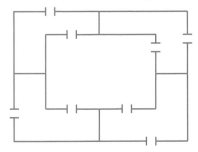

Figure 8.48

In Exercises 15 through 18 (Figures 8.49 through 8.52), no Euler circuit is possible for the graph given. For each graph, show the minimum number of edges that would need to be traveled twice in order to travel every edge and return to the starting vertex.

15.   **16.**

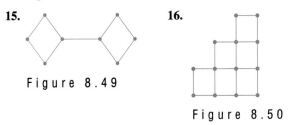

Figure 8.49

Figure 8.50

17.

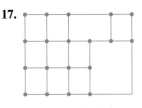

Figure 8.51

18.

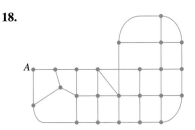

Figure 8.52

19. Modify the graph in Figure 8.51 by adding the minimum number of duplicate edges needed to make all vertices have even degree. Use Fleury's algorithm to find an Euler circuit for the modified version of the graph. Begin at the upper-left corner.

20. Modify the graph in Figure 8.52 by adding the minimum number of duplicate edges needed to make all vertices have even degree. Use Fleury's algorithm to find an Euler circuit for the modified version of the graph. Begin at *A*.

8.3   HAMILTONIAN PATHS AND CIRCUITS

We turn now to the second category of graph problems in which the task is to visit each vertex exactly once, with the exception of the beginning vertex if it must also be the last vertex. For example, such a path would be useful to someone who must service a set of vending machines on a regular basis. Each vending machine could be represented by a vertex.

A **Hamiltonian path** is a path that contains each vertex exactly once. A **Hamiltonian circuit** is a circuit that contains each vertex exactly once except for the first vertex, which is also the last. This sort of path is named for the mathematician Sir William Hamilton, who developed and marketed a game consisting of a wooden graph in the shape of a regular dodecahedron and instructions to find what we have called a Hamiltonian circuit. A planar version of this solid is shown in Figure 8.53(a), with a Hamiltonian circuit (one of many) shown in Figure 8.53(b) by the consecutively numbered vertices.

It is clear that loops and multiple edges are of no use in finding Hamiltonian circuits, since loops could not be used, and only one edge can be used between any two vertices. Thus we will suppose that any graph we mention has no loops or multiple edges.

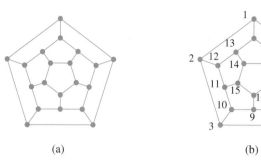

(a)   (b)

Figure 8.53

EXAMPLE 1   Consider the graph in Figure 8.54. The path a, b, c, d, e is a Hamiltonian path because it contains each vertex exactly once. It is not hard to see, however, that there is no Hamiltonian circuit for this graph. For the graph shown in Figure 8.55, the path A, D, C, B, A is a Hamiltonian circuit. In Figures 8.56 and 8.57, no Hamiltonian path is possible. (Verify this.)   ∎

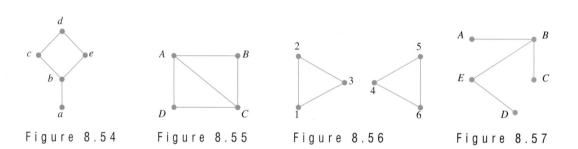

Figure 8.54   Figure 8.55   Figure 8.56   Figure 8.57

EXAMPLE 2   Any complete graph K_n has Hamiltonian circuits. In fact, starting at any vertex, you can visit the other vertices sequentially in any desired order.   ∎

Questions analogous to those about Euler paths and circuits can be asked about Hamiltonian paths and circuits. Is it possible to determine whether a Hamiltonian path or circuit exists? If there must be a Hamiltonian path or circuit, is there an efficient way to find it? Surprisingly, considering Theorems 1 and 2 of Section 8.2, the first question about Hamiltonian paths and circuits has not been completely answered and the second is still unanswered as well. However, we can make several observations based on the examples.

If a graph G on n vertices has a Hamiltonian circuit, then G must have at least n edges.

We now state some partial answers that say if a graph G has "enough" edges, a Hamiltonian circuit can be found. These are again existence statements; no method for constructing a Hamiltonian circuit is given.

Theorem 1 Let G be a connected graph with n vertices, $n > 2$, and no loops or multiple edges. G has a Hamiltonian circuit if for any two vertices u and v of G that are not adjacent, the degree of u plus the degree of v is greater than or equal to n. ●

We omit the proof of this result, but from it we can prove the following:

Corollary 1 G has a Hamiltonian circuit if each vertex has degree greater than or equal to $n/2$. ●

Proof The sum of the degrees of any two vertices is at least $\frac{n}{2} + \frac{n}{2} = n$, so the hypotheses of Theorem 1 hold. ▼

Theorem 2 Let the number of edges of G be m. Then G has a Hamiltonian circuit if $m \geq \frac{1}{2}(n^2 - 3n + 6)$ (recall that n is the number of vertices). ●

Proof Suppose that u and v are any two vertices of G that are not adjacent. We write $\deg(u)$ for the degree of u. Let H be the graph produced by eliminating u and v from G along with any edges that have u or v as end points. Then H has $n - 2$ vertices and $m - \deg(u) - \deg(v)$ edges (one fewer edge would have been removed if u and v had been adjacent). The maximum number of edges that H could possibly have is $_{n-2}C_2$ (see Section 3.2). This happens when there is an edge connecting every distinct pair of vertices. Thus the number of edges of H is at most

$$_{n-2}C_2 = \frac{(n-2)(n-3)}{2} \quad \text{or} \quad \frac{1}{2}(n^2 - 5n + 6).$$

We then have $m - \deg(u) - \deg(v) \leq \frac{1}{2}(n^2 - 5n + 6)$. Therefore, $\deg(u) + \deg(v) \geq m - \frac{1}{2}(n^2 - 5n + 6)$. By the hypothesis of the theorem,

$$\deg(u) + \deg(v) \geq \frac{1}{2}(n^2 - 3n + 6) - \frac{1}{2}(n^2 - 5n + 6) = n.$$

Thus the result follows from Theorem 1. ▼

EXAMPLE 3 The converses of Theorems 1 and 2 given above are not true; that is, the conditions given are sufficient, but not necessary, for the conclusion. Consider the graph represented by Figure 8.58. Here n, the number of vertices, is 8, each vertex has degree 2, and $\deg(u) + \deg(v) = 4$ for every pair of nonadjacent vertices u and v. The total number of edges is also 8. Thus the premises of Theorems 1 and 2 fail to be satisfied, but there are certainly Hamiltonian circuits for this graph. ■

The problem we have been considering has a number of important variations. In one case, the edges may have *weights* representing distance, cost, and the like. The problem is then to find a Hamiltonian circuit (or path) for which the total sum of weights in the path is a minimum. For example, the vertices might represent cities; the edges, lines of transportation; and the weight of an edge, the cost of traveling along that edge. This version of the problem is often called the traveling salesperson problem. Another important category of problems involving graphs with weights assigned to edges is discussed in Section 6.5.

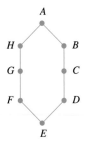

Figure 8.58

8.3 Exercises

In Exercises 1 through 6 (Figures 8.59 through 8.64), determine whether the graph shown has a Hamiltonian circuit, a Hamiltonian path but no Hamiltonian circuit, or neither. If the graph has a Hamiltonian circuit, give the circuit.

1.

Figure 8.59

2.

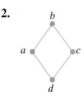

Figure 8.60

3.

Figure 8.61

4.

Figure 8.62

5.

Figure 8.63

6.

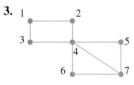

Figure 8.64

In Exercises 7 through 10 (Figures 8.65 through 8.68), find a Hamiltonian circuit for the graph given.

7.

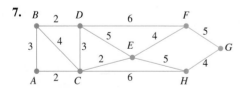

Figure 8.65

8.

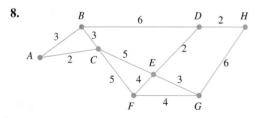

Figure 8.66

9.

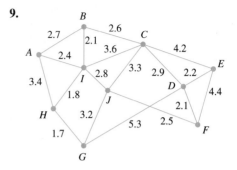

Figure 8.67

10.

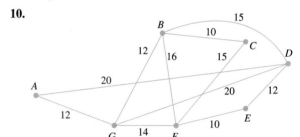

Figure 8.68

In Exercises 11 through 14, find a Hamiltonian circuit of minimal weight for the graph represented by the given figure.

11. Figure 8.65   **12.** Figure 8.66

13. Figure 8.67   **14.** Figure 8.68

15. Find a Hamiltonian circuit of minimal weight for the graph represented by Figure 8.65 if you must begin and end at D.

16. Find a Hamiltonian circuit of minimal weight for the graph represented by Figure 8.66 if you must begin and end at F.

17. Prove that K_n, the complete graph on n vertices with $n \geq 3$, has $(n - 1)!$ Hamiltonian circuits.

18. Give an example of a graph with at least four vertices with a circuit that is both an Euler and a Hamiltonian circuit.

19. Give an example of a graph that has an Euler circuit and a Hamiltonian circuit that are not the same.

20. Let $G = (V, E, \gamma)$ be a graph with $|V| = n$ that has no multiple edges. The relation R on V defined by G can be represented by a matrix, M_R. Explain how to use M_{R^∞} (see Section 4.3) to determine if G is connected.

8.4 TRANSPORT NETWORKS

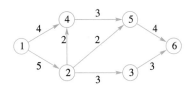

Figure 8.69

We have previously examined several uses of labeled graphs. In this section we return to the idea of a directed graph (digraph). An important use of labeled digraphs is to model what are commonly called transport networks. Consider the labeled digraph shown in Figure 8.69. This might represent a pipeline that carries water from vertex 1 to vertex 6 as part of a municipal water system. The label on an edge represents the maximum flow that can be passed through that edge and is called the **capacity** of the edge. Many situations can be modeled in this way. For instance, Figure 8.69 might as easily represent an oil pipeline, a highway system, a communications network, or an electric power grid. The vertices of a network are usually called nodes and may denote pumping stations, shipping depots, relay stations, or highway interchanges.

More formally, a **transport network**, or a **network**, is a connected digraph N with the following properties:

(a) There is a unique node, the **source**, that has in-degree 0. We generally label the source node 1.

(b) There is a unique node, the **sink**, that has out-degree 0. If N has n nodes, we generally label the sink as node n.

(c) The graph N is labeled. The label, C_{ij}, on edge (i, j) is a nonnegative number called the capacity of the edge.

For simplicity we also assume that all edges carry material in one direction only; that is, if (i, j) is in N, then (j, i) is not.

Flows

The purpose of a network is to implement a flow of water, oil, electricity, traffic, or whatever the network is designed to carry. Mathematically, a **flow** in a network N is a function that assigns to each edge (i, j) of N a nonnegative number F_{ij} that does not exceed C_{ij}. Intuitively, F_{ij} represents the amount of material passing through the edge (i, j) when the flow is F. Informally, we refer to F_{ij} as the flow through edge (i, j). We also require that for each node other than the source and sink, the sum of the F_{ik} on edges entering node k must be equal to the sum of the F_{kj} on edges leaving node k. This means that material cannot accumulate, be created, dissipate, or be lost at any node other than the source or the sink. This is called **conservation of flow**. A consequence of this requirement is that the sum of the flows leaving the source must equal the sum of the flows entering the sink. This sum is called the **value of the flow**, written $value(F)$. We can represent a flow F by labeling each edge (i, j) with the pair (C_{ij}, F_{ij}). A flow F in the network represented by Figure 8.69 is shown in Figure 8.70.

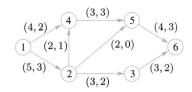

Figure 8.70

EXAMPLE 1

In Figure 8.70, flow is conserved at node 4 since there are input flows of size 2 and 1, and an output flow of size 3. (Verify that flow is conserved properly at the other nodes.) Here $value(F) = 5$. ∎

Maximum Flows

For any network an important problem is to determine the maximum value of a flow through the network and to describe a flow that has the maximum value. For obvious reasons this is commonly referred to the maximum flow problem.

EXAMPLE 2   Figure 8.71(a) shows a flow that has value 8. Three of the five edges are carrying their maximum capacity. This seems to be a good flow function, but Figure 8.71(b) shows a flow with value 10 for the same network.   ■

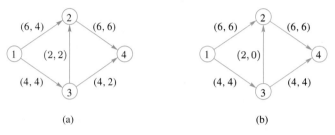

(a)   (b)

Figure 8.71

Example 2 shows that even for a small network, we need a systematic procedure for solving the maximum flow problem. Examining the flow in Figure 8.71(a) shows that using the edge from node 3 to node 2 as we did was a mistake. We should reduce flow in edge (3, 2) so that we can increase it in other edges.

Suppose that in some network N we have an edge (i, j) that is carrying a flow of 5 units. If we want to reduce this flow to 3 units, we can imagine that it is combined with a flow of two units in the opposite direction. Although edge (j, i) is not in N, there is no harm in considering such a *virtual flow* as long as it only has the effect of reducing the existing flow in the actual edge (i, j). Figure 8.72 displays a portion of the flow shown in Figure 8.71(a).

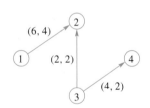

Figure 8.72

The path π: 1, 2, 3, 4 is not an actual path in this network, since (2, 3) is not an actual edge. However, π is a path in the symmetric closure of the network. (Refer to Section 4.7 for symmetric closure.) Moreover, if we consider a virtual flow of two units through π, the effect on the network is to increase the flows through edges (1, 2) and (3, 4) by two units and decrease the flow through edge (3, 2) by two units. Thus, the flow of Figure 8.71(a) becomes the flow of Figure 8.71(b).

We now describe this improvement in general terms. Let N be a network and let G be the symmetric closure of N. Choose a path in G and an edge (i, j) in this path. If (i, j) belongs to N, then we say this edge has positive excess capacity if $e_{ij} = C_{ij} - F_{ij} > 0$. If (i, j) is not an edge of N, then we are traveling this edge in the wrong direction. In this case we say (i, j) has excess capacity $e_{ij} = F_{ji}$ if $F_{ji} > 0$. Then increasing flow through edge (i, j) will have the effect of reducing F_{ji}. We now give a procedure for solving a maximum flow problem.

A Maximum Flow Algorithm

The algorithm we present is due to Ford and Fulkerson and is often called the **labeling algorithm**. The labeling referred to is an additional labeling of nodes. We have used integer capacities for simplicity, but Ford and Fulkerson show that this algorithm will stop in a finite number of steps if the capacities are rational numbers.

Let N be a network with n nodes and G be the symmetric closure of N. All edges and paths used are in G. Begin with all flows set to 0. As we proceed, it will be convenient to track the excess capacities in the edges and how they change rather than tracking the increasing flows. When the algorithm terminates, it is easy to find the maximum flow from the final excess capacities.

◆ THE LABELING ALGORITHM

Step 1 Let N_1 be the set of all nodes connected to the source by an edge with positive excess capacity. Label each j in N_1 with $[E_j, 1]$, where E_j is the excess capacity e_{1j} of edge $(1, j)$. The 1 in the label indicates that j is connected to the source, node 1.

Step 2 Let node j in N_1 be the node with smallest node number and let $N_2(j)$ be the set of all unlabeled nodes, other than the source, that are joined to node j and have positive excess capacity. Suppose that node k is in $N_2(j)$ and (j, k) is the edge with positive excess capacity. Label node k with $[E_k, j]$, where E_k is the minimum of E_j and the excess capacity e_{jk} of edge (j, k). When all the nodes in $N_2(j)$ are labeled in this way, repeat this process for the other nodes in N_1. Let $N_2 = \bigcup_{j \in N_1} N_2(j)$.

Note that after Step 1, we have labeled each node j in N_1 with E_j, the amount of material that can flow from the source to j through one edge and with the information that this flow came from node 1. In Step 2, previously unlabeled nodes k that can be reached from the source by a path $\pi : 1, j, k$ are labeled with $[E_k, j]$. Here E_k is the maximum flow that can pass through π since it is the smaller of the amount that can reach j and the amount that can then pass on to k. Thus when Step 2 is finished, we have constructed two-step paths to all nodes in N_2. The label for each of these nodes records the total flow that can reach the node through the path and its immediate predecessor in the path. We attempt to continue this construction increasing the lengths of the paths until we reach the sink (if possible). Then the total flow can be increased and we can retrace the path used for this increase.

Step 3 Repeat Step 2, labeling all previously unlabeled nodes N_3 that can be reached from a node in N_2 by an edge having positive excess capacity. Continue this process forming sets $N_4, N_5, \ldots$ until after a finite number of steps either

   (i) the sink has not been labeled and no other nodes can be labeled. It can happen that no nodes have been labeled; remember that the source is not labeled.
        or
   (ii) the sink has been labeled.

Step 4 In case (i), the algorithm terminates and the total flow then is a maximum flow. (We show this later.)

Step 5 In case (ii) the sink, node n, has been labeled with $[E_n, m]$ where

E_n is the amount of extra flow that can be made to reach the sink through a path π. We examine π in reverse order. If edge $(i, j) \in N$, then we increase the flow in (i, j) by E_n and decrease the excess capacity e_{ij} by the same amount. Simultaneously, we increase the excess capacity of the (virtual) edge (j, i) by E_n since there is that much more flow in (i, j) to reverse. If, on the other hand, $(i, j) \notin N$, we decrease the flow in (j, i) by E_n and increase its excess capacity by E_n. We simultaneously decrease the excess capacity in (i, j) by the same amount, since there is less flow in (i, j) to reverse. We now have a new flow that is E_n units greater than before and we return to Step 1.

EXAMPLE 3   Use the labeling algorithm to find a maximum flow for the network in Figure 8.69.

Solution   Figure 8.73 shows the network with initial capacities of all edges in G. The initial flow in all edges is zero.

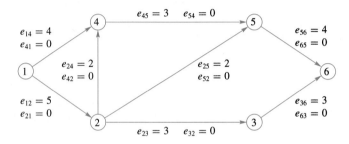

Figure 8.73

STEP 1:   Starting at the source, we can reach nodes 2 and 4 by edges having excess capacity, so $N_1 = \{2, 4\}$. We label nodes 2 and 4 with the labels $[5, 1]$ and $[4, 1]$, respectively, as shown in Figure 8.74.

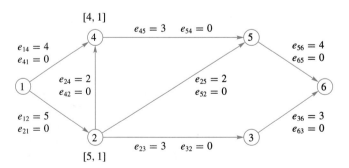

Figure 8.74

STEP 2:   From node 2 we can reach nodes 5 and 3 using edges with positive excess capacity. Node 5 is labeled with $[2, 2]$ since only two additional units of flow can pass through edge $(2, 5)$. Node 3 is labeled with $[3, 2]$ since only

3 additional units of flow can pass through edge (2, 3). The result of this step is shown in Figure 8.75.

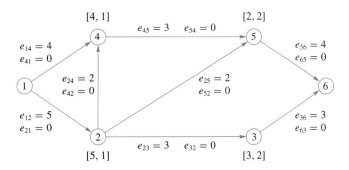

Figure 8.75

We cannot travel from node 4 to any unlabeled node by one edge. Thus, $N_2 = \{3, 5\}$ and Step 2 is complete.

STEP 3:   We repeat Step 2 using N_2. We can reach the sink from node 3 and 3 units through edge (3, 6). Thus the sink is labeled with [3, 3]

STEP 5:   We work backward through the path 1, 2, 3, 6 and subtract 3 from the excess capacity of each edge, indicating an increased flow through that edge, and adding an equal amount to the excess capacities of the (virtual) edges. We now return to Step 1 with the situation shown in Figure 8.76.

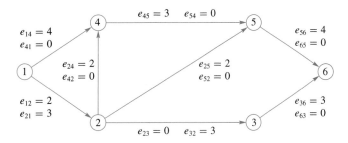

Figure 8.76

Proceeding as before, nodes 2 and 4 are labeled [2, 1] and [4, 1], respectively. Note that E_2 is now only 2 units, the new excess capacity of edge (1, 2). Node 2 can no longer be used to label node 3, since there is no excess capacity in the edge (2, 3). But node 5 now will be labeled [2, 2]. Once again no unlabeled node can be reached from node 4, so we move to Step 3. Here we can reach node 6 from node 5 so node 6 is labeled with [2, 5]. The final result of Step 3 is shown in Figure 8.77, and we have increased the flow by 2 units to a total of 5 units.

We move to Step 5 again and work back along the path 1, 2, 5, 6, subtracting 2 from the excess capacities of these edges and adding 2 to the capacities of the corresponding (virtual) edges. We return to Step 1 with Figure 8.78.

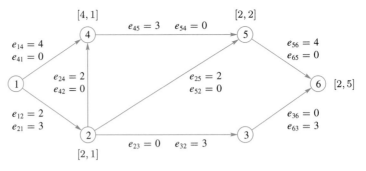

Figure 8.77

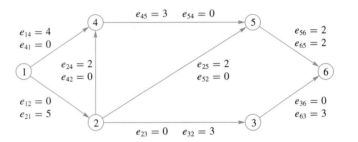

Figure 8.78

This time Steps 1 and 2 produce the following results. Only node 4 is labeled from node 1, with [4, 1]. Node 5 is the only node labeled from node 4, with [3, 4]. Step 3 begins with Figure 8.79.

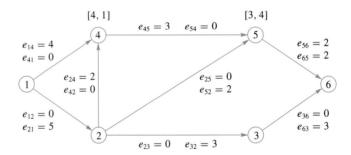

Figure 8.79

At this point, node 5 could label node 2 using the excess capacity of edge $(5, 2)$. (Verify that this would label node 2 with [2, 5].) However, node 5 can also be used to label the sink. The sink is labeled [2, 5] and the total flow is increased to 7 units. In Step 5, we work back along the path $1, 4, 5, 6$, adjusting excess capacities. We return to Step 1 with the configuration shown in Figure 8.80.

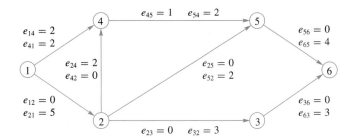

Figure 8.80

Verify that after Steps 1, 2, and 3, nodes 4, 5, and 2 have been labeled as shown in Figure 8.81 and no further labeling is possible. The final labeling of node 2 uses the virtual edge $(5, 2)$.

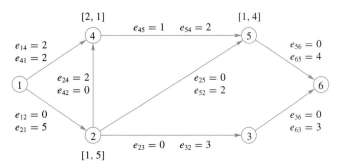

Figure 8.81

Thus, the final overall flow has value 7. By subtracting the final excess capacity e_{ij} of each edge (i, j) in N from the capacity C_{ij}, the flow F that produces the maximum value 7 can be seen in Figure 8.82. ■

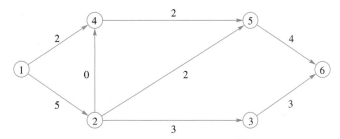

Figure 8.82

There remains the problem of showing that the labeling algorithm produces a maximum flow. First, we define a **cut** in a network N as a set K of edges having the property that every path from the source to the sink contains at least one edge from K. In effect, a cut does "cut" a digraph into two pieces, one containing the source and one containing the sink. If the edges of a cut were removed, nothing could flow from the source to the sink. The **capacity of a cut K**, $c(K)$, is the sum of the capacities of all edges in K.

EXAMPLE 4   Figure 8.83 shows two cuts for the network given by Figure 8.69. Each cut is marked by a jagged line and consists of all edges touched by the jagged line. Verify that $c(K_1) = 10$ and $c(K_2) = 7$.   ■

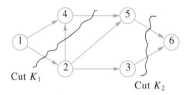

Cut K_1

Cut K_2

Figure 8.83

If F is any flow and K is any cut, then $value(F) \leq c(K)$. This is true because all parts of F must pass through the edges of K, and $c(K)$ is the maximum amount that can pass through the edges of K. Now suppose for some flow F and some cut K, $value(F) = c(K)$; in other words, the flow F uses the full capacity of all edges in K. Then F would be a flow with maximum value, since no flow can have value bigger than $c(K)$. Similarly, K must be a minimum capacity cut, because every cut must have capacity at least equal to $value(F)$. From this discussion we conclude the following.

Theorem 1
The Max Flow
Min Cut Theorem

A maximum flow F in a network has value equal to the capacity of a minimum cut of the network.   ●

We now show that the labeling algorithm results in a maximum flow by finding a minimum cut whose capacity is equal to the value of the flow. Suppose that the algorithm has been run and has stopped at Step 4. Then the sink has not been labeled. Divide the nodes into two sets, M_1 and M_2, where M_1 contains the source and all nodes that have been labeled, and M_2 contains all unlabeled nodes, other than the source. Let K consists of all edges of the network N that connect a node in M_1 with a node in M_2. Any path π in N from the source to the sink begins with a node in M_1 and ends with a node in M_2. If i is the last node in π that belongs to M_1 and j is the node that follows i in the path, then j belongs to M_2 and so by definition (i, j) is in K. Therefore, K is a cut.

Now suppose that (i, j) is an edge in K, so that $i \in M_1$ and $j \in M_2$. The final flow F produced by the algorithm must result in (i, j) carrying its full capacity; otherwise, we could use node i and the excess capacity to label j, which by definition is not labeled. Thus the value of the final flow of the algorithm is equal to the capacity $c(K)$, and so F is a maximum flow.

EXAMPLE 5   The minimum cut corresponding to the maximum flow found in Example 3 is $K = \{(5, 6), (3, 6)\}$ with $c(K) = 7 = value(F)$.   ■

8.4 Exercises

In Exercises 1 through 4 (Figures 8.84 through 8.87), label the network in the given figure with a flow that conserves flow at each node, except the source and the sink. Each edge is labeled with its maximum capacity.

1.

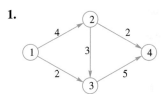

Figure 8.84

2.

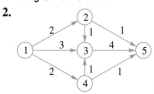

Figure 8.85

3.

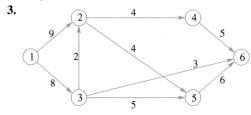

Figure 8.86

4.

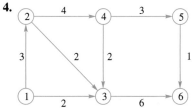

Figure 8.87

In Exercises 5 through 10, find a maximum flow in the given network by using the labeling algorithm.

5. The network shown in Figure 8.84

6. The network shown in Figure 8.85

7. The network shown in Figure 8.86

8. The network shown in Figure 8.87

9. The network shown in Figure 8.88

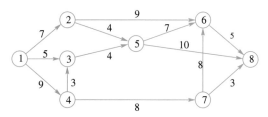

Figure 8.88

10. The network shown in Figure 8.89

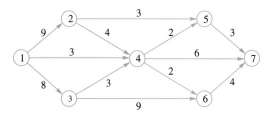

Figure 8.89

In Exercises 11 through 16, find the minimum cut that corresponds to the maximum flow for the given network.

11. The network of Exercise 5

12. The network of Exercise 6

13. The network of Exercise 7

14. The network of Exercise 8

15. The network of Exercise 9

16. The network of Exercise 10

8.5 MATCHING PROBLEMS

The definition of a transport network can be extended, and the concept of a maximal flow in a network can be used to model situations that, at first glance, do not seem to be network problems. We consider two examples in this section.

The first example is to allow a network to have multiple sources or multiple

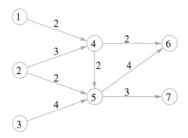

Figure 8.90

sinks as many real networks do. In the network N shown in Figure 8.90, nodes 1, 2, and 3 are all sources, and nodes 6 and 7 are sinks. For example, the sources could carry water from different pumping stations on a lake to two tanks (nodes 6 and 7) that supply two towns' drinking water.

In a case like this, we want to maximize the flow from all sources taken together to all the sinks taken together. As before, a flow F consists of the quantities F_{ij} assigned to each edge (i, j). We require that flow be conserved at each node that is not a source or a sink and that the flow in each edge not exceed the capacity of that edge. The value of the flow F, $value(F)$, is the sum of the flows in all edges that come from any source. It is not hard to show that this value must equal the sum of flows in all edges that lead to any sink. To find the maximal flow in a general network N, we change the problem into a single-source, single-sink network problem by enlarging N to N' as follows. We add two nodes, which we call a and b. Node a is the source for N' and is connected to all nodes that are sources in N. Similarly, node b is the sink for N', and all nodes that were sinks in N are connected to it. Nodes a and b are called, respectively, a **supersource** and a **supersink**. To complete the new network we set the capacities of all new edges to some number C_0 that is so large that no real flow through N' can pass more than C_0 units through any of the new edges. Some authors choose C_0 to be infinity.

EXAMPLE 1   An enlarged network N' for the network of Figure 8.90 is shown in Figure 8.91. We have set C_0 to be 11, the sum of all capacities leaving sources of N.   ∎

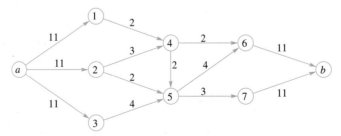

Figure 8.91

By adding a supersource and a supersink (if necessary) to a network, we can apply the labeling algorithm to find a maximal flow for the enlarged network. This flow will also be maximal for the original network.

EXAMPLE 2

Find the maximal flow for the network N given in Figure 8.90.

Solution Applying the labeling algorithm to the enlarged network N' given in Figure 8.91 produces the result shown in Figure 8.92. (Verify.) The value of this flow is 9. This is the maximal flow from nodes 1, 2, and 3 to nodes 6 and 7 in N. ∎

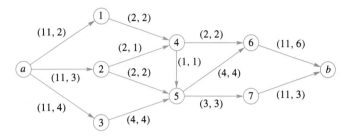

Figure 8.92

The Matching Problem

We consider now an important class of problems, matching problems, that can also be modeled by network flows. We begin with two finite sets A and B and a relation R from A to B. A **matching function** M is a one-to-one function from a subset of A to a subset of B. We say a is matched with b if $M(a) = b$. A matching function M is **compatible with R** if $M \subseteq R$; that is, if $M(a) = b$, then $a\ R\ b$.

EXAMPLE 3

Let A be a set of girls and B a set of boys attending a school dance. Define R by $a\ R\ b$ if and only if a knows b. A matching function M is defined from A to B by $M(a) = b$ if a and b dance the third dance together. M is compatible with R if each girl knows her partner for the third dance. ∎

EXAMPLE 4

Let $A = \{s_1, s_2, s_3, s_4, s_5\}$ be a set of students working on a research project and $B = \{b_1, b_2, b_3, b_4, b_5\}$ be a set of reference books on reserve in the library for the project. Define R by $s_i\ R\ b_k$ if and only if student s_i wants to sign out b_k. A matching of students to books would be compatible with R if each student is matched with a book that he or she wants to sign out. ∎

Given any relation R from A to B, a matching M that is compatible with R is called **maximal** if its domain is as large as possible and is **complete** if its domain is A. In general, a matching problem is, given A, B, and R, find a maximal matching from A to B that is compatible with R. Somewhat surprisingly, matching problems can be solved using networks. We create a network to model the situation by using the elements of A as sources and the elements of B as sinks. There is an directed edge (a, b) if and only if $a\ R\ b$. To complete the network, each edge is assigned capacity 1.

EXAMPLE 5

Let A, B, and R be as in Example 4. Suppose student s_1 wants books b_2 and b_3; s_2 wants b_1, b_2, b_3, b_4; s_3 wants b_2, b_3; s_4 wants b_2, b_3, b_4; s_5 wants b_2, b_3. Then the network N that represents this situation is given in Figure 8.93. ∎

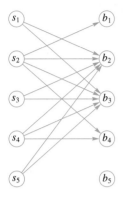

Figure 8.93

For the network in Figure 8.93, one maximal matching M can be found by inspection: $M(s_1) = b_2$, $M(s_2) = b_1$, $M(s_3) = b_3$, $M(s_4) = b_4$. It is easy to see that no complete matching is possible for this case. Usually it can be very difficult to find a maximal matching, so let us examine a network solution to the matching problem. In Figure 8.94, a supersource x and a super sink y have been provided and new edges have been assigned capacity 1 to create N'.

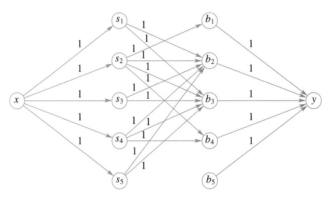

Figure 8.94

Every flow in N provides a matching of students to books that is compatible with R, and every compatible matching arises from a flow in N. To see this, suppose that F is a flow in N. If the flow into node s_m is 1, then the flow out must be 1, so the flow from s_m (if any) can go to only one node, say b_n. Similarly, flow can enter a node b_n from at most one s_m since the flow out of b_n is 1. We match s_m to b_n if and only if there is flow between these two nodes. The matching function M that we construct is clearly compatible with R. Conversely, if we have an R-compatible matching M, we can define a flow F by letting the flow be 1 between any two matched nodes, 1 from x to each student matched with a book, and from each matched book to y, and 0 on all other edges. This flow yields the matching M again. Hence, there is a one-to-one correspondence between flows in N and matching functions compatible with R. This means that we can use the labeling algorithm to solve the matching problem by constructing N and N' as in this example.

We saw that no complete matching is possible in the student-book example, but whether or not there is a complete matching for a particular problem is generally not obvious. A condition that will guarantee the existence of a complete matching was first found for the matching problem for a set of men and a set of women with the relation "suitable for marriage." Because the question posed was, "is it possible for each man to marry a suitable woman", this is often referred to as the marriage problem.

Theorem 1
Hall's Marriage Theorem

Let R be a relation from A to B. Then there exists a complete matching M if and only if for each $X \subseteq A$, $|X| \leq |R(X)|$.   ●

Proof If a complete matching M exists, then $M(X) \subseteq R(X)$ for every subset X of A. (See Section 4.2 for the definition of R-relative sets.) But M is one to one, so $|X| = |M(X)| \leq |R(X)|$.

Conversely, suppose that for any $X \subseteq A$, $|X| \leq |R(X)|$. Construct the network N that corresponds to R. Suppose $|A| = n$. We want to show that there is a flow in N with value n, which will correspond to a complete matching. We know by the max-flow-min-cut theorem that it is sufficient to show that the minimal cut in N has value n. A typical situation is shown in Figure 8.95, with the wavy line representing a cut in N.

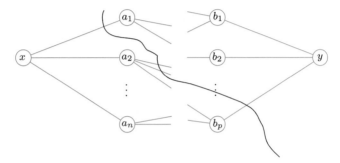

Figure 8.95

Remember that all edges are directed to the right and have capacity 1. We say that two vertices of N are **connected** if there is a path from one to the other; if not, the two vertices are **disconnected**. With this language, a cut in N is a set of edges whose removal will make the supersource x and the supersink y disconnected.

Let $n = |A|$ and $p = |B|$. Suppose K is a minimal cut of N. Divide the edges of K into three sets: S_1 contains the edges that begin at x, S_2 contains the edges that correspond to pairs in R, S_3 contains the edges that end with y. Consider removing the edges of K one set at a time. Suppose that $|S_1| = k$ so x is connected to k elements of A. Let A_1 be this subset of A. When the edges in S_1 are removed, no path from x to y can pass through an element of A_1. Since K is minimal, we can suppose that no edges in K begin with elements in A_1. Let $A_2 = A - A_1$ so $|A_2| = n-k$. Let $B_2 = R(A_2)$. Thus each element of B_2 labels the terminal node of an edge that begins in A_2. Since the supersource x is connected to each element in A_2, x is also connected to each element in B_2. We know that $|A_2| \leq |R(A_2)| = |B_2|$ so there are at least $n - k$ elements in B_2.

Let $|S_2| = r$. Each of these edges connects some element $a \in A_2$ to some element $b \in B_2$. Removing an edge in S_2 may or may not disconnect x from any element of B_2, but it certainly cannot disconnect x from more than one element of B_2. Thus, when the edges in S_2 are removed, x is still connected to at least $(n - k) - r$ elements of B_2. These elements must then be disconnected from y to ensure there is no path from x to y, and so S_3 must contain at least $(n - k) - r$ edges. Hence the capacity of K = the number of edges in $K = |S_1| + |S_2| + |S_3| \geq k + r + ((n - k) - r) = n$.

Since the cut that removes all edges beginning at x has capacity n, we see that the minimal cut, and therefore the maximum flow, is exactly n.   ▼

Note that this is an example of an existence theorem; no method is given for finding a complete matching, if one exists.

EXAMPLE 6   Let $\mathbf{M}_R$ be the matrix of a marriage suitability relation between five men and five women. Can each man marry a suitable woman?

$$\mathbf{M}_R = \begin{bmatrix} 1 & 1 & 0 & 0 & 0 \\ 0 & 0 & 0 & 1 & 1 \\ 1 & 0 & 1 & 0 & 0 \\ 0 & 0 & 1 & 0 & 1 \\ 0 & 1 & 0 & 1 & 0 \end{bmatrix}$$

Solution   We could construct an enlarged network N' to model this problem and apply the labeling algorithm in order to answer the question. Instead, we use Hall's marriage theorem. Note that each man considers exactly two women suitable and each woman is considered suitable by exactly two men. Consider the network that represents this problem. Let S be any subset of the men and E be the set of edges that begin in S. Clearly $|E| = 2|S|$. Each edge in E must terminate in a node of $R(S)$. But we know the number of edges terminating at elements of $R(S)$ is exactly $2|R(S)|$. Thus, $2|S| \leq 2|R(S)|$, and so $|S| \leq |R(S)|$. By Hall's marriage theorem, a complete match is possible.   ■

8.5 Exercises

1. For Example 5, find a different maximal matching. Is it possible to find a maximal matching where s_5 is matched with a book?

2. Verify that the labeling algorithm used on the network of Figure 8.94 gives the maximal matching $M(s_1) = b_2$, $M(s_2) = b_1$, $M(s_3) = s_3$, $M(s_4) = s_4$.

In Exercises 3 through 5 (Figures 8.96 through 8.98), find a maximum flow through the network.

3.

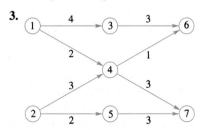

Figure 8.96

4.

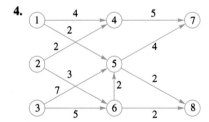

Figure 8.97

5.

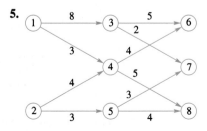

Figure 8.98

In Exercises 6 through 10, the matrix $\mathbf{M}_R$ for a relation from A to B is given. Find a maximal matching for A, B, and R.

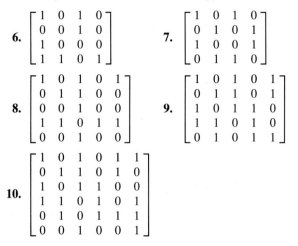

6. $\begin{bmatrix} 1 & 0 & 1 & 0 \\ 0 & 0 & 1 & 0 \\ 1 & 0 & 0 & 0 \\ 1 & 1 & 0 & 1 \end{bmatrix}$

7. $\begin{bmatrix} 1 & 0 & 1 & 0 \\ 0 & 1 & 0 & 1 \\ 1 & 0 & 0 & 1 \\ 0 & 1 & 1 & 0 \end{bmatrix}$

8. $\begin{bmatrix} 1 & 0 & 1 & 0 & 1 \\ 0 & 1 & 1 & 0 & 0 \\ 0 & 0 & 1 & 0 & 0 \\ 1 & 1 & 0 & 1 & 1 \\ 0 & 0 & 1 & 0 & 0 \end{bmatrix}$

9. $\begin{bmatrix} 1 & 0 & 1 & 0 & 1 \\ 0 & 1 & 1 & 0 & 1 \\ 1 & 0 & 1 & 1 & 0 \\ 1 & 1 & 0 & 1 & 0 \\ 0 & 1 & 0 & 1 & 1 \end{bmatrix}$

10. $\begin{bmatrix} 1 & 0 & 1 & 0 & 1 & 1 \\ 0 & 1 & 1 & 0 & 1 & 0 \\ 1 & 0 & 1 & 1 & 0 & 0 \\ 1 & 1 & 0 & 1 & 0 & 1 \\ 0 & 1 & 0 & 1 & 1 & 1 \\ 0 & 0 & 1 & 0 & 0 & 1 \end{bmatrix}$

11. Which of the matchings found in Exercises 6 through 10 are complete matchings?

12. Let R be a relation from A to B with $|A| = |B| = n$. Prove that if the number of ones in each column of $\mathbf{M}_R$ is k and the number of ones in each row of $\mathbf{M}_R$ is k, then there is a complete matching for A, B, and R.

13. Let R be a relation from A to B with $|A| = |B| = n$. Let j be the maximum number of ones in any column of $\mathbf{M}_R$ and k be the minimum number of ones in any row of $\mathbf{M}_R$. Prove that if $j \leq k$, then there is a complete matching for A, B, and R.

14. Let R be a relation from A to B with $|A| = |B| = n$. Prove that a complete matching exists for A, B, and R if each node in the network corresponding to A, B, and R has degree at least $\frac{n}{2}$. (*Hint*: Use a result from Section 8.3.)

8.6 COLORING GRAPHS

Suppose that $G = (V, E, \gamma)$ is a graph with no multiple edges, and $C = \{c_1, c_2, \ldots, c_n\}$ is any set of n "colors." Any function $f: V \to C$ is called a **coloring of the graph G using n colors** (or using the colors of C). For each vertex v, $f(v)$ is the color of v. As we usually present a graph pictorially, so we also think of a coloring in the intuitive way of simply painting each vertex with a color from C. However, graph coloring problems have a wide variety of practical applications in which "color" may have almost any meaning. For example, if the graph represents a connected grid of cities, each city can be marked with the name of the airline having the most flights to and from that city. In this case, the vertices are cities and the colors are airline names. Other examples are given later.

A coloring is **proper** if any two adjacent vertices v and w have different colors.

EXAMPLE 1

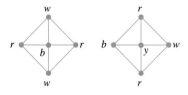

Let $C = \{r, w, b, y\}$ so that $n = 4$. Figure 8.99 shows a graph G properly colored with the colors from C in two different ways, one using three colors from C and one using all four. We show the colors as labels, which helps to explain why we avoid giving names to vertices. There are many ways to color this graph properly with three or four colors, but it is not hard to see that this cannot be done with two or fewer colors. (Experiment to convince yourself that this is true.) ∎

Figure 8.99

The smallest number of colors needed to produce a proper coloring of a graph G is called the **chromatic number of G**, denoted by $\chi(G)$. For the graph G of Figure 8.99, our discussion leads us to believe that $\chi(G) = 3$.

Of the many problems that can be viewed as graph-coloring problems, one of the oldest is the map-coloring problem. Consider the map shown in Figure 8.100.

A coloring of a map is a way to color each region (country, state, county, province, etc.) so that no two distinct regions sharing a common border have the same color. The map-coloring problem is to find the smallest number of colors that can be used. We can view this problem as a proper graph-coloring problem as

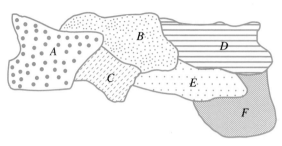

Figure 8.100

follows. Given a map M, construct a graph G_M with one vertex for each region and an edge connecting any two vertices whose corresponding regions share a common boundary. Then the proper colorings of G_M correspond exactly to the colorings of M.

EXAMPLE 2   Consider the map M shown in Figure 8.100. Then G_M is represented by Figure 8.101.   ■

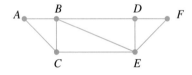

Figure 8.101

The map-coloring problem dates back to the mid-nineteenth century and has been an active object of research at various times since then. A conjecture was that four colors are always enough to color any map drawn on a plane. This conjecture was proved to be true in 1976 with the aid of computer computations performed on almost 2,000 configurations of graphs. There is still no proof known that does not depend on computer checking.

The graph corresponding to a map is an example of a **planar graph**, meaning that it can be drawn in a plane so that no edges cross except at vertices. Figure 8.101 illustrates the planarity of the graph corresponding to the map of Figure 8.100. The complete graph K_5 is not planar, so graph-coloring problems are more general than map-coloring problems. Later, we will see that five colors are required to color K_5.

Graph-coloring problems also arise from counting problems.

EXAMPLE 3   Fifteen different foods are to be held in refrigerated compartments within the same refrigerator. Some of them can be kept together, but other foods must be kept apart. For example, spicy meats and cheeses should be kept separate from bland meats and vegetables. Apples, eggs, and onions should be isolated or they will contaminate many other foods. Butter, margarine, and cream cheese can be kept together, but must be separated from foods with strong odors. We can construct a graph G as follows. Construct one vertex for each food and connect two with an edge if they must be kept in separate compartments in the refrigerator. Then $\chi(G)$ is the smallest number of separate containers needed to store the 15 foods properly.   ■

A similar method could be used to calculate the minimum number of laboratory drawers needed to store chemicals if we need to separate chemicals that will react with one another if stored close to each other.

Chromatic Polynomials

Closely related to the problem of computing $\chi(G)$ is the problem of computing the total number of different proper colorings of a graph G using a set $C = \{c_1, c_2, \ldots, c_n\}$ of colors.

If G is a graph and $n \geq 0$ is an integer, let $P_G(n)$ be the number of ways to color G properly using n or fewer colors. Since $P_G(n)$ is a definite number for each n, we see that P_G is a function. What may not be obvious is that P_G is a polynomial in n. This can be shown in general and is clearly seen in the examples of this section. We call P_G the **chromatic polynomial** of G.

EXAMPLE 4

Consider the linear graph L_4 defined in Section 8.1 and shown in Figure 8.9. Suppose that we have x colors. The first vertex can be colored with any color. No matter how this is done, the second can be colored with any color that was not chosen for vertex 1. Thus there are $x - 1$ choices for vertex 2. Vertex 3 can then be colored with any of the $x - 1$ colors not used for vertex 2. A similar result holds for vertex 4. By the multiplication principle of counting (Section 3.1), the total number of proper colorings is $x(x - 1)^3$. Thus $P_{L_4} = x(x - 1)^3$. ∎

We can see from Example 4 that $P_{L_4}(0) = 0$, $P_{L_4}(1) = 0$, and $P_{L_4}(2) = 2$. Thus there are no proper colorings of L_4 using zero colors (obviously) or one color, and there are two using two colors. From this we see that $\chi(L_4) = 2$. This connection holds in general, and we have the following principle: If G is a graph with no multiple edges, and P_G is the chromatic polynomial of G, then $\chi(G)$ is the smallest positive integer x for which $P_G(x) \neq 0$.

An argument similar to the one given in Example 4 shows that for L_n, $n \geq 1$, $P_{L_n}(x) = x(x - 1)^{n-1}$. Thus, by the above principle, $\chi(L_n) = 2$ for every n.

EXAMPLE 5

For any $n \geq 1$, consider the complete graph K_n defined in Section 8.1. Suppose that we again have x colors to use in coloring K_n. If $x < n$, no proper coloring is possible. So let $x \geq n$. Vertex v_1 can be colored with any of the x colors. For vertex v_2, only $x - 1$ remain since v_2 is connected to v_1. We can only color v_3 with $x - 2$ colors, since v_3 is connected to v_1 and v_2 and so the colors of v_1 and v_2 cannot be used again. Similarly, only $x - 3$ colors remain for v_4 and so on. Again using the multiplication principle of counting, we find that $P_{K_n}(x) = x(x - 1)(x - 2) \cdots (x - n + 1)$. This shows that $\chi(K_n) = n$. Note that if there are at least n colors, then $P_{K_n}(x)$ is the number of permutations of x objects taken n at a time (see Section 3.1). ∎

Suppose that a graph G is not connected and that G_1 and G_2 are two components of G. This means that no vertex in G_1 is connected to any vertex in G_2. Thus any coloring of G_1 can be paired with any coloring of G_2. This can be extended to any number of components, so the multiplication principle of counting gives the following result.

Theorem 1 If G is a disconnected graph with components $G_1, G_2, \ldots, G_m$, then $P_G(x) = P_{G_1}(x)P_{G_2}(x) \cdots P_{G_m}(x)$, the product of the chromatic polynomials for each component. ●

EXAMPLE 6   Let G be the graph shown in Figure 8.6. Then G has two components, each of which is K_3. The chromatic polynomial of K_3 is $x(x-1)(x-2)$, $x \geq 3$. Thus, by Theorem 1, $P_G(x) = x^2(x-1)^2(x-2)^2$. We see that $\chi(G) = 3$ and that the number of distinct ways to color G using three colors is $P_G(3) = 36$. If x is 4, then the total number of proper colorings of G is $4^2 \cdot 3^2 \cdot 2^2$ or 576.   ∎

EXAMPLE 7   Consider the discrete graph U_n of Section 8.1, having n vertices and no edges. All n components are single points. The chromatic polynomial of a single point is x, so, by Theorem 1, $P_{U_n}(x) = x^n$. Thus $\chi(D_n) = 1$ as can also be seen directly.   ∎

There is a useful theorem for computing chromatic polynomials using the subgraph and quotient graph constructions of Section 8.1. Let $G = \{V, E, \gamma\}$ be a graph with no multiple edges, and let $e \in E$, say $e = \{a, b\}$. As in Section 8.1, let G_e be the subgraph of G obtained by deleting e, and let G^e be the quotient graph of G obtained by merging the end points of e. Then we have the following result.

Theorem 2   With the preceding notation and using x colors,

$$P_G(x) = P_{G_e}(x) - P_{G^e}(x).$$ ●

Proof   Consider all the proper colorings of G_e. They are of two types, those for which a and b have different colors and those for which a and b have the same color. Now a coloring of the first type is also a proper coloring for G, since a and b are connected in G, and this coloring gives them different colors. On the other hand, a coloring of G_e of the second type corresponds to a proper coloring of G^e. In fact, since a and b are combined in G^e, they must have the same color there. All other vertices of G_e have the same connections as in G. Thus we have proved that $P_{G_e}(x) = P_G(x) + P_{G^e}(x)$ or $P_G(x) = P_{G_e}(x) - P_{G^e}(x)$.   ▼

EXAMPLE 8   Let us compute $P_G(x)$ for the graph G shown in Figure 8.102, using the edge e. Then G^e is K_3 and G_e has two components, one being a single point and the other being K_3. By Theorem 1,

$$P_{G_e}(x) = x(x(x-1)(x-2)) = x^2(x-1)(x-2),$$

if $x \geq 2$. Also,

$$P_{G^e}(x) = x(x-1)(x-2).$$

Figure 8.102

Thus, by Theorem 2, we see that

$$P_G(x) = x^2(x-1)(x-2) - x(x-1)(x-2)$$

or

$$x(x-1)^2(x-2).$$

Clearly, $P_G(1) = P_G(2) = 0$, and $P_G(3) = 12$. This shows that $\chi(G) = 3$.   ∎

8.6 Exercises

In Exercises 1 through 4 (Figures 8.103 through 8.106), construct a graph for the map given as done in Example 2.

1.

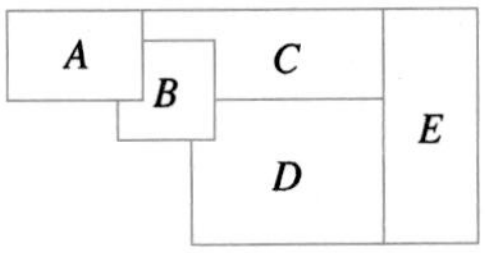

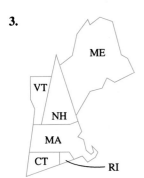

Figure 8.103

2.

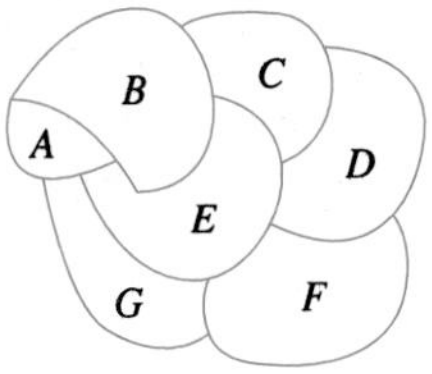

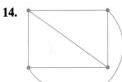

Figure 8.104

3.

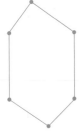

Figure 8.105

4.

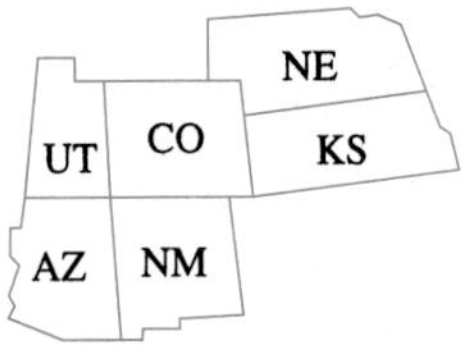

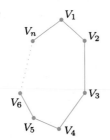

Figure 8.106

In Exercises 5 through 8 (Figures 8.107 through 8.110), determine the chromatic number of the graph by inspection.

5.

Figure 8.107

6.

Figure 8.108

7.

Figure 8.109

8.

Figure 8.110

In Exercises 9 through 12, find the chromatic polynomial for the graph represented by the given figure. Confirm each of these by using the results of Exercises 5 through 8.

9. Figure 8.107

10. Figure 8.108

11. Figure 8.109

12. Figure 8.110

In Exercises 13 through 16 (Figures 8.111 through 8.114), find the chromatic polynomial P_G for the given graph and use P_G to find $\chi(G)$.

13.

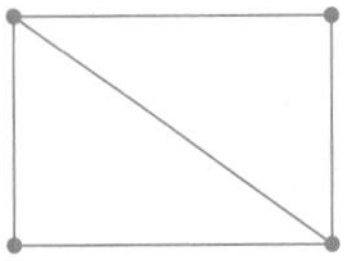

Figure 8.111

14.

Figure 8.112

15.

Figure 8.113

16.

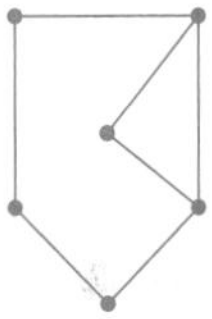

Figure 8.114

17. Find P_G and $\chi(G)$ for the graph G of the map in Exercise 1.

18. Find P_G and $\chi(G)$ for the graph G of the map in Exercise 3.

19. Find P_G and $\chi(G)$ for the graph G of the map in Exercise 4. Consider using Theorem 2 to do this.

20. Let G be the graph represented by Figure 8.63, in Section 8.3 Exercises. Find P_G and compute $\chi(G)$.

21. Prove by mathematical induction that $P_{L_n}(x) = x(x - 1)^{n-1}$, $n \geq 1$.

22. Prove that if G is a graph where the longest cycle has odd length n and n is greater than 2, then $\chi(G) \leq n + 1$.

TIPS FOR PROOFS

No new proof techniques are introduced in this chapter, but most of those used previously make an appearance—direct, indirect by contradiction, mathematical induction. For statements about graphs, a typical structure for a contradiction proof is to examine the largest (or smallest) object with certain properties and show that if the statement were not true, then a larger (or smaller) object with the same properties could be constructed. This is the structure of the proof of Theorem 1(b), Section 8.2.

Often indirect proofs are also existence proofs since they tell us only that an object with certain properties must exist, but not how to construct one. The proofs of Theorems 1 and 2 in Section 8.2 are of this type. Induction proofs are common in graph theory, because many theorems are stated in terms of the number of vertices or the number of edges. Analysis of all possible cases is also common; the proof of the four-color theorem that was mentioned earlier is an extreme case of this technique.

KEY IDEAS FOR REVIEW

- Graph: $G = (V, E, \gamma)$, where V is a finite set of objects, called vertices, E is a set of objects, called edges, and γ is a function that assigns to each edge a two-element subset of V.

- Degree of a vertex: number of edges at the vertex

- Path: see page 281

- Circuit: path that begins and ends at the same vertex

- Simple path or circuit: see page 281

- Connected graph: There is a path from any vertex to any other vertex.

- Subgraph: see page 283

- Euler path (circuit): path (circuit) that contains every edge of the graph exactly once

- Theorem: (a) If a graph G has a vertex of odd degree, there can be no Euler circuit in G. (b) If G is a connected graph and every vertex has even degree, then there is an Euler circuit in G.

- Theorem: (a) If a graph G has more than two vertices of odd degree, then there can be no Euler path in G. (b) If G is connected and has exactly two vertices of odd degree, there is an Euler path in G.

- Bridge: edge whose deletion would cause the graph to become disconnected

- Fleury's algorithm: see pages 290–291

- Hamiltonian path: path that includes each vertex of the graph exactly once

- Hamiltonian circuit: circuit that includes each vertex exactly once except for the first vertex, which is also the last

- Theorem: Let G be a graph on n vertices with no loops or multiple edges, $n > 2$. If for any two vertices u and v of G, the degree of u plus the degree of v is at least n, then G has a Hamiltonian circuit.

- Theorem: Let G be a graph on n vertices that has no loops or multiple edges, $n > 2$. If the number of edges in G is at least $\frac{1}{2}(n^2 - 3n + 6)$, then G has a Hamiltonian circuit.

- Transport network: see page 297

- Capacity: maximum flow that can be passed through an edge

- Flow in a network: a function that assigns a flow to each edge of a network

- Value of a flow: the sum of flows entering the sink

- Labeling algorithm: see page 299

- Cut in a network: set of edges in a network such that every path from the source to the sink contains at least one edge in the set.

- Capacity of a cut: see page 304

- Max Flow Min Cut Theorem: A maximum flow in a network has value equal to the capacity of a minimum cut of the network

- Matching (function) M compatible with a relation R: a one-to-one function such that $M \subseteq R$.

- Hall's marriage theorem: Let R be a relation from A to B. There exists a complete matching M if and only if for each $X \subseteq A$, $|X| \le |R(X)|$.

- Coloring of a graph using n colors: see page 311

- Proper coloring of a graph: Adjacent edges have different colors.

- Chromatic number of a graph G, $\chi(G)$: smallest number of colors needed for a proper coloring of G

- Planar graph: graph that can be drawn in a plane with no crossing edges

- Chromatic polynomial of a graph G, P_G: number of proper colorings of G in terms of the number of colors available

CODING EXERCISES

For each of the following, write the requested program or subroutine in pseudocode (as described in Appendix A) or in a programming language that you know. Test your code either with a paper-and-pencil trace or with a computer run. In each of these exercises assume that a graph is defined by $G = (V, E, \gamma)$.

1. Write a function that given G and an element v of V will return the degree of v.

2. Write a subroutine that will determine if two vertices of G are adjacent.

3. Write code for Fleury's algorithm.

4. Write a subroutine that, with input a list of vertices of G, reports whether or not that list defines a valid path that is a Hamiltonian path.

5. Modify your code for Exercise 4 so that the subroutine checks for Hamiltonian circuits.

CHAPTER 8 SELF-TEST

1. Draw a picture of the graph $G = (V, E, \gamma)$, where $V = \{j, k, l, m, n\}$, $E = \{e_1, e_2, e_3, e_4, e_5\}$, and

$$\gamma(e_1) = \{m, k\}, \quad \gamma(e_2) = \{j, l\}, \quad \gamma(e_3) = \{n, k\},$$
$$\gamma(e_4) = \{m, j\}, \quad \text{and} \quad \gamma(e_5) = \{n, l\}.$$

2. Give an example of a graph with 7 vertices and exactly 3 components.

3. If R is the equivalence relation defined by the partition $\{\{v_2, v_4, v_5\}, \{v_1, v_3\}, \{v_7, v_8\}, \{v_6, v_9\}\}$, find the quotient graph G^R of the graph G represented by Figure 8.115.

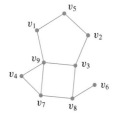

Figure 8.115

Use Figures 8.116 and 8.117 for problems 4 and 5.

4. Tell whether the graph in the specified figure has an Euler circuit, an Euler path but no Euler circuit, or neither. Give reasons for your decision.

(a) Figure 8.116   (b) Figure 8.117

5. Tell whether the graph in the specified figure has a Hamiltonian circuit, a Hamiltonian path but no Hamiltonian circuit, or neither. Give reasons for your decision.

(a) Figure 8.116   (b) Figure 8.117

6. (a) The graph represented by Figure 8.118 does not have an Euler circuit. Mark the minimal number of edges that would have to be traveled twice in order to travel every edge and return to the starting vertex.

(b) For each edge you marked in part (a), add an edge between the same vertices. Use Fleury's algorithm to find an Euler circuit for the modified version of Figure 8.118.

Figure 8.118

7. Give an Euler circuit, by numbering the edges, for the graph represented by Figure 8.119. Begin at A.

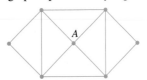

Figure 8.119

Figure 8.116

Figure 8.117

8. Give a Hamiltonian circuit, by numbering the edges, for the graph represented by Figure 8.119.

9. Find a Hamiltonian circuit of minimal weight for the graph represented by Figure 8.120.

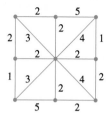

Figure 8.120

10. Find a maximal flow in the network shown in Figure 8.121.

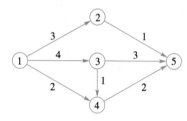

Figure 8.121

11. (a) Construct a graph for the map given in Figure 8.122.

(b) Determine the number of colors required for a proper coloring of the map.

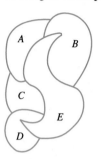

Figure 8.122

12. Compute the chromatic polynomial for the graph constructed in Problem 11(a) and use it to prove the result of Problem 11(b).

9

SEMIGROUPS AND GROUPS

Prerequisites: Chapters 4, 5, and 6

The notion of a mathematical structure was introduced in Section 1.6. In the following chapters, other types of mathematical systems were developed; some, such as [propositions, $\wedge$, $\vee$, $\sim$], were not given specific names; but others, such as B_n, the Boolean algebra on n elements, were named. In this chapter, we identify two more types of mathematical structures, semigroups and groups. Semigroups will be used in our study of finite state machines in Chapter 10. We also develop the basic ideas of group theory, which we will apply to coding theory in Chapter 11.

9.1 BINARY OPERATIONS REVISITED

We defined binary operations earlier (see Section 1.6) and noted in Section 5.2 that a binary operation may be used to define a function. Here we turn the process around and define a binary operation as a function with certain properties.

A **binary operation on a set** A is an everywhere defined function $f : A \times A \rightarrow A$. Observe the following properties that a binary operation must satisfy:

1. Since $\text{Dom}(f) = A \times A$, f assigns an element $f(a, b)$ of A to each ordered pair (a, b) in $A \times A$. That is, the binary operation must be defined for each ordered pair of elements of A.

2. Since a binary operation is a function, only one element of A is assigned to each ordered pair.

Thus we can say that a binary operation is a rule that assigns to each ordered pair of elements of A a unique element of A. The reader should note that this definition is more restrictive than that given in Chapter 1, but we have made the change to simplify the discussion in this chapter. We shall now turn to a number of examples.

It is customary to denote binary operations by a symbol such as $*$, instead of f, and to denote the element assigned to (a, b) by $a * b$ [instead of $*(a, b)$]. It should be emphasized that if a and b are elements in A, then $a * b \in A$, and this property is often described by saying that A is **closed** under the operation $*$.

EXAMPLE 1   Let $A = Z$. Define $a * b$ as $a + b$. Then $*$ is a binary operation on Z. ∎

EXAMPLE 2   Let $A = \mathbb{R}$. Define $a * b$ as a/b. Then $*$ is not a binary operation, since it is not defined for every ordered pair of elements of A. For example, $3 * 0$ is not defined, since we cannot divide by zero. ∎

EXAMPLE 3   Let $A = Z^{+}$. Define $a * b$ as $a - b$. Then $*$ is not a binary operation since it does not assign an element of A to every ordered pair of elements of A; for example, $2 * 5 \notin A$. ∎

EXAMPLE 4   Let $A = Z$. Define $a * b$ as a number less than both a and b. Then $*$ is not a binary operation, since it does not assign a *unique* element of A to each ordered pair of elements of A; for example, $8 * 6$ could be 5, 4, 3, 1, and so on. Thus, in this case, $*$ would be a relation from $A \times A$ to A, but not a function. ∎

EXAMPLE 5   Let $A = Z$. Define $a * b$ as $\max\{a, b\}$. Then $*$ is a binary operation; for example, $2 * 4 = 4$, $-3 * (-5) = -3$. ∎

EXAMPLE 6   Let $A = P(S)$, for some set S. If V and W are subsets of S, define $V * W$ as $V \cup W$. Then $*$ is a binary operation on A. Moreover, if we define $V *' W$ as $V \cap W$, then $*'$ is another binary operation on A. ∎

As Example 6 shows, it is possible to define many binary operations on the same set.

EXAMPLE 7   Let M be the set of all $n \times n$ Boolean matrices for a fixed n. Define $\mathbf{A} * \mathbf{B}$ as $\mathbf{A} \vee \mathbf{B}$ (see Section 1.5). Then $*$ is a binary operation. This is also true of $\mathbf{A} \wedge \mathbf{B}$. ∎

EXAMPLE 8   Let L be a lattice. Define $a * b$ as $a \wedge b$ (the greatest lower bound of a and b). Then $*$ is a binary operation on L. This is also true of $a \vee b$ (the least upper bound of a and b). ∎

Tables

If $A = \{a_1, a_2, \ldots, a_n\}$ is a finite set, we can define a binary operation on A by means of a table as shown in Figure 9.1. The entry in position i, j denotes the element $a_i * a_j$.

EXAMPLE 9   Let $A = \{0, 1\}$. We define binary operations $\vee$ and $\wedge$ by the following tables:

$\vee$	0	1
0	0	1
1	1	1

$\wedge$	0	1
0	0	0
1	0	1

∎

$$
\begin{array}{c|ccccccc}
* & a_1 & a_2 & \cdots & a_j & \cdots & a_n \\
\hline
a_1 & & & & & & \\
a_2 & & & & & & \\
\vdots & & & & & & \\
a_i & & & & a_i * a_j & & \\
\vdots & & & & & & \\
a_n & & & & & &
\end{array}
$$

Figure 9.1

If $A = \{a, b\}$, we shall now determine the number of binary operations that can be defined on A. Every binary operation $*$ on A can be described by a table

$$
\begin{array}{c|cc}
* & a & b \\
\hline
a & & \\
b & &
\end{array}
$$

Since every blank can be filled in with the element a or b, we conclude that there are $2 \cdot 2 \cdot 2 \cdot 2 = 2^4$ or 16 ways to complete the table. Thus, there are 16 binary operations on A.

Properties of Binary Operations

Several of the properties defined for binary operations in Section 1.6 are of particular importance in this chapter. We repeat them here.

A binary operation on a set A is said to be **commutative** if

$$
a * b = b * a
$$

for all elements a and b in A.

EXAMPLE 10 The binary operation of addition on Z (as discussed in Example 1) is commutative.

EXAMPLE 11 The binary operation of subtraction on Z is not commutative, since

$$
2 - 3 \neq 3 - 2.
$$

A binary operation that is described by a table is commutative if and only if the entries in the table are symmetric with respect to the main diagonal.

EXAMPLE 12 Which of the following binary operations on $A = \{a, b, c, d\}$ are commutative?

*	a	b	c	d
a	a	c	b	d
b	b	c	b	a
c	c	d	b	c
d	a	a	b	b

*	a	b	c	d
a	a	c	b	d
b	c	d	b	a
c	b	b	a	c
d	d	a	c	d

(a)   (b)

Solution   The operation in (a) is not commutative, since $a * b$ is c while $b * a$ is b. The operation in (b) is commutative, since the entries in the table are symmetric with respect to the main diagonal.   ■

A binary operation $*$ on a set A is said to be **associative** if

$$a * (b * c) = (a * b) * c$$

for all elements a, b, and c in A.

EXAMPLE 13   The binary operation of addition on Z is associative.   ■

EXAMPLE 14   The binary operation of subtraction on Z is not associative, since

$$2 - (3 - 5) \neq (2 - 3) - 5.$$ ■

EXAMPLE 15   Let L be a lattice. The binary operation defined by $a * b = a \wedge b$ (see Example 8) is commutative and associative. It also satisfies the **idempotent** property $a \wedge a = a$. A partial converse of this example is also true, as shown in Example 16.   ■

EXAMPLE 16   Let $*$ be a binary operation on a set A, and suppose that $*$ satisfies the following properties for any a, b, and c in A.

1. $a = a * a$ Idempotent property
2. $a * b = b * a$ Commutative property
3. $a * (b * c) = (a * b) * c$ Associative property

Define a relation $\leq$ on A by

$$a \leq b \quad \text{if and only if} \quad a = a * b.$$

Show that $(A, \leq)$ is a poset, and for all a, b in A, GLB$(a, b) = a * b$.

Solution   We must show that $\leq$ is reflexive, antisymmetric, and transitive. Since $a = a * a$, $a \leq a$ for all a in A, and $\leq$ is reflexive.

Now suppose that $a \leq b$ and $b \leq a$. Then, by definition and property 2, $a = a * b = b * a = b$, so $a = b$. Thus $\leq$ is antisymmetric.

If $a \leq b$ and $b \leq c$, then $a = a * b = a * (b * c) = (a * b) * c = a * c$, so $a \leq c$ and $\leq$ is transitive.

Finally, we must show that, for all a and b in A, $a*b = a \wedge b$ (the greatest lower bound of a and b with respect to $\leq$). We have $a * b = a * (b * b) = (a * b) * b$, so $a * b \leq b$. In a similar way, we can show that $a * b \leq a$, so $a * b$ is a lower bound for a and b. Now, if $c \leq a$ and $c \leq b$, then $c = c * a$ and $c = c * b$ by definition. Thus $c = (c * a) * b = c * (a * b)$, so $c \leq a * b$. This shows that $a * b$ is the greatest lower bound of a and b. ∎

9.1 Exercises

In Exercises 1 through 8, determine whether the description of $*$ is a valid definition of a binary operation on the set.

1. On $\mathbb{R}$, where $a * b$ is ab (ordinary multiplication).

2. On $\mathbb{Z}^+$, where $a * b$ is a/b.

3. On $\mathbb{Z}$, where $a * b$ is a^b.

4. On $\mathbb{Z}^+$, where $a * b$ is a^b.

5. On $\mathbb{Z}^+$, where $a * b$ is $a - b$.

6. On $\mathbb{R}$, where $a * b$ is $a\sqrt{b}$.

7. On $\mathbb{R}$, where $a * b$ is the largest rational number that is less than ab.

8. On $\mathbb{Z}$, where $a * b$ is $2a + b$.

In Exercises 9 through 17, determine whether the binary operation $*$ is commutative and whether it is associative on the set.

9. On $\mathbb{Z}^+$, where $a * b$ is $a + b + 2$.

10. On $\mathbb{Z}$, where $a * b$ is ab.

11. On $\mathbb{R}$, where $a * b$ is $a \times |b|$.

12. On the set of nonzero real numbers, where $a * b$ is a/b.

13. On $\mathbb{R}$, where $a * b$ is the minimum of a and b.

14. On the set of $n \times n$ Boolean matrices, where $\mathbf{A} * \mathbf{B}$ is $\mathbf{A} \odot \mathbf{B}$ (see Section 1.5).

15. On $\mathbb{R}$, where $a * b$ is $ab/3$.

16. On $\mathbb{R}$, where $a * b$ is $ab + 2b$.

17. On a lattice A, where $a * b$ is $a \vee b$.

18. Fill in the following table so that the binary operation $*$ is commutative.

$*$	a	b	c
a	b		
b	c	b	a
c	a		c

19. Fill in the following table so that the binary operation $*$ is commutative and has the idempotent property.

$*$	a	b	c
a			c
b			
c	c		a

20. Consider the binary operation $*$ defined on the set $A = \{a, b, c\}$ by the following table.

$*$	a	b	c
a	b	c	b
b	a	b	c
c	c	a	b

(a) Is $*$ a commutative operation?

(b) Compute $a * (b * c)$ and $(a * b) * c$.

(c) Is $*$ an associative operation?

21. Consider the binary operation $*$ defined on the set $A = \{a, b, c, d\}$ by the following table.

$*$	a	b	c	d
a	a	c	b	d
b	d	a	b	c
c	c	d	a	a
d	d	b	a	c

Compute

(a) $c * d$ and $d * c$.

(b) $b * d$ and $d * b$.

(c) $a * (b * c)$ and $(a * b) * c$.

(d) Is $*$ commutative? associative?

In Exercises 22 and 23, complete the given table so that the binary operation $*$ is associative.

22.

$*$	a	b	c	d
a	a	b	c	d
b	b	a	d	c
c	c	d	a	b
d				

23.

*	a	b	c	d
a	b	a	c	d
b	b	a	c	d
c				
d	d	c	c	d

24. Let A be a set with n elements. How many binary operations can be defined on A?

25. Let A be a set with n elements. How many commutative binary operations can be defined on A?

26. Let $A = \{a, b\}$.

    (a) Make a table for each of the 16 binary operations that can be defined on A.

    (b) Using part (a), identify the binary operations on A that are commutative.

27. Let $A = \{a, b\}$.

    (a) Using Exercise 26, identify the binary operations on A that are associative.

    (b) Using Exercise 26, identify the binary operations on A that satisfy the idempotent property.

28. Let $*$ be a binary operation on a set A, and suppose that $*$ satisfies the idempotent, commutative, and associative properties, as discussed in Example 16. Define a relation $\leq$ on A by $a \leq b$ if and only if $b = a * b$. Show that $(A, \leq)$ is a poset and, for all a and b, $\mathrm{LUB}(a, b) = a * b$.

29. Describe how the definition of a binary operation on a set A is different from the definition of a binary operation given in Section 1.6. Explain also whether a binary operation on a set is or is not a binary operation according to the earlier definition.

9.2 SEMIGROUPS

In this section we define a simple mathematical structure, consisting of a set together with a binary operation, that has many important applications.

    A **semigroup** is a nonempty set S together with an associative binary operation $*$ defined on S. We shall denote the semigroup by $(S, *)$ or, when it is clear what the operation $*$ is, simply by S. We also refer to $a * b$ as the **product** of a and b. The semigroup $(S, *)$ is said to be commutative if $*$ is a commutative operation.

EXAMPLE 1   It follows from Section 9.1 that $(Z, +)$ is a commutative semigroup.   ∎

EXAMPLE 2   The set $P(S)$, where S is a set, together with the operation of union is a commutative semigroup.   ∎

EXAMPLE 3   The set Z with the binary operation of subtraction is not a semigroup, since subtraction is not associative.   ∎

EXAMPLE 4   Let S be a fixed nonempty set, and let S^S be the set of all functions $f : S \to S$. If f and g are elements of S^S, we define $f * g$ as $f \circ g$, the composite function. Then $*$ is a binary operation on S^S, and it follows from Section 4.7 that $*$ is associative. Hence $(S^S, *)$ is a semigroup. The semigroup S^S is not commutative.   ∎

EXAMPLE 5   Let $(L, \leq)$ be a lattice. Define a binary operation on L by $a * b = a \vee b$. Then L is a semigroup.   ∎

EXAMPLE 6   Let $A = \{a_1, a_2, \ldots, a_n\}$ be a nonempty set. Recall from Section 1.3 that A^* is the set of all finite sequences of elements of A. That is, A^* consists of all words that can be formed from the alphabet A. Let α and β be elements of A^*. Observe that catenation is a binary operation $\cdot$ on A^*. Recall that if $\alpha = a_1 a_2 \cdots a_n$ and

$\beta = b_1b_2\cdots b_k$, then $\alpha\cdot\beta = a_1a_2\cdots a_nb_1b_2\cdots b_k$. It is easy to see that if α, β, and γ are any elements of A^*, then

$$\alpha\cdot(\beta\cdot\gamma) = (\alpha\cdot\beta)\cdot\gamma$$

so that $\cdot$ is an associative binary operation, and $(A^*, \cdot)$ is a semigroup. The semigroup $(A^*, \cdot)$ is called the **free semigroup generated by** A. ∎

In a semigroup $(S, *)$ we can establish the following generalization of the associative property; we omit the proof.

Theorem 1 If $a_1, a_2, \ldots, a_n, n \geq 3$, are arbitrary elements of a semigroup, then all products of the elements $a_1, a_2, \ldots, a_n$ that can be formed by inserting meaningful parentheses arbitrarily are equal. ●

EXAMPLE 7 Theorem 1 shows that the products

$$((a_1 * a_2) * a_3) * a_4, \quad a_1 * (a_2 * (a_3 * a_4)), \quad (a_1 * (a_2 * a_3)) * a_4$$

are all equal. ∎

If $a_1, a_2, \ldots, a_n$ are elements in a semigroup $(S, *)$, we shall write their product as

$$a_1 * a_2 * \cdots * a_n,$$

omitting the parentheses.

An element e in a semigroup $(S, *)$ is called an **identity** element if

$$e * a = a * e = a$$

for all $a \in S$. As shown by Theorem 1, Section 1.6, an identity element must be unique.

EXAMPLE 8 The number 0 is an identity in the semigroup $(Z, +)$. ∎

EXAMPLE 9 The semigroup $(Z^+, +)$ has no identity element. ∎

A **monoid** is a semigroup $(S, *)$ that has an identity.

EXAMPLE 10 The semigroup $P(S)$ defined in Example 2 has the identity $\varnothing$, since

$$\varnothing * A = \varnothing \cup A = A = A \cup \varnothing = A * \varnothing$$

for any element $A \in P(S)$. Hence $P(S)$ is a monoid. ∎

EXAMPLE 11 The semigroup S^S defined in Example 4 has the identity 1_S, since

$$1_S * f = 1_S \circ f = f \circ 1_S = f * 1_S$$

for any element $f \in S^S$, we see that S^S is a monoid. ∎

EXAMPLE 12   The semigroup A^* defined in Example 6 is actually a monoid with identity Λ, the empty sequence, since $\alpha \cdot \Lambda = \Lambda \cdot \alpha = \alpha$ for all $\alpha \in A^*$.   ∎

EXAMPLE 13   The set of all relations on a set A is a monoid under the operation of composition. The identity element is the equality relation Δ (see Section 4.7).   ∎

Let $(S, *)$ be a semigroup and let T be a subset of S. If T is closed under the operation $*$ (that is, $a * b \in T$ whenever a and b are elements of T), then $(T, *)$ is called a **subsemigroup** of $(S, *)$. Similarly, let $(S, *)$ be a monoid with identity e, and let T be a nonempty subset of S. If T is closed under the operation $*$ and $e \in T$, then $(T, *)$ is called a **submonoid** of $(S, *)$.

Observe that the associative property holds in any subset of a semigroup so that a subsemigroup $(T, *)$ of a semigroup $(S, *)$ is itself a semigroup. Similarly, a submonoid of a monoid is itself a monoid.

EXAMPLE 14   If T is the set of all even integers, then $(T, \times)$ is a subsemigroup of the monoid $(Z, \times)$, where $\times$ is ordinary multiplication, but it is not a submonoid since the identity of Z, the number 1, does not belong to T.   ∎

EXAMPLE 15   If $(S, *)$ is a semigroup, then $(S, *)$ is a subsemigroup of $(S, *)$. Similarly, let $(S, *)$ be a monoid. Then $(S, *)$ is a submonoid of $(S, *)$, and if $T = \{e\}$, then $(T, *)$ is also a submonoid of $(S, *)$.   ∎

Suppose that $(S, *)$ is a semigroup, and let $a \in S$. For $n \in Z^+$, we define the powers of a^n recursively as follows:

$$a^1 = a, \quad a^n = a^{n-1} * a, \quad n \geq 2.$$

Moreover, if $(S, *)$ is a monoid, we also define

$$a^0 = e.$$

It can be shown that if m and n are nonnegative integers, then

$$a^m * a^n = a^{m+n}.$$

EXAMPLE 16   (a) If $(S, *)$ is a semigroup, $a \in S$, and

$$T = \{a^i \mid i \in Z^+\},$$

then $(T, *)$ is a subsemigroup of $(S, *)$.

(b) If $(S, *)$ is a monoid, $a \in S$, and

$$T = \{a^i \mid i \in Z^+ \text{ or } i = 0\},$$

then $(T, *)$ is a submonoid of $(S, *)$.   ∎

Isomorphism and Homomorphism

An isomorphism between two posets was defined in Section 6.1 as a one-to-one correspondence that preserved order relations, the distinguishing feature of posets. We now define an isomorphism between two semigroups as a one-to-one correspondence that preserves the binary operations. In general, an isomorphism between two mathematical structures of the same type should preserve the distinguishing features of the structures.

Let $(S, *)$ and $(T, *')$ be two semigroups. A function $f: S \to T$ is called an **isomorphism** from $(S, *)$ to $(T, *')$ if it is a one-to-one correspondence from S to T, and if

$$f(a * b) = f(a) *' f(b)$$

for all a and b in S.

If f is an isomorphism from $(S, *)$ to $(T, *')$, then, since f is a one-to-one correspondence, it follows from Theorem 1 of Section 5.1 that f^{-1} exists and is a one-to-one correspondence from T to S. We now show that f^{-1} is an isomorphism from $(T, *')$ to $(S, *)$. Let a' and b' be any elements of T. Since f is onto, we can find elements a and b in S such that $f(a) = a'$ and $f(b) = b'$. Then $a = f^{-1}(a')$ and $b = f^{-1}(b')$. Now

$$
\begin{aligned}
f^{-1}(a' *' b') &= f^{-1}(f(a) *' f(b)) \\
&= f^{-1}(f(a * b)) \\
&= (f^{-1} \circ f)(a * b) \\
&= a * b = f^{-1}(a') * f^{-1}(b').
\end{aligned}
$$

Hence f^{-1} is an isomorphism.

We now say that the semigroups $(S, *)$ and $(T, *')$ are **isomorphic** and we write $S \simeq T$.

To show that two semigroups $(S, *)$ and $(T, *')$ are isomorphic, we use the following procedure:

STEP 1: Define a function $f: S \to T$ with $\text{Dom}(f) = S$.

STEP 2: Show that f is one-to-one.

STEP 3: Show that f is onto.

STEP 4: Show that $f(a * b) = f(a) *' f(b)$.

EXAMPLE 17

Let T be the set of all even integers. Show that the semigroups $(Z, +)$ and $(T, +)$ are isomorphic.

Solution

STEP 1: We define the function $f: Z \to T$ by $f(a) = 2a$.

STEP 2: We now show that f is one-to-one as follows. Suppose that $f(a_1) = f(a_2)$. Then $2a_1 = 2a_2$, so $a_1 = a_2$. Hence f is one-to-one.

STEP 3: We next show that f is onto. Suppose that b is any even integer. Then $a = b/2 \in Z$ and

$$f(a) = f(b/2) = 2(b/2) = b,$$

so f is onto.

STEP 4:   We have

$$f(a + b) = 2(a + b) = 2a + 2b = f(a) + f(b).$$

Hence $(Z, +)$ and $(T, +)$ are isomorphic semigroups.   ■

In general, it is rather straightforward to verify that a given function $f : S \to T$ is or is not an isomorphism. However, it is generally more difficult to show that two semigroups are isomorphic, because one has to create the isomorphism f.

As in the case of poset or lattice isomorphisms, when two semigroups $(S, *)$ and $(T, *')$ are isomorphic, they can differ only in the nature of their elements; their semigroup structures are identical. If S and T are finite semigroups, their respective binary operations can be given by tables. Then S and T are isomorphic if we can rearrange and relabel the elements of S so that its table is identical with that of T.

EXAMPLE 18   Let $S = \{a, b, c\}$ and $T = \{x, y, z\}$. It is easy to verify that the following operation tables give semigroup structures for S and T, respectively.

*	a	b	c
a	a	b	c
b	b	c	a
c	c	a	b

*	x	y	z
x	z	x	y
y	x	y	z
z	y	z	x

Let

$$f(a) = y$$
$$f(b) = x$$
$$f(c) = z.$$

Replacing the elements in S by their images and rearranging the table, we obtain exactly the table for T. Thus S and T are isomorphic.   ■

Theorem 2   Let $(S, *)$ and $(T, *')$ be monoids with identities e and e', respectively. Let $f : S \to T$ be an isomorphism. Then $f(e) = e'$.   ●

Proof   Let b be any element of T. Since f is onto, there is an element a in S such that $f(a) = b$. Then

$$a = a * e$$
$$b = f(a) = f(a * e) = f(a) *' f(e)$$
$$= b *' f(e).$$

Similarly, since $a = e * a$, $b = f(e) *' b$. Thus for any $b \in T$,

$$b = b *' f(e) = f(e) *' b,$$

which means that $f(e)$ is an identity for T. Thus since the identity is unique, it follows that $f(e) = e'$.   ▽

If $(S, *)$ and $(T, *')$ are semigroups such that S has an identity and T does not, it then follows from Theorem 2 that $(S, *)$ and $(T, *')$ cannot be isomorphic.

EXAMPLE 19 Let T be the set of all even integers and let $\times$ be ordinary multiplication. Then the semigroups $(Z, \times)$ and $(T, \times)$ are not isomorphic, since Z has an identity and T does not. ∎

By dropping the conditions of one to one and onto in the definition of an isomorphism of two semigroups, we get another important method for comparing the algebraic structures of the two semigroups.

Let $(S, *)$ and $(T, *')$ be two semigroups. An everywhere-defined function $f \colon S \to T$ is called a **homomorphism** from $(S, *)$ to $(T, *')$ if

$$f(a * b) = f(a) *' f(b)$$

for all a and b in S. If f is also onto, we say that T is a **homomorphic image** of S.

EXAMPLE 20 Let $A = \{0, 1\}$ and consider the semigroups $(A^*, \cdot)$ and $(A, +)$, where $\cdot$ is the catenation operation and $+$ is defined by the table

+	0	1
0	0	1
1	1	0

Define the function $f \colon A^* \to A$ by

$$f(\alpha) = \begin{cases} 1 & \text{if } \alpha \text{ has an odd number of 1's} \\ 0 & \text{if } \alpha \text{ has an even number of 1's.} \end{cases}$$

It is easy to verify that if α and β are any elements of A^*, then

$$f(\alpha \cdot \beta) = f(\alpha) + f(\beta).$$

Thus f is a homomorphism. The function f is onto since

$$f(0) = 0$$
$$f(1) = 1$$

but f is not an isomorphism, since it is not one to one. ∎

The difference between an isomorphism and a homomorphism is that an isomorphism must be one to one and onto. For both an isomorphism and a homomorphism, the image of a product is the product of the images.

The proof of the following theorem, which is left as an exercise for the reader, is completely analogous to the proof of Theorem 2.

Theorem 3 Let $(S, *)$ and $(T, *')$ be monoids with identities e and e', respectively. Let $f \colon S \to T$ be a homomorphism from $(S, *)$ onto $(T, *')$. Then $f(e) = e'$. ●

Theorem 3 is a stronger, or more general, statement than Theorem 2, because it requires fewer (weaker) conditions for the conclusion.

Theorem 3, together with the following two theorems, shows that, if a semigroup $(T, *')$ is a homomorphic image of the semigroup $(S, *)$, then $(T, *')$ has a strong algebraic resemblance to $(S, *)$.

Theorem 4   Let f be a homomorphism from a semigroup $(S, *)$ to a semigroup $(T, *')$. If S' is a subsemigroup of $(S, *)$, then

$$f(S') = \{t \in T \mid t = f(s) \text{ for some } s \in S'\},$$

the image of S' under f, is a subsemigroup of $(T, *')$.   ●

Proof   If t_1 and t_2 are any elements of $f(S')$, then there exist s_1 and s_2 in S' with

$$t_1 = f(s_1) \quad \text{and} \quad t_2 = f(s_2).$$

Then

$$t_1 *' t_2 = f(s_1) *' f(s_2) = f(s_1 * s_2) = f(s_3),$$

where $s_3 = s_1 * s_2 \in S'$. Hence $t_1 *' t_2 \in f(S')$.

Thus $f(S')$ is closed under the operation $*'$. Since the associative property holds in T, it holds in $f(S')$, so $f(S')$ is a subsemigroup of $(T, *')$.   ▼

Theorem 5   If f is a homomorphism from a commutative semigroup $(S, *)$ onto a semigroup $(T, *')$, then $(T, *')$ is also commutative.   ●

Proof   Let t_1 and t_2 be any elements of T. Then there exist s_1 and s_2 in S with

$$t_1 = f(s_1) \quad \text{and} \quad t_2 = f(s_2).$$

Therefore,

$$t_1 *' t_2 = f(s_1) *' f(s_2) = f(s_1 * s_2) = f(s_2 * s_1) = f(s_2) *' f(s_1) = t_2 *' t_1.$$

Hence $(T, *')$ is also commutative.   ▼

9.2 Exercises

1. Let $A = \{a, b\}$. Which of the following tables define a semigroup on A? Which define a monoid on A?

(a)

*	a	b
a	a	b
b	a	a

(b)

*	a	b
a	a	b
b	b	b

2. Let $A = \{a, b\}$. Which of the following tables define a semigroup on A? Which define a monoid on A?

(a)

*	a	b
a	b	a
b	a	b

(b)

*	a	b
a	a	b
b	b	a

3. Let $A = \{a, b\}$. Which of the following tables define a semigroup on A? Which define a monoid on A?

(a)

*	a	b
a	a	a
b	b	b

(b)

*	a	b
a	b	b
b	a	a

In Exercises 4 through 14, determine whether the set together with the binary operation is a semigroup, a monoid, or neither. If it is a monoid, specify the identity. If it is a semigroup or a monoid, determine if it is commutative.

4. Z^+, where $*$ is defined as ordinary multiplication.

5. Z^+, where $a * b$ is defined as $\max\{a, b\}$.

6. Z^+, where $a * b$ is defined as $\text{GCD}\{a, b\}$.

7. Z^+, where $a * b$ is defined as a.

8. The nonzero real numbers, where $*$ is ordinary multiplication.

9. $P(S)$, with S a set, where $*$ is defined as intersection.

10. A Boolean algebra B, where $a * b$ is defined as $a \wedge b$.

11. $S = \{1, 2, 3, 6, 12\}$, where $a * b$ is defined as GCD(a, b).

12. $S = \{1, 2, 3, 6, 9, 18\}$, where $a * b$ is defined as LCM(a, b).

13. Z, where $a * b = a + b - ab$.

14. The even integers, where $a * b$ is defined as $\dfrac{ab}{2}$.

15. Does the following table define a semigroup?

*	a	b	c
a	c	b	a
b	b	c	b
c	a	b	c

16. Does the following table define a semigroup?

*	a	b	c
a	a	c	b
b	c	b	a
c	b	a	c

17. Complete the following table to obtain a semigroup.

*	a	b	c
a	c	a	b
b	a	b	c
c			a

18. Complete the following table so that it defines a monoid.

*	a	b	c	d
a	c	d	a	b
b		a	b	
c			c	
d	b		d	a

19. Let $S = \{a, b\}$. Write the operation table for the semigroup S^S. Is the semigroup commutative?

20. Let $S = \{a, b\}$. Write the operation table for the semigroup $(P(S), \cup)$.

21. Let $A = \{a, b, c\}$ and consider the semigroup $(A^*, \cdot)$, where $\cdot$ is the operation of catenation. If $\alpha = abac$, $\beta = cba$, and $\gamma = babc$, compute

    (a) $(\alpha \cdot \beta) \cdot \gamma$ (b) $\gamma \cdot (\alpha \cdot \alpha)$ (c) $(\gamma \cdot \beta) \cdot \alpha$

22. Prove or disprove that the intersection of two subsemigroups of a semigroup $(S, *)$ is a subsemigroup of $(S, *)$.

23. Prove or disprove that the intersection of two submonoids of a monoid $(S, *)$ is a submonoid of $(S, *)$.

24. Let $A = \{0, 1\}$, and consider the semigroup $(A^*, \cdot)$, where $\cdot$ is the operation of catenation. Let T be the subset of A^* consisting of all sequences having an odd number of 1's. Is $(T, \cdot)$ a subsemigroup of $(A, \cdot)$?

25. Let $A = \{a, b\}$. Are there two semigroups $(A, *)$ and $(A, *')$ that are not isomorphic?

26. An element x in a monoid is called an **idempotent** if $x^2 = x * x = x$. Show that the set of all idempotents in a commutative monoid S is a submonoid of S.

27. Let $(S_1, *_1)$, $(S_2, *_2)$, and $(S_3, *_3)$ be semigroups and $f: S_1 \to S_2$ and $g: S_2 \to S_3$ be homomorphisms. Prove that $g \circ f$ is a homomorphism from S_1 to S_3.

28. Let $(S_1, *)$, $(S_2, *')$, and $(S_3, *'')$ be semigroups, and let $f: S_1 \to S_2$ and $g: S_2 \to S_3$ be isomorphisms. Show that $g \circ f: S_1 \to S_3$ is an isomorphism.

29. Which properties of f are used in the proof of Theorem 2?

30. Explain why the proof of Theorem 1 can be used as a proof of Theorem 3.

31. Let R^+ be the set of all positive real numbers. Show that the function $f: R^+ \to R$ defined by $f(x) = \ln x$ is an isomorphism of the semigroup $(R^+, \times)$ to the semigroup $(R, +)$, where $\times$ and $+$ are ordinary multiplication and addition, respectively.

9.3 PRODUCTS AND QUOTIENTS OF SEMIGROUPS

In this section we shall obtain new semigroups from existing semigroups.

Theorem 1   If $(S, *)$ and $(T, *')$ are semigroups, then $(S \times T, *'')$ is a semigroup, where $*''$ is defined by $(s_1, t_1) *'' (s_2, t_2) = (s_1 * s_2, t_1 *' t_2)$.

    Proof   The proof is left as an exercise.

It follows at once from Theorem 1 that if S and T are monoids with identities e_S and e_T, respectively, then $S \times T$ is a monoid with identity (e_S, e_T).

We now turn to a discussion of equivalence relations on a semigroup $(S, *)$. Since a semigroup is not merely a set, we shall find that certain equivalence relations on a semigroup give additional information about the structure of the semigroup.

An equivalence relation R on the semigroup $(S, *)$ is called a **congruence relation** if

$$a \, R \, a' \quad \text{and} \quad b \, R \, b' \quad \text{imply} \quad (a * b) \, R \, (a' * b').$$

EXAMPLE 1   Consider the semigroup $(Z, +)$ and the equivalence relation R on Z defined by

$$a \, R \, b \quad \text{if and only if} \quad a \equiv b \pmod 2.$$

Recall that we discussed this equivalence relation in Section 4.5. Remember that if $a \equiv b \pmod 2$, then $2 \mid (a - b)$. We now show that this relation is a congruence relation as follows.

If

$$a \equiv b \pmod 2 \quad \text{and} \quad c \equiv d \pmod 2,$$

then 2 divides $a - b$ and 2 divides $c - d$, so

$$a - b = 2m \quad \text{and} \quad c - d = 2n,$$

where m and n are in Z. Adding, we have

$$(a - b) + (c - d) = 2m + 2n$$

or

$$(a + c) - (b + d) = 2(m + n),$$

so

$$a + c \equiv b + d \pmod 2.$$

Hence the relation is a congruence relation.   ∎

EXAMPLE 2   Let $A = \{0, 1\}$ and consider the free semigroup $(A^*, \cdot)$ generated by A. Define the following relation on A:

$$\alpha \, R \, \beta \quad \text{if and only if} \quad \alpha \text{ and } \beta \text{ have the same number of 1's.}$$

Show that R is a congruence relation on $(A^*, \cdot)$.

Solution   We first show that R is an equivalence relation. We have

1. $\alpha \, R \, \alpha$ for any $\alpha \in A^*$.
2. If $\alpha \, R \, \beta$, then α and β have the same number of 1's, so $\beta \, R \, \alpha$.
3. If $\alpha \, R \, \beta$ and $\beta \, R \, \gamma$, then α and β have the same number of 1's and β and γ have the same number of 1's, so α and γ have the same number of 1's. Hence $\alpha \, R \, \gamma$.

We next show that R is a congruence relation. Suppose that $\alpha \, R \, \alpha'$ and $\beta \, R \, \beta'$. Then α and α' have the same number of 1's and β and β' have the same number of 1's. Since the number of 1's in $\alpha \cdot \beta$ is the sum of the number of 1's in α and the number of 1's in β, we conclude that the number of 1's in $\alpha \cdot \beta$ is the same as the number of 1's in $\alpha' \cdot \beta'$. Hence

$$(\alpha \cdot \beta) \; R \; (\alpha' \cdot \beta')$$

and thus R is a congruence relation. ∎

EXAMPLE 3

Consider the semigroup $(Z, +)$, where $+$ is ordinary addition. Let $f(x) = x^2 - x - 2$. We now define the following relation on Z:

$$a \; R \; b \quad \text{if and only if} \quad f(a) = f(b).$$

It is straightforward to verify that R is an equivalence relation on Z. However, R is not a congruence relation since we have

$$-1 \; R \; 2 \qquad \text{since } f(-1) = f(2) = 0$$

and

$$-2 \; R \; 3 \qquad \text{since } f(-2) = f(3) = 4$$

but

$$(-1 + -2) \; \not{R} \; (2 + 3)$$

since $f(-3) = 10$ and $f(5) = 18$. ∎

Recall from Section 4.5 that an equivalence relation R on the semigroup $(S, *)$ determines a partition of S. We let $[a] = R(a)$ be the equivalence class containing a and S/R denote the set of all equivalence classes. The notation $[a]$ is more traditional in this setting and produces less confusing computations.

Theorem 2   Let R be a congruence relation on the semigroup $(S, *)$. Consider the relation $\circledast$ from $S/R \times S/R$ to S/R in which the ordered pair $([a], [b])$ is, for a and b in S, related to $[a * b]$.

(a) $\circledast$ is a function from $S/R \times S/R$ to S/R, and as usual we denote $\circledast([a], [b])$ by $[a] \circledast [b]$. Thus $[a] \circledast [b] = [a * b]$.

(b) $(S/R, \circledast)$ is a semigroup. ●

Proof   Suppose that $([a], [b]) = ([a'], [b'])$. Then $a \; R \; a'$ and $b \; R \; b'$, so we must have $a * b \; R \; a' * b'$, since R is a congruence relation. Thus $[a * b] = [a' * b']$; that is, $\circledast$ is a function. This means that $\circledast$ is a binary operation on S/R.

Next, we must verify that $\circledast$ is an associative operation. We have

$$[a] \circledast ([b] \circledast [c]) = [a] \circledast [b * c]$$
$$= [a * (b * c)]$$
$$= [(a * b) * c] \quad \text{by the associative property of } * \text{ in } S$$
$$= [a * b] \circledast [c]$$
$$= ([a] \circledast [b]) \circledast [c].$$

Hence S/R is a semigroup. We call S/R the **quotient semigroup** or **factor semigroup**. Observe that $\circledast$ is a type of "quotient binary relation" on S/R that is constructed from the original binary relation $*$ on S by the congruence relation R. ▼

Corollary 1   Let R be a congruence relation on the monoid $(S, *)$. If we define the operation $\circledast$ in S/R by $[a] \circledast [b] = [a * b]$, then $(S/R, \circledast)$ is a monoid.   ●

Proof   If e is the identity in $(S, *)$, then it is easy to verify that $[e]$ is the identity in $(S/R, \circledast)$.   ▼

EXAMPLE 4   Consider the situation in Example 2. Since R is a congruence relation on the monoid $S = (A^*, \cdot)$, we conclude that $(S/R, \odot)$ is a monoid, where

$$[\alpha] \odot [\beta] = [\alpha \cdot \beta].$$ ■

EXAMPLE 5   As has already been pointed out in Section 4.5, we can repeat Example 4 of that section with the positive integer n instead of 2. That is, we define the following relation on the semigroup $(Z, +)$:

$$a \; R \; b \quad \text{if and only if} \quad a \equiv b \pmod{n}.$$

Using exactly the same method as in Example 4 in Section 4.5, we show that R is an equivalence relation and, as in the case of $n = 2$, $a \equiv b \pmod{n}$ implies $n \mid (a - b)$. Thus, if n is 4, then

$$2 \equiv 6 \pmod{4}$$

and 4 divides $(2 - 6)$. We also leave it for the reader to show that $\equiv \pmod{n}$ is a congruence relation on Z.

We now let $n = 4$ and we compute the equivalence classes determined by the congruence relation $\equiv \pmod{4}$ on Z. We obtain

$$[0] = \{\dots, -8, -4, 0, 4, 8, 12, \dots, \} = [4] = [8] = \cdots$$
$$[1] = \{\dots, -7, -3, 1, 5, 9, 13, \dots\} = [5] = [9] = \cdots$$
$$[2] = \{\dots, -6, -2, 2, 6, 10, 14, \dots\} = [6] = [10] = \cdots$$
$$[3] = \{\dots, -5, -1, 3, 7, 11, 15, \dots\} = [7] = [11] = \cdots.$$

These are all the distinct equivalence classes that form the quotient set $Z/\equiv \pmod{4}$. It is customary to denote the quotient set $Z/\equiv \pmod{n}$ by Z_n; Z_n is a monoid with operation $\oplus$ and identity $[0]$. We now determine the addition table for the semigroup Z_4 with operation $\oplus$.

$\oplus$	[0]	[1]	[2]	[3]
[0]	[0]	[1]	[2]	[3]
[1]	[1]	[2]	[3]	[0]
[2]	[2]	[3]	[0]	[1]
[3]	[3]	[0]	[1]	[2]

The entries in this table are obtained from

$$[a] \oplus [b] = [a + b].$$

Thus

$$[1] \oplus [2] = [1 + 2] = [3]$$
$$[1] \oplus [3] = [1 + 3] = [4] = [0]$$
$$[2] \oplus [3] = [2 + 3] = [5] = [1]$$
$$[3] \oplus [3] = [3 + 3] = [6] = [2].$$

It can be shown that, in general, Z_n has the n equivalence classes

$$[0], [1], [2], \ldots, [n - 1]$$

and that

$$[a] \oplus [b] = [r],$$

where r is the remainder when $a + b$ is divided by n. Thus, if n is 6,

$$[2] \oplus [3] = [5]$$
$$[3] \oplus [5] = [2]$$
$$[3] \oplus [3] = [0].$$ ∎

We shall now examine the connection between the structure of a semigroup $(S, *)$ and the quotient semigroup $(S/R, \circledast)$, where R is a congruence relation on $(S, *)$.

Theorem 3   Let R be a congruence relation on a semigroup $(S, *)$, and let $(S/R, \circledast)$ be the corresponding quotient semigroup. Then the function $f_R: S \to S/R$ defined by

$$f_R(a) = [a]$$

is an onto homomorphism, called the **natural homomorphism**.   ●

Proof   If $[a] \in S/R$, then $f_R(a) = [a]$, so f_R is an onto function. Moreover, if a and b are elements of S, then

$$f_R(a * b) = [a * b] = [a] \circledast [b] = f_R(a) \circledast f_R(b),$$

so f_R is a homomorphism.   ▼

Theorem 4
Fundamental
Homomorphism Theorem
Let $f: S \to T$ be a homomorphism of the semigroup $(S, *)$ onto the semigroup $(T, *')$. Let R be the relation on S defined by $a \; R \; b$ if and only if $f(a) = f(b)$, for a and b in S. Then

(a) R is a congruence relation.

(b) $(T, *')$ and the quotient semigroup $(S/R, \circledast)$ are isomorphic.   ●

Proof

(a) We show that R is an equivalence relation. First, $a \ R \ a$ for every $a \in S$, since $f(a) = f(a)$. Next, if $a \ R \ b$, then $f(a) = f(b)$, so $b \ R \ a$. Finally, if $a \ R \ b$ and $b \ R \ c$, then $f(a) = f(b)$ and $f(b) = f(c)$, so $f(a) = f(c)$ and $a \ R \ c$. Hence R is an equivalence relation. Now suppose that $a \ R \ a_1$ and $b \ R \ b_1$. Then

$$f(a) = f(a_1) \quad \text{and} \quad f(b) = f(b_1).$$

Multiplying in T, we obtain

$$f(a) *' f(b) = f(a_1) *' f(b_1).$$

Since f is a homomorphism, this last equation can be rewritten as

$$f(a * b) = f(a_1 * b_1).$$

Hence

$$(a * b) \ R \ (a_1 * b_1)$$

and R is a congruence relation.

(b) We now consider the relation $\overline{f}$ from S/R to T defined as follows:

$$\overline{f} = \{([a], f(a)) \mid [a] \in S/R\}.$$

We first show that $\overline{f}$ is a function. Suppose that $[a] = [a']$. Then $a \ R \ a'$, so $f(a) = f(a')$, which implies that $\overline{f}$ is a function. We may now write $\overline{f} : S/R \to T$, where $\overline{f}([a]) = f(a)$ for $[a] \in S/R$.

We next show that $\overline{f}$ is one to one. Suppose that $\overline{f}([a]) = \overline{f}([a'])$. Then

$$f(a) = f(a').$$

So $a \ R \ a'$, which implies that $[a] = [a']$. Hence $\overline{f}$ is one to one.

Now we show that $\overline{f}$ is onto. Suppose that $b \in T$. Since f is onto, $f(a) = b$ for some element a in S. Then

$$\overline{f}([a]) = f(a) = b.$$

So $\overline{f}$ is onto.

Finally,

$$\overline{f}([a] \circledast [b]) = \overline{f}([a * b])$$
$$= f(a * b) = f(a) *' f(b)$$
$$= \overline{f}([a]) *' \overline{f}([b]).$$

Hence $\overline{f}$ is an isomorphism.   ▼

EXAMPLE 6   Let $A = \{0, 1\}$, and consider the free semigroup A^* generated by A under the operation of catenation. Note that A^* is a monoid with the empty string Λ as its identity. Let N be the set of all nonnegative integers. Then N is a semigroup under the operation of ordinary addition, denoted by $(N, +)$. The function $f : A^* \to N$ defined by

$$f(\alpha) = \text{the number of 1's in } \alpha$$

is readily checked to be a homomorphism. Let R be the following relation on A^*:

$$\alpha \; R \; \beta \quad \text{if and only if} \quad f(\alpha) = f(\beta).$$

That is, $\alpha \; R \; \beta$ if and only if α and β have the same number of 1's. Theorem 4 implies that $A^*/R \simeq N$ under the isomorphism $\overline{f} \colon A^*/R \to N$ defined by

$$\overline{f}([\alpha]) = f(\alpha) = \text{the number of 1's in } \alpha. \qquad \blacksquare$$

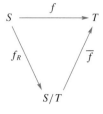

S —f→ T

f_R ↘ ↗ $\overline{f}$

S/T

Figure 9.2

Theorem 4(b) can be described by the diagram shown in Figure 9.2. Here f_R is the natural homomorphism. It follows from the definitions of f_R and $\overline{f}$ that

$$\overline{f} \circ f_R = f$$

since

$$(\overline{f} \circ f_R)(a) = \overline{f}(f_R(a)) = \overline{f}([a]) = f(a).$$

9.3 Exercises

1. Let $(S, *)$ and $(T, *')$ be commutative semigroups. Show that $S \times T$ (see Theorem 1) is also a commutative semigroup.

2. Let $(S, *)$ and $(T, *')$ be monoids. Show that $S \times T$ is also a monoid. Show that the identity of $S \times T$ is (e_S, e_T).

3. Let $(S, *)$ and $(T, *')$ be semigroups. Show that the function $f \colon S \times T \to S$ defined by $f(s, t) = s$ is a homomorphism of the semigroup $S \times T$ onto the semigroup S.

4. Let $(S, *)$ and $(T, *')$ be semigroups. Show that $S \times T$ and $T \times S$ are isomorphic semigroups.

5. Prove Theorem 1.

In Exercises 6 through 15, determine whether the relation R on the semigroup S is a congruence relation.

6. $S = Z$ under the operation of ordinary addition; $a \; R \; b$ if and only if 2 does not divide $a - b$.

7. $S = Z$ under the operation of ordinary addition; $a \; R \; b$ if and only if $a + b$ is even.

8. $S =$ any semigroup; $a \; R \; b$ if and only if $a = b$.

9. $S =$ the set of all rational numbers under the operation of addition; $a/b \; R \; c/d$ if and only if $ad = bc$.

10. $S =$ the set of all rational numbers under the operation of multiplication; $a/b \; R \; c/d$ if and only if $ad = bc$.

11. $S = Z$ under the operation of ordinary addition; $a \; R \; b$ if and only if $a \equiv b$ (mod 3).

12. $S = Z$ under the operation of ordinary addition; $a \; R \; b$ if and only if a and b are both even or a and b are both odd.

13. $S = Z^+$ under the operation of ordinary multiplication; $a \; R \; b$ if and only if $|a - b| \leq 2$.

14. $A = \{0, 1\}$ and $S = A^*$, the free semigroup generated by A under the operation of catenation; $\alpha \; R \; \beta$ if and only if α and β both have an even number of 1's or both have an odd number of 1's.

15. $S = \{0, 1\}$ under the operation $*$ defined by the table

$*$	0	1
0	0	1
1	1	0

$a \; R \; b$ if and only if $a * a = b * b$. (*Hint*: Observe that if x is any element in S, then $x * x = 0$.)

16. Show that the intersection of two congruence relations on a semigroup is a congruence relation.

17. Show that the composition of two congruence relations on a semigroup need not be a congruence relation.

18. Describe the quotient semigroup for S and R given in Exercise 10.

19. Describe the quotient semigroup for S and R given in Exercise 11.

20. Describe the quotient semigroup for S and R given in Exercise 12.

21. Describe the quotient semigroup for $S = Z$ with ordinary addition and R defined by $a \; R \; b$ if and only if $a \equiv b$ (mod 5).

22. Consider the semigroup $S = \{a, b, c, d\}$ with the following operation table.

$*$	a	b	c	d
a	a	b	c	d
b	b	a	d	c
c	c	d	a	b
d	d	c	b	a

Consider the congruence relation $R = \{(a, a), (a, b), (b, a), (b, b), (c, c), (c, d), (d, c), (d, d)\}$ on S.

(a) Write the operation table of the quotient semigroup S/R.

(b) Describe the natural homomorphism $f_R \colon S \to S/R$.

23. Consider the monoid $S = \{e, a, b, c\}$ with the following operation table.

$*$	e	a	b	c
e	e	a	b	c
a	a	e	b	c
b	b	c	b	c
c	c	b	b	c

Consider the congruence relation $R = \{(e, e), (e, a), (a, e), (a, a), (b, b), (b, c), (c, b), (c, c)\}$ on S.

(a) Write the operation table of the quotient monoid S/R.

(b) Describe the natural homomorphism $f_R \colon S \to S/R$.

24. Let $A = \{0, 1\}$ and consider the free semigroup A^* generated by A under the operation of catenation. Let N be the semigroup of all nonnegative integers under the operation of ordinary addition.

(a) Verify that the function $f \colon A^* \to N$, defined by $f(\alpha) = $ the number of digits in α, is a homomorphism.

(b) Let R be the following relation on A^*: $\alpha\, R\, \beta$ if and only if $f(\alpha) = f(\beta)$. Show that R is a congruence relation on A^*.

(c) Show that A^*/R and N are isomorphic.

25. Describe the strategy of the proof of Theorem 4. Outline the proof.

9.4 GROUPS

In this section we examine a special type of monoid, called a group, that has applications in every area where symmetry occurs. Applications of groups can be found in mathematics, physics, and chemistry, as well as in less obvious areas such as sociology. Recent and exciting applications of group theory have arisen in fields such as particle physics and in the solutions of puzzles such as Rubik's cube. In this book, we shall present an important application of group theory to binary codes in Section 11.2.

A **group** $(G, *)$ is a monoid, with identity e, that has the additional property that for every element $a \in G$ there exists an element $a' \in G$ such that $a * a' = a' * a = e$. Thus a group is a set together with a binary operation $*$ on G such that

1. $(a * b) * c = a * (b * c)$ for any elements a, b, and c in G.
2. There is a unique element e in G such that

$$a * e = e * a \qquad \text{for any } a \in G.$$

3. For every $a \in G$, there is an element $a' \in G$, called an **inverse** of a, such that

$$a * a' = a' * a = e.$$

Observe that if $(G, *)$ is a group, then $*$ is a binary operation, so G must be closed under $*$; that is,

$$a * b \in G \qquad \text{for any elements } a \text{ and } b \text{ in } G.$$

To simplify our notation, from now on when only one group $(G, *)$ is under consideration and there is no possibility of confusion, we shall write the product $a * b$ of the elements a and b in the group $(G, *)$ simply as ab, and we shall also refer to $(G, *)$ simply as G.

A group G is said to be **Abelian** if $ab = ba$ for all elements a and b in G.

EXAMPLE 1 The set of all integers Z with the operation of ordinary addition is an Abelian group. If $a \in Z$, then an inverse of a is its opposite $-a$. ∎

EXAMPLE 2 The set Z^+ under the operation of ordinary multiplication is not a group since, for example, the element 2 in Z^+ has no inverse. However, this set together with the given operation is a monoid. ∎

EXAMPLE 3 The set of all nonzero real numbers under the operation of ordinary multiplication is a group. An inverse of $a \neq 0$ is $1/a$. ∎

EXAMPLE 4 Let G be the set of all nonzero real numbers and let

$$a * b = \frac{ab}{2}.$$

Show that $(G, *)$ is an Abelian group.

Solution We first verify that $*$ is a binary operation. If a and b are elements of G, then $ab/2$ is a nonzero real number and hence is in G. We next verify associativity. Since

$$(a * b) * c = \left(\frac{ab}{2}\right) * c = \frac{(ab)c}{4}$$

and since

$$a * (b * c) = a * \left(\frac{bc}{2}\right) = \frac{a(bc)}{4} = \frac{(ab)c}{4},$$

the operation $*$ is associative.

The number 2 is the identity in G, for if $a \in G$, then

$$a * 2 = \frac{(a)(2)}{2} = a = \frac{(2)(a)}{2} = 2 * a.$$

Finally, if $a \in G$, then $a' = 4/a$ is an inverse of a, since

$$a * a' = a * \frac{4}{a} = \frac{a(4/a)}{2} = 2 = \frac{(4/a)(a)}{2} = \frac{4}{a} * a = a' * a.$$

Since $a * b = b * a$ for all a and b in G, we conclude that G is an Abelian group. ∎

Before proceeding with additional examples of groups, we develop several important properties that are satisfied in any group G.

Theorem 1 Let G be a group. Each element a in G has only one inverse in G. ●

Proof   Let a' and a'' be inverses of a. Then

$$a'(aa'') = a'e = a'$$

and

$$(a'a)a'' = ea'' = a''.$$

Hence, by associativity,

$$a' = a''.$$ ▼

From now on we shall denote the inverse of a by a^{-1}. Thus in a group G we have

$$aa^{-1} = a^{-1}a = e.$$

Theorem 2   Let G be a group and let a, b, and c be elements of G. Then
  (a) $ab = ac$ implies that $b = c$ (**left cancellation property**).
  (b) $ba = ca$ implies that $b = c$ (**right cancellation property**).   ●

Proof
  (a) Suppose that

$$ab = ac.$$

Multiplying both sides of this equation by a^{-1} on the left, we obtain

$$a^{-1}(ab) = a^{-1}(ac)$$
$$(a^{-1}a)b = (a^{-1}a)c \qquad \text{by associativity}$$
$$eb = ec \qquad \text{by the definition of an inverse}$$
$$b = c \qquad \text{by definition of an identity.}$$

  (b) The proof is similar to that of part (a).   ▼

Theorem 3   Let G be a group and let a and b be elements of G. Then
  (a) $(a^{-1})^{-1} = a$.
  (b) $(ab)^{-1} = b^{-1}a^{-1}$.   ●

Proof
  (a) We show that a acts as an inverse for a^{-1}:

$$a^{-1}a = aa^{-1} = e.$$

Since the inverse of an element is unique, we conclude that $(a^{-1})^{-1} = a$.
  (b) We easily verify that

$$(ab)(b^{-1}a^{-1}) = a(b(b^{-1}a^{-1})) = a((bb^{-1})a^{-1}) = a(ea^{-1}) = aa^{-1} = e$$

and, similarly,

$$(b^{-1}a^{-1})(ab) = e,$$

so

$$(ab)^{-1} = b^{-1}a^{-1}.$$ ▼

Theorem 4 Let G be a group, and let a and b be elements of G. Then

    (a) The equation $ax = b$ has a unique solution in G.

    (b) The equation $ya = b$ has a unique solution in G.   ●

Proof

    (a) The element $x = a^{-1}b$ is a solution of the equation $ax = b$, since

$$a(a^{-1}b) = (aa^{-1})b = eb = b.$$

        Suppose now that x_1 and x_2 are two solutions of the equation $ax = b$. Then

$$ax_1 = b \quad \text{and} \quad ax_2 = b.$$

        Hence

$$ax_1 = ax_2.$$

        Theorem 2 implies that $x_1 = x_2$.

    (b) The proof is similar to that of part (a).   ▼

From our discussion of monoids, we know that if a group G has a finite number of elements, then its binary operation can be given by a table, which is generally called a **multiplication table**. The multiplication table of a group $G = \{a_1, a_2, \ldots, a_n\}$ under the binary operation $*$ must satisfy the following properties:

    1. The row labeled by e must be

$$a_1, a_2, \ldots, a_n$$

    and the column labeled by e must be

$$\begin{array}{c} a_1 \\ a_2 \\ \vdots \\ a_n. \end{array}$$

    2. From Theorem 4, it follows that each element b of the group must appear exactly once in each row and column of the table. Thus each row and column is a permutation of the elements $a_1, a_2, \ldots, a_n$ of G, and each row (and each column) determines a different permutation.

If G is a group that has a finite number of elements, we say that G is a **finite group**, and the **order** of G is the number of elements $|G|$ in G. We shall now determine the multiplication tables of all nonisomorphic groups of orders 1, 2, 3, and 4.

If G is a group of order 1, then $G = \{e\}$, and we have $ee = e$. Now let $G = \{e, a\}$ be a group of order 2. Then we obtain a multiplication table (Table 9.1) where we need to fill in the blank. The blank can be filled in by e or by a. Since there can be no repeats in any row or column, we must write e in the blank. The multiplication table shown in Table 9.2 satisfies the associativity property and the other properties of a group, so it is the multiplication table of a group of order 2.

Table 9.1

	e	a
e	e	a
a	a	

Table 9.2

	e	a
e	e	a
a	a	e

Table 9.3

	e	a	b
e	e	a	b
a	a		
b	b		

Table 9.4

	e	a	b
e	e	a	b
a	a	b	e
b	b	e	a

Next, let $G = \{e, a, b\}$ be a group of order 3. We have a multiplication table (Table 9.3) where we must fill in four blanks. A little experimentation shows that we can only complete the table as shown in Table 9.4. It can be shown (a tedious task) that Table 9.4 satisfies the associative property and the other properties of a group. Thus it is the multiplication table of a group of order 3. Observe that the groups of orders 1, 2, and 3 are also Abelian and that there is just one group of each order for a fixed labeling of the elements.

We next come to a group $G = \{e, a, b, c\}$ of order 4. It is not difficult to show that the possible multiplication table for G can be completed as shown in Tables 9.5 through 9.8. It can be shown that each of these tables satisfies the associative property and the other properties of a group. Thus there are four possible multiplication tables for a group of order 4. Again, observe that a group of order 4 is Abelian. We shall return to groups of order 4 toward the end of this section, where we shall see that there are only two and not four different nonisomorphic groups of order 4.

Table 9.5

	e	a	b	c
e	e	a	b	c
a	a	e	c	b
b	b	c	e	a
c	c	b	a	e

Table 9.6

	e	a	b	c
e	e	a	b	c
a	a	e	c	b
b	b	c	a	e
c	c	b	e	a

Table 9.7

	e	a	b	c
e	e	a	b	c
a	a	b	c	e
b	b	c	e	a
c	c	e	a	b

Table 9.8

	e	a	b	c
e	e	a	b	c
a	a	c	e	b
b	b	e	c	a
c	c	b	a	e

EXAMPLE 5   Let $B = \{0, 1\}$, and let $+$ be the operation defined on B as follows:

+	0	1
0	0	1
1	1	0

Then B is a group. In this group, every element is its own inverse.   ■

We next turn to an important example of a group.

EXAMPLE 6   Consider the equilateral triangle shown in Figure 9.3 with vertices 1, 2, and 3. A **symmetry** of the triangle (or of any geometrical figure) is a one-to-one correspondence from the set of points forming the triangle (the geometrical figure) to itself that preserves the distance between adjacent points. Since the triangle is determined by its vertices, a symmetry of the triangle is merely a permutation of the vertices that preserves the distance between adjacent points. Let l_1, l_2, and l_3 be the angle bisectors of the corresponding angles as shown in Figure 9.3, and let O be their point of intersection.

We now describe the symmetries of this triangle. First, there is a counterclockwise rotation f_2 of the triangle about O through $120°$. Then f_2 can be written (see Section 5.3) as the permutation

$$f_2 = \begin{pmatrix} 1 & 2 & 3 \\ 2 & 3 & 1 \end{pmatrix}.$$

We next obtain a counterclockwise rotation f_3 about O through $240°$, which can be written as the permutation

$$f_3 = \begin{pmatrix} 1 & 2 & 3 \\ 3 & 1 & 2 \end{pmatrix}.$$

Finally, there is a counterclockwise rotation f_1 about O through $360°$, which can be written as the permutation

$$f_1 = \begin{pmatrix} 1 & 2 & 3 \\ 1 & 2 & 3 \end{pmatrix}.$$

Of course, f_1 can also be viewed as the result of rotating the triangle about O through $0°$.

We may also obtain three additional symmetries of the triangle, g_1, g_2, and g_3, by reflecting about the lines l_1, l_2, and l_3, respectively. We may denote these reflections as the following permutations:

$$g_1 = \begin{pmatrix} 1 & 2 & 3 \\ 1 & 3 & 2 \end{pmatrix}, \quad g_2 = \begin{pmatrix} 1 & 2 & 3 \\ 3 & 2 & 1 \end{pmatrix}, \quad g_3 = \begin{pmatrix} 1 & 2 & 3 \\ 2 & 1 & 3 \end{pmatrix}.$$

Observe that the set of all symmetries of the triangle is described by the set of permutations of the set $\{1, 2, 3\}$, which is considered in Section 5.3 and is denoted by S_3. Thus

$$S_3 = \{f_1, f_2, f_3, g_1, g_2, g_3\}.$$

We now introduce the operation $*$, *followed by*, on the set S_3, and we obtain the multiplication table shown in Table 9.9.

Table 9.9

$*$	f_1	f_2	f_3	g_1	g_2	g_3
f_1	f_1	f_2	f_3	g_1	g_2	g_3
f_2	f_2	f_3	f_1	g_3	g_1	g_2
f_3	f_3	f_1	f_2	g_2	g_3	g_1
g_1	g_1	g_2	g_3	f_1	f_2	f_3
g_2	g_2	g_3	g_1	f_3	f_1	f_2
g_3	g_3	g_1	g_2	f_2	f_3	f_1

Each of the entries in this table can be obtained in one of two ways: algebraically or geometrically. For example, suppose that we want to compute $f_2 * g_2$. Geometrically, we proceed as in Figure 9.4. Note that "followed by" here refers to the geometric order. To compute $f_2 * g_2$ algebraically, we compute $f_2 \circ g_2$.

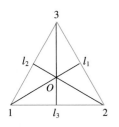

Figure 9.3

$$\begin{pmatrix} 1 & 2 & 3 \\ 2 & 3 & 1 \end{pmatrix} \circ \begin{pmatrix} 1 & 2 & 3 \\ 3 & 2 & 1 \end{pmatrix} = \begin{pmatrix} 1 & 2 & 3 \\ 1 & 3 & 2 \end{pmatrix} = g_1$$

and find that $f_2 * g_2 = g_1$.

Since composition of functions is always associative, we see that $*$ is an associative operation on S_3. Observe that f_1 is the identity in S_3 and that every element of S_3 has a unique inverse in S_3. For example, $f_2^{-1} = f_3$. Hence S_3 is a group called the **group of symmetries of the triangle**. Observe that S_3 is the first example that we have given of a group that is not Abelian. ■

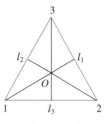 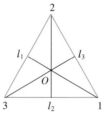 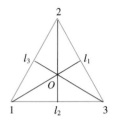

Given triangle   Triangle resulting after   Triangle resulting after applying
                        applying f_2 g_2 to the triangle at the left

Figure 9.4

EXAMPLE 7   The set of all permutations of n elements is a group of order $n!$ under the operation of composition. This group is called the **symmetric group on n letters** and is denoted by S_n. We have seen that S_3 also represents the group of symmetries of the equilateral triangle. ■

As in Example 6, we can also consider the group of symmetries of a square. However, it turns out that this group is of order 8, so it does not agree with the group S_4, whose order is $4! = 24$.

EXAMPLE 8   In Section 9.3 we discussed the monoid Z_n. We now show that Z_n is a group as follows. Let $[a] \in Z_n$. Then we may assume that $0 \le a < n$. Moreover, $[n - a] \in Z_n$ and since

$$[a] \oplus [n - a] = [a + n - a] = [n] = [0],$$

we conclude that $[n - a]$ is the inverse of $[a]$. Thus, if n is 6, then [2] is the inverse of [4]. Observe that Z_n is an Abelian group. ■

We next turn to a discussion of important subsets of a group. Let H be a subset of a group G such that

(a) The identity e of G belongs to H.

(b) If a and b belong to H, then $ab \in H$.

(c) If $a \in H$, then $a^{-1} \in H$.

Then H is called a **subgroup** of G. Part (b) says that H is a subsemigroup of G. Thus a subgroup of G can be viewed as a subsemigroup having properties (a) and (c).

Observe that if G is a group and H is a subgroup of G, then H is also a group with respect to the operation in G, since the associative property in G also holds in H.

EXAMPLE 9   Let G be a group. Then G and $H = \{e\}$ are subgroups of G, called the **trivial subgroups of** G.   ∎

EXAMPLE 10   Consider S_3, the group of symmetries of the equilateral triangle, whose multiplication table is shown in Table 9.9. It is easy to verify that $H = \{f_1, f_2, f_3\}$ is a subgroup of S_3.   ∎

EXAMPLE 11   Let A_n be the set of all even permutations (see Section 5.4) in the group S_n. It can be shown from the definition of even permutation that A_n is a subgroup of S_n, called the **alternating group on n letters**.   ∎

EXAMPLE 12   Let G be a group and let $a \in G$. Since a group is a monoid, we have already defined, in Section 9.2, a^n for $n \in Z^+$ as $aa \cdots a$ (n factors), and a^0 as e. If n is a negative integer, we now define a^{-n} as $a^{-1}a^{-1} \cdots a^{-1}$ (n factors). Then, if n and m are any integers, we have

$$a^n a^m = a^{n+m}.$$

It is easy to show that

$$H = \{a^i \mid i \in Z\}$$

is a subgroup of G.   ∎

Let $(G, *)$ and $(G', *')$ be two groups. Since groups are also semigroups, we can consider isomorphisms and homomorphisms from $(G, *)$ to $(G', *')$.

Since an isomorphism must be a one-to-one and onto function, it follows that two groups whose orders are unequal cannot possibly be isomorphic.

EXAMPLE 13   Let G be the group of real numbers under addition, and let G' be the group of positive real numbers under multiplication. Let $f: G \to G'$ be defined by $f(x) = e^x$. We now show that f is an isomorphism.

If $f(a) = f(b)$, so that $e^a = e^b$, then $a = b$. Thus f is one to one. If $c \in G'$, then $\ln c \in G$ and

$$f(\ln c) = e^{\ln c} = c,$$

so f is onto. Finally,

$$f(a + b) = e^{a+b} = e^a e^b = f(a)f(b).$$

Hence f is an isomorphism.   ∎

EXAMPLE 14

Let G be the symmetric group of n letters, and let G' be the group B defined in Example 5. Let $f : G \to G'$ be defined as follows: for $p \in G$,

$$f(p) = \begin{cases} 0 & \text{if } p \in A_n \quad \text{(the subgroup of all even permutations in } G) \\ 1 & \text{if } p \notin A_n. \end{cases}$$

Then f is a homomorphism.   ∎

EXAMPLE 15

Let G be the group of integers under addition, and let G' be the group Z_n as discussed in Example 8. Let $f : G \to G'$ be defined as follows: If $m \in G$, then $f(m) = [r]$, where r is the remainder when m is divided by n. We now show that f is a homomorphism of G onto G'.

Let $[r] \in Z_n$. Then we may assume that $0 \leq r < n$, so

$$r = 0 \cdot n + r,$$

which means that the remainder when r is divided by n is r. Hence

$$f(r) = [r]$$

and thus f is onto.

Next, let a and b be elements of G expressed as

$$a = q_1 n + r_1, \qquad \text{where } 0 \leq r_1 < n, \text{ and } r_1 \text{ and } q_1 \text{ are integers} \qquad (1)$$
$$b = q_2 n + r_2, \qquad \text{where } 0 \leq r_2 < n, \text{ and } r_2 \text{ and } q_2 \text{ are integers} \qquad (2)$$

so that

$$f(a) = [r_1] \quad \text{and} \quad f(b) = [r_2].$$

Then

$$f(a) + f(b) = [r_1] + [r_2] = [r_1 + r_2].$$

To find $[r_1 + r_2]$, we need the remainder when $r_1 + r_2$ is divided by n. Write

$$r_1 + r_2 = q_3 n + r_3, \qquad \text{where } 0 \leq r_3 < n, \text{ and } r_3 \text{ and } q_3 \text{ are integers.}$$

Thus

$$f(a) + f(b) = [r_3].$$

Adding, we have

$$a + b = q_1 n + q_2 n + r_1 + r_2 = (q_1 + q_2 + q_3)n + r_3,$$

so

$$f(a + b) = [r_1 + r_2] = [r_3].$$

Hence

$$f(a + b) = f(a) + f(b),$$

which implies that f is a homomorphism.

Note that when n is 2, f assigns each even integer to [0] and each odd integer to [1].   ∎

Theorem 5 Let $(G, *)$ and $(G', *')$ be two groups, and let $f: G \rightarrow G'$ be a homomorphism from G to G'.

(a) If e is the identity in G and e' is the identity in G', then $f(e) = e'$.
(b) If $a \in G$, then $f(a^{-1}) = (f(a))^{-1}$.
(c) If H is a subgroup of G, then

$$f(H) = \{f(h) \mid h \in H\}$$

is a subgroup of G'.   ●

Proof

(a) Let $x = f(e)$. Then

$$x *' x = f(e) *' f(e) = f(e * e) = f(e) = x,$$

so $x *' x = x$. Multiplying both sides by x^{-1} on the right, we obtain

$$x = x *' x *' x^{-1} = x *' x^{-1} = e'.$$

Thus $f(e) = e'$.

(b) $a * a^{-1} = e$, so

$$f(a * a^{-1}) = f(e) = e' \quad \text{by part (a)}$$

or

$$f(a) *' f(a^{-1}) = e' \quad \text{since } f \text{ is a homomorphism.}$$

Similarly,

$$f(a^{-1}) *' f(a) = e'.$$

Hence $f(a^{-1}) = (f(a))^{-1}$.

(c) This follows from Theorem 4 of Section 9.2 and parts (a) and (b).   ▼

EXAMPLE 16 The groups S_3 and Z_6 are both of order 6. However, S_3 is not Abelian and Z_6 is Abelian. Hence they are not isomorphic. Remember that an isomorphism preserves all properties defined in terms of the group operations.   ■

EXAMPLE 17 Earlier in this section we found four possible multiplication tables (Tables 9.5 through 9.8) for a group or order 4. We now show that the groups with multiplication Tables 9.6, 9.7, and 9.8 are isomorphic as follows. Let $G = \{e, a, b, c\}$ be the group whose multiplication table is Table 9.6, and let $G' = \{e', a', b', c'\}$ be the group whose multiplication table is Table 9.7, where we put primes on every entry in this last table. Let $f: G \rightarrow G'$ be defined by $f(e) = e'$, $f(a) = b'$, $f(b) = a'$, $f(c) = c'$. We can then verify that under this renaming of elements the two tables become identical, so the corresponding groups are isomorphic. Similarly, let $G'' = \{e'', a'', b'', c''\}$ be the group whose multiplication table is Table 9.8, where we put double primes on every entry in this last table. Let $g: G \rightarrow G''$ be defined by $g(e) = e''$, $g(a) = c''$, $g(b) = b''$, $g(c) = a''$. We can then verify that under this renaming of elements the two tables become identical, so the corresponding groups are isomorphic. That is, the groups given by Tables 9.6, 9.7, and 9.8 are isomorphic.

Now, how can we be sure that Tables 9.5 and 9.6 do not yield isomorphic groups? Observe that if x is any element in the group determined by Table 9.5, then $x^2 = e$. If the groups were isomorphic, then the group determined by Table 9.6 would have the same property. Since it does not, we conclude that these groups are not isomorphic. Thus there are exactly two nonisomorphic groups of order 4.

The group with multiplication Table 9.5 is called the **Klein 4 group** and it is denoted by V. The one with multiplication Table 9.6, 9.7, or 9.8 is denoted by Z_4, since a relabeling of the elements of Z_4 results in this multiplication table.   ■

9.4 Exercises

In Exercises 1 through 11, determine whether the set together with the binary operation is a group. If it is a group, determine if it is Abelian; specify the identity and the inverse of a generic element.

1. Z, where $*$ is ordinary multiplication.

2. Z, where $*$ is ordinary subtraction.

3. Q, the set of all rational numbers under the operation of addition.

4. Q, the set of all rational numbers under the operation of multiplication.

5. $\mathbb{R}$, under the operation of multiplication.

6. $\mathbb{R}$, where $a * b = a + b + 2$.

7. Z^+, under the operation of addition.

8. The real numbers that are not equal to -1, where $a * b = a + b + ab$.

9. The set of odd integers under the operation of multiplication.

10. The set of all $m \times n$ matrices under the operation of matrix addition.

11. If S is a nonempty set, the set $P(S)$, where $A * B = A \oplus B$. (See Section 1.2.)

12. Let $S = \{x \mid x$ is a real number and $x \neq 0, x \neq -1\}$. Consider the following functions $f_i \colon S \to S, i = 1, 2, \ldots, 6$:

$$f_1(x) = x, \quad f_2(x) = 1 - x, \quad f_3(x) = \frac{1}{x},$$

$$f_4(x) = \frac{1}{1-x}, \quad f_5(x) = 1 - \frac{1}{x}, \quad f_6(x) = \frac{x}{x-1}.$$

Show that $G = \{f_1, f_2, f_3, f_4, f_5, f_6\}$ is a group under the operation of composition. Give the multiplication table of G.

13. Consider S_3, the group of symmetries of the equilateral triangle, and the group in Exercise 12. Prove or disprove that these two groups are isomorphic.

14. Show that the mapping in Example 14 is a homomorphism.

15. Let G be the group defined in Example 4. Solve the following equations:

   (a) $3 * x = 4$ (b) $y * 5 = -2$

16. Let G be a group with identity e. Show that if $a^2 = e$ for all a in G, then G is Abelian.

17. Consider the square shown in Figure 9.5.

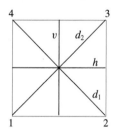

Figure 9.5

The symmetries of the square are as follows:

Rotations f_1, f_2, f_3, and f_4 through $0°$, $90°$, $180°$, and $270°$, respectively

f_5 and f_6, reflections about the lines v and h, respectively

f_7 and f_8, reflections about the diagonals d_1 and d_2, respectively

Write the multiplication table of D_4, the group of symmetries of the square.

18. Let G be a group. Show by mathematical induction that if $ab = ba$, then $(ab)^n = a^n b^n$ for $n \in Z^+$.

19. Let G be a finite group with identity e, and let a be an arbitrary element of G. Prove that there exists a nonnegative integer n such that $a^n = e$.

20. Let G be the nonzero integers under the operation of multiplication, and let $H = \{3^n \mid n \in Z\}$. Is H a subgroup of G?

21. Let G be the group of integers under the operation of addition, and let $H = \{3k \mid k \in Z\}$. Is H a subgroup of G?

22. Let G be an Abelian group with identity e, and let $H = \{x \mid x^2 = e\}$. Show that H is a subgroup of G.

23. Let G be a group, and let $H = \{x \mid x \in G$ and $xy = yx$ for all $y \in G\}$. Prove that H is a subgroup of G.

24. Let G be a group and let $a \in G$. Define $H_a = \{x \mid x \in G$ and $xa = ax\}$. Prove that H_a is a subgroup of G.

25. Let A_n be the set of all even permutations in S_n. Show that A_n is a subgroup of S_n.

26. Let H and K be subgroups of a group G.
    (a) Prove that $H \cap K$ is a subgroup of G.
    (b) Show that $H \cup K$ need not be a subgroup of G.

27. Find all subgroups of the group given in Exercise 17.

28. Let G be an Abelian group and n a fixed integer. Prove that the function $f: G \to G$ defined by $f(a) = a^n$, for $a \in G$, is a homomorphism.

29. Prove that the function $f(x) = |x|$ is a homomorphism from the group G of nonzero real numbers under multiplication to the group G' of positive real numbers under multiplication.

30. Let G be a group with identity e. Show that the function $f: G \to G$ defined by $f(a) = e$ for all $a \in G$ is a homomorphism.

31. Let G be a group. Show that the function $f: G \to G$ defined by $f(a) = a^2$ is a homomorphism if and only if G is Abelian.

32. Let G be a group. Show that the function $f: G \to G$ defined by $f(a) = a^{-1}$ is an isomorphism if and only if G is Abelian.

33. Let G be a group and let a be a fixed element of G. Show that the function $f_a: G \to G$ defined by $f_a(x) = axa^{-1}$, for $x \in G$, is an isomorphism.

34. Let $G = \{e, a, a^2, a^3, a^4, a^5\}$ be a group under the operation of $a^i a^j = a^r$, where $i + j \equiv r \pmod 6$. Prove that G and Z_6 are isomorphic.

9.5 PRODUCTS AND QUOTIENTS OF GROUPS

In this section, we shall obtain new groups from other groups by using the ideas of product and quotient. Since a group has more structure than a semigroup, our results will be deeper than analogous results for semigroups as discussed in Section 9.3.

Theorem 1 If G_1 and G_2 are groups, then $G = G_1 \times G_2$ is a group with binary operation defined by

$$(a_1, b_1)(a_2, b_2) = (a_1 a_2, b_1 b_2).$$ ●

Proof By Theorem 1, Section 9.3, we have that G is a semigroup. The existence of an identity and inverses is easy to verify. ▼

EXAMPLE 1 Let G_1 and G_2 be the group Z_2. For simplicity of notation, we shall write the elements of Z_2 as $\bar{0}$ and $\bar{1}$, respectively, instead of $[0]$ and $[1]$. Then the multiplication table of $G = G_1 \times G_2$ is given in Table 9.10.

Since G is a group of order 4, it must be isomorphic to V or to Z_4 (see Section 9.4), the only groups of order 4. By looking at the multiplication tables, we see that the function $f: V \to Z_2 \times Z_2$ defined by $f(e) = (\bar{0}, \bar{0})$, $f(a) = (\bar{1}, \bar{0})$, $f(b) = (\bar{0}, \bar{1})$, and $f(c) = (\bar{1}, \bar{1})$ is an isomorphism. ■

Table 9.10 Multiplication Table of $Z_2 \times Z_2$

	$(\bar{0}, \bar{0})$	$(\bar{1}, \bar{0})$	$(\bar{0}, \bar{1})$	$(\bar{1}, \bar{1})$
$(\bar{0}, \bar{0})$	$(\bar{0}, \bar{0})$	$(\bar{1}, \bar{0})$	$(\bar{0}, \bar{1})$	$(\bar{1}, \bar{1})$
$(\bar{1}, \bar{0})$	$(\bar{1}, \bar{0})$	$(\bar{0}, \bar{0})$	$(\bar{1}, \bar{1})$	$(\bar{0}, \bar{1})$
$(\bar{0}, \bar{1})$	$(\bar{0}, \bar{1})$	$(\bar{1}, \bar{1})$	$(\bar{0}, \bar{0})$	$(\bar{1}, \bar{0})$
$(\bar{1}, \bar{1})$	$(\bar{1}, \bar{1})$	$(\bar{0}, \bar{1})$	$(\bar{1}, \bar{0})$	$(\bar{0}, \bar{0})$

If we repeat Example 1 with Z_2 and Z_3, we find that $Z_2 \times Z_3 \simeq Z_6$. It can be shown, in general, that $Z_m \times Z_n \simeq Z_{mn}$ if and only if $\text{GCD}(m, n) = 1$, that is, if and only if m and n are relatively prime.

Theorem 1 can obviously be extended to show that if $G_1, G_2, \ldots, G_n$ are groups, then $G = G_1 \times G_2 \times \cdots G_n$ is also a group.

EXAMPLE 2   Let $B = \{0, 1\}$ be the group defined in Example 5 of Section 9.4, where $+$ is defined as follows:

$$
\begin{array}{c|cc}
+ & 0 & 1 \\
\hline
0 & 0 & 1 \\
1 & 1 & 0
\end{array}
$$

Then $B^n = B \times B \times \cdots \times B$ (n factors) is a group with operation $\oplus$ defined by

$$(x_1, x_2, \ldots, x_n) \oplus (y_1, y_2, \ldots, y_n) = (x_1 + y_1, x_2 + y_2, \ldots, x_n + y_n).$$

The identity of B^n is $(0, 0, \ldots, 0)$, and every element is its own inverse. This group is essentially the same as the Boolean algebra B_n defined in Section 6.4, but the binary operation is very different from $\wedge$ and $\vee$.   ∎

A congruence relation on a group is simply a congruence relation on the group when it is viewed as a semigroup. We now discuss quotient structures determined by a congruence relation on a group.

Theorem 2   Let R be a congruence relation on the group $(G, *)$. Then the semigroup $(G/R, \circledast)$ is a group, where the operation $\circledast$ is defined on G/R by

$$[a] \circledast [b] = [a * b] \quad \text{(see Section 9.3)}. \qquad \bullet$$

Proof   Since a group is a monoid, we know from Corollary 1 of Section 9.3 that G/R is a monoid. We need to show that each element of G/R has an inverse. Let $[a] \in G/R$. Then $[a^{-1}] \in G/R$, and

$$[a] \circledast [a^{-1}] = [a * a^{-1}] = [e].$$

So $[a]^{-1} = [a^{-1}]$. Hence $(G/R, \circledast)$ is a group.   ∎

Since the definitions of homomorphism, isomorphism, and congruence for groups involve only the semigroup and monoid structure of groups, the following corollary is an immediate consequence of Theorems 3 and 4 of Section 9.3.

Corollary 1   (a) If R is a congruence relation on a group G, then the function $f_R: G \to G/R$, given by $f_R(a) = [a]$, is a group homomorphism.

(b) If $f: G \to G'$ is a homomorphism from the group $(G, *)$ onto the group $(G', *')$, and R is the relation defined on G by $a \; R \; b$ if and only if $f(a) = f(b)$, for a and b in G, then

    1. R is a congruence relation.

    2. The function $\overline{f}: G/R \to G'$, given by $\overline{f}([a]) = f(a)$, is an isomorphism from the group $(G/R, \circledast)$ onto the group $(G', *')$. ●

Congruence relations on groups have a very special form, which we will now develop. Let H be a subgroup of a group G, and let $a \in G$. The **left coset** of H in G determined by a is the set $aH = \{ah \mid h \in H\}$. The **right coset** of H in G determined by a is the set $Ha = \{ha \mid h \in H\}$. Finally, we will say that a subgroup H of G is **normal** if $aH = Ha$ for all a in G.

    Warning: If $Ha = aH$, it does *not* follow that, for $h \in H$ and $a \in G$, $ha = ah$. It does follows that $ha = ah'$, where h' is some element in H.

If H is a subgroup of G, we shall need in some applications to compute all the left cosets of H in G. First, suppose that $a \in H$. Then $aH \subseteq H$, since H is a subgroup of G; moreover, if $h \in H$, then $h = ah'$, where $h' = a^{-1}h \in H$, so that $H \subseteq aH$. Thus, if $a \in H$, then $aH = H$. This means that, when finding all the cosets of H, we need not compute aH for $a \in H$, since it will always be H.

EXAMPLE 3   Let G be the symmetric group S_3 discussed in Example 6 of Section 9.4. The subset $H = \{f_1, g_2\}$ is a subgroup of G. Compute all the distinct left cosets of H in G.

    Solution   If $a \in H$, then $aH = H$. Thus

$$f_1 H = g_2 H = H.$$

Also,

$$f_2 H = \{f_2, g_1\}$$
$$f_3 H = \{f_3, g_3\}$$
$$g_1 H = \{g_1, f_2\} = f_2 H$$
$$g_3 H = \{g_3, f_3\} = f_3 H.$$

The distinct left cosets of H in G are H, $f_2 H$, and $f_3 H$. ■

EXAMPLE 4   Let G and H be as in Example 3. Then the right coset $Hf_2 = \{f_2, g_3\}$. In Example 3 we saw that $f_2 H = \{f_2, g_1\}$. It follows that H is not a normal subgroup of G. ■

EXAMPLE 5   Show that if G is an Abelian group, then every subgroup of G is a normal subgroup.

    Solution   Let H be a subgroup of G and let $a \in G$ and $h \in H$. Then $ha = ah$, so $Ha = aH$, which implies that H is a normal subgroup of G. ■

Theorem 3   Let R be a congruence relation on a group G, and let $H = [e]$, the equivalence class containing the identity. Then H is a normal subgroup of G and, for each $a \in G$, $[a] = aH = Ha$.   ●

Proof   Let a and b be any elements in G. Since R is an equivalence relation, $b \in [a]$ if and only if $[b] = [a]$. Also, G/R is a group by Theorem 2. Therefore, $[b] = [a]$ if and only if $[e] = [a]^{-1}[b] = [a^{-1}b]$. Thus $b \in [a]$ if and only if $H = [e] = [a^{-1}b]$. That is, $b \in [a]$ if and only if $a^{-1}b \in H$ or $b \in aH$. This proves that $[a] = aH$ for every $a \in G$. We can show similarly that $b \in [a]$ if and only if $H = [e] = [b][a]^{-1} = [ba^{-1}]$. This is equivalent to the statement $[a] = Ha$. Thus $[a] = aH = Ha$, and H is normal.   ▼

Combining Theorem 3 with Corollary 1, we see that in this case the quotient group G/R consists of all the left cosets of $N = [e]$. The operation in G/R is given by

$$(aN)(bN) = [a] \circledast [b] = [ab] = abN$$

and the function $f_R: G \to G/R$, defined by $f_R(a) = aN$, is a homomorphism from G onto G/R. For this reason, we will often write G/R as G/N.

We next consider the question of whether every normal subgroup of a group G is the equivalence class of the identity of G for some congruence relation.

Theorem 4   Let N be a normal subgroup of a group G, and let R be the following relation on G:

$$a \mathrel{R} b \quad \text{if and only if} \quad a^{-1}b \in N.$$

Then

(a) R is a congruence relation on G.

(b) N is the equivalence class $[e]$ relative to R, where e is the identity of G.

Proof

(a) Let $a \in G$. Then $a \mathrel{R} a$, since $a^{-1}a = e \in N$, so R is reflexive. Next, suppose that $a \mathrel{R} b$, so that $a^{-1}b \in N$. Then $(a^{-1}b)^{-1} = b^{-1}a \in N$, so $b \mathrel{R} a$. Hence R is symmetric. Finally, suppose that $a \mathrel{R} b$ and $b \mathrel{R} c$. Then $a^{-1}b \in N$ and $b^{-1}c \in N$. Then $(a^{-1}b)(b^{-1}c) = a^{-1}c \in N$, so $a \mathrel{R} c$. Hence R is transitive. Thus R is an equivalence relation on G.

Next we show that R is a congruence relation on G. Suppose that $a \mathrel{R} b$ and $c \mathrel{R} d$. Then $a^{-1}b \in N$ and $c^{-1}d \in N$. Since N is normal, $Nd = dN$; that is, for any $n_1 \in N$, $n_1 d = dn_2$ for some $n_2 \in N$. In particular, since $a^{-1}b \in N$, we have $a^{-1}bd = dn_2$ for some $n_2 \in N$. Then $(ac)^{-1}bd = (c^{-1}a^{-1})(bd) = c^{-1}(a^{-1}b)d = (c^{-1}d)n_2 \in N$, so $ac \mathrel{R} bd$. Hence R is a congruence relation on G.

(b) Suppose that $x \in N$. Then $x^{-1}e = x^{-1} \in N$ since N is a subgroup, so $x \mathrel{R} e$ and therefore $x \in [e]$. Thus $N \subseteq [e]$. Conversely, if $x \in [e]$, then $x \mathrel{R} e$, so $x^{-1}e = x^{-1} \in N$. Then $x \in N$ and $[e] \subseteq N$. Hence $N = [e]$. ▼

We see, thanks to Theorems 3 and 4, that if G is any group, then the equivalence classes with respect to a congruence relation on G are always the cosets of some normal subgroup of G. Conversely, the cosets of any normal subgroup of G are just the equivalence classes with respect to some congruence relation on G. We may now, therefore, translate Corollary 1(b) as follows: Let f be a homomorphism from a group $(G, *)$ onto a group $(G', *')$, and let the **kernel** of f, $\ker(f)$, be defined by

$$\ker(f) = \{a \in G \mid f(a) = e'\}.$$

Then

(a) $\ker(f)$ is a normal subgroup of G.

(b) The quotient group $G/\ker(f)$ is isomorphic to G'.

This follows from Corollary 1 and Theorem 3, since if R is the congruence relation on G given by

$$a \, R \, b \quad \text{if and only if} \quad f(a) = f(b),$$

then it is easy to show that $\ker(f) = [e]$.

EXAMPLE 6   Consider the homomorphism f from Z onto Z_n defined by

$$f(m) = [r],$$

where r is the remainder when m is divided by n. (See Example 15 of Section 9.4.) Find $\ker(f)$.

Solution   An integer m in Z belongs to $\ker(f)$ if and only if $f(m) = [0]$, that is, if and only if m is a multiple of n. Hence $\ker(f) = nZ$. ∎

9.5 Exercises

1. Write the multiplication table for the group $Z_2 \times Z_3$.

2. Prove that if G and G' are Abelian groups, then $G \times G'$ is an Abelian group.

3. Let G_1 and G_2 be groups. Prove that $G_1 \times G_2$ and $G_2 \times G_1$ are isomorphic.

4. Let G_1 and G_2 be groups. Show that the function $f : G_1 \times G_2 \to G_1$ defined by $f(a, b) = a$, for $a \in G_1$ and $b \in G_2$, is a homomorphism.

5. Determine the operational table of the quotient group $Z/3Z$, where Z has operation $+$.

6. Let Z be the group of integers under the operation of addition. Prove that the function $f : Z \times Z \to Z$ defined by $f(a, b) = a + b$ is a homomorphism.

7. Let $G = Z_4$. Determine all the left cosets of $H = \{[0]\}$ in G.

8. Let $G = Z_4$. Determine all the left cosets of $H = \{[0], [2]\}$ in G.

9. Let $G = Z_4$. Determine all the left cosets of $H = \{[0], [1], [2], [3]\}$ in G.

10. Let $G = S_3$. Determine all the left cosets of $H = \{f_1, g_1\}$ in G.

11. Let $G = S_3$. Determine all the left cosets of $H = \{f_1, g_3\}$ in G.

12. Let $G = S_3$. Determine all the left cosets of $H = \{f_1, f_2, f_3\}$ in G.

13. Let $G = S_3$. Determine all the left cosets of $H = \{f_1\}$ in G.

14. Let $G = S_3$. Determine all the left cosets of $H = \{f_1, f_2, f_3, g_1, g_2, g_3\}$ in G.

15. Let $G = Z_8$. Determine all the left cosets of $H = \{[0], [4]\}$ in G.

16. Let $G = Z_8$. Determine all the left cosets of $H = \{[0], [2], [4], [6]\}$ in G.

17. Let Z be the group of integers under the operation of addition, and let $G = Z \times Z$. Consider the subgroup $H = \{(x, y) \mid x = y\}$ of G. Describe the left cosets of H in G.

18. Let N be a subgroup of a group G, and let $a \in G$. Define

$$a^{-1}Na = \{a^{-1}na \mid n \in N\}.$$

Prove that N is a normal subgroup of G if and only if $a^{-1}Na = N$ for all $a \in G$.

19. Let N be a subgroup of group G. Prove that N is a normal subgroup of G if and only if $a^{-1}Na \subseteq N$ for all $a \in G$.

20. Find all the normal subgroups of S_3.

21. Find all the normal subgroups of D_4. (See Exercise 17 of Section 9.4.)

22. Let G be a group, and let $H = \{x \mid x \in G \text{ and } xa = ax \text{ for all } a \in G\}$. Show that H is a normal subgroup of G.

23. Let H be a subgroup of a group G. Prove that every left coset aH of H has the same number of elements as H by showing that the function $f_a : H \to aH$ defined by $f_a(h) = ah$, for $h \in H$, is one to one and onto.

24. Let H and K be normal subgroups of G. Show that $H \cap K$ is a normal subgroup of G.

25. Let G be a group and H a subgroup of G. Let S be the set of all left cosets of H in G, and let T be the set of all right cosets of H in G. Prove that the function $f : S \to T$ defined by $f(aH) = Ha^{-1}$ is one to one and onto.

26. Let G_1 and G_2 be groups. Let $f : G_1 \times G_2 \to G_2$ be the homomorphism from $G_1 \times G_2$ onto G_2 given by $f((g_1, g_2)) = g_2$. Compute $\ker(f)$.

27. Let f be a homomorphism from a group G_1 onto a group G_2, and suppose that G_2 is Abelian. Show that $\ker(f)$ contains all elements of G_1 of the form $aba^{-1}b^{-1}$, where a and b are arbitrary in G_1.

28. Let G be an Abelian group and N a subgroup of G. Prove that G/N is an Abelian group.

29. Let H be a subgroup of the finite group G and suppose that there are only two left cosets of H in G. Prove that H is a normal subgroup of G.

30. Let H and N be subgroups of the group G. Prove that if N is a normal subgroup of G, then $H \cap N$ is a normal subgroup of H.

31. Let $f : G \to G'$ be a group homomorphism. Prove that f is one to one if and only if $\ker(f) = \{e\}$.

TIPS FOR PROOFS

The proofs in this chapter are mostly simple direct proofs, in part because we have introduced several new mathematical structures (semigroup, monoids, groups, Abelian groups). With a new structure we first explore the simple consequences of the definitions; for example, Theorem 1, Section 9.2. However, proofs of uniqueness are frequently indirect as in Theorems 1 and 4 in Section 9.4.

The idea of a substructure appears several times in this chapter. In general, to prove that a subset forms a substructure of a mathematical structure, we show that the subset together with the operation(s) satisfy the definition of this type of structure. But any global property such as associativity is inherited by the subset so we need only check closure properties and properties involving special elements. Thus, to show that a subset is a subgroup, we check closure for the multiplication, that the identity belongs to the subset, and that the inverse of each element in the subset belongs to the subset.

Isomorphism is a powerful tool for proving statements, since, roughly speaking, establishing an isomorphism between two structures allows us to transfer knowledge about one structure to the other. This can be seen in Theorem 4, Section 9.2.

KEY IDEAS FOR REVIEW

- Binary operation on A: everywhere defined function $f : A \times A \to A$
- Commutative binary operation: $a * b = b * a$
- Associative binary operation: $a * (b * c) = (a * b) * c$
- Semigroup: nonempty set S together with an associative binary operation $*$ defined on S
- Monoid: semigroup that has an identity
- Subsemigroup $(T, *)$ of semigroup $(S, *)$: T is a nonempty subset of S and $a * b \in T$ whenever a and b are in T.
- Submonoid $(T, *)$ of monoid $(S, *)$: T is a nonempty subset of S, $e \in T$, and $a * b \in T$ whenever a and b are in T.

- Isomorphism: see page 327
- Homomorphism: see page 329
- Theorem: Let $(S, *)$ and $(T, *')$ be monoids with identities e and e', respectively, and suppose that $f : S \to T$ is an isomorphism. Then $f(e) = e'$.
- Theorem: If $(S, *)$ and $(T, *')$ are semigroups, then $(S \times T, *'')$ is a semigroup, where $*''$ is defined by

$$(s_1, t_1) *'' (s_2, t_2) = (s_1 * s_2, t_1 *' t_2).$$

- Congruence relation R on semigroup $(S, *)$: equivalence relation R such that $a \, R \, a'$ and $b \, R \, b'$ imply that $(a * b) \, R \, (a' * b')$
- Theorem: Let R be a congruence relation on the semigroup $(S, *)$. Define the operation $\circledast$ in S/R as follows:

$$[a] \circledast [b] = [a * b].$$

Then $(S/R, \circledast)$ is a semigroup.
- Quotient semigroup or factor semigroup S/R: see page 333
- Z_n: see page 334
- Theorem (fundamental homomorphism theorem): Let $f : S \to T$ be a homomorphism of the semigroup $(S, *)$ onto the semigroup $(T, *')$. Let R be the relation on S defined by $a \, R \, b$ if and only if $f(a) = f(b)$, for a and b in S. Then
  (a) R is a congruence relation.
  (b) T is isomorphic to S/R.
- Group $(G, *)$: monoid with identity e such that for every $a \in G$ there exists $a' \in G$ with the property that $a * a' = a' * a = e$.
- Theorem: Let G be a group, and let a, b, and c be elements of G. Then

(a) $ab = ac$ implies that $b = c$ (left cancellation property).
(b) $ba = ca$ implies that $b = c$ (right cancellation property).
- Theorem: Let G be a group, and let a and b be elements of G. Then
(a) $(a^{-1})^{-1} = a$.
(b) $(ab)^{-1} = b^{-1}a^{-1}$.
- Order of a group G: $|G|$, the number of elements in G
- S_n: the symmetric group on n letters
- Subgroup: see page 344
- Theorem: Let R be a congruence relation on the group $(G, *)$. Then the semigroup $(G/R, \circledast)$ is a group, where the operation $\circledast$ is defined in G/R by

$$[a] \circledast [b] = [a * b].$$

- Left coset aH of H in G determined by a: $\{ah \mid h \in H\}$
- Normal subgroup: subgroup H such that $aH = Ha$ for all a in G
- Theorem: Let R be a congruence relation on a group G, and let $H = [e]$, the equivalence class containing the identity. Then H is a normal subgroup of G and, for each $a \in G$, $[a] = aH = Ha$.
- Theorem: Let N be a normal subgroup of a group G, and let R be the following relation on G:

$$a \, R \, b \quad \text{if and only if} \quad a^{-1}b \in N.$$

Then
(a) R is a congruence relation on G.
(b) N is the equivalence class $[e]$ relative to R, where e is the identity of G.

CODING EXERCISES

For each of the following, write the requested program or subroutine in pseudocode (as described in Appendix A) or in a programming language that you know. Test your code either with a paper-and-pencil trace or with a computer run.

Let Z_n be as defined in Section 9.3.

1. Write a function SUM that takes two elements of Z_n, $[x]$ and $[y]$ and returns their sum $[x] \oplus [y]$. The user should be able to input a choice for n.

2. Let $H = \{[0], [2]\}$. Write a subroutine that computes the left cosets of H in Z_6.

3. Let $H = \{[0], [2], [4], [6]\}$. Write a subroutine that computes the right cosets of H in Z_8.

4. Write a program that given a finite operation table will determine if the operation satisfies the associative property.

5. Write a program that given a finite group G and a subgroup H determines if H is a normal subgroup of G.

CHAPTER 9 SELF-TEST

1. For each of the following, determine whether the description of $*$ is a valid definition of a binary operation on the given set.

   (a) On the set of 2×2 Boolean matrices, where $\mathbf{A} * \mathbf{B} = \left[(a_{ij} + b_{ij}) \pmod 2 \right]$

   (b) On the set of even integers, where $a * b = a + b$

   (c) On Z^+, where $a * b = 2^{ab}$

2. Complete the operational table so that $*$ is a commutative and idempotent binary operation.

$*$	a	b	c
a		c	
b			
c		b	

3. Let Q be the set of rational numbers and define $a * b = a + b - ab$

   (a) Is $(Q, *)$ a monoid? Justify your answer.

   (b) If $(Q, *)$ is a monoid, which elements of Q have an inverse?

4. Determine whether the set together with the operation is a semigroup, a monoid, or neither for each of the pairs given in Exercise 1.

5. Let $A = \{0, 1\}$, and consider the semigroup $(A^*, \cdot)$, where $\cdot$ is the operation of catenation. Define a relation R on this semigroup by $\alpha \, R \, \beta$ if and only if α and β have the same length. Prove that R is a congruence relation.

6. Let G be a group and define $f: G \to G$ by $f(a) = a^{-1}$. Is f a homomorphism? Justify your answer.

7. Let G be the group whose multiplication table is given below and let H be the subgroup $\{c, d, e\}$.

$*$	e	a	b	c	d	f
e	e	a	b	c	d	f
a	a	e	c	b	f	d
b	b	d	e	f	a	c
c	c	f	a	d	e	b
d	d	b	f	e	c	a
f	f	c	d	a	b	e

   Find the right cosets of H in G.

8. Let $f: G_1 \to G_2$ be a homomorphism from the group $(G_1, *_1)$ onto the group $(G_2, *_2)$. If N is a normal subgroup of G_1, show that its image $f(N)$ is a normal subgroup of G_2.

9. Let G be a group with identity e. Show that if $x^2 = x$ for some x in G, then $x = e$.

10. Let G be the group of integers under the operation of addition and G' be the group of all even integers under the operation of addition. Show that the function $f: G \to G'$ defined by $f(a) = 2a$ is an isomorphism.

11. Let $H_1, H_2, \ldots, H_k$ be subgroups of a group G. Prove that $\bigcap_{i=1}^{k} H_i$ is also a subgroup of G.

12. Let G be the group of all nonzero real numbers under the operation of multiplication and consider the subgroup $H = \{3^n \mid n \in Z\}$ of G. Determine all the left cosets of H in G.

10

LANGUAGES AND FINITE-STATE MACHINES

Prerequisites: Chapters 7 and 9

In this chapter we introduce to the study of formal languages and develop another mathematical structure, phrase structure grammars, a simple device for the construction of useful formal languages. We also examine several popular methods for representing these grammars.

10.1 LANGUAGES

In Section 1.3, we considered the set S^* consisting of all finite strings of elements from the set S. There are many possible interpretations of the elements of S^*, depending on the nature of S. If we think of S as a set of "words," then S^* may be regarded as the collection of all possible "sentences" formed from words in S. Of course, such "sentences" do not have to be meaningful or even sensibly constructed. We may think of a language as a complete specification, at least in principle, of three things. First, there must be a set S consisting of all "words" that are to be regarded as being part of the language. Second, a subset of S^* must be designated as the set of "properly constructed sentences" in the language. The meaning of this term will depend very much on the language being constructed. Finally, it must be determined which of the properly constructed sentences have meaning and what the meaning is.

Suppose, for example, that S consists of all English words. The specification of a properly constructed sentence involves the complete rules of English grammar; the meaning of a sentence is determined by this construction and by the meaning of the words. The sentence

"Going to the store John George to sing."

is a string in S^*, but is not a properly constructed sentence. The arrangement of nouns and verb phrases is illegal. On the other hand, the sentence

"Noiseless blue sounds sit cross-legged under the mountaintop."

is properly constructed, but completely meaningless.

For another example, S may consist of the integers, the symbols $+$, $-$, $\times$, and $\div$, and left and right parentheses. We will obtain a language if we designate

357

as proper those strings in S^* that represent unambiguously parenthesized algebraic expressions. Thus

$$((3-2)+(4 \times 7)) \div 9 \quad \text{and} \quad (7-(8-(9-10)))$$

are properly constructed "sentences" in this language. On the other hand, $(2-3))+4$, $4-3-2$, and $)2+(3-) \times 4$ are not properly constructed. The first has too many parentheses, the second has too few (we do not know which subtraction to perform first), and the third has parentheses and numbers completely out of place. All properly constructed expressions have meaning except those involving division by zero. The meaning of an expression is the rational number it represents. Thus the meaning of $((2-1) \div 3) + (4 \times 6)$ is 73/3, while $2+(3 \div 0)$ and $(4+2)-(0 \div 0)$ are not meaningful.

The specification of the proper construction of sentences is called the **syntax** of a language. The specification of the meaning of sentences is called the **semantics** of a language. Among the languages that are of fundamental importance in computer science are the programming languages. These include BASIC, FORTRAN, JAVA, PASCAL, C^{++}, LISP, ADA, FORTH, and many other general and special-purpose languages. When they are taught to program in some programming language, people are actually taught the syntax of the language. In a compiled language such as FORTRAN, most mistakes in syntax are detected by the compiler, and appropriate error messages are generated. The semantics of a programming language forms a much more difficult and advanced topic of study. The meaning of a line of programming is taken to be the entire sequence of events that takes place inside the computer as a result of executing or interpreting that line.

We will not deal with semantics at all. We will study the syntax of a class of languages called phrase structure grammars. Although these are not nearly complex enough to include natural languages such as English, they are general enough to encompass many languages of importance in computer science. This includes most aspects of programming languages, although the complete specification of some higher-level programming languages exceeds the scope of these grammars. On the other hand, phrase structure grammars are simple enough to be studied precisely, since the syntax is determined by substitution rules. The grammars that will occupy most of our attention lead to interesting examples of labeled trees.

Grammars

A **phrase structure grammar** G is defined to be a 4-tuple $(V, S, v_0, \mapsto)$, where V is a finite set, S is a subset of V, $v_0 \in V - S$, and $\mapsto$ is a finite relation on V^*. The idea here is that S is, as discussed previously, the set of all allowed "words" in the language, and V consists of S together with some other symbols. The element v_0 of V is a starting point for the substitutions, which will shortly be discussed. Finally, the relation $\mapsto$ on V^* specifies allowable replacements, in the sense that, if $w \mapsto w'$, we may replace w by w' whenever the string w occurs, either alone or as a substring of some other string. Traditionally, the statement $w \mapsto w'$ is called a **production** of G. Then w and w' are termed the **left** and **right** sides of the production, respectively. We assume that no production of G has the empty string Λ as its left side. We will call $\mapsto$ the **production relation** of G.

With these ingredients, we can introduce a substitution relation, denoted by $\Rightarrow$, on V^*. We let $x \Rightarrow y$ mean that $x = l \cdot w \cdot r$, $y = l \cdot w' \cdot r$, and $w \mapsto w'$, where l and r are completely arbitrary strings in V^*. In other words, $x \Rightarrow y$ means that y results from x by using one of the allowed productions to replace part or all of x. The relation $\Rightarrow$ is usually called **direct derivability**. Finally, we let $\Rightarrow^\infty$ be the transitive closure of $\Rightarrow$ (see Section 4.3), and we say that a string w in S^* is a syntactically correct sentence if and only if $v_0 \Rightarrow^\infty w$. In more detail, this says that a string w is a properly constructed sentence if w is in S^*, not just in V^*, and if we can get from v_0 to w by making a finite number of substitutions. This may seem complicated, but it is really a simple idea, as the following examples will show.

If $G = (V, S, v_0, \mapsto)$ is a phrase structure grammar, we will call S the set of **terminal symbols** and $N = V - S$ the set of **nonterminal symbols**. Note that $V = S \cup N$.

The reader should be warned that other texts have slight variations of the definitions and notations that we have used for phrase structure grammars.

EXAMPLE 1 Let $S = \{$John, Jill, drives, jogs, carelessly, rapidly, frequently$\}$, $N = \{$sentence, noun, verphrase, verb, adverb$\}$, and let $V = S \cup N$. Let $v_0 = $ sentence, and suppose that the relation $\mapsto$ on V^* is described by

$$
\begin{aligned}
\text{sentence} &\mapsto \text{noun verbphrase} \\
\text{noun} &\mapsto \text{John} \\
\text{noun} &\mapsto \text{Jill} \\
\text{verbphrase} &\mapsto \text{verb adverb} \\
\text{verb} &\mapsto \text{drives} \\
\text{verb} &\mapsto \text{jogs} \\
\text{adverb} &\mapsto \text{carelessly} \\
\text{adverb} &\mapsto \text{rapidly} \\
\text{adverb} &\mapsto \text{frequently}
\end{aligned}
$$

The set S contains all the allowed words in the language; N consists of words that describe parts of sentences but that are not actually contained in the language.

We claim that the sentence "Jill drives frequently," which we will denote by w, is an allowable or syntactically correct sentence in this language. To prove this, we consider the following sequence of strings in V^*.

sentence		
noun	verbphrase	
Jill	verbphrase	
Jill	verb	adverb
Jill	drives	adverb
Jill	drives	frequently

Now each of these strings follows from the preceding one by using a production to make a partial or complete substitution. In other words, each string is related to the following string by the relation $\Rightarrow$, so sentence $\Rightarrow^\infty w$. By definition then, w is syntactically correct since, for this example, v_0 is sentence. In phrase structure grammars, correct syntax simply refers to the process by which a sentence is formed, nothing else. ∎

It should be noted that the sequence of substitutions that produces a valid sentence, a sequence that will be called a **derivation** of the sentence, is not unique.

The following derivation produces the sentence w of Example 1 but is not identical with the derivation given there.

sentence		
noun	verbphrase	
noun	verb	adverb
noun	verb	frequently
noun	drives	frequently
Jill	drives	frequently

The set of all properly constructed sentences that can be produced using a grammar G is called the **language** of G and is denoted by $L(G)$. The language of the grammar given in Example 1 is a somewhat simple-minded sublanguage of English, and it contains exactly 12 sentences. The reader can verify that "John jogs carelessly" is in the language $L(G)$ of this grammar, while "Jill frequently jogs" is not in $L(G)$.

It is also true that many different phrase structure grammars may produce the same language; that is, they have exactly the same set of syntactically correct sentences. Thus a grammar cannot be reconstructed from its language. In Section 10.2 we will give examples in which different grammars are used to construct the same language.

Example 1 illustrates the process of derivation of a sentence in a phrase structure grammar. Another method that may sometimes be used to show the derivation process is the construction of a **derivation tree** for the sentence. The starting symbol, v_0, is taken as the label for the root of this tree. The level-1 vertices correspond to and are labeled in order by the various words involved in the first substitution for v_0. Then the offspring of each vertex, at every succeeding level, are labeled by the various words (if any) that are substituted for that vertex the next time it is subjected to substitution. Consider, for example, the first derivation of sentence w in Example 1. Its derivation tree begins with "sentence," and the next-level vertices correspond to "noun" and "verbphrase" since the first substitution replaces the word "sentence" with the string "noun verbphrase." This part of the tree is shown in Figure 10.1(a). Next, we substitute "Jill" for "noun," and the tree becomes as shown in Figure 10.1(b). The next two substitutions, "verb adverb" for "verbphrase" and "drives" for "verb," extend the tree as shown in Figures 10.1(c) and (d). Finally, the tree is completed with the substitution of "frequently" for "adverb." The finished derivation tree is shown in Figure 10.1(e).

The second derivation sequence, given in Example 1 for the sentence w, yields a derivation tree in the stages shown in Figure 10.2 on page 362. Notice that the same tree results in both figures. Thus these two derivations yield the same tree, and the differing orders of substitution simply create the tree in different ways. The sentence being derived labels the leaves of the resulting tree.

EXAMPLE 2   Let $V = \{v_0, w, a, b, c\}$, $S = \{a, b, c\}$, and let $\mapsto$ be the relation on V^* given by

$$1.\ v_0 \mapsto aw. \qquad 2.\ w \mapsto bbw. \qquad 3.\ w \mapsto c.$$

Consider the phrase structure grammar $G = (V, S, v_0, \mapsto)$. To derive a sentence of $L(G)$, it is necessary to perform successive substitutions, using (1), (2),

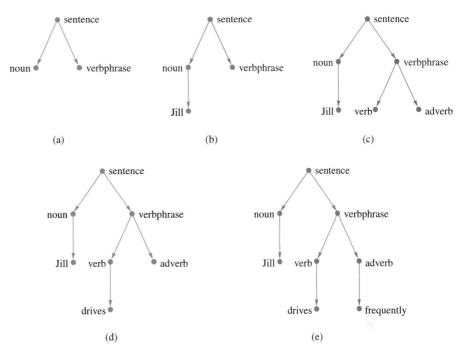

Figure 10.1

and (3), until all symbols are eliminated other than the terminal symbols a, b, and c. Since we begin with the symbol v_0, we must first use production (1), or we could never eliminate v_0. This first substitution results in the string aw. We may now use (2) or (3) to substitute for w. If we use production (2), the result will contain a w. Thus one application of (2) to aw produces the string ab^2w (here b^n means n consecutive b's). If we use (2) again, we will have the string ab^4w. We may use production (2) any number of times, but we will finally have to use production 3 to eliminate the symbol w. Once we use (3), only terminal symbols remain, so the process ends. We may summarize this analysis by saying that $L(G)$ is the subset of S^* corresponding to the regular expression $a(bb)^*c$ (see Section 1.3). Thus the word ab^6c is in the language of G, and its derivation tree is shown in Figure 10.3. Note that, unlike the tree of Example 1, the derivation tree for ab^6c is not a binary tree. ■

EXAMPLE 3   Let $V = \{v_0, w, a, b, c\}$, $S = \{a, b, c\}$, and let $\mapsto$ be a relation on V^* given by

1. $v_0 \mapsto av_0b$.   2. $v_0b \mapsto bw$.   3. $abw \mapsto c$.

Let $G = (V, S, v_0, \mapsto)$ be the corresponding phrase structure grammar. As we did in Example 2, we determine the form of allowable sentences in $L(G)$.

Since we must begin with the symbol v_0 alone, we must use production (1) first. We may continue to use (1) any number of times, but we must eventually use production (2) to eliminate v_0. Repeated use of (1) will result in a string of the form $a^n v_0 b^n$; that is, there are equal numbers of a's and b's. When (2) is used repeatedly, the result is a string of the form $a^m(abw)b^m$ with $m \geq 0$. At this point the only

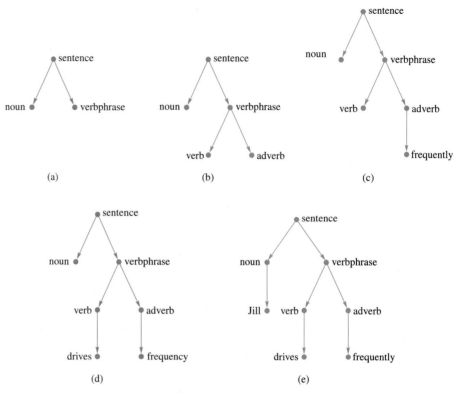

Figure 10.2

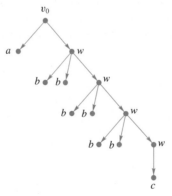

Figure 10.3

production that can be used is (3), and we must use it to remove the nonterminal symbol w. The use of (3) finishes the substitution process and produces a string in S^*. Thus the allowable sentences $L(G)$ of the grammar G all have the form $w = a^n c b^n$, where $n \geq 0$. In this case it can be shown that $L(G)$ does not correspond to a regular expression over S. ∎

Another interesting feature of the grammar in Example 3 is that the derivations of the sentences cannot be expressed as trees. Our construction of derivation trees works only when the left-hand sides of all productions used consist of single, nonterminal symbols. The left-hand sides of the productions in Example 3 do not have this simple form. Although it is possible to construct a graphical representation of these derivations, the resulting digraph would not be a tree. Many other problems can arise if no restrictions are placed on the productions. For this reason, a classification of phrase structure grammars has been devised.

Let $G = (V, S, v_0, \mapsto)$ be a phrase structure grammar. Then we say that G is

TYPE 0:   if no restrictions are placed on the productions of G

TYPE 1:   if for any production $w_1 \mapsto w_2$, the length of w_1 is less than or equal to the length of w_2 (where the **length** of a string is the number of words in that string)

TYPE 2:   if the left-hand side of each production is a single, nonterminal symbol and the right-hand side consists of one or more symbols

TYPE 3:   if the left-hand side of each production is a single, nonterminal symbol and the right-hand side has one or more symbols, including at most one nonterminal symbol, which must be at the extreme right of the string

In each of the preceding types, we permit the inclusion of the trivial production $v_0 \mapsto \Lambda$, where Λ represents the empty string. This is an exception to the defining rule for types 1, 2, and 3, but it is included so that the empty string can be made part of the language. This avoids constant consideration of unimportant special cases.

It follows from the definition that each type of grammar is a special case of the type preceding it. Example 1 is a type 2 grammar, Example 2 is type 3, and Example 3 is type 0. Grammars of types 0 or 1 are quite difficult to study and little is known about them. They include many pathological examples that are of no known practical use. We will restrict further consideration of grammars to types 2 and 3. These types have derivation trees for the sentences of their languages, and they are sufficiently complex to describe many aspects of actual programming languages. Type 2 grammars are sometimes called **context-free grammars**, since the symbols on the left of the productions are substituted for wherever they occur. On the other hand, a production of the type $l \cdot w \cdot r \mapsto l \cdot w' \cdot r$ (which could not occur in a type 2 grammar) is called **context sensitive**, since w' is substituted for w only in the context where it is surrounded by the strings l and r. Type 3 grammars have a very close relationship with finite-state machines. (See Section 10.3.) Type 3 grammars are also called **regular grammars**.

A language will be called **type 2** or **type 3** if there is a grammar of type 2 or type 3 that produces it. This concept can cause problems. Even if a language is produced by a non-type-2 grammar, it is possible that some type 2 grammar also produces this same language. In this case, the language is type 2. The same situation may arise in the case of type 3 grammars.

The process we have considered in this section, namely deriving a sentence within a grammar, has a converse process. The converse process involves taking a sentence and verifying that it is syntactically correct in some grammar G by constructing a derivation tree that will produce it. This process is called **parsing** the sentence, and the resulting derivation tree is often called the **parse tree** of the sentence. Parsing is of fundamental importance for compilers and other forms of language translation. A sentence in one language is parsed to show its structure, and a tree is constructed. The tree is then searched and, at each step, corresponding sentences are generated in another language. In this way a C^{++} program, for example, is compiled into a machine-language program. The contents of this section and the next two sections are essential to the compiling process, but the complete details must be left to a more advanced course.

10.1 Exercises

In Exercises 1 through 7, a grammar G is specified. In each case describe precisely the language, $L(G)$, produced by this grammar; that is, describe all syntactically correct "sentences."

1. $G = (V, S, v_0, \mapsto)$
   $V = \{v_0, v_1, x, y, z\}, \ S = \{x, y, z\}$
   $\mapsto : \ v_0 \mapsto xv_0$
   $\quad\quad v_0 \mapsto yv_1$
   $\quad\quad v_1 \mapsto yv_1$
   $\quad\quad v_1 \mapsto z$

2. $G = (V, S, v_0, \mapsto)$
   $V = \{v_0, a\}, \ S = \{a\}$
   $\mapsto : \ v_0 \mapsto aav_0$
   $\quad\quad v_0 \mapsto aa$

3. $G = (V, S, v_0, \mapsto)$
   $V = \{v_0, a, b\}, \ S = \{a, b\}$
   $\mapsto : \ v_0 \mapsto aav_0$
   $\quad\quad v_0 \mapsto a$
   $\quad\quad v_0 \mapsto b$

4. $G = (V, S, v_0, \mapsto)$
   $V = \{v_0, x, y, z\}, \ S = \{x, y, z\}$
   $\mapsto : \ v_0 \mapsto xv_0$
   $\quad\quad v_0 \mapsto yv_0$
   $\quad\quad v_0 \mapsto z$

5. $G = (V, S, v_0, \mapsto)$
   $V = \{v_0, v_1, v_2, a, +, (,)\},$
   $S = \{(,), a, +\}$
   $\mapsto : \ v_0 \mapsto (v_0)$ (where left and right parentheses
   $\quad\quad\quad\quad$ are symbols from S)
   $\quad\quad v_0 \mapsto a + v_1$
   $\quad\quad v_1 \mapsto a + v_2$
   $\quad\quad v_2 \mapsto a + v_2$
   $\quad\quad v_2 \mapsto a$

6. $G = (V, S, v_0, \mapsto)$
   $V = \{v_0, v_1, a, b\}, \ S = \{a, b\}$
   $\mapsto : \ v_0 \mapsto av_1$
   $\quad\quad v_1 \mapsto bv_0$
   $\quad\quad v_1 \mapsto a$

7. $G = (V, S, v_0, \mapsto)$
   $V = \{v_0, v_1, v_2, x, y, z\}, \ S = \{x, y, z\}$
   $\mapsto : \ v_0 \mapsto v_0v_1$
   $\quad\quad v_0v_1 \mapsto v_2v_0$
   $\quad\quad v_2v_0 \mapsto xy$
   $\quad\quad v_2 \mapsto x$
   $\quad\quad v_1 \mapsto z$

8. For each grammar in Exercises 1 through 7, state whether the grammar is type 1, 2, or 3.

9. Let $G = (V, S, I, \mapsto)$, where

$$V = \{I, L, D, W, a, b, c, 0, 1, 2, 3, 4, 5, 6, 7, 8, 9\}$$
$$S = \{a, b, c, 0, 1, 2, 3, 4, 5, 6, 7, 8, 9\}$$

$\mapsto$ is given by

1. $I \mapsto L$	8. $L \mapsto b$
2. $I \mapsto LW$	9. $L \mapsto c$
3. $W \mapsto LW$	10. $D \mapsto 0$
4. $W \mapsto DW$	11. $D \mapsto 1$
5. $W \mapsto L$	$\vdots$
6. $W \mapsto D$	19. $D \mapsto 9$
7. $L \mapsto a$	

Which of the following statements are true for this grammar?

(a) $ab092 \in L(G)$

(b) $2a3b \in L(G)$

(c) $aaaa \in L(G)$

(d) $I \Rightarrow a$

(e) $I \Rightarrow^\infty ab$

(f) $DW \Rightarrow 2$

(g) $DW \Rightarrow^\infty 2$

(h) $W \Rightarrow^\infty 2abc$

(i) $W \Rightarrow^\infty ba2c$

10. Draw a derivation tree for $ab3$ in the grammar of Exercise 9.

11. If G is the grammar of Exercise 9, describe $L(G)$.

12. Let $G = (V, S, v_0, \mapsto)$ where
$$V = \{v_0, v_1, v_2, a, b, c\}, \quad S = \{a, b, c\}$$
$$\mapsto \ : \ v_0 \mapsto aav_0$$
$$v_0 \mapsto bv_1$$
$$v_1 \mapsto cv_2b$$
$$v_1 \mapsto cb$$
$$v_2 \mapsto bbv_2$$
$$v_2 \mapsto bb$$
State which of the following are in $L(G)$.

(a) *aabcb*

(b) *abbcb*

(c) *aaaabcbb*

(d) *aaaabcbbb*

(e) *abcbbbbb*

13. If G is the grammar of Exercise 12, describe $L(G)$.

14. Draw a derivation tree for the string a^8 in the grammar of Exercise 2.

15. Give two distinct derivations (sequences of substitutions that start at v_0) for the string $xyz \in L(G)$, where G is the grammar of Exercise 7.

16. Let G be the grammar of Exercise 5. Can you give two distinct derivations (see Exercise 15) for the string $((a + a + a))$?

17. Let G be the grammar of Exercise 9. Give two distinct derivations (see Exercise 15) of the string $a100$.

In Exercises 18 through 24, construct a phrase structure grammar G such that the language, L(G), of G is equal to the language L.

18. $L = \{a^n b^n \mid n \geq 1\}$

19. $L = \{$strings of 0's and 1's with an equal number $n \geq 0$ of 0's and 1's$\}$

20. $L = \{a^n b^m \mid n \geq 1, m \geq 1\}$

21. $L = \{a^n b^n \mid n \geq 3\}$

22. $L = \{a^n b^m \mid n \geq 1, m \geq 3\}$

23. $L = \{x^n y^m \mid n \geq 2, m \text{ nonnegative and even}\}$

24. $L = \{x^n y^m \mid n \text{ even}, m \text{ positive and odd}\}$

In Exercises 25 and 26, let $V = \{v_0, v_1, v_2, v_3\}$, $S = \{a\}$, and let $\mapsto$ be the relation on V^* given by

1. $v_0 \mapsto av_1$ 2. $v_1 \mapsto av_2$ 3. $v_2 \mapsto av_3$

4. $v_2 \mapsto a$ 5. $v_3 \mapsto av_1$

Let $G = (V, S, v_0, \mapsto)$.

25. Complete the following proof that each $(aaa)^*$, $n \geq 1$ belongs to $L(G)$.

Proof: Basis Step: For $n = 1$, aaa is produced by _____. Hence $aaa \in L(G)$.

Induction Step: We use $P(k)$: $(aaa)^k$ is in $L(G)$ to show $P(k+1)$: $(aaa)^{k+1}$ is in $L(G)$. If $(aaa)^k = a^{3k}$ is in $L(G)$, then we must have applied _____ to _____. Instead we may use production rule _____ to produce $a^{3k}av_3$. (Complete the derivation of aaa^{3k+3}.) Hence $(aaa)^{3(k+1)}$ is in $L(G)$.

26. Complete the following proof that any string in $L(G)$ is of the form $(aaa)^n$.

Proof: Clearly all strings in $L(G)$ are of the form a^i, $1 \leq i$, because _____. Let $P(n)$: If a^i is in $L(G)$ and $3n < i \leq 3(n + 1)$, then $i = 3(n + 1)$.

Basis Step: For $n = 0$, suppose a^i is in $L(G)$ and _____. It is easy to see that the smallest possible i with these production rules is 3. Hence $P(0)$ is true.

Induction Step: Let $k \geq 1$. We use $P(k)$: _____ to show $P(k + 1)$: _____. Suppose $3(k + 1) < i \leq 3(k + 2)$. We know that a^i must have been produced from _____ by production rule _____. But at the step just before that we must have used production rule _____ on $a^{i-2}v_1$. The string $a^{i-2}v_1$ could not have been produced using production rule 1, because _____, but was produced from _____. Since $3(k+1) < i \leq 3(k+2)$, we have $3k < i \leq 3(k+1)$. By $P(k)$, _____.

27. Let G_1 and G_2 be regular grammars with languages L_1 and L_2 respectively. Define a new language $L_3 = \{w_1 \cdot w_2 \mid w_1 \in L_1 \text{ and } w_2 \in L_2\}$. Describe how to create a regular grammar for L_3 from G_1 and G_2.

28. Complete the following proof that no regular grammar G can produce the language of Exercise 18.

Proof: Suppose that there is a regular grammar $G = (V, S, v_0, \mapsto)$ with n nonterminals. $L(G)$ must contain exactly the strings of the form $a^i b^i$. But since G is regular, in the derivation of $a^i b^i$ we must have _____ with some nonterminal v_j. But with only n nonterminals, there must be at least two strings ___(1)___ and ___(2)___ with the same nonterminals. (Why?) There is no way to guarantee that using the production rules on (1) and (2) will produce exactly _____b's from (1) and exactly _____b's from (2). Hence there is no such regular grammar G.

10.2   REPRESENTATIONS OF SPECIAL GRAMMARS AND LANGUAGES

BNF Notation

For type 2 grammars (which include type 3 grammars), there are some useful, alternative methods of displaying the productions. A commonly encountered alternative is called the **BNF notation** (for Backus–Naur form). We know that the left-hand sides of all productions in a type 2 grammar are single, nonterminal symbols. For any such symbol w, we combine all productions having w as the left-hand side. The symbol w remains on the left, and all right-hand sides associated with w are listed together, separated by the symbol |. The relational symbol $\mapsto$ is replaced by the symbol ::=. Finally, the nonterminal symbols, wherever they occur, are enclosed in pointed brackets ⟨ ⟩. This has the additional advantage that nonterminal symbols may be permitted to have embedded spaces. Thus ⟨word1 word2⟩ shows that the string between the brackets is to be treated as one "word," not as two words. That is, we may use the space as a convenient and legitimate "letter" in a word, as long as we use pointed brackets to delimit the words.

EXAMPLE 1   In BNF notation, the productions of Example 1 of Section 10.1 appear as follows.

$$
\begin{aligned}
\langle \text{sentence} \rangle &\ ::= \langle \text{noun} \rangle \langle \text{verbphrase} \rangle \\
\langle \text{noun} \rangle &\ ::= \text{John} \mid \text{Jill} \\
\langle \text{verbphrase} \rangle &\ ::= \langle \text{verb} \rangle \langle \text{adverb} \rangle \\
\langle \text{verb} \rangle &\ ::= \text{drives} \mid \text{jogs} \\
\langle \text{adverb} \rangle &\ ::= \text{carelessly} \mid \text{rapidly} \mid \text{frequently}
\end{aligned}
$$

■

EXAMPLE 2   In BNF notation, the productions of Example 2 of Section 10.1 appear as follows.

$$
\begin{aligned}
\langle v_0 \rangle &\ ::= a \langle w \rangle \\
\langle w \rangle &\ ::= bb \langle w \rangle \mid c
\end{aligned}
$$

■

Note that the left-hand side of a production may also appear in one of the strings on the right-hand side. Thus, in the second line of Example 2, $\langle w \rangle$ appears on the left, and it appears in the string $bb\langle w \rangle$ on the right. When this happens, we say that the corresponding production $w \mapsto bbw$ is **recursive**. If a recursive production has w as left-hand side, we will say that the production is **normal** if w appears only once on the right-hand side and is the rightmost symbol. Other nonterminal symbols may also appear on the right side. The recursive production $w \mapsto bbw$ given in Example 2 is normal. Note that any recursive production that appears in a type 3 (regular) grammar is normal, by the definition of type 3.

EXAMPLE 3   BNF notation is often used to specify actual programming languages. PASCAL and many other languages had their grammars given in BNF initially. In this example, we consider a small subset of PASCAL's grammar. This subset describes the syntax of decimal numbers and can be viewed as a mini-grammar whose corresponding language consists precisely of all properly formed decimal numbers.

Let $S = \{0, 1, 2, 3, 4, 5, 6, 7, 8, 9, .\}$. Let V be the union of S with the set

$N = \{$decimal-number, decimal-fraction, unsigned-integer, digit$\}$.

Then let G be the grammar with symbol sets V and S, with starting symbol "decimal-number" and with productions given in BNF form as follows:

1. ⟨decimal-number⟩ ::= ⟨unsigned-integer⟩ | ⟨decimal-fraction⟩ | ⟨unsigned-integer⟩⟨decimal-fraction⟩
2. ⟨decimal-fraction⟩ ::= .⟨unsigned-integer⟩
3. ⟨unsigned-integer⟩ ::= ⟨digit⟩ | ⟨digit⟩⟨unsigned-integer⟩
4. ⟨digit⟩ ::= 0 | 1 | 2 | 3 | 4 | 5 | 6 | 7 | 8 | 9

Figure 10.4 shows a derivation tree, in this grammar, for the decimal number 23.14. Notice that the BNF statement numbered 3 is recursive in the second part of its right-hand side. That is, the production "unsigned-integer ↦ digit unsigned-integer" is recursive, and it is also normal. In general, we know that many different grammars may produce the same language. If the line numbered 3 were replaced by the line

3′. ⟨unsigned-integer⟩ ::= ⟨digit⟩ | ⟨unsigned-integer⟩⟨digit⟩

we would have a different grammar that produced exactly the same language, namely the correctly formed decimal numbers. However, this grammar contains a production that is recursive but not normal. ∎

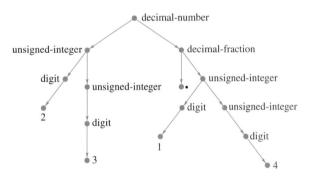

Figure 10.4

EXAMPLE 4

As in Example 3, we give a grammar that specifies a piece of several actual programming languages. In these languages, an identifier (a name for a variable, function, subroutine, and so on) must be composed of letters and digits and must begin with a letter. The following grammar, with productions given in BNF, has precisely these identifiers as its language.

$$G = (V, S, \text{identifier}, \mapsto)$$
$$N = \{\text{identifier, remaining, digit, letter}\}$$
$$S = \{a, b, c, \ldots, z, 0, 1, 2, 3, \ldots, 9\},$$
$$V = N \cup S$$

1. $\langle$identifier$\rangle$ ::= $\langle$letter$\rangle$ | $\langle$letter$\rangle$$\langle$remaining$\rangle$
2. $\langle$remaining$\rangle$ ::= $\langle$letter$\rangle$ | $\langle$digit$\rangle$ | $\langle$letter$\rangle$$\langle$remaining$\rangle$ | $\langle$digit$\rangle$$\langle$remaining$\rangle$
3. $\langle$letter$\rangle$ ::= a | b | $c \cdots$ | z
4. $\langle$digit$\rangle$ ::= 0 | 1 | 2 | 3 | 4 | 5 | 6 | 7 | 8 | 9

Again we see that the productions "remaining $\mapsto$ letter remaining" and "remaining $\mapsto$ digit remaining," occurring in BNF statement 2, are recursive and normal.   ■

Syntax Diagrams

A second alternative method for displaying the productions in some type 2 grammars is the **syntax diagram**. This is a pictorial display of the productions that allows the user to view the substitutions dynamically, that is, to view them as movement through the diagram. We will illustrate, in Figure 10.5, the diagrams that result from translating typical sets of productions, usually all the productions appearing on the right-hand side of some BNF statement.

A BNF statement that involves just a single production, such as $\langle w \rangle$ ::= $\langle w_1 \rangle \langle w_2 \rangle \langle w_3 \rangle$, will result in the diagram shown in Figure 10.5(a). The symbols (words) that make up the right-hand side of the production are drawn in sequence from left to right. The arrows indicate the direction in which to move to accomplish a substitution, while the label w indicates that we are substituting for the symbol w. Finally, the rectangles enclosing w_1, w_2, and w_3 denote the fact that these are non-terminal symbols. If terminal symbols were present, they would instead be enclosed in circles or ellipses. Figure 10.5(b) shows the situation when there are several productions with the same left-hand side. This figure is a syntax diagram translation of the following BNF specification:

$$\langle w \rangle ::= \langle w_1 \rangle \langle w_2 \rangle \mid \langle w_1 \rangle a \mid bc \langle w_2 \rangle$$

(where a, b, and c are terminal symbols). Here the diagram shows that when we substitute for w, by moving through the figure in the direction of the arrows, we may

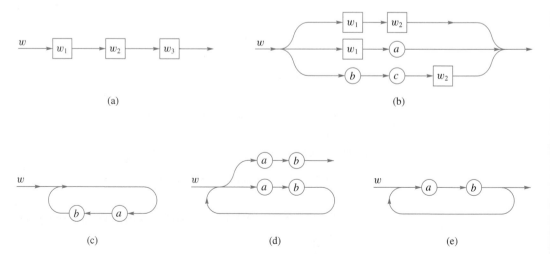

Figure 10.5

take any one of three paths. This corresponds to the three alternative substitutions for the symbol w. Now consider the following normal, recursive production, in BNF form:

$$\langle w \rangle ::= ab\langle w \rangle.$$

The syntax diagram for this production is shown in Figure 10.5(c). If we go through the loop once, we encounter a, then b, and we then return to the starting point designated by w. This represents the recursive substitution of abw for w. Several trips around the diagram represent several successive substitutions. Thus, if we traverse the diagram three times and return to the starting point, we see that w will be replaced by $abababw$ in three successive substitutions. This is typical of the way in which movement through a syntax diagram represents the substitution process.

The preceding remarks show how to construct a syntax diagram for a normal recursive production. Nonnormal recursive productions do not lead to the simple diagrams discussed, but we may sometimes replace nonnormal, recursive productions by normal recursive productions and obtain a grammar that produces the same language. Since recursive productions in regular grammars must be normal, syntax diagrams can always be used to represent regular grammars.

We also note that syntax diagrams for a language are by no means unique. They will not only change when different, equivalent productions are used, but they may be combined and simplified in a variety of ways. Consider the following BNF specification:

$$\langle w \rangle ::= ab \mid ab\langle w \rangle.$$

If we construct the syntax diagram for w using exactly the rules presented, we will obtain the diagram of Figure 10.5(d). This shows that we can "escape" from w, that is, eliminate w entirely, only by passing through the upper path. On the other hand, we may first traverse the lower loop any number of times. Thus any movement through the diagram that eventually results in the complete elimination of w by successive substitutions will produce a string of terminal symbols of the form $(ab)^n$, $n \geq 1$.

It is easily seen that in the simpler diagram of Figure 10.5(e), produced by combining the paths of Figure 10.5(d) in an obvious way, is an entirely equivalent syntax diagram. These types of simplifications are performed whenever possible.

EXAMPLE 5 The syntax diagrams of Figure 10.6(a) represent the BNF statements of Example 2, constructed with our original rules for drawing syntax diagrams. A slightly more aesthetic version is shown in Figure 10.6(b). ■

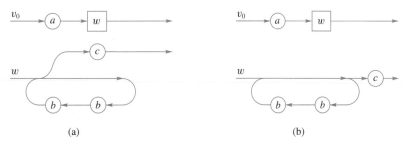

(a)   (b)

Figure 10.6

EXAMPLE 6   Consider the BNF statements 1, 2, 3, and 4 of Example 4. The direct translation into syntax diagrams is shown in Figure 10.7. In Figure 10.8 we combine the first two diagrams of Figure 10.7 and simplify the result. We thus eliminate the symbol "remaining," and we arrive at the customary syntax diagrams for identifiers.   ■

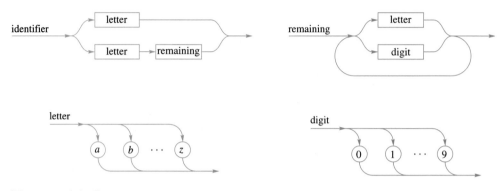

Figure 10.7

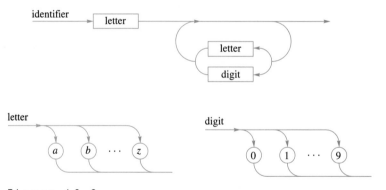

Figure 10.8

EXAMPLE 7   The productions of Example 3, for well-formed decimal numbers, are shown in syntax diagram form in Figure 10.9. Figure 10.10 shows the result of substituting the diagram for "unsigned-integer" into that for "decimal-number" and "decimal-fraction." In Figure 10.11 (see p. 372) the process of substitution is carried one step further. Although this is not usually done, it does illustrate the fact that we can be very flexible in designing syntax diagrams.   ■

If we were to take the extreme case and combine the diagrams of Figure 10.11 into one huge diagram, that diagram would contain only terminal symbols. In that case a valid "decimal-number" would be any string that resulted from moving through the diagram, recording each symbol encountered in the order in which it was encountered, and eventually exiting to the right.

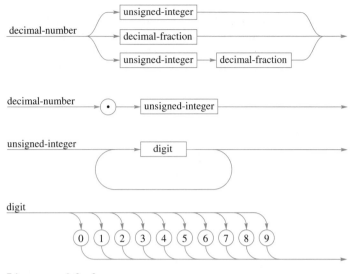

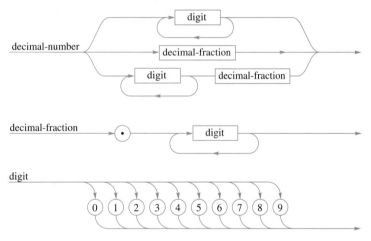

Figure 10.9

Figure 10.10

Regular Grammars and Regular Expressions

There is a close connection between the language of a regular grammar and a regular expression (see Section 1.3). We state the following theorem without proof.

Theorem 1 Let S be a finite set, and $L \subseteq S^*$. Then L is a regular set if and only if $L = L(G)$ for some regular grammar $G = (V, S, v_0, \mapsto)$. ●

Theorem 1 tells us that the language $L(G)$ of a regular grammar G must be the set corresponding to some regular expression over S, but it does not tell us how to find such a regular expression. If the relation $\mapsto$ of G is specified in BNF or syntax diagram form, we may compute the regular expression desired in a reasonably

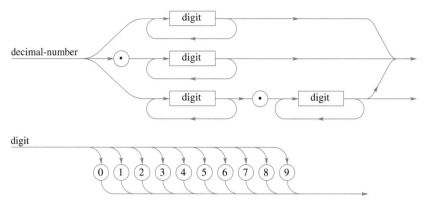

decimal-number

digit

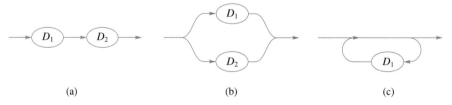

Figure 10.11

straightforward way. Suppose, for example, that $G = (V, S, v_0, \mapsto)$ and that $\mapsto$ is specified by a set of syntax diagrams. As we previously mentioned, it is possible to combine all the syntax diagrams into one large diagram that represents v_0 and involves only terminal symbols. We will call the result the **master diagram** of G. Consider the following rules of correspondence between regular expressions and parts, or segments, of the master diagram of G.

1. Terminal symbols of the diagram correspond to themselves, as regular expressions.
2. If a segment D of the diagram is composed of two segments D_1 and D_2 in sequence, as shown in Figure 10.12(a), and if D_1 and D_2 correspond to regular expressions α_1 and α_2, respectively, then D corresponds to $\alpha_1 \cdot \alpha_2$.
3. If a segment D of the diagram is composed of alternative segments D_1 and D_2, as shown in Figure 10.12(b), and if D_1 and D_2 correspond to regular expressions α_1 and α_2, respectively, then D corresponds to $\alpha_1 \vee \alpha_2$.
4. If a segment D of the diagram is a loop through a segment D_1, as shown in Figure 10.12(c), and if D_1 corresponds to the regular expression α, then D corresponds to α^*.

D_1 — D_2	D_1 / D_2	D_1
(a)	(b)	(c)

Figure 10.12

Rules 2 and 3 extend to any finite number of segments D_i of the diagram. Using the foregoing rules, we may construct the single expression that corresponds to the master diagram as a whole. This expression is the regular expression that corresponds to $L(G)$.

EXAMPLE 8 Consider the syntax diagram shown in Figure 10.13(a). It is composed of three alternative segments, the first corresponding to the expression a, the second to the expression b, and the third, a loop, corresponding to the expression c^*. Thus the entire diagram corresponds to the regular expression $a \lor b \lor c^*$.

The diagram shown in Figure 10.13(b) is composed of three sequential segments. The first segment is itself composed of two alternative subsegments, and it corresponds to the regular expression $a \lor b$. The second component segment of the diagram corresponds to the regular expression c, and the third component, a loop, corresponds to the regular expression d^*. Thus the overall diagram corresponds to the regular expression $(a \lor b)cd^*$.

Finally, consider the syntax diagram shown in Figure 10.13(c). This is one large loop through a segment that corresponds to the regular expression $a \lor bc$. Thus the entire diagram corresponds to the regular expression $(a \lor bc)^*$. ■

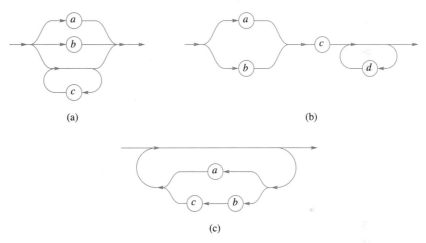

(a)   (b)

(c)

Figure 10.13

EXAMPLE 9 Consider the grammar G given in BNF in Example 2. Syntax diagrams for this grammar were discussed in Example 5 and shown in Figure 10.6(b). If we substitute the diagram representing w into the diagram that represents v_0, we get the master diagram for this grammar. This is easily visualized, and it shows that $L(G)$ corresponds to the regular expression $a(bb)^*c$, as we stated in Example 2 of Section 10.1. ■

EXAMPLE 10 Consider the grammar G of Examples 4 and 6. Then $L(G)$ is the set of legal identifiers, whose syntax diagrams are shown in Figure 10.8. In Figure 10.14 we show the master diagram that results from combining the diagrams of Figure 10.7. It follows that a regular expression corresponding to $L(G)$ is

$$(a \lor b \lor \cdots \lor z)(a \lor b \lor \cdots \lor z \lor 0 \lor 1 \lor \cdots \lor 9)^*.$$ ■

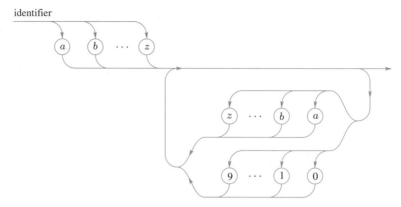

Figure 10.14

The type of diagram segments illustrated in Figure 10.12 can be combined to produce syntax diagrams for any regular grammar. Thus we may always proceed as just illustrated to find the corresponding regular expression. With practice one can learn to compute this expression directly from multiple syntax diagrams or BNF, thus avoiding the need to make a master diagram. In any event, complex cases may prove too cumbersome for hand analysis.

10.2 Exercises

In each of Exercises 1 through 5, we have referenced a grammar described in the exercises of a previous section. In each case, give the BNF *and corresponding syntax diagrams for the productions of the grammar.*

1. Exercise 1 of Section 10.1

2. Exercise 2 of Section 10.1

3. Exercise 6 of Section 10.1

4. Exercise 9 of Section 10.1

5. Exercise 12 of Section 10.1

6. Give the BNF for the productions of Exercise 3 of Section 10.1.

7. Give the BNF for the productions of Exercise 4 of Section 10.1.

8. Give the BNF for the productions of Exercise 5 of Section 10.1.

9. Give the BNF for the productions of Exercise 6 of Section 10.1.

10. Let $G = (V, S, v_0, \mapsto)$, where $V = \{v_0, v_1, 0, 1\}$, $S = \{0, 1\}$, and
    $\mapsto : v_0 \mapsto 0v_1$
    $\quad\quad v_1 \mapsto 11v_1$
    $\quad\quad v_1 \mapsto 010v_1$
    $\quad\quad v_1 \mapsto 1$

Give the BNF representation for the productions of G.

In Exercises 11 through 14, give the BNF *representation for the syntax diagram shown. The symbols a, b, c, and d are terminal symbols of some grammar. You may provide nonterminal symbols as needed (in addition to v_0), to use in the* BNF *productions. You may use several* BNF *statements if needed.*

11.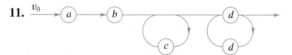

Figure 10.15

12.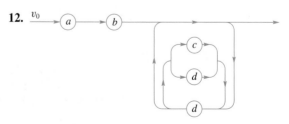

Figure 10.16

13.

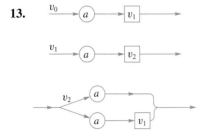

Figure 10.17

14.

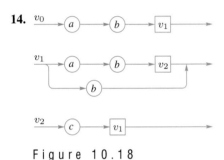

Figure 10.18

In each of Exercises 15 *through* 19, *we have referenced a*

grammar G, described in the exercises of a previous section. In each case find a regular expression that corresponds to the language $L(G)$.

15. Exercise 2 of Section 10.1

16. Exercise 3 of Section 10.1

17. Exercise 5 of Section 10.1

18. Exercise 6 of Section 10.1

19. Exercise 9 of Section 10.1

20. Find the regular expression that corresponds to the syntax diagram of Exercise 11.

21. Find the regular expression that corresponds to the syntax diagram of Exercise 12.

22. Find the regular expression that corresponds to the syntax diagram of Exercise 13.

23. Find the regular expression that corresponds to the syntax diagram of Exercise 14.

24. Find the regular expression that corresponds to $L(G)$ for G given in Exercise 10.

10.3 FINITE-STATE MACHINES

The question of whether a certain string belongs to the language of a given grammar is, in general, a difficult one to answer. In fact, in some cases it cannot be answered at all. Regular grammars and regular languages, though, have properties that enable us to construct a "recognizer" (or acceptor) for strings that belong to a given regular grammar. We lay the foundation for this construction in this section.

We think of a machine as a system that can accept **input**, possibly produce **output**, and have some sort of internal memory that can keep track of certain information about previous inputs. The complete internal condition of the machine and all of its memory, at any particular time, is said to constitute the **state** of the machine at that time. The state in which a machine finds itself at any instant summarizes its memory of past inputs and determines how it will react to subsequent input. When more input arrives, the given state of the machine determines (with the input) the next state to be occupied, and any output that may be produced. If the number of states is finite, the machine is a finite-state machine.

Suppose that we have a finite set $S = \{s_0, s_1, \ldots, s_n\}$, a finite set I, and for each $x \in I$, a function $f_x : S \to S$. Let $\mathcal{F} = \{f_x \mid x \in I\}$. The triple $(S, I, \mathcal{F})$ is called a **finite-state machine**, S is called the **state set** of the machine, and the elements of S are called **states**. The set I is called the **input set** of the machine. For any input $x \in I$, the function f_x describes the effect that this input has on the states of the machine and is called a **state transition function**. Thus, if the machine is in state s_i and input x occurs, the next state of the machine will be $f_x(s_i)$.

Since the next state $f_x(s_i)$ is uniquely determined by the pair (s_i, x), there is a function $F: S \times I \to S$ given by

$$F(s_i, x) = f_x(s_i).$$

The individual functions f_x can all be recovered from a knowledge of F. Many authors use a function $F: S \times I \to S$, instead of a set $\{f_x \mid x \in I\}$, to define a finite-state machine. The definitions are completely equivalent.

EXAMPLE 1   Let $S = \{s_0, s_1\}$ and $I = \{0, 1\}$. Define f_0 and f_1 as follows:

$$f_0(s_0) = s_0, \quad f_1(s_0) = s_1,$$
$$f_0(s_1) = s_1, \quad f_1(s_1) = s_0.$$

This finite-state machine has two states, s_0 and s_1, and accepts two possible inputs, 0 and 1. The input 0 leaves each state fixed, and the input 1 reverses states.   ■

We can think of the machine in Example 1 as a model for a circuit (or logical) device and visualize such a device as in Figure 10.19. The output signals will, at any given time, consist of two voltages, one higher than the other. Either line 1 will be at the higher voltage and line 2 at the lower, or the reverse. The first set of output conditions will be denoted s_0 and the second will be denoted s_1. An input pulse, represented by the symbol 1, will reverse output voltages. The symbol 0 represents the absence of an input pulse and so results in no change of output. This device is often called a **T flip-flop** and is a concrete realization of the machine in this example.

	0	1
s_0	s_0	s_1
s_1	s_1	s_0

Figure 10.19   Figure 10.20

We summarize this machine in Figure 10.20. The table shown there lists the states down the side and inputs across the top. The column under each input gives the values of the function corresponding to that input at each state shown on the left.

The arrangement illustrated in Figure 10.20 for summarizing the effect of inputs on states is called the **state transition table** of the finite-state machine. It can be used with any machine of reasonable size and is a convenient method of specifying the machine.

	a	b
s_0	s_0	s_1
s_1	s_2	s_0
s_2	s_1	s_2

Figure 10.21

EXAMPLE 2   Consider the state transition table shown in Figure 10.21. Here a and b are the possible inputs, and there are three states, s_0, s_1, and s_2. The table shows us that

$$f_a(s_0) = s_0, \quad f_a(s_1) = s_2, \quad f_a(s_2) = s_1$$

and

$$f_b(s_0) = s_1, \quad f_b(s_1) = s_0, \quad f_b(s_2) = s_2.$$ ■

If M is a finite-state machine with states S, inputs I, and state transition functions $\{f_x \mid x \in I\}$, we can determine a relation R_M on S in a natural way. If s_i, $s_j \in S$, we say that $s_i \, R_M \, s_j$ if there is an input x so that $f_x(s_i) = s_j$.

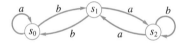

Figure 10.22

Thus $s_i \, R_M \, s_j$ means that if the machine is in state s_i, there is some input $x \in I$ that, if received next, will put the machine in state s_j. The relation R_M permits us to describe the machine M as a labeled digraph of the relation R_M on S, where each edge is labeled by the set of all inputs that cause the machine to change states as indicated by that edge.

EXAMPLE 3 Consider the machine of Example 2. Figure 10.22 shows the digraph of the relation R_M, with each edge labeled appropriately. Notice that the entire structure of M can be recovered from this digraph, since edges and their labels indicate where each input sends each state. ■

EXAMPLE 4 Consider the machine M whose table is shown in Figure 10.23(a). The digraph of R_M is then shown in Figure 10.23(b), with edges labeled appropriately. ■

	a	b	c
s_0	s_0	s_0	s_0
s_1	s_2	s_3	s_2
s_2	s_1	s_0	s_3
s_3	s_3	s_2	s_3

(a)

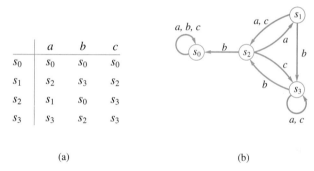

(b)

Figure 10.23

Note that an edge may be labeled by more than one input, since several inputs may cause the same change of state. The reader will observe that every input must be part of the label of exactly one edge out of each state. This is a general property that holds for the labeled digraphs of all finite-state machines. For brevity, we will refer to the labeled digraph of a machine M simply as the **digraph** of M.

It is possible to add a variety of extra features to a finite-state machine in order to increase the utility of the concept. A simple, yet very useful extension results in what is often called a **Moore machine**, or **recognition machine**, which is defined as a sequence $(S, I, \mathcal{F}, s_0, T)$, where $(S, I, \mathcal{F})$ constitutes a finite-state machine, $s_0 \in S$ and $T \subseteq S$. The state s_0 is called the **starting state** of M, and it will be used to represent the condition of the machine before it receives any input. The set T is called the set of **acceptance states** of M. These states will be used in Section 10.4 in connection with language recognition.

When the diagraph of a Moore machine is drawn, the acceptance states are indicated with two concentric circles, instead of one. No special notation will be used on these digraphs for the starting state, but unless otherwise specified, this state will be named s_0.

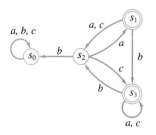

Figure 10.24

EXAMPLE 5

Let M be the Moore machine $(S, I, \mathcal{F}, s_0, T)$, where $(S, I, \mathcal{F})$ is the finite-state machine of Figure 10.23 and $T = \{s_1, s_3\}$. Figure 10.24 shows the digraph of M. ∎

Machine Congruence and Quotient Machines

Let $M = (S, I, \mathcal{F})$ be a finite-state machine, and suppose that R is an equivalence relation on S. We say that R is a **machine congruence** on M if, for any $s, t \in S$, $s\ R\ t$ implies that $f_x(s)\ R\ f_x(t)$ for all $x \in I$. In other words, R is a machine congruence if R-equivalent pairs of states are always taken into R-equivalent pairs of states by every input in I. If R is a machine congruence on $M = (S, I, \mathcal{F})$, we let $\overline{S} = S/R$ be the partition of S corresponding to R (see Section 4.5). Then $\overline{S} = \{[s] \mid s \in S\}$.

For any input $x \in I$, consider the relation $\overline{f}_x$ on $\overline{S}$ defined by

$$\overline{f}_x = \{([s], [f_x(s)])\}.$$

If $[s] = [t]$, then $s\ R\ t$; therefore, $f_x(s)\ R\ f_x(t)$, so $[f_x(s)] = [f_x(t)]$. This shows that the relation $\overline{f}_x$ is a function from $\overline{S}$ to $\overline{S}$, and we may write $\overline{f}_x([s]) = [f_x(s)]$ for all equivalence classes $[s]$ in $\overline{S}$. If we let $\overline{\mathcal{F}} = \{\overline{f}_x \mid x \in I\}$, then the triple $\overline{M} = (\overline{S}, I, \overline{\mathcal{F}})$ is a finite-state machine called the **quotient of M corresponding to R**. We will also denote $\overline{M}$ by M/R.

Generally, a quotient machine will be simpler than the original machine. We will show in Section 10.6 that it is often possible to find a simpler quotient machine that will replace the original machine for certain purposes.

EXAMPLE 6

Let M be the finite-state machine whose state transition table is shown in Figure 10.25. Then $S = \{s_0, s_1, s_2, s_3, s_4, s_5\}$. Let R be the equivalence relation on S whose matrix is

$$\mathbf{M}_R = \begin{bmatrix} 1 & 0 & 1 & 0 & 0 & 0 \\ 0 & 1 & 0 & 1 & 0 & 1 \\ 1 & 0 & 1 & 0 & 0 & 0 \\ 0 & 1 & 0 & 1 & 0 & 1 \\ 0 & 0 & 0 & 0 & 1 & 0 \\ 0 & 1 & 0 & 1 & 0 & 1 \end{bmatrix}.$$

Then we have $S/R = \{[s_0], [s_1], [s_4]\}$, where

$$[s_0] = \{s_0, s_2\} = [s_2],$$
$$[s_1] = \{s_1, s_3, s_5\} = [s_3] = [s_5],$$

and

$$[s_4] = \{s_4\}.$$

We check that R is a machine congruence. The state transition table in Figure 10.25 shows that f_a takes each element of $[s_i]$ to an element of $[s_i]$ for $i = 0, 1, 4$. Also, f_b takes each element of $[s_0]$ to an element of $[s_4]$, each element of $[s_1]$ to an element of $[s_0]$, and each element of $[s_4]$ to an element of $[s_1]$. These observations show that R is a machine congruence; the state transition table of the quotient machine M/R is shown in Figure 10.26. ∎

	a	b
s_0	s_0	s_4
s_1	s_1	s_0
s_2	s_2	s_4
s_3	s_5	s_2
s_4	s_4	s_3
s_5	s_3	s_2

Figure 10.25

	a	b
$[s_0]$	$[s_0]$	$[s_4]$
$[s_1]$	$[s_1]$	$[s_0]$
$[s_4]$	$[s_4]$	$[s_2]$

Figure 10.26

EXAMPLE 7

Let $I = \{0, 1\}$, $S = \{s_0, s_1, s_2, s_3, s_4, s_5, s_6, s_7\}$, and $M = (S, I, \mathcal{F})$, the finite-state machine whose digraph is shown in Figure 10.27.

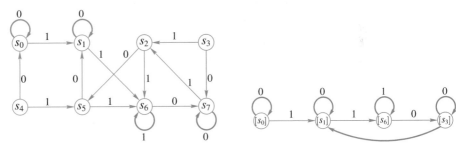

Figure 10.27   Figure 10.28

Suppose that R is the equivalence relation whose corresponding partition of S, S/R, is $\{\{s_0, s_4\}, \{s_1, s_2, s_5\}, \{s_6\}, \{s_3, s_7\}\}$. Then it is easily checked, from the digraph of Figure 10.27, that R is a machine congruence. To obtain the digraph of the quotient machine $\overline{M}$, draw a vertex for each equivalence class, $[s_0] = \{s_0, s_4\}$, $[s_1] = \{s_1, s_2, s_5\}$, $[s_6] = \{s_6\}$, $[s_3] = \{s_3, s_7\}$, and construct an edge from $[s_i]$ to $[s_j]$ if there is, in the original digraph, an edge from some vertex in $[s_i]$ to some vertex in $[s_j]$. In this case, the constructed edge is labeled with all inputs that take some vertex in $[s_i]$ to some vertex in $[s_j]$. Figure 10.28 shows the result. The procedure illustrated in this example works in general. ∎

If $M = (S, I, \mathcal{F}, s_0, T)$ is a Moore machine, and R is a machine congruence on M, then we may let $\overline{T} = \{[t] \mid t \in T\}$. Here, the sequence $\overline{M} = (\overline{S}, I, \overline{\mathcal{F}}, [s_0], \overline{T})$ is a Moore machine. In other words, we compute the usual quotient machine M/R; then we designate $[s_0]$ as a starting state, and let $\overline{T}$ be the set of equivalence classes of acceptance states. The resulting Moore machine $\overline{M}$, constructed this way, will be called the **quotient Moore machine** of M.

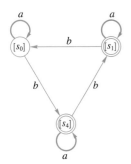

Figure 10.29

EXAMPLE 8

Consider the Moore machine $(S, I, \mathcal{F}, s_0, T)$, where $(S, I, \mathcal{F})$ is the finite-state machine of Example 6 and T is the set $\{s_1, s_3, s_4\}$. The digraph of the resulting quotient Moore machine is shown in Figure 10.29. ∎

10.3 Exercises

In Exercises 1 through 6, draw the digraph of the machine whose state transition table is shown. Remember to label the edges with the appropriate inputs.

1.

	0	1
s_0	s_0	s_1
s_1	s_1	s_2
s_2	s_2	s_0

2.

	0	1	2
s_0	s_1	s_0	s_2
s_1	s_0	s_0	s_1
s_2	s_2	s_0	s_2

3.

	a	b
s_0	s_1	s_0
s_1	s_2	s_0
s_2	s_2	s_0

4.

	a	b
s_0	s_1	s_0
s_1	s_2	s_1
s_2	s_3	s_2
s_3	s_3	s_3

5.

	a	b	c
s_0	s_0	s_1	s_2
s_1	s_2	s_1	s_1
s_2	s_1	s_1	s_2
s_3	s_2	s_0	s_1

6.

	0	1	2
s_0	s_0	s_2	s_1
s_1	s_1	s_3	s_2
s_2	s_2	s_1	s_3
s_3	s_3	s_3	s_2

In Exercises 7 through 12 (Figures 10.30 through 10.35), construct the state transition table of the finite-state machine whose digraph is shown.

7.

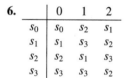

Figure 10.30

8.

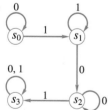

Figure 10.31

9.

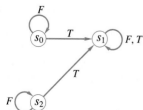

Figure 10.32

10.

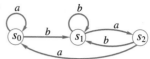

Figure 10.33

11.

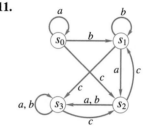

Figure 10.34

12.

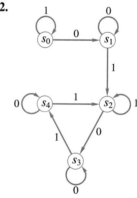

Figure 10.35

13. Let $M = (S, I, \mathcal{F})$ be a finite-state machine. Define a relation R on I as follows: $x_1 \, R \, x_2$ if and only if $f_{x_1}(s) = f_{x_2}(s)$ for every s in S. Show that R is an equivalence relation on I.

14. Let $(S, *)$ be a finite semigroup. Then we may consider the machine $(S, S, \mathcal{F})$, where $\mathcal{F} = \{f_x \mid x \in S\}$, and $f_x(y) = x * y$ for all $x, y \in S$. Thus we have a finite-state machine in which the state set and the input are the same. Define a relation R on S as follows: $x \, R \, y$ if and only if there is some $z \in S$ such that $f_z(x) = y$. Show that R is transitive.

15. Consider a finite group $(S, *)$ and let $(S, S, \mathcal{F})$ be the finite-state machine constructed in Exercise 14. Show that if R is the relation defined in Exercise 14, then R is an equivalence relation.

16. Let $I = \{0, 1\}$ and $S = \{a, b\}$. Construct all possible state transition tables of finite-state machines that have S as state set and I as input set.

17. Consider the machine whose state transition table is

	0	1
1	1	4
2	3	2
3	2	3
4	4	1

Here $S = \{1, 2, 3, 4\}$.

(a) Show that $R = \{(1, 1), (1, 4), (4, 1), (4, 4), (2, 2), (2, 3), (3, 2), (3, 3)\}$ is a machine congruence.

(b) Construct the state transition table for the corresponding quotient machine.

18. Consider the machine whose state transition table is

	a	b	c
s_0	s_0	s_1	s_3
s_1	s_0	s_1	s_2
s_2	s_2	s_3	s_0
s_3	s_2	s_2	s_0

Let $R = \{(s_0, s_1), (s_0, s_0), (s_1, s_1), (s_1, s_0), (s_3, s_2), (s_2, s_2), (s_3, s_3), (s_2, s_3)\}$.

(a) Show that R is a machine congruence.

(b) Construct the digraph for the corresponding quotient machine.

19. Consider the Moore machine whose digraph is shown in Figure 10.36. Show that the relation R on S whose matrix is

$$\mathbf{M}_R = \begin{bmatrix} 1 & 0 & 0 & 0 & 1 & 0 \\ 0 & 1 & 0 & 0 & 0 & 1 \\ 0 & 0 & 1 & 1 & 0 & 0 \\ 0 & 0 & 1 & 1 & 0 & 0 \\ 1 & 0 & 0 & 0 & 1 & 0 \\ 0 & 1 & 0 & 0 & 0 & 1 \end{bmatrix}$$

is a machine congruence. Draw the digraph of the corresponding quotient Moore machine.

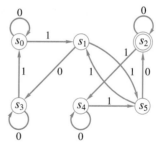

Figure 10.36

20. Consider the Moore machine whose digraph is shown in Figure 10.37. Show that the relation R on S whose matrix is

$$\mathbf{M}_R = \begin{bmatrix} 1 & 0 & 0 & 0 & 0 \\ 0 & 1 & 0 & 0 & 0 \\ 0 & 0 & 1 & 1 & 1 \\ 0 & 0 & 1 & 1 & 1 \\ 0 & 0 & 1 & 1 & 1 \end{bmatrix}$$

is a machine congruence. Draw the digraph of the corresponding quotient Moore machine.

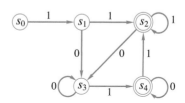

Figure 10.37

10.4 SEMIGROUPS, MACHINES, AND LANGUAGES

Let $M = (S, I, \mathcal{F})$ be a finite-state machine with state set $S = \{s_0, s_1, \ldots, s_n\}$, input set I, and state transition functions $\mathcal{F} = \{f_x \mid x \in I\}$.

We will associate with M two monoids, whose construction we recall from Section 9.2. First, there is the free monoid I^* on the input set I. This monoid consists of all finite sequences (or "strings" or "words") from I, with catenation as its binary operation. The identity is the empty string Λ. Second, we have the monoid S^S, which consists of all functions from S to S and which has function composition as its binary operation. The identity in S^S is the function 1_S defined by $1_S(s) = s$, for all s in S.

If $w = x_1 x_2 \cdots x_n \in I^*$, we let $f_w = f_{x_n} \circ f_{x_{n-1}} \circ \cdots \circ f_{x_1}$, the composition of the functions $f_{x_n}, f_{x_{n-1}}, \ldots, f_{x_1}$. Also we define f_Λ to be 1_S. In this way we assign an element f_w of S^S to each element w of I^*. If we think of each f_x as the effect of the input x on the states of the machine M, then f_w represents the combined effect of all the input letters in the word w, received in the sequence specified by w. We call f_w the **state transition function corresponding to w**.

EXAMPLE 1

Let $M = (S, I, \mathcal{F})$, where $S = \{s_0, s_1, s_2\}$, $I = \{0, 1\}$, and $\mathcal{F}$ is given by the following state transition table.

	0	1
s_0	s_0	s_1
s_1	s_2	s_2
s_2	s_1	s_0

Let $w = 011 \in I^*$. Then

$$f_w(s_0) = (f_1 \circ f_1 \circ f_0)(s_0) = f_1(f_1(f_0(s_0)))$$
$$= f_1(f_1(s_0)) = f_1(s_1) = s_2.$$

Similarly,

$$f_w(s_1) = f_1(f_1(f_0(s_1))) = f_1(f_1(s_2)) = f_1(s_0) = s_1$$

and

$$f_w(s_2) = f_1(f_1(f_0(s_2))) = f_1(f_1(s_1)) = f_1(s_2) = s_0. \qquad \blacksquare$$

EXAMPLE 2

Let us consider the same machine M as in Example 1 and examine the problem of computing f_w a little differently. In Example 1 we used the definition directly, and for a large machine we would program an algorithm to compute the values of f_w in just that way. However, if the machine is of moderate size, we may find another procedure to be preferable.

We begin by drawing the digraph of the machine M as shown in Figure 10.38. We may use this digraph to compute word transition functions by just following the edges corresponding to successive input letter transitions. Thus, to compute $f_w(s_0)$, we start at state s_0 and see that input 0 takes us to state s_0. The input 1 that follows takes us on to state s_1, and the final input of 1 takes us to s_2. Thus $f_w(s_0) = s_2$, as before.

Let us compute $f_{w'}$, where $w' = 01011$. The successive transitions of s_0 are

$$s_0 \xrightarrow{0} s_0 \xrightarrow{1} s_1 \xrightarrow{0} s_2 \xrightarrow{1} s_0 \xrightarrow{1} s_1,$$

so $f_{w'}(s_0) = s_1$. Similar displays show that $f_{w'}(s_1) = s_2$ and $f_{w'}(s_2) = s_0$. $\qquad \blacksquare$

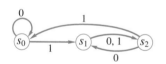

Figure 10.38

This method of interpreting word transition functions such as f_w and $f_{w'}$ is useful in designing machines that have word transitions possessing certain desired properties. This is a crucial step in the practical application of the theory and we will consider it in the next section.

Let $M = (S, I, \mathcal{F})$ be a finite-state machine. We define a function T from I^* to S^S. If w is a string in I^*, let $T(w) = f_w$ as defined previously. Then we have the following result.

Theorem 1   (a) If w_1 and w_2 are in I^*, then $T(w_1 \cdot w_2) = T(w_2) \circ T(w_1)$.
(b) If $\mathcal{M} = T(I^*)$, then $\mathcal{M}$ is a submonoid of S^S.

Proof

(a) Let $w_1 = x_1 x_2 \cdots x_k$ and $w_2 = y_1 y_2 \cdots y_m$ be two strings in I^*. Then

$$T(w_1 \cdot w_2) = T(x_1 x_2 \cdots x_k y_1 y_2 \cdots y_m)$$
$$= (f_{y_m} \circ f_{y_{m-1}} \circ \cdots \circ f_{y_1}) \circ (f_{x_k} \circ f_{x_{k-1}} \circ \cdots \circ f_{x_1})$$
$$= T(w_2) \circ T(w_1).$$

Also, $T(\Lambda) = 1_S$ by definition. Thus T is a monoid homomorphism.

(b) Part (a) shows that if f and g are in $\mathcal{M}$, then $f \circ g$ and $g \circ f$ are in $\mathcal{M}$. Thus $\mathcal{M}$ is a subsemigroup of S^S. Since $1_S = T(\Lambda)$, $1_S \in \mathcal{M}$. Thus $\mathcal{M}$ is a submonoid of S^S. The monoid $\mathcal{M}$ is called the **monoid of the machine $\mathcal{M}$**.

▼

EXAMPLE 3   Let $S = \{s_0, s_1, s_2\}$ and $I = \{a, b, d\}$. Consider the finite-state machine $M = (S, I, \mathcal{F})$ defined by the digraph shown in Figure 10.39. Compute the functions f_{bad}, f_{add}, and f_{badadd}, and verify that

$$f_{add} \circ f_{bad} = f_{badadd}.$$

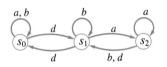

Figure 10.39

Solution f_{bad} is computed by the following sequence of transitions:

$$s_0 \xrightarrow{b} s_0 \xrightarrow{a} s_0 \xrightarrow{d} s_1$$
$$s_1 \xrightarrow{b} s_1 \xrightarrow{a} s_2 \xrightarrow{d} s_1$$
$$s_2 \xrightarrow{b} s_1 \xrightarrow{a} s_2 \xrightarrow{d} s_1.$$

Thus $f_{bad}(s_0) = s_1$, $f_{bad}(s_1) = s_1$, and $f_{bad}(s_2) = s_1$.
Similarly, for f_{add},

$$s_0 \xrightarrow{a} s_0 \xrightarrow{d} s_1 \xrightarrow{d} s_0$$
$$s_1 \xrightarrow{a} s_2 \xrightarrow{d} s_1 \xrightarrow{d} s_0$$
$$s_2 \xrightarrow{a} s_2 \xrightarrow{d} s_1 \xrightarrow{d} s_0,$$

so $f_{add}(s_i) = s_0$ for $i = 0, 1, 2$. A similar computation shows that

$$f_{badadd}(s_0) = s_0, \qquad f_{badadd}(s_1) = s_0, \qquad f_{badadd}(s_2) = s_0$$

and the same results hold for $f_{add} \circ f_{bad}$. In fact,

$$(f_{add} \circ f_{bad})(s_0) = f_{add}(f_{bad}(s_0)) = f_{add}(s_1) = s_0$$
$$(f_{add} \circ f_{bad})(s_1) = f_{add}(f_{bad}(s_1)) = f_{add}(s_1) = s_0$$
$$(f_{add} \circ f_{bad})(s_2) = f_{add}(f_{bad}(s_2)) = f_{add}(s_1) = s_0.$$

■

EXAMPLE 4   Consider the machine whose graph is shown in Figure 10.40. Show that $f_w(s_0) = s_0$ if and only if w has $3n$ 1's for some $n \geq 0$.

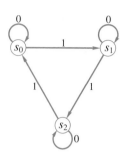

Figure 10.40

Solution   From Figure 10.40 we see that $f_0 = 1_S$, so the 0's in a string $w \in I^*$ have no effect on f_w. Thus, if $\overline{w}$ is w with all 0's removed, then $f_w = f_{\overline{w}}$. Let $l(w)$ denote the **length** of w, that is, the number of digits in w. Then $l(\overline{w})$ is the number if 1's in w, for all $w \in I^*$. For each $n \geq 0$, consider the statement

$$P(n): \text{Let } w \in I^* \text{ and let } l(\overline{w}) = m.$$
(a) If $m = 3n$, then $f_w(s_0) = s_0$.
(b) If $m = 3n + 1$, then $f_w(s_0) = s_1$.
(c) If $m = 3n + 2$, then $f_w(s_0) = s_2$.

We prove by mathematical induction that $P(n)$ is true for all $n \geq 0$.

Basis Step: Suppose that $n = 0$. In case (a), $m = 0$; therefore, w has no 1's and $f_w(s_0) = 1_S(s_0) = s_0$. In case (b), $m = 1$, so $\overline{w} = 1$ and $f_w(s_0) = f_{\overline{w}}(s_0) = f_1(s_0) = s_1$. Finally, in case (c), $m = 2$, so $\overline{w} = 11$, and $f_w(s_0) = f_{\overline{w}}(s_0) = f_{11}(s_0) = f_1(s_1) = s_2$.

Induction Step: We must use $P(k)$ to show $P(k+1)$. Let $w \in I^*$, and denote $l(\overline{w})$ by m. In case (a), $m = 3(k + 1) = 3k + 3$; therefore, $\overline{w} = w' \cdot 111$, where $l(w') = 3k$. Then $f_{w'}(s_0) = s_0$ by $P(k)$, part (a), and $f_{111}(s_0) = s_0$ by direct computation, so $f_{\overline{w}}(s_0) = f_{w'}(f_{111}(s_0)) = f_{w'}(s_0) = s_0$. Cases (b) and (c) are handled in the same way. Thus $P(k + 1)$ is true.

By mathematical induction, $P(n)$ is true for all $n \geq 0$, so $f_w(s_0) = s_0$ if and only if the number of 1's in w is a multiple of 3.   ∎

Suppose now that $(S, I, \mathcal{F}, s_0, T)$ is a Moore machine. As in Section 10.1, we may think of certain subsets of I^* as "languages" with "words" from I. Using M, we can define such a subset, which we will denote by $L(M)$, and call the **language of the machine** M. Define $L(M)$ to be the set of all $w \in I^*$ such that $f_w(s_0) \in T$. In other words, $L(M)$ consists of all strings that, when used as input to the machine, cause the starting state s_0 to move to an acceptance state in T. Thus, in this sense, M accepts the string. It is for this reason that the states in T were named acceptance states in Section 10.3.

EXAMPLE 5   Let $M = (S, I, \mathcal{F}, s_0, T)$ be the Moore machine in which $(S, I, \mathcal{F})$ is the finite-state machine whose digraph is shown in Figure 10.40, and $T = \{s_1\}$. The discussion of Example 4 shows that $f_w(s_0) = s_1$ if and only if the number of 1's in w is of the form $3n + 1$ for some $n \geq 0$. Thus $L(M)$ is exactly the set of all strings with $3n + 1$ 1's for some $n \geq 0$.   ∎

EXAMPLE 6   Consider the Moore machine M whose digraph is shown in Figure 10.41. Here state s_0 is the starting state, and $T = \{s_2\}$. What is $L(M)$? Clearly, the input set is $I = \{a, b\}$. Observe that, in order for a string w to cause a transition from s_0 to s_2, w must contain at least two b's. After reaching s_2, any additional letters have no effect. Thus $L(M)$ is the set of all strings having two or more b's. We see, for example, that $f_{aabaa}(s_0) = s_1$, so *aabaa* is rejected. On the other hand, $f_{abaab}(s_0) = s_2$, so *abaab* is accepted.   ∎

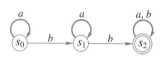

Figure 10.41

10.4 Exercises

In Exercises 1 through 5, we refer to the finite-state machine whose state transition table is

	0	1
s_0	s_0	s_1
s_1	s_1	s_2
s_2	s_2	s_3
s_3	s_3	s_0

1. List the values of the transition function f_w for $w = 01001$.

2. List the values of the transition function f_w for $w = 11100$.

3. Describe the set of binary words (sequences of 0's and 1's) w having the property that $f_w(s_0) = s_0$.

4. Describe the set of binary words w having the property that $f_w = f_{010}$.

5. Describe the set of binary words w having the property that $f_w(s_0) = s_2$.

In Exercises 6 through 10, we refer to the finite-state machine whose digraph is shown in Figure 10.42.

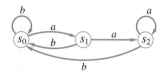

Figure 10.42

6. List the values of the transition function f_w for $w = abba$.

7. List the values of the transition function f_w for $w = babab$.

8. Describe the set of words w having the property that $f_w(s_0) = s_2$.

9. Describe the set of words w having the property that $f_w(s_0) = s_0$.

10. Describe the set of words w having the property that $f_w = f_{aba}$.

In Exercises 11 through 15, describe (in words) the language accepted by the Moore machines whose digraphs are given in Figures 10.43 through 10.47.

11.

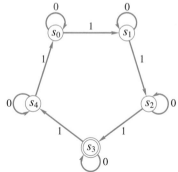

Figure 10.43

12.

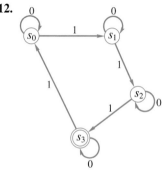

Figure 10.44

13.

Figure 10.45

14.

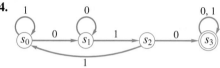

Figure 10.46

15.

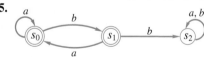

Figure 10.47

In Exercises 16 through 20, describe (in words) the language accepted by the Moore machine whose state table is given. The starting state is s_0, and the set T of acceptance states is shown.

16.

	0	1
s_0	s_1	s_2
s_1	s_1	s_2
s_2	s_2	s_1

$T = \{s_2\}$

17.

	0	1
s_0	s_1	s_0
s_1	s_1	s_2
s_2	s_1	s_0

$T = \{s_2\}$

18.

	0	1
s_0	s_0	s_1
s_1	s_0	s_1

$T = \{s_1\}$

19.

	x	y	z
s_0	s_1	s_3	s_4
s_1	s_4	s_2	s_4
s_2	s_4	s_4	s_4
s_3	s_4	s_4	s_2
s_4	s_4	s_4	s_4

$T = \{s_2\}$

20.

	x	y
s_0	s_1	s_2
s_1	s_3	s_2
s_2	s_1	s_2
s_3	s_3	s_3

$T = \{s_1, s_2\}$

21. Let $M = \{S, I, \mathcal{F}, s_0, T\}$ be a Moore machine. Suppose that if $s \in T$ and $w \in I^*$, then $f_w(s) \in T$. Prove that $L(M)$ is a subsemigroup of $(I^*, \cdot)$, where $\cdot$ is catenation.

10.5   MACHINES AND REGULAR LANGUAGES

Let $M = (S, I, \mathcal{F}, s_0, T)$ be a Moore machine. In Section 10.4 we defined the language $L(M)$ of the machine M. It is natural to ask if there is a connection between such a language and the languages of phrase structure grammars, discussed in Section 10.1. The following theorem, due to S. Kleene, describes the connection.

Theorem 1   Let I be a set and let $L \subseteq I^*$. Then L is a type 3 language; that is, $L = L(G)$, where G is a type 3 grammar having I as its set of terminal symbols, if and only if $L = L(M)$ for some Moore machine $M = (S, I, \mathcal{F}, s_0, T)$.   ●

We stated in Section 10.2 that a set $L \subseteq I^*$ is a type 3 language if and only if L is a regular set, that is, if and only if L corresponds to some regular expression over I. This leads to the following corollary of Theorem 1.

Corollary 1   Let I be a set and let $L \subseteq I^*$. Then $L = L(M)$ for some Moore machine $M = (S, I, \mathcal{F}, s_0, T)$ if and only if L is a regular set.   ●

We will not give a complete and detailed proof of Theorem 1. However, it is easy to give a construction that produces a type 3 grammar from a given Moore machine. This is done in such a way that the grammar and the machine have the same language. Let $M = (S, I, \mathcal{F}, s_0, T)$ be a given Moore machine. We construct a type 3 grammar $G = (V, I, s_0, \mapsto)$ as follows. Let $V = I \cup S$; that is, I will be the set of terminal symbols for G, while S will be the set of nonterminal symbols. Let s_i and s_j be in S, and $x \in I$. We write $s_i \mapsto x s_j$, if $f_x(s_i) = s_j$, that is, if the input x takes state s_i to s_j. We also write $s_i \mapsto x$ if $f_x(s_i) \in T$, that is, if the input x takes the state s_i to some acceptance state. Now let $\mapsto$ be the relation determined by the preceding two conditions and take this relation as the production relation of G.

The grammar G constructed above has the same language as M. Suppose, for example, that $w = x_1 x_2 x_3 \in I^*$. The string w is in $L(M)$ if and only if $f_w(s_0) =$

$f_{x_3}(f_{x_2}(f_{x_1}(s_0))) \in T$. Let $a = f_{x_1}(s_0)$, $b = f_{x_2}(a)$, and $c = f_{x_3}(b)$, where $c = f_w(s_0)$ is in T. Then the rules given for constructing $\mapsto$ tell us that

1. $s_0 \mapsto x_1 a$
2. $a \mapsto x_2 b$
3. $b \mapsto x_3$

are all productions in G. The last one occurs because $c \in T$. If we begin with s_0 and substitute, using (1), (2), and (3) in succession, we see that $s_0 \Rightarrow^* x_1 x_2 x_3 = w$ (see Section 10.1), so $w \in L(G)$. A similar argument works for any string in $L(M)$, so $L(M) \subseteq L(G)$. If we reverse the argument, we can see that we also have $L(G) \subseteq L(M)$. Thus M and G have the same language.

EXAMPLE 1   Consider the Moore machine M shown in Figure 10.41. Construct a type 3 grammar G such that $L(G) = L(M)$. Also, find a regular expression over $I = \{a, b\}$ that corresponds to $L(M)$.

Solution   Let $I = \{a, b\}$, $S = \{s_0, s_1, s_2\}$, and $V = I \cup S$. We construct the grammar $(V, I, s_0, \mapsto)$, where $\mapsto$ is described as follows:

$$\mapsto: \quad \begin{array}{ll} s_0 \mapsto as_0 & s_2 \mapsto bs_2 \\ s_0 \mapsto bs_1 & s_1 \mapsto b \\ s_1 \mapsto as_1 & s_2 \mapsto a \\ s_1 \mapsto bs_2 & s_2 \mapsto b \\ s_2 \mapsto as_2. & \end{array}$$

The production relation $\mapsto$ is constructed as we indicated previously; therefore, $L(M) = L(G)$.

If we consult Figure 10.41, we see that a string $w \in L(M)$ has the following properties. Any number $n \geq 0$ of a's can occur at the beginning of w. At some point, a b must occur in order to cause the transition from s_0 to s_1. After this b, any number $k \geq 0$ of a's may occur, followed by another b to cause transition to s_2. The remainder of w, if any, is completely arbitrary, since the machine cannot leave s_2 after once entering this state. From this description we can readily see that $L(M)$ corresponds to the regular expression

$$a^*ba^*b(a \vee b)^*.$$

EXAMPLE 2   Consider the Moore machine whose digraph is shown in Figure 10.48. Describe in words the language $L(M)$. Then construct the regular expression that corresponds to $L(M)$ and describe the production of the corresponding grammar G in BNF form.

Solution   It is clear that 0's in the input string have no effect on the states. If an input string w has an odd number of 1's, then $f_w(s_0) = s_1$. If w has an even number of 1's, then $f_w(s_0) = s_0$. Since $T = \{s_1\}$, we see that $L(M)$ consists of all w in I^* that have an odd number of 1's.

We now find the regular expression corresponding to $L(M)$. Any input string corresponding to the expression 0^*10^* will be accepted, since it will have exactly one 1. If an input w begins in this way, but has more 1's, then the additional ones must come in pairs, with any number of 0's allowed between,

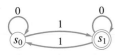

Figure 10.48

or after each pair of 1's. The previous sentence describes the set of strings corresponding to the expression $(10^*10^*)^*$. Thus $L(M)$ corresponds to the regular expression

$$0^*10^*(10^*10^*)^*.$$

Finally, the type 3 grammar constructed from M is $G = (V, I, s_0, \mapsto)$ with $V = I \cup S$. The BNF of the relation $\mapsto$ is

$$\langle s_0 \rangle ::= 0\langle s_0 \rangle \mid 1\langle s_1 \rangle \mid 1$$
$$\langle s_1 \rangle ::= 0\langle s_1 \rangle \mid 1\langle s_0 \rangle \mid 0$$ ■

Occasionally, we may need to determine the function performed by a given Moore machine, as we did in the preceding examples. More commonly, however, it is necessary to construct a machine that will perform a given task. This task may be defined by giving a verbal description, a regular expression, or an equivalent type 3 grammar, perhaps in BNF or with a syntax diagram. There are systematic, almost mechanical ways to construct such a machine. Most of these use the concept of *nondeterministic machines* and employ a tedious translation process from such machines to the Moore machines that we have discussed. If the task of the machine is not too complex, we may use simple reasoning to construct the machine in steps, usually in the form of its digraph. Whichever method is used, the resulting machine may be quite inefficient; for example, it may have unneeded states. In Section 10.6, we will give a procedure for constructing an equivalent machine that may be much more efficient.

EXAMPLE 3   Construct a Moore machine M that will accept exactly the string 001 from input strings of 0's and 1's. In other words, $I = \{0, 1\}$ and $L(M) = \{001\}$.

Solution   We must begin with a starting state s_0. If w is an input string of 0's and 1's and if w begins with a 0, then w *may* be accepted (depending on the remainder of its components). Thus one step toward acceptance has been taken, and there needs to be a state s_1 that corresponds to this step. We therefore begin as in Figure 10.49(a). If we next receive another 0, we have progressed one more step toward acceptance. We therefore construct another state s_2 and let 0 give a transition from s_1 to s_2. State s_1 represents the condition "first symbol of input is a 0," whereas state s_2 represents the condition "first two symbols of the input are 00." This situation is shown in Figure 10.49(b). Finally, if the third input symbol is a 1, we move to an acceptance state, as shown in Figure 10.49(c). Any other beginning sequence of input digits or any additional digits will move us to a "failure state" s_4 from which there is no escape. Thus Figure 10.49(d) shows the completed machine.   ■

The process illustrated in Example 3 is difficult to describe precisely or to generalize. We try to construct states representing each successive stage of input complexity leading up to an acceptable string. There must also be states indicating the ways in which a promising input pattern may be destroyed when a certain symbol is received. If the machine is to recognize several, essentially different types of input, then we will need to construct separate branches corresponding to each

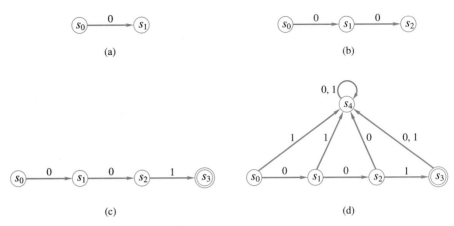

Figure 10.49

type of input. This process may result is some redundancy, but the machine can be simplified later.

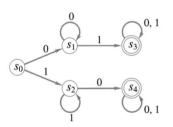

Figure 10.50

EXAMPLE 4

Let $I = \{0, 1\}$. Construct a Moore machine that accepts those input sequences w that contain the string 01 or the string 10 anywhere within them. In other words, we are to accept exactly those strings that do not consist entirely of 0's or entirely of 1's.

Solution   This is a simple example in which, whatever input digit is received first, a string will be accepted if and only if the other digit is eventually received. There must be a starting state s_0, states s_1 and s_2 corresponding respectively to first receiving a 0 or 1, and (acceptance) states s_3 and s_4, which will be reached if and when the other digit is received. Having once reached an acceptance state, the machine stays in that state. Thus we construct the digraph of this machine as shown in Figure 10.50. ∎

In Example 3, once an acceptance state is reached, any additional input will cause a permanent transition to a nonaccepting state. In Example 4, once an acceptance state is reached, any additional input will have no effect. Sometimes the situation is between these two extremes. As input is received, the machine may repeatedly enter and leave acceptance states. Consider the Moore machine M whose digraph is shown in Figure 10.51. This machine is a slight modification of the finite-state machine given in Example 4 of Section 10.4. We know from that example that $w \in L(M)$ if and only if the number of 1's in w is of the form $3n$, $n \geq 0$. As input symbols are received, M may enter and leave s_0 repeatedly. The conceptual states "one 1 has been received" and "four 1's have been received" may both be represented by s_1. When constructing machines, we should keep in mind the fact that a state, previously defined to represent one conceptual input condition, may be used for a new input condition if these two conditions represent the same degree of progress of the input stream toward acceptance. The next example illustrates this fact.

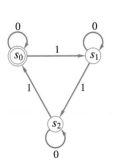

Figure 10.51

EXAMPLE 5   Construct a Moore machine that accepts exactly those input strings of x's and y's that end in yy.

> *Solution*   Again we need a starting state s_0. If the input string begins with a y, we progress one step to a new state s_1 ("last input component received is a y"). On the other hand, if the input begins with an x, we have made no progress toward acceptance. Thus we may suppose that M is again in state s_0. This situation is shown in Figure 10.52(a). If, while in state s_1, a y is received, we progress to an acceptance state s_2 ("last two symbols of input received were y's"). If instead the input received is an x, we must again receive two y's in order to be in an acceptance state. Thus we may again regard this as a return to state s_0. The situation at this point is shown in Figure 10.52(b). Having reached state s_2, an additional input of y will have no effect, but an input of x will necessitate two more y's for acceptance. Thus we can again regard M as being in state s_0. The final Moore machine is shown in Figure 10.52(c). ■

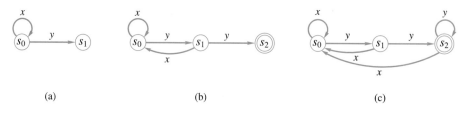

(a)   (b)   (c)

F i g u r e   1 0 . 5 2

We have not mentioned the question of implementation of finite-state machines. Indeed, many such machines, including all digital computers, are implemented as hardware devices, that is, as electronic circuitry. There are, however, many occasions when finite-state machines are simulated in software. This is frequently seen in compilers and interpreters, for which Moore machines may be programmed to retrieve and interpret words and symbols in an input string. We provide just a hint of the techniques available by simulating the machine of Example 2 in pseudocode. The reader should refer back to Example 2 and Figure 10.48 for the details of the machine. The following subroutine gives a pseudocode program for this machine.

This program uses a subroutine INPUT to get the next 0 or 1 in variable X and assumes that a logical variable EOI will be set true when no further input is available. The variable RESULT will be true if the input string contains an odd number of 1's; otherwise, it will be false.

SUBROUTINE ODDONES (RESULT)
1.   EOI $\leftarrow F$
2.   RESULT $\leftarrow F$
3.   STATE $\leftarrow 0$
4.   **UNTIL** (EOI)
    a.   **CALL** INPUT $(X,$ EOI$)$
        1.   **IF** (EOI $= F$) **THEN**
            a.   **IF** (STATE $= 0$) **THEN**

    1. **IF** $(X = 1)$ **THEN**
        a. RESULT $\leftarrow T$
        b. STATE $\leftarrow 1$
b. **ELSE**
    1. **IF** $(X = 1)$ **THEN**
        a. RESULT $\leftarrow F$
        b. STATE $\leftarrow 0$
5. **RETURN**
END OF SUBROUTINE ODDONES

    In this particular coding technique, a state is denoted by a variable that may be assigned different values depending on input and whose values then determine other effects of the input. An alternative procedure is to represent a state by a particular location in code. This location then determines the effect of input and the branch to a new location (subsequent state). The following subroutine shows the same subroutine ODDONES coded in this alternative way.

    SUBROUTINE ODDONES (RESULT) version 2
      1. RESULT $\leftarrow F$
S0: 2. **CALL** INPUT (X, EOI)
      3. **IF** (EOI) **THEN**
        a. **RETURN**
      4. **ELSE**
        a. **IF** $(X = 1)$ **THEN**
           1. RESULT $\leftarrow T$
           2. **GO TO** $S1$
        b. **ELSE**
           1. **GO TO** $S0$
S1: 5. **CALL** INPUT (X, EOI)
      6. **IF** (EOI) **THEN**
        a. **RETURN**
      7. **ELSE**
        a. **IF** $(X = 1)$ **THEN**
           1. RESULT $\leftarrow F$
           2. **GO TO** $S0$
        b. **ELSE**
           1. **GO TO** $S1$
END OF SUBROUTINE ODDONES version 2

    It is awkward to avoid **GO TO** statements in this approach, and we have used them. In languages with multiple **GO TO** statements, such as FORTRAN's indexed **GO TO** or PASCAL's CASE statement, this method may be particularly efficient for finite-state machines with a fairly large number of states. In such cases, the first method may become quite cumbersome.

10.5 Exercises

1. Let M be the Moore machine of Figure 10.53. Construct a type 3 grammar $G = (V, I, s_0, \mapsto)$, such that $L(M) = L(G)$.

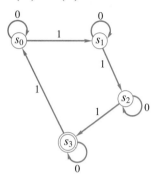

Figure 10.53

2. Let M be the Moore machine of Figure 10.54. Give a regular expression over $I = \{0, 1\}$, which corresponds to the language $L(M)$.

Figure 10.54

3. Let M be the Moore machine of Exercise 18, Section 10.4. Give a regular expression over $I = \{0, 1\}$, which corresponds to the language $L(M)$.

4. Let M be the Moore machine of Figure 10.55. Construct a type 3 grammar $G = (V, I, s_0, \mapsto)$, such that $L(M) = L(G)$. Describe $\mapsto$ in BNF.

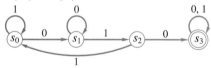

Figure 10.55

5. Let M be the Moore machine of Figure 10.56. Construct a type 3 grammar $G = (V, I, s_0, \mapsto)$, such that $L(M) = L(G)$. Describe $\mapsto$ in BNF.

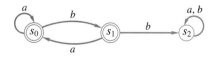

Figure 10.56

In Exercises 6 through 20, construct the digraph of a Moore machine that accepts the input strings described and no others.

6. Inputs a, b: strings where the number of a's and the number of b's are the same

7. Inputs a, b: strings where the number of b's is divisible by 3

8. Inputs a, b: strings where the number of a's is even and the number of b's is a multiple of 3

9. Inputs x, y: strings that have an even number of y's

10. Inputs 0, 1: strings that contain 0011

11. Inputs 0, 1: strings that end with 0011

12. Inputs $\square$, $\triangle$: strings that contain $\square\triangle$ or $\triangle\square$

13. Inputs $+$, $\times$: strings that contain $+ \times \times$ or $\times + +$

14. Inputs w, z: strings that contain wz or zzw

15. Inputs a, b: strings that contain ab and end in bbb

16. Inputs $+$, $\times$: strings that end in $+ \times \times$

17. Inputs w, z: strings that end in wz or zzw

18. Inputs 0, 1, 2: the string 0120

19. Inputs a, b, c: the strings aab or abc

20. Inputs x, y, z: the strings xzx or yx or zyx

In Exercises 21 through 25, construct the state table of a Moore machine that recognizes the given input strings and no others.

21. Inputs 0, 1: strings ending in 0101

22. Inputs a, b: strings where the number of b's is divisible by 4

23. Inputs x, y: strings having exactly two x's

24. Inputs a, b: strings that do not have two successive b's

25. Let $\mathcal{M} = (S, I, \mathcal{F}, s_0, T)$ be a Moore machine. Define a relation R on S as follows: $s_i\ R\ s_j$ if and only if $f_w(s_i)$ and $f_w(s_j)$ either both belong to T or neither does, for every $w \in I^*$. Show that R is an equivalence relation on S.

10.6 SIMPLIFICATION OF MACHINES

As we have seen, the method in Section 10.5 for the construction of a finite-state machine to perform a given task is as much an art as a science. Generally, graphical methods are first used, and states are constructed for all intermediate steps in the process. Not surprisingly, a machine constructed in this way may not be efficient, and we need to find a method for obtaining an equivalent, more efficient machine. Fortunately, a method is available that is systematic (and can be computerized), and this method will take any correct machine, however redundant it is, and produce an equivalent machine that is usually more efficient. Here we will use the number of states as our measure of efficiency. We will demonstrate this technique for Moore machines, but the principles extend, with small changes, to various other types of finite-state machines. Let $(S, I, \mathcal{F}, s_0, T)$ be a Moore machine. We define a relation R on S as follows: For any $s, t \in S$ and $w \in I^*$, we say that s and t are w-**compatible** if $f_w(s)$ and $f_w(t)$ both belong to T, or neither does. Let $s \ R \ t$ mean that s and t are w-compatible for all $w \in I^*$.

Theorem 1   Let $(S, I, \mathcal{F}, s_0, T)$ be a Moore machine, and let R be the relation defined previously.

   (a) R is an equivalence relation on S.
   (b) R is a machine congruence (see Section 10.3).   ●

Proof

   (a) R is clearly reflexive and symmetric. Suppose now that $s \ R \ t$ and $t \ R \ u$ for s, t, and u in S, and let $w \in I^*$. Then s and t are w-compatible, as are t and u, so if we consider $f_w(s), f_w(t), f_w(u)$, it follows that either all belong to T or all belong to $\overline{T}$, the complement of T. Thus s and u are w-compatible, so R is transitive, and therefore R is an equivalence relation.

   (b) We must show that if s and t are in S and $x \in I$, then $s \ R \ t$ implies that $f_x(s) \ R \ f_x(t)$. To show this, let $w \in I^*$, and let $w' = x \cdot w$ ($\cdot$ is the operation of catenation). Since $s \ R \ t$, $f_{w'}(s)$ and $f_{w'}(t)$ are both in T or both in $\overline{T}$. But $f_{w'}(s) = f_{x \cdot w}(s) = f_w(f_x(s))$ and $f_{w'}(t) = f_{x \cdot w}(t) = f_w(f_x(t))$, so $f_x(s)$ and $f_x(t)$ are w-compatible. Since w is arbitrary in I^*, $f_x(s) \ R \ f_x(t)$.   ▼

Since R is a machine congruence, we may form the quotient Moore machine $\overline{M} = (S/R, I, \overline{\mathcal{F}}, [s_0], T/R)$ as in Section 10.3. The machine $\overline{M}$ is the efficient version of M that we have promised. We will show that $\overline{M}$ is **equivalent** to M, meaning that $L(\overline{M}) = L(M)$.

EXAMPLE 1   Consider the Moore machine whose digraph is shown in Figure 10.57. In this machine $I = \{0, 1\}$. The starting state is s_0, and $T = \{s_2, s_3\}$. Let us compute the quotient machine $\overline{M}$. First, we see that $s_0 \ R \ s_1$. In fact, $f_w(s_0) \in T$ if and only if w contains at least one 1, and $f_w(s_1) \in T$ under precisely the same condition. Thus s_0 and s_1 are w-compatible for all $w \in I^*$. Now $s_2 \ \mathcal{R} \ s_0$ and $s_3 \ \mathcal{R} \ s_0$, since $f_0(s_2) \in T$, $f_0(s_3) \in T$, but $f_0(s_0) \notin T$. This implies that $\{s_0, s_1\}$ is one R-equivalence class.

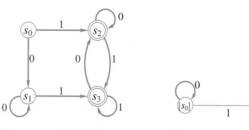

Figure 10.57   Figure 10.58

Also s_2 R s_3, since $f_w(s_2) \in T$ and $f_w(s_3) \in T$ for all $w \in I^*$. This proves that

$$S/R = \{\{s_0, s_1\}, \{s_2, s_3\}\} = \{[s_0], [s_2]\}.$$

Also note that $T/R = \{[s_2]\}$. The resulting quotient Moore machine $\overline{M}$ is equivalent to M and its digraph is shown in Figure 10.58. ∎

In this case it is clear that M and $\overline{M}$ are equivalent since each accepts a word w if and only if w has at least one 1. In general, we have the following result.

Theorem 2   Let $M = (S, I, \mathcal{F}, s_0, T)$ be a Moore machine, let R be the equivalence relation defined above, and let $\overline{M} = (S/R, I, \overline{\mathcal{F}}, [s_0], T/R)$ be the corresponding quotient Moore machine. Then $L(\overline{M}) = L(M)$. ●

Proof   Suppose that w is accepted by M so that $f_w(s_0) \in T$. Then $\overline{f}_w([s_0]) = [f_w(s_0)] \in T/R$; that is, $\overline{M}$ also accepts w.

Conversely, suppose that $\overline{M}$ accepts w so that $\overline{f}_w([s_0]) = [f_w(s_0)]$ is in T/R. This means that t R $f_w(s_0)$ for some element t in T. By definition of R, we know that t and $f_w(s_0)$ are w'-compatible for every $w' \in I^*$. When w' is Λ, the empty string, then $f_{w'} = 1_S$, so $t = f_{w'}(t)$ and $f_w(s_0) = f_{w'}(f_w(s_0))$ are both in T or both in $\overline{T}$. Since $t \in T$, we must have $f_w(s_0) \in T$, so M accepts w. ▼

Thus we see that after initially designing the Moore machine M, we may compute R and pass to the quotient machine $\overline{M} = M/R$, thereby obtaining an equivalent machine that may be considerably more efficient, in the sense that it may have many fewer states. Often the quotient machine is one that would have been difficult to discover at the outset.

We now need an algorithm for computing the relation R. In Example 1 we found R by direct analysis of input, but this example was chosen to be particularly simple. In general, a direct analysis will be very difficult. We now define and investigate a set of relations that provides an effective method for computing R.

If k is a nonnegative integer, we define a relation R_k on S, the state set of a Moore machine $(S, I, \mathcal{F}, s_0, T)$. If $w \in I^*$, recall that $l(w)$ is the length of the string w, that is, the number of symbols in w. Note that $l(\Lambda) = 0$. Now, if s and $t \in S$, we let s R_k t mean that s and t are w-compatible for all $w \in I^*$ with $l(w) \leq k$. The relations R_k are not machine congruences but are successive approximations to the desired congruence R.

Theorem 3 (a) $R_{k+1} \subseteq R_k$ for all $k \geq 0$.

(b) Each R_k is an equivalence relation.

(c) $R \subseteq R_k$ for all $k \geq 0$. ●

Proof If $s, t \in S$, and s and t are w-compatible for all $w \in I^*$ or for all w with $l(w) \leq k + 1$, then in either case s and t are compatible for all w with $l(w) \leq k$. This proves parts (a) and (c). The proof of part (b) is similar to the proof of Theorem 1(a), and we omit it. ▼

The key result for computing the relations R_k recursively is the following theorem.

Theorem 4 (a) $S/R_0 = \{T, \overline{T}\}$, where $\overline{T}$ is the complement of T.

(b) Let k be a nonnegative integer and $s, t \in S$. Then $s \ R_{k+1} \ t$ if and only if

(1) $s \ R_k \ t$.

(2) $f_x(s) \ R_k \ f_x(t)$ for all $x \in I$. ●

Proof

(a) Since only Λ has length 0, it follows that $s \ R_0 \ t$ if and only if both s and t are in T or both are in $\overline{T}$. This proves that $S/R_0 = \{T, \overline{T}\}$.

(b) Let $w \in I^*$ be such that $l(w) \leq k + 1$. Then $w = x \cdot w'$, for some $x \in I$ and for some $w' \in I^*$ with $l(w') \leq k$. Conversely, if any $x \in I$ and $w' \in I^*$ with $l(w') \leq k$ are chosen, the resulting string $w = x \cdot w'$ has length less than or equal to $k + 1$.

Now $f_w(s) = f_{x \cdot w'}(s) = f_{w'}(f_x(s))$ and $f_w(t) = f_{w'}(f_x(t))$ for any s, t in S. This shows that s and t are w-compatible for any $w \in I^*$ with $l(w) \leq k + 1$ if and only if $f_x(s)$ and $f_x(t)$ are, for all $x \in I$, w'-compatible, for any w' with $l(w') \leq k$. That is, $s \ R_{k+1} \ t$ if and only if $f_x(s) \ R_k \ f_x(t)$ for all $x \in I$.

Now either of these equivalent conditions implies that $s \ R_k \ t$, since $R_{k+1} \subseteq R_k$, so we have proved the theorem. ▼

This result says that we may find the partitions P_k, corresponding to the relations R_k, by the following recursive method:

STEP 1: Begin with $P_0 = \{T, \overline{T}\}$.

STEP 2: Having reached partition $P_k = \{A_1, A_2, \ldots, A_m\}$, examine each equivalence class A_i and break it into pieces where two elements s and t of A_i fall into the same piece if all inputs x take both s and t into the same subset A_j (depending on x).

STEP 3: The new partition of S, obtained by taking all pieces of all the A_i, will be P_{k+1}.

The final step in this method, telling us when to stop, is given by the following result.

___

Theorem 5   If $R_k = R_{k+1}$ for any nonnegative integer k, then $R_k = R$.   ●

Proof   Suppose that $R_k = R_{k+1}$. Then, by Theorem 4, $s\ R_{k+2}\ t$ if and only if $f_x(s)\ R_{k+1}\ f_x(t)$ for all $x \in I$, or (since $R_k = R_{k+1}$) if and only if $f_x(s)\ R_k\ f_x(t)$ for all $x \in I$. This happens if and only if $s\ R_{k+1}\ t$. Thus $R_{k+2} = R_{k+1} = R_k$. By induction, it follows that $R_k = R_n$ for all $n \geq k$. Now it is easy to see that $R = \bigcap_{n=0}^{\infty} R_n$, since every string w in I^* must have some finite length. Since $R_1 \supseteq R_2 \cdots \supseteq R_k = R_{k+1} = \cdots$, the intersection of the R_n's is exactly R_k, so $R = R_k$.   ▼

A procedure for reducing a given Moore machine to an equivalent machine is as follows.

STEP 1:   Start with the partition $P_0 = \{T, \overline{T}\}$.

STEP 2:   Construct successive partitions $P_1, P_2, \ldots$ corresponding to the equivalence relations $R_1, R_2, \ldots$ by using the method outlined after Theorem 4.

STEP 3:   Whenever $P_k = P_{k+1}$, stop. The resulting partition $P = P_k$ corresponds to the relation R.

STEP 4:   The resulting quotient machine is equivalent to the given Moore machine.

EXAMPLE 2   Consider the machine of Example 1. Here $S = \{s_0, s_1, s_2, s_3\}$ and $T = \{s_2, s_3\}$. We use the preceding method to compute an equivalent quotient machine. First, $P_0 = \{\{s_0, s_1\}, \{s_2, s_3\}\}$. We must decompose this partition in order to find P_1. Consider first the set $\{s_0, s_1\}$. Input 0 takes each of these states into $\{s_0, s_1\}$. Input 1 takes both s_0 and s_1 into $\{s_2, s_3\}$. Thus the equivalence class $\{s_0, s_1\}$ does not decompose in passing to P_1. We also see that input 0 takes both s_2 and s_3 into $\{s_2, s_3\}$ and input 1 takes both s_2 and s_3 into $\{s_2, s_3\}$. Again, the equivalence class $\{s_2, s_3\}$ does not decompose in passing to P_1. This means that $P_1 = P_0$, so P_0 corresponds to the congruence R. We found this result directly in Example 1.   ■

EXAMPLE 3   Let M be the Moore machine shown in Figure 10.59. Find the relation R and draw the digraph of the corresponding quotient machine $\overline{M}$.

Solution   The partition $P_0 = \{T, \overline{T}\} = \{\{s_0, s_5\}, \{s_1, s_2, s_3, s_4\}\}$. Consider first the set $\{s_0, s_5\}$. Input 0 carries both s_0 and s_5 into T, and input 1 carries both into $\overline{T}$. Thus $\{s_0, s_5\}$ does not decompose further in passing to P_1. Next consider the set $\overline{T} = \{s_1, s_2, s_3, s_4\}$. State s_1 is carried to $\overline{T}$ by input 0 and to T by input 1. This is also true for state s_4, but not for s_2 and s_3; so the equivalence class of s_1 in P_1 will be $\{s_1, s_4\}$. Since states s_2 and s_3 are carried into $\overline{T}$ by inputs 0 and 1, they will also form an equivalence class in P_1. Thus $\overline{T}$ has decomposed into the subsets $\{s_1, s_4\}$ and $\{s_2, s_3\}$ in passing to P_1, and $P_1 = \{\{s_0, s_5\}, \{s_1, s_4\}, \{s_2, s_3\}\}$.

To find P_2, we must examine each subset of P_1 in turn. Consider $\{s_0, s_5\}$. Input 0 takes s_0 and s_5 to $\{s_0, s_5\}$, and input 1 takes each of them to $\{s_1, s_4\}$. This means that $\{s_0, s_5\}$ does not further decompose in passing to P_2. A similar

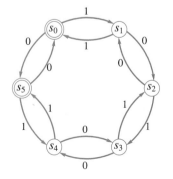

Figure 10.59

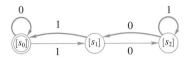

Figure 10.60

argument shows that neither of the sets $\{s_1, s_4\}$ and $\{s_2, s_3\}$ decomposes, so that $P_2 = P_1$. Hence P_1 corresponds to R. The resulting quotient machine is shown in Figure 10.60. It can be shown (we omit the proof) that each of these machines will accept a string $w = b_1 b_2 \cdots b_n$ in $\{0, 1\}^*$ if and only if w is the binary representation of a number that is divisible by 3. ∎

10.6 Exercises

In Exercises 1 through 8, find the specified relation R_k for the Moore machine whose digraph is given.

1. Find R_0.

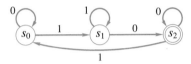

Figure 10.61

2. Find R_1 for the Moore machine depicted by Figure 10.61.

3. Find R_1 for the Moore machine depicted by Figure 10.62.

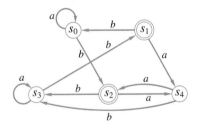

Figure 10.62

4. Find R_2 for the machine of Exercise 3.

5. Find R_{127} for the machine of Exercise 3.

6. Find R_1 for the Moore machine depicted by Figure 10.63.

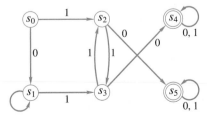

Figure 10.63

7. Find R_2 for the machine of Exercise 6.

8. Find R_1 for the Moore machine depicted by Figure 10.64.

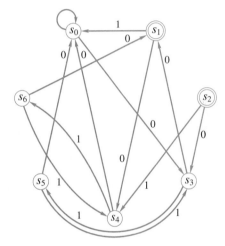

Figure 10.64

9. Find R for the machine of Exercise 1.

10. Find R for the machine of Exercise 3.

11. Find R for the machine of Exercise 6.

12. Find R for the Moore machine depicted by Figure 10.65.

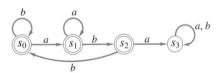

Figure 10.65

13. Find the relation R and construct the digraph of the corresponding equivalent quotient machine for the Moore machine whose digraph is shown in Figure 10.64.

In Exercises 14 through 17, draw the digraph of the quotient machine $\overline{M}$ for the given machine.

14. The machine of Exercise 1

15. The machine of Exercise 3

16. The machine of Exercise 6

17. The machine of Exercise 12

In Exercises 18 and 19, find the partition corresponding to the relation R, and construct the state table of the corresponding quotient machine that is equivalent to the Moore machine whose state table is shown.

18.

	0	1
s_0	s_5	s_2
s_1	s_6	s_2
s_2	s_0	s_4
s_3	s_3	s_5
s_4	s_6	s_2
s_5	s_3	s_0
s_6	s_3	s_1

$T = \{s_2\}$

19.

	0	1
a	a	c
b	g	d
c	f	e
d	a	d
e	a	d
f	g	f
g	g	c

$s_0 = a$
$T = \{d, e\}$

20. Draw the digraph of the Moore machine whose state table is given in Exercise 18. Also, draw the digraph of the quotient machine $\overline{M}$.

21. Draw the digraph of the Moore machine whose state table is given in Exercise 19. Also, draw the digraph of the quotient machine $\overline{M}$.

TIPS FOR PROOFS

The proof techniques of Chapter 2 are sufficient to handle all the proofs we have presented up to this point. These techniques should be very familiar to you by now. Direct proofs using definitions and previously proven results are usually the first strategy to try. Example 4, Section 10.4, uses an induction proof, because the statement to be proved is of the form $P(n), n \geq 0$. The form of the statement is often a clue to what proof strategy to choose. As another example, you should recognize that phrases such as "at least two" in the theorem or statement may make it a good candidate for a pigeonhole proof. (See Exercise 28, Section 10.1.) Even when you have a good idea which proof technique will be useful, there is still likely to be some trial and error involved in developing the proof. Just how to use a definition or which previous theorem is applicable is not always obvious, but persistence is also a powerful tool for producing a proof.

KEY IDEAS FOR REVIEW

- Phrase structure grammar $(V, S, v_0, \mapsto)$: see page 358
- Production: a statement $w \mapsto w'$, where $(w, w') \in \mapsto$
- Direct derivability: see page 359
- Terminal symbols: the elements of S
- Nonterminal symbols: the elements of $V - S$
- Derivation of a sentence: substitution process that produces a valid sentence
- Language of a grammar G: set of all properly constructed sentences that can be produced from G
- Derivation tree for a sentence: see page 360
- Types 0, 1, 2, 3 phrase structure grammars: see page 363
- Context-free grammar: type 2 grammar
- Regular grammar: type 3 grammar
- Parsing: process of obtaining a derivation tree that will produce a given sentence

- BNF notation: see page 366
- Syntax diagram: see page 368
- Theorem: Let S be a finite set, and $L \subseteq S^*$. Then L is a regular set if and only if $L = L(G)$ for some regular grammar $G = (V, S, v_0, \mapsto)$.
- Finite-state machine: $(S, I, \mathcal{F})$, where S is a finite set of states, I is a set of inputs, and $\mathcal{F} = \{f_x \mid x \in I\}$
- State transition tables: see page 376
- R_M: $s_i \ R_M \ s_j$, if there is an input x so that $f_x(s_i) = s_j$
- Moore machine: $M = (S, I, \mathcal{F}, s_0, T)$, where $s_0 \in S$ is the starting state and $T \subseteq S$ is the set of acceptance states
- Machine congruence R on M: For any $s, t \in S$, $s \ R \ t$ implies that $f_x(s) \ R \ f_x(t)$ for all $x \in I$.
- Quotient of M corresponding to R: see page 378

- State transition function f_w, $w = x_1 x_2 \cdots x_n$:
  $f_w = f_{x_n} \circ f_{x_{n-1}} \circ \cdots \circ f_{x_1}$, $f_\Lambda = 1_S$
- Theorem: Let $M = (S, I, \mathcal{F})$ be a finite-state machine. Define $T : I^* \to S^S$ by $T(w) = f_w$, $w \neq \Lambda$, and $T(\Lambda) = 1_S$. Then
  (a) If w_1 and w_2 are in I^*, then $T(w_1 \cdot w_2) = T(w_2) \circ T(w_1)$.
  (b) If $\mathcal{M} = T(I^*)$, then $\mathcal{M}$ is a submonoid of S^S.
- Monoid of a machine: $\mathcal{M}$ in the preceding theorem

- w-compatible: see page 393
- Equivalent machines M and N: $L(M) = L(N)$
- $l(w)$: length of the string w
- Language accepted by M: $L(M) = \{w \in I^* \mid f_w(s_0) \in T\}$
- Theorem: Let I be a set and $L \subseteq I^*$. Then L is a type 3 language, that is, $L = L(G)$ if and only if $L = L(M)$ for some Moore machine M.

CODING EXERCISES

For each of the following, write the requested program or subroutine in pseudocode (as described in Appendix A) or in a programming language that you know. Test your code either with a paper-and-pencil trace or with a computer run.

1. Let $M = (S, I, \mathcal{F})$ be a finite state machine where $S = \{s_0, s_1\}$, $I = \{0, 1\}$, and $\mathcal{F}$ is given by the following state transition table:

	0	1
s_0	s_0	s_1
s_1	s_1	s_0

Write a subroutine that given a state and an input returns the next state of the machine.

2. Write a function ST_TRANS that takes a word w, a string of 0's and 1's, and a state s and returns $f_w(s)$, the state transition function corresponding to w evaluated at s.

3. Let $M = (S, I, \mathcal{F})$ be a Moore machine where $S = \{s_0, s_1, s_2\}$, $I = \{0, 1\}$, $T = \{s_2\}$ and $\mathcal{F}$ is given by the following state transition table:

	0	1
s_0	s_0	s_1
s_1	s_2	s_2
s_2	s_1	s_0

Write a program that determines if a given word w is in $L(M)$.

4. Write a subroutine that simulates the Moore machine given in Exercise 2, Section 10.5.

5. Write a subroutine that simulates the Moore machine given in Exercise 4, Section 10.5.

CHAPTER 10 SELF-TEST

1. Let $G = (V, S, v_0, \mapsto)$ be a phrase structure grammar with $V = \{v_0, v_1, a, b, c, d\}$, $S = \{a, b, c, d\}$, and

$$\mapsto : v_0 \mapsto a v_0 b \quad v_0 \mapsto v_1 \quad v_1 \mapsto c v_1 \quad v_1 \mapsto d.$$

Tell whether each of the following is true or false.

(a) $abcd \in L(G)$
(b) $aaav_1 bbb \Rightarrow^* aaacc v_1 bbb$
(c) $ccccd \in L(G)$
(d) $av_1 b \Rightarrow acdb$

2. Describe precisely $L(G)$, the language of G as given in Problem 1.

3. Let $G = (V, S, v_0, \mapsto)$ be a phrase structure grammar with

$V = \{v_0, v_1, 1, 2, 3, 4, 5\}$, $S = \{1, 2, 3, 4, 5\}$ and

$$\mapsto : \quad v_0 \mapsto 1 v_1 \quad v_0 \mapsto 3 v_1 \quad v_0 \mapsto 5 v_1$$
$$v_1 \mapsto 1 v_1 \quad v_1 \mapsto 2 v_1 \quad v_1 \mapsto 3 v_1$$
$$v_1 \mapsto 4 v_1 \quad v_1 \mapsto 5 v_1 \quad v_1 \mapsto 2$$

(a) Draw the syntax diagram for the productions of G.
(b) Give the BNF for the productions of G.

4. Describe the language of G as given in Problem 3.

5. Construct a phrase structure grammar G such that $L(G) = \{0^n 10^m \mid n \geq 0, m \geq 1\}$.

6. Consider the finite-state machine whose state transition

table is

	a	b
s_0	s_0	s_1
s_1	s_1	s_2
s_2	s_2	s_3
s_3	s_3	s_0

Construct the digraph of this machine.

7. Describe the language accepted by the Moore machine whose digraph is given in Figure 10.66.

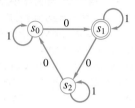

Figure 10.66

8. Consider the Moore machine M whose digraph is given in Figure 10.67.

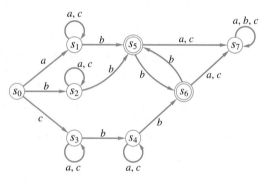

Figure 10.67

(a) Show that $R = \{(s_0, s_0), (s_1, s_1), (s_2, s_2), (s_3, s_3), (s_4, s_4), (s_5, s_5), (s_6, s_6), (s_7, s_7), (s_1, s_2), (s_1, s_4), (s_2, s_1), (s_2, s_4), (s_4, s_1), (s_4, s_2), (s_5, s_6), (s_6, s_5)\}$ is a machine congruence.

(b) Draw the digraph of the corresponding quotient Moore machine.

9. For the machine described in Problem 6, describe all words w in $\{a, b\}^*$ such that $f_w(s_0) = s_0$.

10. Construct a Moore machine that accepts a string of 0's and 1's if and only if the string has exactly two 1's.

11. Consider the Moore machine M whose digraph is given in Figure 10.68. Define R_k, $k = 0, 1, 2, \ldots$, as follows: $s \, R_k \, t$ if and only if s and t are w-compatible for all $w \in \{0, 1\}^*$ with length of $w \le k$.

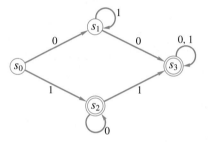

Figure 10.68

(a) Give the matrix of R_0.

(b) List the elements of R_1.

12. Using the machine M and the definitions in Problem 11,

(a) Find the smallest k such that $R_k = R_{k+1}$.

(b) Let $R = R_k$ for the k found in part (a) and draw the digraph of M/R.

11

GROUPS AND CODING

Prerequisites: *Chapter 9*

In today's modern world of communication, data items are constantly being transmitted from point to point. This transmission may range from the simple task of a computer terminal interacting with the mainframe computer located 200 feet away, to the more complicated task of sending a signal thousands of miles away via a satellite that is parked in an orbit 20,000 miles from the earth, or to a telephone call or letter to another part of the country. The basic problem in transmission of data is that of receiving the data as sent and not receiving a distorted piece of data. Distortion can be caused by a number of factors.

Coding theory has developed techniques for introducing redundant information in transmitted data that help in detecting, and sometimes in correcting, errors. Some of these techniques make use of group theory. We present a brief introduction to these ideas in this chapter.

11.1 CODING OF BINARY INFORMATION AND ERROR DETECTION

The basic unit of information, called a **message**, is a finite sequence of characters from a finite alphabet. We shall choose as our alphabet the set $B = \{0, 1\}$. Every character or symbol that we want to transmit is now represented as a sequence of m elements from B. That is, every character or symbol is represented in binary form. Our basic unit of information, called a **word**, is a sequence of m 0's and 1's.

The set B is a group under the binary operation $+$ whose table is shown in Table 11.1. (See Example 5 of Section 9.4.)

Table 11.1

+	0	1
0	0	1
1	1	0

If we think of B as the group Z_2, then $+$ is merely mod 2 addition. It follows from Theorem 1 of Section 9.5 that $B^m = B \times B \times \cdots \times B$ (m factors) is a group under

the operation $\oplus$ defined by

$$(x_1, x_2, \ldots, x_m) \oplus (y_1, y_2, \ldots, y_m) = (x_1 + y_1, x_2 + y_2, \ldots, x_m + y_m).$$

This group has been introduced in Example 2 of Section 9.5. Its identity is $\overline{0} = (0, 0, \ldots, 0)$ and every element is its own inverse. An element in B^m will be written as $(b_1, b_2, \ldots, b_m)$ or more simply as $b_1 b_2 \cdots b_m$. Observe that B^m has 2^m elements. That is, the order of the group B^m is 2^m.

Figure 11.1 shows the basic process of sending a word from one point to another point over a transmission channel. An element $x \in B^m$ is sent through the transmission channel and is received as an element $x_t \in B^m$. In actual practice, the transmission channel may suffer disturbances, which are generally called **noise**, due to weather interference, electrical problems, and so on, that may cause a 0 to be received as a 1, or vice versa. This erroneous transmission of digits in a word being sent may give rise to the situation where the word received is different from the word that was sent; that is, $x \neq x_t$. If an error does occur, then x_t could be any element of B^m.

Figure 11.1

The basic task in the transmission of information is to reduce the likelihood of receiving a word that differs from the word that was sent. This is done as follows. We first choose an integer $n > m$ and a one-to-one function $e: B^m \to B^n$. The function e is called an (m, n) **encoding function**, and we view it as a means of representing every word in B^m as a word in B^n. If $b \in B^m$, then $e(b)$ is called the **code word** representing b. The additional 0's and 1's can provide the means to detect or correct errors produced in the transmission channel.

We now transmit the code words by means of a transmission channel. Then each code word $x = e(b)$ is received as the word x_t in B^n. This situation is illustrated in Figure 11.2.

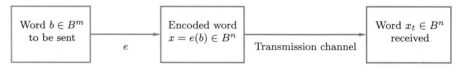

Figure 11.2

Observe that we want an encoding function e to be one to one so that different words in B^m will be assigned different code words.

If the transmission channel is noiseless, then $x_t = x$ for all x in B^n. In this case $x = e(b)$ is received for each $b \in B^m$, and since e is a known function, b may be identified.

In general, errors in transmission do occur. We will say that the code word $x = e(b)$ has been transmitted with **k or fewer errors** if x and x_t differ in at least 1 but no more than k positions.

Let $e: B^m \to B^n$ be an (m, n) encoding function. We say that e **detects k or fewer errors** if whenever $x = e(b)$ is transmitted with k or fewer errors, then x_t is not a code word (thus x_t could not be x and therefore could not have been correctly transmitted). For $x \in B^n$, the number of 1's in x is called the **weight** of x and is denoted by $|x|$.

EXAMPLE 1

Find the weight of each of the following words in B^5:

(a) $x = 01000$; (b) $x = 11100$; (c) $x = 00000$; (d) $x = 11111$.

Solution

(a) $|x| = 1$. (b) $|x| = 3$. (c) $|x| = 0$. (d) $|x| = 5$. ■

EXAMPLE 2

Parity Check Code

The following encoding function $e: B^m \to B^{m+1}$ is called the **parity $(m, m + 1)$ check code**: If $b = b_1 b_2 \cdots b_m \in B^m$, define

$$e(b) = b_1 b_2 \cdots b_m b_{m+1},$$

where

$$b_{m+1} = \begin{cases} 0 & \text{if } |b| \text{ is even} \\ 1 & \text{if } |b| \text{ is odd.} \end{cases}$$

Observe that b_{m+1} is zero if and only if the number of 1's in b is an even number. It then follows that every code word $e(b)$ has even weight. A single error in the transmission of a code word will change the received word to a word of odd weight and therefore can be detected. In the same way we see that any odd number of errors can be detected.

For a concrete illustration of this encoding function, let $m = 3$. Then

$$\left. \begin{array}{l} e(000) = 0000 \\ e(001) = 0011 \\ e(010) = 0101 \\ e(011) = 0110 \\ e(100) = 1001 \\ e(101) = 1010 \\ e(110) = 1100 \\ e(111) = 1111 \end{array} \right\} \text{ code words.}$$

Suppose now that $b = 111$. Then $x = e(b) = 1111$. If the transmission channel transmits x as $x_t = 1101$, then $|x_t| = 3$, and we know that an odd number of errors (at least one) has occurred. ■

It should be noted that if the received word has even weight, then we cannot conclude that the code word was transmitted correctly, since this encoding function does not detect an even number of errors. Despite this limitation, the parity check code is widely used.

EXAMPLE 3

Consider the following $(m, 3m)$ encoding function $e: B^m \to B^{3m}$. If

$$b = b_1 b_2 \cdots b_m \in B^m,$$

define

$$e(b) = e(b_1 b_2 \cdots b_m) = b_1 b_2 \cdots b_m b_1 b_2 \cdots b_m b_1 b_2 \cdots b_m.$$

That is, the encoding function e repeats each word of B^m three times. For a concrete example, let $m = 3$. Then

$$\left.\begin{array}{l} e(000) = 000000000 \\ e(001) = 001001001 \\ e(010) = 010010010 \\ e(011) = 011011011 \\ e(100) = 100100100 \\ e(101) = 101101101 \\ e(110) = 110110110 \\ e(111) = 111111111 \end{array}\right\} \text{ code words.}$$

Suppose now that $b = 011$. Then $e(011) = 011011011$. Assume now that the transmission channel makes an error in the underlined digit and that we receive the word 011111011. This is not a code word, so we have detected the error. It is not hard to see that any single error and any two errors can be detected. ∎

Let x and y be words in B^m. The **Hamming distance** $\delta(x, y)$ between x and y is the weight, $|x \oplus y|$, of $x \oplus y$. Thus the distance between $x = x_1 x_2 \cdots x_m$ and $y = y_1 y_2 \cdots y_m$ is the number of values of i such that $x_i \neq y_i$, that is, the number of positions in which x and y differ. Using the weight of $x \oplus y$ is a convenient way to count the number of different positions.

EXAMPLE 4   Find the distance between x and y:

(a) $x = 110110$, $y = 000101$.
(b) $x = 001100$, $y = 010110$.

Solution

(a) $x \oplus y = 110011$, so $|x \oplus y| = 4$.
(b) $x \oplus y = 011010$, so $|x \oplus y| = 3$. ∎

Theorem 1
Properties of the
Distance Function

Let x, y, and z be elements of B^m. Then
(a) $\delta(x, y) = \delta(y, x)$.
(b) $\delta(x, y) \geq 0$.
(c) $\delta(x, y) = 0$ if and only if $x = y$.
(d) $\delta(x, y) \leq \delta(x, z) + \delta(z, y)$.   ●

Proof   Properties (a), (b), and (c) are simple to prove and are left as exercises.
(d) For a and b in B^m,

$$|a \oplus b| \leq |a| + |b|,$$

since at any position where a and b differ one of them must contain a 1. Also, if $a \in B^m$, then $a \oplus a = \bar{0}$, the identity element in B^m. Then

$$\delta(x, y) = |x \oplus y| = |x \oplus \bar{0} \oplus y| = |x \oplus z \oplus z \oplus y|$$
$$\leq |x \oplus z| + |z \oplus y|$$
$$= \delta(x, z) + \delta(z, y).$$ ▼

The **minimum distance** of an encoding function $e \colon B^m \to B^n$ is the minimum of the distances between all distinct pairs of code words; that is,

$$\min\{\delta(e(x), e(y)) \mid x, y \in B^m\}.$$

EXAMPLE 5

Consider the following $(2, 5)$ encoding function e:

$$\left. \begin{array}{l} e(00) = 00000 \\ e(10) = 00111 \\ e(01) = 01110 \\ e(11) = 11111 \end{array} \right\} \quad \text{code words.}$$

The minimum distance is 2, as can be checked by computing the minimum of the distances between all six distinct pairs of code words.   ■

Theorem 2   An (m, n) encoding function $e \colon B^m \to B^n$ can detect k or fewer errors if and only if its minimum distance is at least $k + 1$.   ●

Proof   Suppose that the minimum distance between any two code words is at least $k + 1$. Let $b \in B^m$, and let $x = e(b) \in B^n$ be the code word representing b. Then x is transmitted and is received as x_t. If x_t were a code word different from x, then $\delta(x, x_t) \geq k + 1$, so x would be transmitted with $k + 1$ or more errors. Thus, if x is transmitted with k or fewer errors, then x_t cannot be a code word. This means that e can detect k or fewer errors.

Conversely, suppose that the minimum distance between code words is $r \leq k$, and let x and y be code words with $\delta(x, y) = r$. If $x_t = y$, that is, if x is transmitted and is mistakenly received as y, then $r \leq k$ errors have been committed and have not been detected. Thus it is not true that e can detect k or fewer errors.   ▼

EXAMPLE 6

Consider the $(3, 8)$ encoding function $e \colon B^3 \to B^8$ defined by

$$\left. \begin{array}{l} e(000) = 00000000 \\ e(001) = 10111000 \\ e(010) = 00101101 \\ e(011) = 10010101 \\ e(100) = 10100100 \\ e(101) = 10001001 \\ e(110) = 00011100 \\ e(111) = 00110001 \end{array} \right\} \quad \text{code words.}$$

How many errors will e detect?

Solution   The minimum distance of e is 3, as can be checked by computing the minimum of the distances between all 28 distinct pairs of code words. By Theorem 2, the code will detect k or fewer errors if and only if its minimum distance is at least $k + 1$. Since the minimum distance is 3, we have $3 \geq k + 1$ or $k \leq 2$. Thus the code will detect two or fewer errors.   ∎

Group Codes

So far, we have not made use of the fact that $(B^n, \oplus)$ is a group. We shall now consider an encoding function that makes use of this property of B^n.

An (m, n) encoding function $e \colon B^m \to B^n$ is called a **group code** if

$$e(B^m) = \{e(b) \mid b \in B^m\} = \text{Ran}(e)$$

is a subgroup of B^n.

Recall from the definition of subgroup given in Section 9.4 that N is a subgroup of B^n if (a) the identity of B^n is in N, (b) if x and y belong to N, then $x \oplus y \in N$, and (c) if x is in N, then its inverse is in N. Property (c) need not be checked here, since every element in B^n is its own inverse. Moreover, since B^n is Abelian, every subgroup of B^n is a normal subgroup.

EXAMPLE 7   Consider the $(3, 6)$ encoding function $e \colon B^3 \to B^6$ defined by

$$
\left. \begin{array}{l}
e(000) = 000000 \\
e(001) = 001100 \\
e(010) = 010011 \\
e(011) = 011111 \\
e(100) = 100101 \\
e(101) = 101001 \\
e(110) = 110110 \\
e(111) = 111010
\end{array} \right\} \text{code words.}
$$

Show that this encoding function is a group code.

Solution   We must show that the set of all code words

$$N = \{000000, 001100, 010011, 011111, 100101, 101001, 110110, 111010\}$$

is a subgroup of B^6. This is done by first noting that the identity of B^6 belongs to N. Next we verify, by trying all possibilities, that if x and y are elements in N, then $x \oplus y$ is in N. Hence N is a subgroup of B^6, and the given encoding function is a group code.   ∎

The strategy of the next proof is similar to the way we often show two sets A and B are the same by showing that $A \subseteq B$ and $B \subseteq A$. Here we show that $\delta = \eta$ by proving $\delta \leq \eta$ and $\eta \leq \delta$.

Theorem 3   Let $e \colon B^m \to B^n$ be a group code. The minimum distance of e is the minimum weight of a nonzero code word.   ●

Proof Let δ be the minimum distance of the group code, and suppose that $\delta = \delta(x, y)$, where x and y are distinct code words. Also, let η be the minimum weight of a nonzero code word and suppose that $\eta = |z|$ for a code word z. Since e is a group code, $x \oplus y$ is a nonzero code word. Thus

$$\delta = \delta(x, y) = |x \oplus y| \geq \eta.$$

On the other hand, since 0 and z are distinct code words,

$$\eta = |z| = |z \oplus 0| = \delta(z, 0) \geq \delta.$$

Hence $\eta = \delta$. ▼

One advantage of a group code is given in the following example.

EXAMPLE 8   The minimum distance of the group code in Example 7 is 2, since by Theorem 3 this distance is equal to the smallest number of 1's in any of the seven nonzero code words. To check this directly would require 28 different calculations.   ∎

We shall now take a brief look at a procedure for generating group codes. First, we need several additional results on Boolean matrices. Consider the set B with operation $+$ defined in Table 11.1. Now let $\mathbf{D} = [d_{ij}]$ and $\mathbf{E} = [e_{ij}]$ be $m \times n$ Boolean matrices. We define the **mod-2 sum** $\mathbf{D} \oplus \mathbf{E}$ as the $m \times n$ Boolean matrix $\mathbf{F} = [f_{ij}]$, where

$$f_{ij} = d_{ij} + e_{ij}, \qquad 1 \leq i \leq m, \qquad 1 \leq j \leq n. \qquad \text{(Here } + \text{ is addition in } B.)$$

EXAMPLE 9   We have

$$
\begin{bmatrix} 1 & 0 & 1 & 1 \\ 0 & 1 & 1 & 0 \\ 1 & 0 & 0 & 1 \end{bmatrix} \oplus \begin{bmatrix} 1 & 1 & 0 & 1 \\ 1 & 1 & 0 & 1 \\ 0 & 1 & 1 & 1 \end{bmatrix} = \begin{bmatrix} 1+1 & 0+1 & 1+0 & 1+1 \\ 0+1 & 1+1 & 1+0 & 0+1 \\ 1+0 & 0+1 & 0+1 & 1+1 \end{bmatrix}
$$

$$
= \begin{bmatrix} 0 & 1 & 1 & 0 \\ 1 & 0 & 1 & 1 \\ 1 & 1 & 1 & 0 \end{bmatrix}.
$$

Observe that if $\mathbf{F} = \mathbf{D} \oplus \mathbf{E}$, then f_{ij} is zero when *both* d_{ij} and e_{ij} are zero or both are one.   ∎

Table 11.2

·	0	1
0	0	0
1	0	1

Next, consider the set $B = \{0, 1\}$ with the binary operation given in Table 11.2. This operation has been seen earlier in a different setting and with a different symbol. In Chapter 6 it was shown that B is the unique Boolean algebra with two elements. In particular, B is a lattice with partial order $\leq$ defined by $0 \leq 0, 0 \leq 1$, $1 \leq 1$. Then the reader may easily check that if a and b are any two elements of B,

$$a \cdot b = a \wedge b \qquad \text{(the greatest lower bound of } a \text{ and } b).$$

Thus Table 11.2 for · is just a copy of the table for $\wedge$.

Let $\mathbf{D} = [\, d_{ij} \,]$ be an $m \times p$ Boolean matrix, and let $\mathbf{E} = [\, e_{ij} \,]$ be a $p \times n$ Boolean matrix. We define the **mod-2 Boolean product** $\mathbf{D} * \mathbf{E}$ as the $m \times n$ matrix $\mathbf{F} = [\, f_{ij} \,]$, where

$$f_{ij} = d_{i1} \cdot e_{1j} + d_{i2} \cdot e_{2j} + \cdots + d_{ip} \cdot e_{pj}, \qquad 1 \le i \le m, \quad 1 \le j \le n.$$

This type of multiplication is illustrated in Figure 11.3. Compare this with similar figures in Section 1.5.

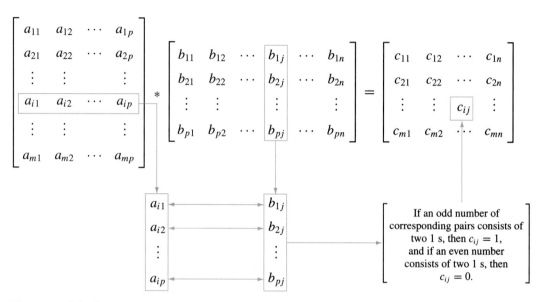

Figure 11.3

EXAMPLE 10   We have

$$\begin{bmatrix} 1 & 1 & 0 \\ 0 & 1 & 1 \end{bmatrix} * \begin{bmatrix} 1 & 0 \\ 1 & 1 \\ 0 & 1 \end{bmatrix} = \begin{bmatrix} 1 \cdot 1 + 1 \cdot 1 + 0 \cdot 0 & 1 \cdot 0 + 1 \cdot 1 + 0 \cdot 1 \\ 0 \cdot 1 + 1 \cdot 1 + 1 \cdot 0 & 0 \cdot 0 + 1 \cdot 1 + 1 \cdot 1 \end{bmatrix}$$

$$= \begin{bmatrix} 0 & 1 \\ 1 & 0 \end{bmatrix}. \qquad \blacksquare$$

The proof of the following theorem is left as an exercise.

Theorem 4   Let $\mathbf{D}$ and $\mathbf{E}$ be $m \times p$ Boolean matrices, and let $\mathbf{F}$ be a $p \times n$ Boolean matrix. Then

$$(\mathbf{D} \oplus \mathbf{E}) * \mathbf{F} = (\mathbf{D} * \mathbf{F}) \oplus (\mathbf{E} * \mathbf{F}).$$

That is, a distributive property holds for $\oplus$ and $*$.   ●

We shall now consider the element $x = x_1 x_2 \cdots x_n \in B^n$ as the $1 \times n$ matrix $[\, x_1 \quad x_2 \quad \cdots \quad x_n \,]$.

Theorem 5 Let m and n be nonnegative integers with $m < n$, $r = n - m$, and let $\mathbf{H}$ be an $n \times r$ Boolean matrix. Then the function $f_H \colon B^n \to B^r$ defined by

$$f_H(x) = x * \mathbf{H}, \qquad x \in B^n$$

is a homomorphism from the group B^n to the group B^r.

Proof If x and y are elements in B^n, then

$$
\begin{aligned}
f_H(x \oplus y) &= (x \oplus y) * \mathbf{H} \\
&= (x * \mathbf{H}) \oplus (y * \mathbf{H}) \qquad \text{by Theorem 4} \\
&= f_H(x) \oplus f_H(y).
\end{aligned}
$$

Hence f_H is a homomorphism from B^n to B^r.

Corollary 1 Let $m, n, r, \mathbf{H}$, and f_H be as in Theorem 5. Then

$$N = \{x \in B^n \mid x * \mathbf{H} = \bar{0}\}$$

is a normal subgroup of B^n.

Proof It follows from the results in Section 9.5 that N is the kernel of the homomorphism f_H, so it is a normal subgroup of B^n.

Let $m < n$ and $r = n - m$. An $n \times r$ Boolean matrix

$$
\mathbf{H} =
\begin{bmatrix}
h_{11} & h_{12} & \cdots & h_{1r} \\
h_{21} & h_{22} & \cdots & h_{2r} \\
\vdots & \vdots & & \vdots \\
h_{m1} & h_{m2} & \cdots & h_{mr} \\
1 & 0 & \cdots & 0 \\
0 & 1 & \cdots & 0 \\
\vdots & \vdots & & \vdots \\
0 & 0 & \cdots & 1
\end{bmatrix}
\Biggr\} \, n - m = r \text{ rows}
$$

whose last r rows form the $r \times r$ identity matrix, is called a **parity check matrix**. We use $\mathbf{H}$ to define an encoding function $e_H \colon B^m \to B^n$. If $b = b_1 b_2 \cdots b_m$, let $x = e_H(b) = b_1 b_2 \cdots b_m x_1 x_2 \cdots x_r$, where

$$
\begin{aligned}
x_1 &= b_1 \cdot h_{11} + b_2 \cdot h_{21} + \cdots + b_m \cdot h_{m1} \\
x_2 &= b_1 \cdot h_{12} + b_2 \cdot h_{22} + \cdots + b_m \cdot h_{m2} \\
&\vdots \\
x_r &= b_1 \cdot h_{1r} + b_2 \cdot h_{2r} + \cdots + b_m \cdot h_{mr}.
\end{aligned}
\tag{1}
$$

Theorem 6 Let $x = y_1 y_2 \cdots y_m x_1 \cdots x_r \in B^n$. Then $x * \mathbf{H} = \bar{0}$ if and only if $x = e_H(b)$ for some $b \in B^m$.

Proof   Suppose that $x * \mathbf{H} = \bar{0}$. Then

$$y_1 \cdot h_{11} + y_2 \cdot h_{21} + \cdots + y_m \cdot h_{m1} + x_1 = 0$$
$$y_1 \cdot h_{12} + y_2 \cdot h_{22} + \cdots + y_m \cdot h_{m2} + x_2 = 0$$
$$\vdots$$
$$y_1 \cdot h_{1r} + y_2 \cdot h_{2r} + \cdots + y_m \cdot h_{mr} + x_r = 0.$$

The first equation is of the form

$$a + x_1 = 0, \quad \text{where } a = y_1 \cdot h_{11} + y_2 \cdot h_{21} + \cdots + y_m \cdot h_{m1}.$$

Adding a to both sides, we obtain

$$a + (a + x_1) = a + 0 = a$$
$$(a + a) + x_1 = a$$
$$0 + x_1 = a \quad \text{since } a + a = 0$$
$$x_1 = a.$$

This can be done for each row; therefore,

$$x_i = y_1 \cdot h_{1i} + y_2 \cdot h_{2i} + \cdots + y_m \cdot h_{mi}, \qquad 1 \le i \le r.$$

Letting $b_1 = y_1$, $b_2 = y_2$, ..., $b_m = y_m$, we see that $x_1, x_2, \ldots, x_r$ satisfy the equations in (1). Thus $b = b_1 b_2 \cdots b_m \in B^m$ and $x = e_H(b)$.

Conversely, if $x = e_H(b)$, the equations in (1) can be rewritten by adding x_i to both sides of the ith equation, $i = 1, 2, \ldots, n$, as

$$b_1 \cdot h_{11} + b_2 \cdot h_{21} + \cdots + b_m \cdot h_{m1} + x_1 = 0$$
$$b_1 \cdot h_{12} + b_2 \cdot h_{22} + \cdots + b_m \cdot h_{m2} + x_2 = 0$$
$$\vdots$$
$$b_1 \cdot h_{1r} + b_2 \cdot h_{2r} + \cdots + b_m \cdot h_{mr} + x_r = 0,$$

which shows $x * \mathbf{H} = \bar{0}$.   ▼

Corollary 2 $e_H(B^m) = \{e_H(b) \mid b \in B^m\}$ is a subgroup of B^n.   ●

Proof   The result follows from the observation that

$$e_H(B^m) = \ker(f_H)$$

and from Corollary 1. Thus e_H is a group code.   ▼

EXAMPLE 11   Let $m = 2$, $n = 5$, and

$$\mathbf{H} = \begin{bmatrix} 1 & 1 & 0 \\ 0 & 1 & 1 \\ 1 & 0 & 0 \\ 0 & 1 & 0 \\ 0 & 0 & 1 \end{bmatrix}.$$

Determine the group code $e_H \colon B^2 \to B^5$.

Solution   We have $B^2 = \{00, 10, 01, 11\}$. Then

$$e(00) = 00x_1x_2x_3,$$

where x_1, x_2, and x_3 are determined by the equations in (1). Thus

$$x_1 = x_2 = x_3 = 0$$

and

$$e(00) = 00000.$$

Next

$$e(10) = 10x_1x_2x_3.$$

Using the equations in (1) with $b_1 = 1$ and $b_2 = 0$, we obtain

$$x_1 = 1 \cdot 1 + 0 \cdot 0 = 1$$
$$x_2 = 1 \cdot 1 + 0 \cdot 1 = 1$$
$$x_3 = 1 \cdot 0 + 0 \cdot 1 = 0.$$

Thus $x_1 = 1$, $x_2 = 1$, and $x_3 = 0$, so

$$e(10) = 10110.$$

Similarly (verify),

$$e(01) = 01011$$
$$e(11) = 11101.$$ ■

11.1 Exercises

Find the weights of the given words.

1. (a) 1011   (b) 0110   (c) 1110

2. (a) 011101   (b) 11111   (c) 010101

3. Consider the $(3, 4)$ parity check code. For each of the received words, determine whether an error will be detected.

   (a) 0100   (b) 1100

4. Consider the $(3, 4)$ parity check code. For each of the received words, determine whether an error will be detected.

   (a) 0010   (b) 1001

5. Consider the $(6, 7)$ parity check code. For each of the received words, determine whether an error will be detected.

   (a) 1101010   (b) 1010011

   (c) 0011111   (d) 1001101

6. Consider the $(m, 3m)$ encoding function of Example 3, where $m = 4$. For each of the received words, determine whether an error will be detected.

   (a) 011010011111   (b) 110110010110

7. Consider the $(m, 3m)$ encoding function of Example 3, where $m = 4$. For each of the received words, determine whether an error will be detected.

   (a) 010010110010   (b) 001001111001

8. Explain how $|x \oplus y|$ counts the number of positions in which x and y differ.

9. Find the distance between x and y.

   (a) $x = 1100010$, $y = 1010001$

   (b) $x = 0100110$, $y = 0110010$

10. Find the distance between x and y.

    (a) $x = 00111001$, $y = 10101001$

    (b) $x = 11010010$, $y = 00100111$

11. (a) Prove Theorem 1(a).

(b) Prove Theorem 1(b).

12. Prove Theorem 1(c).

13. Find the minimum distance of the $(2, 4)$ encoding function e.

$$e(00) = 0000 \qquad e(10) = 0110$$
$$e(01) = 1011 \qquad e(11) = 1100$$

14. Find the minimum distance of the $(3, 8)$ encoding function e.

$$e(000) = 00000000 \qquad e(100) = 01100101$$
$$e(001) = 01110010 \qquad e(101) = 10110000$$
$$e(010) = 10011100 \qquad e(110) = 11110000$$
$$e(011) = 01110001 \qquad e(111) = 00001111$$

15. Consider the $(2, 6)$ encoding function e.

$$e(00) = 000000 \qquad e(10) = 101010$$
$$e(01) = 011110 \qquad e(11) = 111000$$

(a) Find the minimum distance of e.

(b) How many errors will e detect?

16. Consider the $(3, 9)$ encoding function e.

$$e(000) = 000000000 \qquad e(100) = 010011010$$
$$e(001) = 011100101 \qquad e(101) = 111101011$$
$$e(010) = 010101000 \qquad e(110) = 001011000$$
$$e(011) = 110010001 \qquad e(111) = 110000111$$

(a) Find the minimum distance of e.

(b) How many errors will e detect?

17. Show that the $(2, 5)$ encoding function $e: B^2 \to B^5$ defined by

$$e(00) = 00000 \qquad e(10) = 10101$$
$$e(01) = 01110 \qquad e(11) = 11011$$

is a group code.

18. Show that the $(3, 7)$ encoding function $e: B^3 \to B^7$ defined by

$$e(000) = 0000000 \qquad e(100) = 1000101$$
$$e(001) = 0010110 \qquad e(101) = 1010011$$
$$e(010) = 0101000 \qquad e(110) = 1101101$$
$$e(011) = 0111110 \qquad e(111) = 1111011$$

is a group code.

19. Find the minimum distance of the group code defined in Exercise 17.

20. Find the minimum distance of the group code defined in Exercise 18.

21. Compute

$$\begin{bmatrix} 1 & 1 & 0 \\ 0 & 1 & 1 \end{bmatrix} \oplus \begin{bmatrix} 1 & 1 & 1 \\ 0 & 1 & 1 \end{bmatrix}.$$

22. Compute

$$\begin{bmatrix} 1 & 0 & 1 \\ 1 & 1 & 0 \\ 0 & 1 & 0 \\ 0 & 1 & 1 \end{bmatrix} \oplus \begin{bmatrix} 1 & 0 & 1 \\ 0 & 1 & 0 \\ 1 & 1 & 1 \\ 0 & 0 & 1 \end{bmatrix}.$$

23. Compute

$$\begin{bmatrix} 1 & 0 \\ 1 & 1 \\ 0 & 1 \end{bmatrix} * \begin{bmatrix} 1 & 1 & 0 \\ 0 & 1 & 1 \end{bmatrix}.$$

24. Compute

$$\begin{bmatrix} 1 & 0 & 1 \\ 0 & 1 & 1 \\ 1 & 0 & 1 \end{bmatrix} * \begin{bmatrix} 1 & 1 & 0 \\ 0 & 1 & 1 \\ 1 & 0 & 1 \end{bmatrix}.$$

25. Let

$$\mathbf{H} = \begin{bmatrix} 0 & 1 & 1 \\ 0 & 1 & 1 \\ 1 & 0 & 0 \\ 0 & 1 & 0 \\ 0 & 0 & 1 \end{bmatrix}$$

be a parity check matrix. Determine the $(2, 5)$ group code function $e_H: B^2 \to B^5$.

26. Let

$$\mathbf{H} = \begin{bmatrix} 1 & 0 & 0 \\ 0 & 1 & 1 \\ 1 & 1 & 1 \\ 1 & 0 & 0 \\ 0 & 1 & 0 \\ 0 & 0 & 1 \end{bmatrix}$$

be a parity check matrix. Determine the $(3, 6)$ group code $e_H: B^3 \to B^6$.

27. Prove Theorem 4.

11.2 DECODING AND ERROR CORRECTION

Consider an (m, n) encoding function $e \colon B^m \to B^n$. Once the encoded word $x = e(b) \in B^n$, for $b \in B^m$, is received as the word x_t, we are faced with the problem of identifying the word b that was the original message.

An onto function $d \colon B^n \to B^m$ is called an (n, m) **decoding function associated with e** if $d(x_t) = b' \in B^m$ is such that when the transmission channel has no noise, then $b' = b$, that is,

$$d \circ e = 1_{B^m},$$

where 1_{B^m} is the identity function on B^m. The decoding function d is required to be onto so that every received word can be decoded to give a word in B^m. It decodes properly received words correctly, but the decoding of improperly received words may or may not be correct.

Consider the parity check code that is defined in Example 2 of Section 11.1. We now define the decoding function $d \colon B^{m+1} \to B^m$. If $y = y_1 y_2 \cdots y_m y_{m+1} \in B^{m+1}$, then

$$d(y) = y_1 y_2 \cdots y_m.$$

Observe that if $b = b_1 b_2 \cdots b_m \in B^m$, then

$$(d \circ e)(b) = d(e(b)) = b,$$

so $d \circ e = 1_{B^m}$.

For a concrete example, let $m = 4$. Then we obtain $d(10010) = 1001$ and $d(11001) = 1100$. ∎

Let e be an (m, n) encoding function and let d be an (n, m) decoding function associated with e. We say that the pair (e, d) **corrects k or fewer errors** if whenever $x = e(b)$ is transmitted correctly or with k or fewer errors and x_t is received, then $d(x_t) = b$. Thus x_t is decoded as the correct message b.

Consider the $(m, 3m)$ encoding function defined in Example 3 of Section 11.1. We now define the decoding function $d \colon B^{3m} \to B^m$. Let

$$y = y_1 y_2 \cdots y_m y_{m+1} \cdots y_{2m} y_{2m+1} \cdots y_{3m}.$$

Then

$$d(y) = z_1 z_2 \cdots z_m,$$

where

$$z_i = \begin{cases} 1 & \text{if } \{y_i, y_{i+m}, y_{i+2m}\} \text{ has at least two 1's} \\ 0 & \text{if } \{y_i, y_{i+m}, y_{i+2m}\} \text{ has less than two 1's.} \end{cases}$$

That is, the decoding function d examines the ith digit in each of the three blocks transmitted. The digit that occurs at least twice in these three blocks is chosen as the decoded ith digit. For a concrete example, let $m = 3$. Then

$$e(100) = 100100100$$
$$e(011) = 011011011$$
$$e(001) = 001001001.$$

Suppose now that $b = 011$. Then $e(011) = 011\underline{0}11011$. Assume now that the transmission channel makes an error in the underlined digit and that we receive the word $x_t = 011111011$. Then, since the first digits in two out of the three blocks are 0, the first digit is decoded as 0. Similarly, the second digit is decoded as 1, since all three second digits in the three blocks are 1. Finally, the third digit is also decoded as 1, for the analogous reason. Hence $d(x_t) = 011$; that is, the decoded word is 011, which is the word that was sent. Therefore, the single error has been corrected. A similar analysis shows that, if e is this $(m, 3m)$ code for any value of m and d is as defined, then (e, d) corrects any single error.   ■

Given an (m, n) encoding function $e: B^m \to B^n$, we often need to determine an (n, m) decoding function $d: B^n \to B^m$ associated with e. We now discuss a method, called the **maximum likelihood technique**, for determining a decoding function d for a given e.

Since B^m has 2^m elements, there are 2^m code words in B^n. We first list the code words in a fixed order:

$$x^{(1)}, x^{(2)}, \ldots, x^{(2^m)}.$$

If the received word is x_t, we compute $\delta(x^{(i)}, x_t)$ for $1 \le i \le 2^m$ and choose the first code word, say it is $x^{(s)}$, such that

$$\min_{1 \le i \le 2^m} \{\delta(x^{(i)}, x_t)\} = \delta(x^{(s)}, x_t).$$

That is, $x^{(s)}$ is a code word that is closest to x_t and the first in the list. If $x^{(s)} = e(b)$, we define the **maximum likelihood decoding function** d associated with e by

$$d(x_t) = b.$$

Observe that d depends on the particular order in which the code words in $e(B^n)$ are listed. If the code words are listed in a different order, we may obtain a different maximum likelihood decoding function d associated with e.

Theorem 1   Suppose that e is an (m, n) encoding function and d is a maximum likelihood decoding function associated with e. Then (e, d) can correct k or fewer errors if and only if the minimum distance of e is at least $2k + 1$.   ●

Proof   Assume that the minimum distance of e is at least $2k + 1$. Let $b \in B^m$ and $x = e(b) \in B^n$. Suppose that x is transmitted with k or fewer errors, and x_t is received. This means that $\delta(x, x_t) \le k$. If z is any other code word, then

$$2k + 1 \le \delta(x, z) \le \delta(x, x_t) + \delta(x_t, z) \le k + d(x_t, z).$$

Thus $\delta(x_t, z) \ge 2k + 1 - k = k + 1$. This means that x is the unique code word that is closest to x_t, so $d(x_t) = b$. Hence (e, d) corrects k or fewer errors.

Conversely, assume that the minimum distance between code words is $r \le 2k$, and let $x = e(b)$ and $x' = e(b')$ be code words with $\delta(x, x') = r$. Suppose that x' precedes x in the list of code words used to define d. Write $x = b_1 b_2 \cdots b_n$, $x' = b_1' b_2' \cdots b_n'$. Then $b_i \ne b_i'$ for exactly r integers i between 1 and n. Assume, for simplicity, that $b_1 \ne b_1', b_2 \ne b_2', \ldots, b_r \ne b_r'$, but $b_i = b_i'$ when $i > r$. Any other case is handled in the same way.

(a) Suppose that $r \leq k$. If x is transmitted as $x_t = x'$, then $r \leq k$ errors have been committed, but $d(x_t) = b'$; so (e, d) has not corrected the r errors.

(b) Suppose that $k + 1 \leq r \leq 2k$, and let

$$y = b'_1 b'_2 \cdots b'_k b_{k+1} \cdots b_n.$$

If x is transmitted as $x_t = y$, then $\delta(x_t, x') = r - k \leq k$ and $\delta(x_t, x) \geq k$. Thus x' is at least as close to x_t as x is, and x' precedes x in the list of code words; so $d(x_t) \neq b$. Then we have committed k errors, which (e, d) has not corrected. ▼

EXAMPLE 3 Let e be the $(3, 8)$ encoding function defined in Example 6 of Section 11.1, and let d be an $(8, 3)$ maximum likelihood decoding function associated with e. How many errors can (e, d) correct?

Solution Since the minimum distance of e is 3, we have $3 \geq 2k + 1$, so $k \leq 1$. Thus (e, d) can correct one error. ∎

We now discuss a simple and effective technique for determining a maximum likelihood decoding function associated with a given group code. First, we prove the following result.

Theorem 2 If K is a finite subgroup of a group G, then every left coset of K in G has exactly as many elements as K. ●

Proof Let aK be a left coset of K in G, where $a \in G$. Consider the function $f : K \to aK$ defined by

$$f(k) = ak, \qquad \text{for } k \in K.$$

We show that f is one to one and onto.

To show that f is one to one, we assume that

$$f(k_1) = f(k_2), \qquad k_1, k_2 \in K.$$

Then

$$ak_1 = ak_2.$$

By Theorem 2 of Section 9.4, $k_1 = k_2$. Hence f is one to one.

To show that f is onto, let b be an arbitrary element in aK. Then $b = ak$ for some $k \in K$. We now have

$$f(k) = ak = b,$$

so f is onto. Since f is one to one and onto, K and aK have the same number of elements. ▼

Let $e : B^m \to B^n$ be an (m, n) encoding function that is a group code. Thus the set N of code words in B^n is a subgroup of B^n whose order is 2^m, say $N = \{x^{(1)}, x^{(2)}, \ldots, x^{(2^m)}\}$.

Suppose that the code word $x = e(b)$ is transmitted and that the word x_t is received. The left coset of x_t is

$$x_t \oplus N = \{\epsilon_1, \epsilon_2, \ldots, \epsilon_{2^m}\},$$

where $\epsilon_i = x_t \oplus x^{(i)}$. The distance from x_t to code word $x^{(i)}$ is just $|\epsilon_i|$, the weight of ϵ_i. Thus, if ϵ_j is a coset member with smallest weight, then $x^{(j)}$ must be a code word that is closest to x_t. In this case, $x^{(j)} = \overline{0} \oplus x^{(j)} = x_t \oplus x_t \oplus x^{(j)} = x_t \oplus \epsilon_j$. An element ϵ_j, having smallest weight, is called a **coset leader**. Note that a coset leader need not be unique.

If $e \colon B^m \to B^n$ is a group code, we now state the following procedure for obtaining a maximum likelihood decoding function associated with e.

STEP 1:   Determine all the left cosets of $N = e(B^m)$ in B^n.

STEP 2:   For each coset, find a coset leader (a word of least weight). Steps 1 and 2 can be carried out in a systematic tabular manner that will be described later.

STEP 3:   If the word x_t is received, determine the coset of N to which x_t belongs. Since N is a normal subgroup of B^n, it follows from Theorems 3 and 4 of Section 9.5 that the cosets of N form a partition of B^n, so each element of B^n belongs to one and only one coset of N in B^n. Moreover, there are $2^n/2^m$ or 2^r distinct cosets of N in B^n.

STEP 4:   Let ϵ be a coset leader for the coset determined in Step 3. Compute $x = x_t \oplus \epsilon$. If $x = e(b)$, we let $d(x_t) = b$. That is, we decode x_t as b.

To implement the foregoing procedure, we must keep a complete list of all the cosets of N in B^n, usually in tabular form, with each row of the table containing one coset. We identify a coset leader in each row. Then, when a word x_t is received, we locate the row that contains it, find the coset leader for that row, and add it to x_t. This gives us the code word closest to x_t. We can eliminate the need for these additions if we construct a more systematic table.

Before illustrating with an example, we make several observations. Let

$$N = \{x^{(1)}, x^{(2)}, \ldots, x^{(2^m)}\},$$

where $x^{(1)}$ is $\overline{0}$, the identity of B^n.

Steps 1 and 2 in the preceding decoding algorithm are carried out as follows. First, list all the elements of N in a row, starting with the identity $\overline{0}$ at the left. Thus we have

$$\overline{0} \quad x^{(2)} \quad x^{(3)} \quad \cdots \quad x^{(2^m)}.$$

This row is the coset $[\overline{0}]$, and it has $\overline{0}$ as its coset leader. For this reason we will also refer to $\overline{0}$ as ϵ_1. Now choose any word y in B^n that has not been listed in the first row. List the elements of the coset $y \oplus N$ as the second row. This coset also has 2^m elements. Thus we have the two rows

$$\begin{array}{ccccc} \overline{0} & x^{(2)} & x^{(3)} & \cdots & x^{(2^m)} \\ y \oplus \overline{0} & y \oplus x^{(2)} & y \oplus x^{(3)} & \cdots & y \oplus x^{(2^m)}. \end{array}$$

In the coset $y \oplus N$, pick an element of least weight, a coset leader, which we denote by $\epsilon^{(2)}$. In case of ties, choose any element of least weight. Recall from Section 9.5 that, since $\epsilon^{(2)} \in y \oplus N$, we have $y \oplus N = \epsilon^{(2)} \oplus N$. This means that every word in the second row can be written as $\epsilon^{(2)} \oplus \nu$, $\nu \in N$. Now rewrite the second row as follows:

$$\epsilon^{(2)} \quad \epsilon^{(2)} \oplus x^{(2)} \quad \epsilon^{(2)} \oplus x^{(3)} \quad \cdots \quad \epsilon^{(2)} \oplus x^{(2^m)}$$

with $\epsilon^{(2)}$ in the leftmost position.

Next, choose another element z in B^n that has not yet been listed in either of the first two rows and form the third row $(z \oplus x^{(j)})$, $1 \leq j \leq 2^m$ (another coset of N in B^n). This row can be rewritten in the form

$$\epsilon^{(3)} \quad \epsilon^{(3)} \oplus x^{(2)} \quad \epsilon^{(3)} \oplus x^{(3)} \quad \cdots \quad \epsilon^{(3)} \oplus x^{(2^m)},$$

where $\epsilon^{(3)}$ is a coset leader for the row.

Continue this process until all elements of B^n have been listed. The resulting Table 11.3 is called a **decoding table**. Notice that it contains 2^r rows, one for each coset of N. If we receive the word x_t, we locate it in the table. If x is the element of N that is at the top of the column containing x_t, then x is the code word closest to x_t. Thus, if $x = e(b)$, we let $d(x_t) = b$.

Table 11.3

$\overline{0}$	$x^{(2)}$	$x^{(3)}$	$\cdots$	$x^{(2^m-1)}$
$\epsilon^{(2)}$	$\epsilon^{(2)} \oplus x^{(2)}$	$\epsilon^{(2)} \oplus x^{(3)}$	$\cdots$	$\epsilon^{(2)} \oplus x^{(2^m-1)}$
$\vdots$	$\vdots$	$\vdots$		$\vdots$
$\epsilon^{(2^r)}$	$\epsilon^{(2^r)} \oplus x^{(2)}$	$\epsilon^{(2^r)} \oplus x^{(3)}$	$\cdots$	$\epsilon^{(2^r)} \oplus x^{(2^m-1)}$

EXAMPLE 4 Consider the $(3, 6)$ group code defined in Example 7 of Section 11.1. Here

$$N = \{000000, 001100, 010011, 011111, 100101, 101001, 110110, 111010\}$$
$$= \{x^{(1)}, x^{(2)}, \ldots, x^{(8)}\}$$

as defined in Example 1. We now implement the decoding procedure for e as follows.

STEPS 1 and 2: Determine all the left cosets of N in B^6, as rows of a table. For each row i, locate the coset leader ϵ_i, and rewrite the row in the order

$$\epsilon_i, \quad \epsilon_i \oplus 001100, \quad \epsilon_i \oplus 010011, \quad \ldots, \quad \epsilon_i \oplus 111010.$$

The result is shown in Table 11.4.

Table 11.4

000000	001100	010011	011111	100101	101001	110110	111010
000001	001101	010010	011110	100100	101000	110111	111011
000010	001110	010001	011101	100111	101011	110100	111000
000100	001000	010111	011011	100001	101101	110010	111110
010000	011100	000011	001111	110101	111001	100110	101010
100000	101100	110011	111111	<u>000101</u>	001001	010110	011010
000110	001010	<u>010101</u>	011001	100011	101111	110000	111100
010100	011000	000111	001011	110001	111101	100010	101110

STEPS 3 and 4:   If we receive the word 000101, we decode it by first locating it in the decoding table: it appears in the fifth column, where it is underlined. The word at the top of the fifth column is 100101. Since $e(100) = 100101$, we decode 000101 as 100. Similarly, if we receive the word 010101, we first locate it in the third column of the decoding table, where it is underlined twice. The word at the top of the third column is 010011. Since $e(010) = 010011$, we decode 010101 as 010.

We make the following observations for this example. In determining the decoding table in Steps 1 and 2, there was more than one candidate for coset leader of the last two cosets. In row 7 we chose 00110 as coset leader. If we had chosen 001010 instead, row 7 would have appeared in the rearranged form

$$001010 \quad 001010 \oplus 001100 \quad \cdots \quad 001010 \oplus 111010$$

or

$$001010 \quad 000110 \quad 011001 \quad 010101 \quad 101111 \quad 100011 \quad 111100 \quad 110000.$$

The new decoding table is shown in Table 11.5.

Table 11.5

000000	001100	010011	011111	100101	101001	110110	111010
000001	001101	010010	011110	100100	101000	110111	111011
000010	001110	010001	011101	100111	101011	110100	111000
000100	001000	010111	011011	100001	101101	110010	111110
010000	011100	000011	001111	110101	111001	100110	101010
100000	101100	110011	111111	000101	001001	010110	011010
001010	000110	011001	<u>010101</u>	101111	100011	111100	110000
010100	011000	000111	001011	110001	111101	100010	101110

Now, if we receive the word 010101, we first locate it in the *fourth* column of Table 11.5. The word at the top of the fourth column is 011111. Since $e(011) = 011111$, we decode 010101 as 011. ∎

Suppose that the (m, n) group code is $e_H: B^m \to B^n$, where **H** is a given parity check matrix. In this case, the decoding technique above can be simplified. We now turn to a discussion of this situation.

Recall from Section 11.1 that $r = n - m$,

$$\mathbf{H} = \begin{bmatrix} h_{11} & h_{12} & \cdots & h_{1r} \\ h_{21} & h_{22} & \cdots & h_{2r} \\ \vdots & \vdots & & \vdots \\ h_{m1} & h_{m2} & \cdots & h_{mr} \\ 1 & 0 & \cdots & 0 \\ 0 & 1 & \cdots & 0 \\ \vdots & \vdots & & \vdots \\ 0 & 0 & \cdots & 1 \end{bmatrix}$$

and the function $f_H: B^n \to B^r$ defined by

$$f_H(x) = x * \mathbf{H}$$

is a homomorphism from the group B^n to the group B^r.

Theorem 3   If $m, n, r, \mathbf{H}$, and f_H are as defined, then f_H is onto.   ●

Proof   Let $b = b_1 b_2 \cdots b_r$ be any element in B^r. Letting

$$x = \underbrace{00 \cdots 0}_{m \text{ 0's}} b_1 b_2 \cdots b_r$$

we obtain $x * \mathbf{H} = b$. Thus $f_H(x) = b$, so f_H is onto.   ▼

It follows from Corollary 1 of Section 9.5 that B^r and B^n/N are isomorphic, where $N = \ker(f_H) = e_H(B^m)$, under the isomorphism $g: B^n/N \to B^r$ defined by

$$g(xN) = f_H(x) = x * \mathbf{H}.$$

The element $x * \mathbf{H}$ is called the **syndrome** of x. We now have the following result.

Theorem 4   Let x and y be elements in B^n. Then x and y lie in the same left coset of N in B^n if and only if $f_H(x) = f_H(y)$, that is, if and only if they have the same syndrome.   ●

Proof   It follows from Theorem 4 of Section 9.5 that x and y lie in the same left coset of N in B^n if and only if $x \oplus y = (-x) \oplus y \in N$. Since $N = \ker(f_H)$, $x \oplus y \in N$ if and only if

$$f_H(x \oplus y) = \bar{0}_{B^r}$$
$$f_H(x) \oplus f_H(y) = \bar{0}_{B^r}$$
$$f_H(x) = f_H(y).$$ ▼

In this case, the decoding procedure given previously can be modified as follows. Suppose that we compute the syndrome of each coset leader. If the word x_t is received, we also compute $f_H(x_t)$, the syndrome of x_t. By comparing $f_H(x_t)$ and the syndromes of the coset leaders, we find the coset in which x_t lies. Suppose that a coset leader of this coset is ϵ. We now compute $x = x_t \oplus \epsilon$. If $x = e(b)$, we then decode x_t as b. Thus we need only the coset leaders and their syndromes in order to decode. We state the new procedure in detail.

STEP 1:   Determine all left cosets of $N = e_H(B^m)$ in B^n.

STEP 2:   For each coset, find a coset leader, and compute the syndrome of each leader.

STEP 3:   If x_t is received, compute the syndrome of x_t and find the coset leader ϵ having the same syndrome. Then $x_t \oplus \epsilon = x$ is a code word $e_H(b)$, and $d(x_t) = b$.

For this procedure, we do not need to keep a table of cosets, and we can avoid the work of computing a decoding table. Simply list all cosets once, in any order, and select a coset leader from each coset. Then keep a table of these coset leaders and their syndromes. The foregoing procedure is easily implemented with such a table.

EXAMPLE 5   Consider the parity check matrix

$$\mathbf{H} = \begin{bmatrix} 1 & 1 & 0 \\ 1 & 0 & 1 \\ 0 & 1 & 1 \\ 1 & 0 & 0 \\ 0 & 1 & 0 \\ 0 & 0 & 1 \end{bmatrix}$$

and the $(3, 6)$ group $e_H : B^3 \to B^6$. Then

$$\left. \begin{array}{l} e(000) = 000000 \\ e(001) = 001011 \\ e(010) = 010101 \\ e(011) = 011110 \\ e(100) = 100110 \\ e(101) = 101101 \\ e(110) = 110011 \\ e(111) = 111000 \end{array} \right\} \quad \text{code words.}$$

Table 11.6

Syndrome of Coset Leader	Coset Leader
000	000000
001	000001
010	000010
011	001000
100	000100
101	010000
110	100000
111	001100

Thus

$$N = \{000000, 001011, 010101, 011110, 100110, 101101, 110011, 111000\}.$$

We now implement the decoding procedure as follows.

In Table 11.6 we give only the coset leaders together with their syndromes. Suppose now that we receive the word 001110. We compute the syndrome of $x_t = 001110$, obtaining $f_H(x_t) = x_t * \mathbf{H} = 101$, which is the sixth entry in the first column of Table 11.6. This means that x_t lies in the coset whose leader is $\epsilon = 010000$. We compute $x = x_t \oplus \epsilon = 001110 \oplus 010000 = 011110$. Since $e(011) = 011110$, we decode 001110 as 011.   ■

11.2 Exercises

1. Let d be the $(4, 3)$ decoding function defined by letting m be 3 in Example 1. Determine $d(y)$ for the word y in B^4.

(a) $y = 0110$ (b) $y = 1011$

2. Let d be the $(6, 5)$ decoding function defined by letting m be 5 in Example 1. Determine $d(y)$ for the word y in B^6.

(a) $y = 001101$ (b) $y = 110100$

3. Let d be the $(6, 2)$ decoding function defined in Example 2. Determine $d(y)$ for the word y in B^6.

(a) $y = 111011$ (b) $y = 010100$

4. Let d be the $(9, 3)$ decoding function defined in the same way as the decoding function in Example 2. Determine $d(y)$ for the word y in B^9.

(a) $y = 101111101$ (b) $y = 100111100$

In Exercises 5 through 10, let e be the indicated encoding function and let d be an associated maximum likelihood decoding function. Determine the number of errors that (e, d) will correct.

5. *e* is the encoding function in Exercise 13 of Section 11.1.

6. *e* is the encoding function in Exercise 14 of Section 11.1.

7. *e* is the encoding function in Exercise 15 of Section 11.1.

8. *e* is the encoding function in Exercise 16 of Section 11.1.

9. *e* is the encoding function in Exercise 17 of Section 11.1.

10. *e* is the encoding function in Exercise 18 of Section 11.1.

11. Consider the group code defined in Exercise 17 of Section 11.1 Decode the following words relative to a maximum likelihood decoding function.
    (a) 11110   (b) 10011   (c) 10100

12. Consider the (2, 4) group encoding function $e: B^2 \to B^4$ defined by

$$e(00) = 0000 \qquad e(10) = 1001$$
$$e(01) = 0111 \qquad e(11) = 1111.$$

Decode the following words relative to a maximum likelihood decoding function.
    (a) 0011   (b) 1011   (c) 1111

13. Consider the (3, 5) group encoding function $e: B^3 \to B^5$ defined by

$$e(000) = 00000 \qquad e(100) = 10011$$
$$e(001) = 00110 \qquad e(101) = 10101$$
$$e(010) = 01001 \qquad e(110) = 11010$$
$$e(011) = 01111 \qquad e(111) = 11100.$$

Decode the following words relative to a maximum likelihood decoding function.
    (a) 11001   (b) 01010   (c) 00111

14. Consider the (3, 6) group encoding function $e: B^3 \to B^6$ defined by

$$e(000) = 000000 \qquad e(100) = 100101$$
$$e(001) = 000110 \qquad e(101) = 100011$$
$$e(010) = 010010 \qquad e(110) = 110111$$
$$e(011) = 010100 \qquad e(111) = 110001.$$

Decode the following words relative to a maximum likelihood decoding function.
    (a) 011110   (b) 101011   (c) 110010

15. Let G be a group and H a subgroup of G.

(a) Prove that if g_1 and g_2 are elements of G, then either $g_1 H = g_2 H$ or $g_1 H \cap g_2 H = \{ \ \}$.

(b) Use the result of part (a) to show that the left cosets of H form a partition of G.

In Exercises 16 through 18, determine the coset leaders for $N = e_H(B^m)$ for the given parity check matrix $\mathbf{H}$.

16. $\mathbf{H} = \begin{bmatrix} 1 & 1 \\ 1 & 0 \\ 1 & 0 \\ 0 & 1 \end{bmatrix}$

17. $\mathbf{H} = \begin{bmatrix} 0 & 1 & 1 \\ 1 & 0 & 1 \\ 1 & 0 & 0 \\ 0 & 1 & 0 \\ 0 & 0 & 1 \end{bmatrix}$

18. $\mathbf{H} = \begin{bmatrix} 1 & 0 & 0 \\ 1 & 1 & 0 \\ 0 & 1 & 1 \\ 1 & 0 & 0 \\ 0 & 1 & 0 \\ 0 & 0 & 1 \end{bmatrix}$

In Exercises 19 through 21, compute the syndrome for each coset leader found in the specified exercise.

19. Exercise 16.

20. Exercise 17.

21. Exercise 18.

22. Let

$$\mathbf{H} = \begin{bmatrix} 1 & 1 \\ 1 & 0 \\ 1 & 0 \\ 0 & 1 \end{bmatrix}$$

be a parity check matrix. Decode the following words relative to a maximum likelihood decoding function.
    (a) 0101   (b) 1010   (c) 1101

23. Let

$$\mathbf{H} = \begin{bmatrix} 0 & 1 & 1 \\ 1 & 0 & 1 \\ 1 & 0 & 0 \\ 0 & 1 & 0 \\ 0 & 0 & 1 \end{bmatrix}$$

be a parity check matrix. Decode the following words relative to a maximum likelihood decoding function associated with e_H.
    (a) 10100   (b) 01101   (c) 11011

24. Let

$$\mathbf{H} = \begin{bmatrix} 1 & 0 & 0 \\ 1 & 1 & 0 \\ 0 & 1 & 1 \\ 1 & 0 & 0 \\ 0 & 1 & 0 \\ 0 & 0 & 1 \end{bmatrix}$$

be a parity check matrix. Decode the following words relative to a maximum likelihood decoding function associated with e_H.

(a) 011001   (b) 101011   (c) 111010

TIPS FOR PROOFS

The proofs in this chapter rely heavily on earlier results. Many of the concepts developed throughout the book are applied here to the problems of coding and decoding. In Section 11.1, we pointed out the similarity of proving two numbers equal to proving two sets are the same. Analogous proofs could be developed for any relation that has the antisymmetric property as

"is less than" and "is a subset of" do.

In Section 11.2, Theorem 2 we use a one-to-one, onto function to "match" the elements of two sets in order to show that they have the same number of elements. This is also a technique that can be used in solving counting problems if the cardinality of one of the sets used is known.

KEY IDEAS FOR REVIEW

- Message: finite sequence of characters from a finite alphabet
- Word: sequence of 0's and 1's
- (m, n) encoding function: one-to-one function $e: B^m \rightarrow B^n, m < n$
- Code word: element in Ran(e)
- Weight of x, $|x|$: number of 1's in x
- Parity check code: see page 403
- Hamming distance between x and y, $\delta(x, y)$: $|x \oplus y|$
- Theorem (Properties of the Distance Function): Let x, y, and z be elements of B^m. Then

  (a) $\delta(x, y) = \delta(y, x)$.

  (b) $\delta(x, y) \geq 0$.

  (c) $\delta(x, y) = 0$ if and only if $x = y$.

  (d) $\delta(x, y) \leq \delta(x, z) + \delta(z, y)$.

- Minimum distance of an (m, n) encoding function: minimum of the distances between all distinct pairs of code words
- Theorem: An (m, n) encoding function $e: B^m \rightarrow B^n$ can detect k or fewer errors if and only if its minimum distance is at least $k + 1$.
- Group code: (m, n) encoding function $e: B^m \rightarrow B^n$ such that $e(B^m) = \{e(b) \mid b \in B^m\}$ is a subgroup of B^n

- Theorem: The minimum distance of a group code is the minimum weight of a nonzero code word.
- Mod-2 sum of Boolean matrices $\mathbf{D}$ and $\mathbf{E}$, $\mathbf{D} \oplus \mathbf{E}$: see page 407
- Mod-2 Boolean product of Boolean matrices $\mathbf{D}$ and $\mathbf{E}$, $\mathbf{D} * \mathbf{E}$: see page 408
- Theorem: Let m and n be nonnegative integers with $m < n$, $r = n - m$, and let $\mathbf{H}$ be an $n \times r$ Boolean matrix. Then the function $f_H: B^n \rightarrow B^r$ defined by

  $$f_H(x) = x * \mathbf{H}, \qquad x \in B^n$$

  is a homomorphism from the group B^n to the group B^r.
- Group code e_H corresponding to parity check matrix $\mathbf{H}$: see page 409
- (n, m) decoding function: see page 413
- Maximum likelihood decoding function associated with e: see page 414
- Theorem: Suppose that e is an (m, n) encoding function and d is a maximum likelihood decoding function associated with e. Then (e, d) can correct k or fewer errors if and only if the minimum distance is at least $2k + 1$.
- Decoding procedure for a group code: see page 416
- Decoding procedure for a group code given by a parity check matrix: see page 419

CODING EXERCISES

For each of the following, write the requested program or subroutine in pseudocode (as described in Appendix A) or in a programming language that you know. Test your code either with a paper-and-pencil trace or with a computer run.

1. Write a function that finds the weight of a word in B^n.

2. Write a subroutine that computes the Hamming distance between two words in B^n.

3. Let M and N be Boolean matrices of size $n \times n$. Write a

program that given M and N returns their mod-2 Boolean product.

4. Write a subroutine to simulate the $(m, 3m)$-encoding function $e: B^m \to B^{3m}$ described in Example 3, Section 11.1.

5. Write a subroutine to simulate the decoding function d for the encoding function of Exercise 4 as described in Example 2, Section 11.2.

CHAPTER 11 SELF-TEST

1. Consider the (3, 4) parity check code. For each of the received words, determine whether an error will be detected.

    (a) 1101   (b) 1010   (c) 1111   (d) 0011

2. Consider the $(m, 3m)$ encoding function with $m = 4$. For each of the received words, determine whether an error will be detected.

    (a) 001100100011   (b) 110111001101

    (c) 010111010011

3. Let e be the (3, 5) encoding function defined by

$$
\begin{array}{ll}
e(000) = 0000 & e(100) = 01010 \\
e(001) = 11110 & e(101) = 10100 \\
e(010) = 01101 & e(110) = 00111 \\
e(011) = 10011 & e(111) = 11001.
\end{array}
$$

How many errors will e detect?

4. Show that the (3, 5) encoding function in Problem 3 is a group code.

5. Let e be the encoding function defined in Problem 3 and let d be an associated maximum likelihood decoding function. Determine the number of errors that (e, d) will correct.

6. Let

$$
\mathbf{H} = \begin{bmatrix} 1 & 1 \\ 0 & 1 \\ 1 & 0 \\ 0 & 1 \end{bmatrix}
$$

be a parity check matrix. Decode 0110 relative to a maximum likelihood decoding function associated with e_H.

Appendix A

ALGORITHMS
AND PSEUDOCODE

ALGORITHMS

An **algorithm** is a complete list of the steps necessary to perform a task or computation. The steps in an algorithm may be general descriptions, leaving much detail to be filled in, or they may be totally precise descriptions of every detail.

> **EXAMPLE 1**

A recipe for baking a cake can be viewed as an algorithm. It might be written as follows.

1 ADD MILK TO CAKE MIX
2 ADD EGG TO CAKE MIX AND MILK.
3 BEAT MIXTURE FOR 2 MINUTES.
4 POUR MIXTURE INTO PAN AND COOK IN OVEN FOR 40 MINUTES AT 350°F.

End of Algorithm

It is a good idea to add the last line so that there can be no mistake about where the algorithm ends. The preceding algorithm is fairly general and assumes that the user understands how to pour milk, break an egg, set controls on an oven, and perform a host of other unspecified actions. If these steps were all included, the algorithm would be much more detailed, but long and unwieldy. One possible solution, if the added detail is necessary, is to group collections of related steps into other algorithms that we call **subroutines** and simply refer to these subroutines at appropriate points in the main algorithm. We hasten to point out that we are using the term "subroutine" in the general sense of an algorithm whose primary purpose is to form part of a more general algorithm. We do not give the term the precise meaning that it would have in a computer programming language. Subroutines are given names, and when an algorithm wishes the steps in a subroutine to be performed, it signifies this by *calling* the subroutine. We will specify this by a statement **CALL** NAME, where NAME is the name of the subroutine. ∎

425

EXAMPLE 2

Consider the following version of Example 1, which uses subroutines to add detail. Let us title this algorithm BAKECAKE.

◆ **ALGORITHM** BAKECAKE

    1 **CALL** ADDMILK

    2 **CALL** ADDEGG

    3 **CALL** BEAT(2)

    4 **CALL** COOK(OVEN, 40, 350)

End of Algorithm BAKECAKE

The subroutines of this example will give the details of each step. For example, subroutine ADDEGG might consist of the following general steps.

◆ **SUBROUTINE** ADDEGG

    1 Remove egg from carton.

    2 Break egg on edge of bowl.

    3 Drop egg, without shell, into bowl.

    4 **RETURN**

End of Subroutine ADDEGG

Of course, these steps could be broken into substeps, which themselves could be implemented as subroutines. The purpose of Step 4, the "return" statement, is to signify that one should continue with the original algorithm that "called" the subroutine. ■

Our primary concern is with algorithms to implement mathematical computations, investigate mathematical questions, manipulate strings or sequences of symbols and numbers, move data from place to place in arrays, and so on. Sometimes the algorithms will be of a general nature, suitable for human use, and sometimes they will be stated in a formal, detailed way suitable for programming in a computer language. Later in this appendix we will describe a reasonable language for stating algorithms.

It often happens that a test is performed at some point in an algorithm, and the result of this test determines which of two sets of steps will be performed next. Such a test and the resulting decision to begin performing a certain set of instructions will be called a **branch**.

EXAMPLE 3

Consider the following algorithm for deciding whether to study for a "discrete structures" test.

◆ **ALGORITHM** FLIP

    1 Toss a coin.

2 **IF** the result is "heads," **GO TO** 5.

3 Study for test.

4 **GO TO** 6.

5 See a show.

6 Take test next day.

End of ALGORITHM FLIP   ■

Note that the branching is accomplished by **GO TO** statements, which direct the user to the next instruction to be performed, in case it is not the next instruction in sequence. In the past, especially for algorithms written in computer programming languages such as FORTRAN, the **GO TO** statement was universally used to describe branches. Since then there have been many advances in the art of algorithm and computer program design. Out of this experience has come the view that the indiscriminate use of **GO TO** statements to branch from one instruction to any other instruction leads to algorithms (and computer programs) that are difficult to understand, hard to modify, and prone to error. Also, recent techniques for actually proving that an algorithm or program does what it is supposed to do will not work in the presence of unrestricted **GO TO** statements.

In light of the foregoing remarks, it is a widely held view that algorithms should be **structured**. This term refers to a variety of restrictions on branching, which help to overcome difficulties posed by the **GO TO** statement. In a structured branch, the test condition follows an **IF** statement. When the test is true, the instructions following a **THEN** statement are performed. Otherwise, the instructions following an **ELSE** statement are performed.

EXAMPLE 4 Consider again the algorithm FLIP described in Example 3. The following is a structured version of FLIP.

◆ **ALGORITHM** FLIP

1 Toss a coin.

2 **IF** (heads) **THEN**
   a Study for test.

3 **ELSE**
   a See a show.

4 Take test next day.

End of ALGORITHM FLIP

This algorithm is easy to read and is formulated without **GO TO** statements. In fact, it does not require numbering or lettering of the steps, but we keep these to set off and emphasize the instructions. Of course, the algorithm FLIP of Example 3 is not very different from that of Example 4. The point is that the **GO TO** statement has the potential for abuse, which is eliminated in the structured form.   ■

Another commonly encountered situation that calls for a branch is the **loop**, in which a set of instructions is repeatedly followed either for a definite number of times or until some condition is encountered. In structured algorithms, a loop may be formulated as shown in the following example.

EXAMPLE 5   The following algorithm describes the process of mailing 50 invitations.

◆ **ALGORITHM** INVITATIONS

   1 COUNT ← 50

   2 **WHILE** (COUNT > 0)

      a Address envelope.

      b Insert invitation in envelope.

      c Place stamp on envelope.

      d COUNT ← COUNT − 1

   3 Place envelopes in mailbox.

End of ALGORITHM INVITATIONS

In this algorithm, the variable COUNT is first assigned the value 50. The symbol ← may be read "is assigned." The loop is handled by the **WHILE** statement. The condition COUNT > 0 is checked, and as long as it is true, statements **a** through **d** are performed. When COUNT = 0 (after 50 steps), the looping stops. ∎

Later in this appendix we will give the details of this and other methods of looping, which are generally considered to be structured. In structured algorithms, the only deviations permitted from a normal, sequential execution of steps are those given by loops or iterations and those resulting from the use of the **IF-THEN-ELSE** construction. Use of the latter construction for branching is called **selection**.

In this book we will need to describe numerous algorithms, many of which are highly technical. Descriptions of these algorithms in ordinary English may be feasible, and in many cases we will give such descriptions. However, it is often easier to get an overview of an algorithm if it is presented in a concise, symbolic form. Some authors use diagrammatic representations called **flow charts** for this purpose. Figure A.1 shows a flow chart for the algorithm given in Example 4. These diagrams have a certain appeal and are still used in the computer programming field, but many believe that they are undesirable since they are more in accord with older programming practice than with structured programming ideas.

The other alternative is to express algorithms in a way that resembles a computer programming language or to use an actual programming language such as PASCAL. We choose to use a **pseudocode** language rather than an actual programming language, and the earlier examples of this section provide a hint as to the structure of this pseudocode form. There are several reasons for making this choice. First, knowledge of a programming language is not necessary for the understanding of the contents of this book. The fine details of a programming language are necessary for communication with a computer but may serve only to obscure the description of an algorithm. Moreover, we feel that the algorithms should be expressed in

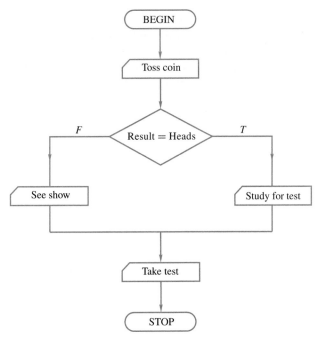

Figure A.1

such a way that an easy translation to any desired computer programming language is possible. Pseudocode is very simple to learn and easy to use, and it in no way interferes with one's learning of an actual programming language.

The second reason for using pseudocode is the fact that many professional programmers believe that developing and maintaining pseudocode versions of a program, before and after translation to an actual programming language, encourage good programming practice and aid in developing and modifying programs. We feel that the student should see a pseudocode in use for this reason. The pseudocode described is largely taken from Rader (*Advanced Software Design Techniques*, Petrocelli, New York, 1968) and has seen service in a practical programming environment. We have made certain cosmetic changes in the interest of pedagogy.

One warning is in order. An algorithm written in pseudocode may, if it is finely detailed, be very reminiscent of a computer program. This is deliberate, even to the use of terms like **SUBROUTINE** and the statement **RETURN** at the end of a subroutine to signify that we should return to the steps of the main algorithm. Also, the actual programming of algorithms is facilitated by the similarity of pseudocode to a programming language. However, always remember that a pseudocode algorithm is *not* a computer program. It is meant for humans, not machines, and we are only obliged to include sufficient detail to make the algorithm clear to human readers.

PSEUDOCODE

In pseudocode, successive steps are usually labeled with consecutive numbers. If a step begins a selection or a loop, several succeeding steps may be considered subordinate to this step (for example, the body of a loop). Subordinate lines are indented several spaces and labeled with consecutive letters instead of numbers. If these steps had subordinates, they in turn would be indented and labeled with numbers. We use only consecutive numbers or letters as labels, and we alternate them in succeeding levels of subordination. A typical structuring of steps with subordinate steps is illustrated in the following.

1. line 1
   a. line 2
   b. line 3
      1. line 4
      2. line 5
   c. line 6
2. line 7
3. line 8
   a. line 9
      1. line 10
   b. line 11
4. line 12

Steps that have the same degree of indentation will be said to be at the **same level**. Thus the next line at the level of line 1 is line 7, while the next line at the level of line 3 is line 6, and so on.

Selection in pseudocode is expressed with the form **IF-THEN-ELSE**, as follows:

1. **IF** (CONDITION) **THEN**
     | true-block |

2. **ELSE**
     | false-block |

The true- and false-blocks (to be executed respectively when CONDITION is true and CONDITION is false) may contain any legitimate pseudocode including selections or iterations. Sometimes we will omit statement 2, the **ELSE** statement, and the false-block. In this case, the true-block is executed only when CONDITION is true and then, whether CONDITION is true or false, control passes to the next statement that is at the same level as statement 1.

EXAMPLE 6

Consider the following statements in pseudocode. Assume that X is a rational number.

1. **IF** $(X > 13{,}000)$ **THEN**
   a. $Y \leftarrow X + 0.02(X - 13{,}000)$
2. **ELSE**
   a. $Y \leftarrow X + 0.03X$

In statement 1, CONDITION is: $X > 13{,}000$. If X is greater than 13,000, then Y is computed by the formula

$$X + 0.02(X - 13{,}000),$$

while if $X \leq 13,000$, then Y is computed by the formula

$$X + 0.03X.$$
∎

We will use the ordinary symbols of mathematics to express algebraic relationships and conditions in pseudocode. The symbols $+$, $-$, $\times$, and $/$ will be used for the basic arithmetic operations, and the symbols $<$, $>$, $\leq$, $\geq$, $=$, and $\neq$ will be used for testing conditions. The number X raised to the power Y will be denoted by X^Y, the square of a number A will be denoted by A^2, and so on. Moreover, products such as $3 \times A$ will be denoted by $3A$, and so on, when no confusion is possible.

We will use a left arrow, $\leftarrow$, rather than the equal sign, for assignments of values to variables. Thus, as in Example 6, the expression $Y \leftarrow X + 0.03X$ means that Y is assigned the value specified by the right-hand side. The use of $=$ for this purpose conflicts with the use of this symbol for testing conditions. Thus $X = X+1$ could either be an assignment or a question about the number X. The use of $\leftarrow$ avoids this problem.

A fundamental way to express iteration expressions in pseudocode is the **WHILE** form:

1. **WHILE** (CONDITION)
    | repeat-block |

Here CONDITION is tested and, if true, the block of pseudocode following it is executed. This process is repeated until CONDITION becomes false, after which control passes to the next statement that is at the same level as statement 1.

EXAMPLE 7

Consider the following algorithm in pseudocode; N is assumed to be a positive integer.

1 $X \leftarrow 0$

2 $Y \leftarrow 0$

3 **WHILE** $(X < N)$

    a $X \leftarrow X + 1$

    b $Y \leftarrow Y + X$

4 $Y \leftarrow Y/2$

End of ALGORITHM

In this algorithm, CONDITION is $X < N$. As long as CONDITION is true, that is, as long as $X < N$, statements a and b will be executed repeatedly. As soon as CONDITION is false, that is, as soon as $X = N$, statement 4 will be executed. This means that the **WHILE** loop is executed N times and the algorithm computes

$$\frac{1 + 2 + \cdots + N}{2},$$

which is the value of variable Y at the completion of the algorithm.
∎

A simple modification of the **WHILE** form called the **UNTIL** form is useful and we include it, although it could be replaced by completely equivalent statements using **WHILE**. This construction is

1. **UNTIL** (CONDITION)
   $\boxed{\text{repeat-block}}$

Here the loop continues to be executed *until* the condition is true; that is, continues only as long as the condition is false. Also, CONDITION is tested *after* the repeat-block rather than *before*, so the block must be repeated at least once.

EXAMPLE 8   The algorithm given in Example 7 could also be written with an **UNTIL** statement as follows:

1 $X \leftarrow 0$

2 $Y \leftarrow 0$

3   **UNTIL** $(X \geq N)$
      a $X \leftarrow X + 1$
      b $Y \leftarrow Y + X$

4 $Y \leftarrow Y/2$

End of ALGORITHM

In this algorithm, the CONDITION $X \geq N$ is tested at the completion of Step 3. If it is false, the body of Step 3 is repeated. This process continues until the test reveals that CONDITION is true (when $X = N$). At that time Step 4 is immediately executed. ■

The **UNTIL** form of iteration is a convenience and could be formulated with a **WHILE** statement. The form

1. **UNTIL** (CONDITION)
   $\boxed{\text{block 1}}$

is actually equivalent to the form

1. $\boxed{\text{block 1}}$

2. **WHILE** (CONDITION = FALSE)
   $\boxed{\text{block 1}}$

In each case, the instructions in block 1 are followed once, regardless of CONDITION. After this, CONDITION is checked, and, if it is true, the process stops; otherwise, block 1 instructions are followed again. This procedure of checking CONDITION and then repeating instructions in block 1 if CONDITION is false is continued until CONDITION is true. Since both forms produce the same results, they are equivalent.

The other form of iteration is the one most like a traditional DO loop, and we express it as a **FOR** statement:

1.   **FOR** VAR $= X$ **THRU** Y **[BY** Z**]**
     repeat-block

In this form, VAR is an integer variable used to count the number of times the instructions in repeat-block have been followed. X, Y, and Z, if desired, are either integers or expressions whose computed values are integers. The variable VAR begins at X and increases Z units at a time (Z is 1 if not specified). After each increase in X, the repeat-block is executed as long as the new value of X is not greater than Y. The conditions on VAR, specified by X, Y, and Z, are checked before each repetition of the instructions in the block. The block is repeated only if those conditions are true. The brackets around **BY** Z are not part of the statement, but simply mean that this part is optional and may be omitted. Note that the repeat block is always executed at least once, since no check is made until X is changed.

EXAMPLE 9

The pseudocode statement

FOR VAR $= 2$ **THRU** 10 **BY** 3

will cause the repeat-block to be executed three times, corresponding to VAR $= 2$, 5, 8. The process ends then, since the next value of VAR would be 11, which is greater than 10. ■

We will use lines of pseudocode by themselves to illustrate different parts of a computation. However, when the code represents a complete thought, we may choose to designate it as an algorithm, a subroutine, or a function.

A set of instructions that will primarily be used at various places by other algorithms is often designated as a **subroutine**. A subroutine is given a name for reference, a list of input variables, which it will receive from other algorithms, and output variables, which it will pass on to the algorithms that use it. A typical title of a subroutine is

SUBROUTINE NAME $(A, B, \ldots ; X, Y, \ldots)$

The values of the input variables are assumed to be supplied to the subroutine when it is used. Here NAME is a name generally chosen as a memory aid for the task performed by the subroutine; A, B, and so on, are input variables; and X, Y, and so on, are output variables. The semicolon is used to separate input variables from output variables.

A subroutine will end with the statement **RETURN**. As we remarked earlier in this section, this simply reminds us to return to the algorithm (if any) that is using the subroutine.

An algorithm uses a subroutine by including the statement

CALL NAME $(A, B, \ldots ; X, Y, \ldots)$

where NAME is a subroutine and the input variables A, B, and so on have all been assigned values. This process was also illustrated in earlier examples.

EXAMPLE 10

The following subroutine computes the square of a positive integer N by successive additions.

◆ **SUBROUTINE** SQR(N; X)

   1 $X \leftarrow N$

   2 $Y \leftarrow 1$

   3 **WHILE** $(Y \neq N)$

      a $X \leftarrow X + N$

      b $Y \leftarrow Y + 1$

   4 **RETURN**

End of SUBROUTINE SQR

If the result of the steps performed by a subroutine is a single number, we may call the subroutine **FUNCTION**. In this case, we title such a program as follows:

    FUNCTION NAME $(A, B, C, \ldots)$

where NAME is the name of the function and $A, B, C, \ldots$ are input variables. We also specify the value to be returned as follows:

    RETURN (Y)

where Y is the value to be returned.

The name **FUNCTION** is used because such subroutines remind us of familiar functions such as $\sin(x)$, $\log(x)$, and so on. When an algorithm requires the use of a function defined elsewhere, it simply uses the function in the familiar way and does not use the phrase **CALL**. Thus, if a function FN1 has been defined, the following steps of pseudocode will compute 1 plus the value of the function FN1 at $3X + 1$.

1. $Y \leftarrow 3X + 1$
2. $Y \leftarrow 1 + \text{FN1}(Y)$

EXAMPLE 11   The program given in Example 10 can be written as a function as follows:

◆ **FUNCTION** SQR(N)

   1 $X \leftarrow N$

   2 $Y \leftarrow 1$

   3 **WHILE** $(Y \neq N)$

      a $X \leftarrow X + N$

      b $Y \leftarrow Y + 1$

   4 **RETURN** (X)

End of FUNCTION SQR

Variables such as Y in Examples 10 and 11 are called **local variables**, since they are used only by the algorithm in its computations and are not part of input or output.

We will have many occasions to use linear arrays, as we need to be able to incorporate them into algorithms written in pseudocode. An array A will have locations indicated by $A[1]$, $A[2]$, $A[3]$, ... (as we noted in Section 1.3) and we will use this notation in pseudocode statements. Later, we will introduce arrays with more dimensions. In most actual programming languages, such arrays must be introduced by dimension statements or declarations, which indicate the maximum number of locations that may be used in the array and the nature of the data to be stored. In pseudocode we will not require such statements, and the presence of brackets after a variables will indicate that the variable names an array.

EXAMPLE 12   Suppose that $X[1]$, $X[2]$, ..., $X[N]$ contain real numbers and that we want to exhibit the maximum such number. The following instructions will do that.

1. MAX $\leftarrow$ $X[1]$
2. **FOR** $I = 2$ **THRU** N
    a. **IF** (MAX $<$ $X[I]$) **THEN**
        1. MAX $\leftarrow$ $X[I]$
3. **RETURN** (MAX)   ∎

EXAMPLE 13   Suppose that $A[1]$, $A[2]$, ..., $A[N]$ contain 0's and 1's so that A represents a subset (which we will also call A) of a universal set U with N elements (see Section 1.3). Similarly, a subset B of U is represented by another array, $B[1]$, $B[2]$, ..., $B[N]$. The following pseudocode will compute the representation of the union $C = A \cup B$ and store it in locations $C[1]$, $C[2]$, ..., $C[N]$ of an array C.

1. **FOR** $I = 1$ **THRU** N
    a. **IF** $((A[I] = 1)$ **OR** $(B[I] = 1))$ **THEN**
        1. $C[I] \leftarrow 1$
    b. **ELSE**
        1. $C[I] \leftarrow 0$ ∎

We will find it convenient to include a **PRINT** statement in the pseudocode. The construction is

1. **PRINT** ('message')

This statement will cause 'message' to be printed. Here we do not specify whether the printing is done on the computer screen or on paper.

Finally, we do include a **GO TO** statement to direct attention to some other point in the algorithm. The usage would be **GO TO** LABEL, where LABEL is a name assigned to some line of the algorithm. If that line had the number 1, for example, then the line would have to begin

LABEL: 1 ...

We avoid the **GO TO** statement when possible, but there are times when the **GO TO** statement is extremely useful, for example, to exit a loop prematurely if certain conditions are detected.

Exercises

In Exercises 1 through 8, write the steps in pseudocode needed to perform the task described.

1. In a certain country, the tax structure is as follows. An income of $30,000 or more results in $6000 tax, an income of $20,000 to $30,000 pays $2500 tax, and an income of less than $20,000 pays a 10% tax. Write a function TAX that accepts a variable INCOME and outputs the tax appropriate to that income.

2. Table A.1 shows brokerage commissions for firm X based on both price per share and number of shares purchased.

Table A.1 Commission Schedule (per share)

	Less Than $150/Share	$150/Share or More
Less than 100 shares	$3.25	$2.75
100 shares or more	$2.75	$2.50

Write a subroutine COMM with input variables NUMBER and PRICE (giving number of shares purchased and price per share) and output variable FEE giving the total commission for the transaction (not the per share commission).

3. Let $X_1, X_2, \ldots, X_N$ be a set of numbers. Write the steps needed to compute the sum and the average of the numbers.

4. Write an algorithm to compute the sum of cubes of all numbers from 1 to N (that is, $1^3 + 2^3 + 3^3 + \cdots + N^3$).

5. Suppose that the array X consists of real numbers $X[1]$, $X[2]$, $X[3]$ and the array Y consists of real numbers $Y[1]$, $Y[2]$, $Y[3]$. Write an algorithm to compute

$$X[1]Y[1] + X[2]Y[2] + X[3]Y[3].$$

6. Let the array $A[1], A[2], \ldots, A[N]$ contain the coefficients $a_1, a_2, \ldots, a_N$ of a polynomial $\sum_{i=1}^{N} a_i x^i$. Write a subroutine that has the array A and variables N and X as inputs and has the value of the polynomial at X as output.

7. Let $A[1], A[2], A[3]$ be the coefficients of a quadratic equation $ax^2 + bx + c = 0$ (that is, $A[1]$ contains a, $A[2]$ contains b, and $A[3]$ contains c). Write an algorithm that computes the roots $R1$ and $R2$ of the equation if they are real and distinct. If the roots are real and equal, the value should be assigned to $R1$ and a message printed. If the roots are not real, an appropriate message should be printed and computation halted. You may use the function SQRT (which returns the square root of any nonnegative number X).

8. Let $[a_1, a_2), [a_2, a_3), \ldots, [a_{N-1}, a_N]$ be N adjacent intervals on the real line. If $A[1], \ldots, A[N]$ contain the numbers $a_1, \ldots, a_n$, respectively, and X is a real number, write an algorithm that computes a variable INTERVAL as follows: If X is not between a_1 and a_N, INTERVAL $= 0$; however, if X is in the ith interval, then INTERVAL $= i$. Thus INTERVAL specifies which interval (if any) contains the number X.

In Exercises 9 through 12, let A and B be arrays of length N that contain 0's and 1's, and suppose they represent subsets (which we also call A and B) of some universal set U with N elements. Write algorithms that specify an array C representing the set indicated.

9. $C = A \oplus B$

10. $C = A \cap \overline{B}$

11. $C = \overline{A} \cap \overline{B}$

12. $C = A \cap (A \oplus \overline{B})$

In Exercises 13 through 20, write pseudocode programs to compute the quantity specified. Here N is a positive integer.

13. The sum of the first N nonnegative even integers

14. The sum of the first N nonnegative odd integers

15. The product of the first N positive even integers

16. The product of the first N positive odd integers

17. The sum of the squares of the first 77 positive integers

18. The sum of the cubes of the first 23 positive integers

19. The sum of the first 10 terms of the series

$$\sum_{n=1}^{\infty} \frac{1}{3n+1}$$

20. The smallest number of terms of the series

$$\sum_{n=1}^{\infty} \frac{1}{n+1}$$

whose sum exceeds 5

In Exercises 21 through 25, describe what is accomplished by the pseudocode. Unspecified inputs or variables X and Y represent rational numbers, while N and M represent integers.

21. SUBROUTINE MAX $(X, Y; Z)$
  1. $Z \leftarrow X$
  2. **IF** $(X < Y)$ **THEN**
      a. $Z \leftarrow Y$
  3. **RETURN**
  END OF SUBROUTINE MAX

22. 1. $X \leftarrow 0$
     2. $I \leftarrow 1$
     3. **WHILE** $(X < 10)$
        a. $X \leftarrow X + (1/I)$
        b. $I \leftarrow I + 1$

23. FUNCTION $F(X)$
  1. **IF** $(X < 0)$ **THEN**
      a. $R \leftarrow -X$
  2. **ELSE**
      a. $R \leftarrow X$
  3. **RETURN** (R)
  END OF FUNCTION F

24. FUNCTION $F(X)$
  1. **IF** $(X < 0)$ **THEN**
      a. $R \leftarrow X^2 + 1$
  2. **ELSE**
      a. **IF** $(X < 3)$ **THEN**
         1. $R \leftarrow 2X + 6$
      b. **ELSE**
         1. $R \leftarrow X + 7$
  3. **RETURN** (R)
  END OF FUNCTION F

25. 1. **IF** $(M < N)$ **THEN**
       a. $R \leftarrow 0$
     2. **ELSE**
       a. $K \leftarrow N$
       b. **WHILE** $(K < M)$
          1. $K \leftarrow K + N$
       c. **IF** $(K = M)$ **THEN**
          1. $R \leftarrow 1$
     3. **ELSE**
          1. $R \leftarrow 0$

In Exercises 26 through 30, give the value of all variables at the time when the given set of instructions terminates. N always represents a positive integer.

26. 1. $I \leftarrow 1$
     2. $X \leftarrow 0$
     3. **WHILE** $(I \leq N)$
        a. $X \leftarrow X + 1$
        b. $I \leftarrow I + 1$

27. 1. $I \leftarrow 1$
     2. $X \leftarrow 0$
     3. **WHILE** $(I \leq N)$
        a. $X \leftarrow X + I$
        b. $I \leftarrow I + 1$

28. 1. $A \leftarrow 1$
     2. $B \leftarrow 1$
     3. **UNTIL** $(B > 100)$
        a. $B \leftarrow 2A - 2$
        b. $A \leftarrow A + 3$

29. 1. **FOR** $I = 1$ **THRU** 50 **BY 2**
       a. $X \leftarrow 0$
       b. $X \leftarrow X + I$
     2. **IF** $(X < 50)$ **THEN**
       a. $X \leftarrow 25$
     3. **ELSE**
       a. $X \leftarrow 0$

30. 1. $X \leftarrow 1$
     2. $Y \leftarrow 100$
     3. **WHILE** $(X < Y)$
        a. $X \leftarrow X + 2$
        b. $Y \leftarrow \frac{1}{2}Y$

Appendix B

EXPERIMENTS IN DISCRETE MATHEMATICS

EXPERIMENT 1

In this experiment you will investigate a family of mathematical structures and classify family members according to certain properties that they have or do not have. In Section 1.4, we define $x \equiv r \pmod{n}$ if $x = kn + r$ with $0 \le r \le n - 1$. This idea is used to define operations in the family of structures to be studied. There will be one member of the family for each positive integer n. Each member of the family has two operations defined as follows:

$$a \oplus_n b = a + b \pmod{n}, \qquad a \otimes_n b = ab \pmod{n}.$$

For example, $5 \oplus_3 8 = 13 \pmod{3} = 1$, because $13 = 4 \cdot 3 + 1$ and $4 \otimes_5 8 = 32 \pmod{5} = 2$. The result of each operation mod n must be a number between 0 and $n - 1$ (inclusive), so to satisfy the closure property for each operation we restrict the objects in the structure based on mod n to $0, 1, 2, \ldots, n - 1$. Let $Z_n = [\{1, 2, 3, \ldots, n - 1\}, \oplus_n, \otimes_n]$. The Z_n are the family of structures to be studied.

Part I.   Some examples need to be collected to begin the investigation.

    1. Compute each of the following.

        (a) $7 \oplus_8 5$ (b) $4 \oplus_6 2$

        (c) $2 \oplus_4 3$ (d) $1 \oplus_5 3$

        (e) $6 \oplus_7 6$ (f) $7 \otimes_8 5$

        (g) $4 \otimes_6 2$ (h) $2 \otimes_4 3$

        (i) $1 \otimes_5 3$ (j) $6 \otimes_7 6$

    2. Construct an operation table for $\oplus_n$ and an operation table for $\otimes_n$ for $n = 2, 3, 4, 5, 6$. There will be a total of 10 tables. These will be used in Part II.

Part II.   Properties that a mathematical structure can have are presented in Section 1.6. In this part you will see if some of these properties are satisfied by Z_n for selected values of n.

    1. Is $\oplus_n$ commutative for $n = 2, 3, 4, 5, 6$? Explain how you made your decisions.

2. Is $\oplus_n$ associative for $n = 2, 3, 4, 5, 6$? Explain how you made your decisions.

3. Is there an identity for $\oplus_n$ in Z_n for $n = 2, 3, 4, 5, 6$? If so, give the identity.

4. Does each element of Z_n have an $\oplus_n$-inverse for $n = 4, 5, 6$? If so, let $-z$ denote the $\oplus_n$-inverse of z and define $a \ominus_n b = a \oplus_n (-b)$ and construct a $\ominus_n$ table.

5. Solve each of the following equations.
   (a) $3 \oplus_4 x = 2$ (b) $3 \oplus_5 x = 2$ (c) $3 \oplus_6 x = 2$

6. Is $\otimes_n$ commutative for $n = 2, 3, 4, 5, 6$? Explain how you made your decisions.

7. Is $\otimes_n$ associative for $n = 2, 3, 4, 5, 6$? Explain how you made your decisions.

8. Is there an identity for $\otimes_n$ in Z_n for $n = 2, 3, 4, 5, 6$? If so, give the identity.

9. Does each element of Z_n have an $\otimes_n$-inverse for $n = 4, 5, 6$? If so, let $1/z$ denote the $\otimes_n$-inverse of z and define $a \oslash_n b = a \otimes_n (1/b)$ and construct a $\oslash_n$ table.

10. Solve each of the following equations.
    (a) $2 \otimes_n x = 0$ for $n = 3, 4, 5, 6$
    (b) $x \otimes_n 3 = 2$ for $n = 4, 5, 6, 7$
    (c) $2 \otimes_n x = 1$ for $n = 3, 4, 5, 6$

Part III. Here you will develop some general conclusions about the family of Z_n.

1. Let $a \in Z_n$ and $a \neq 0$. Tell how to compute $-a$ using n and a.

2. For which positive integers k does $a \otimes_k x = 1$ have a unique solution for each a, $0 < a < k - 1$?

3. For which positive integers k does $a \otimes_k x = 1$ not have a unique solution for each a, $0 < a < k - 1$?

4. Test your conjectures from questions 2 and 3 for $k = 9, 10$, and 11. If necessary, revise your answers for questions 2 and 3.

5. If $a \otimes_k x = 1$ does not have a unique solution for each a, $0 < a < k - 1$, describe the relationship between a and k that guarantees that
   (a) There are no solutions to $a \otimes_k x = 1$.
   (b) There is more than one solution to $a \otimes_k x = 1$.

6. Describe k such that the following statement is true for Z_k.

$$a \otimes_k b = 0 \quad \text{only if} \quad a = 0 \text{ or } b = 0$$

EXPERIMENT 2

Many games and puzzles use strategies based on the rules of mathematical logic developed in Chapter 2. We begin here with a simple puzzle situation: Construct an object from beads and wires that satisfies some given conditions. After investigating this object, you will prove that it satisfies certain properties.

Part I. Here are the conditions for the first object.
  (a) You must use exactly three beads.

  (b) There is exactly one wire between every pair of beads.

  (c) Not all beads can be on the same wire.

  (d) Any pair of wires has at least one bead in common.

  1. Draw a picture of the object.
  2. Your object might not be the only one possible, so the following statements are to be proved referring only to the conditions and not to your object.
  T1. Any two wires have at most one bead in common.
  T2. There are exactly three wires.
  T3. No bead is on all the wires.

Part II.   Here are the conditions for the second object.
  (a) You must use at least one bead.

  (b) Every wire has exactly two beads on it.

  (c) Every bead is on exactly two wires.

  (d) Given a wire, there are exactly three other distinct wires that have no beads in common with the given wire.

  1. Draw a picture of the object.
  2. Your object might not be the only one possible, so the following statements are to be proved referring only to the conditions and not to your object.
  T1. There is at least one wire.

  T2. Given a wire there are exactly two other wires that have a bead in common with the given wire.

  T3. There are exactly _____ wires.

  T4. There are exactly _____ beads.

Part III.   The two objects you created in Parts I and II can be viewed in a number of ways. Instead of beads and wires, consider players and two-person teams, or substitute the words point and line for bead and wire.

  1. Translate the statements T1, T2, and T3 in Part I into statements about players and two-person teams.
  2. Translate the conditions (a)–(d) given in Part II into statements about points and lines.
  3. The type of object created here is often called a finite geometry, because each has a finite number of points and lines. What common geometric concept is described in condition (d) of Part II?

4. The Acian Bolex Tournament will be played soon. Determine the number of players needed and the number of teams that will be formed according to these ancient rules for bolex.

   (a) A team must consist of exactly three players.

   (b) Two players may be on at most one team in common.

   (c) Each player must be on at least three teams.

   (d) Not all the players can be on the same team.

   (e) At least one team must be formed.

   (f) If a player is not on a given team, then the player must be on exactly one team that has no members in common with the given team.

EXPERIMENT 3

An old folktale says that in a faraway monastery there is a platform with three large posts on it and when the world began there were 64 golden disks stacked on one of the posts. The disks are all of different sizes arranged in order from largest at the bottom to smallest at the top. The disks are being moved from the original post to another according to the following rules:

1. One disk at a time is moved from one post to another.
2. A larger disk may never rest on top of a smaller disk.
3. A disk is either on a post or in motion from one post to another.

When the monks have finished moving the disks from the original post to one of the others, the world will end. How long will the world exist?

A useful strategy is to try out smaller cases and look for patterns. Let N_k be the minimum number of moves that are needed to move k disks from one post to another. Then N_1 is 1 and N_2 is 3. (Verify this.)

1. By experimenting, find N_3, N_4, N_5.
2. Describe a recursive process for transferring k disks from post 1 to post 3. Write an algorithm to carry out your process.
3. Use the recursive process in part 2 to develop a recurrence relation for N_k.
4. Solve the recurrence relation in part 3 and verify the solution by comparing the results produced by the solution and the values found in part 1.
5. From part 4 you have an explicit formula for N_k. Use mathematical induction to prove that this statement is true.
6. If the monks move one disk per second and never make a mistake, how long (to the nearest year) will the world exist?

EXPERIMENT 4

Equivalence relations and partial orders are defined as relations with certain properties. In this experiment, you will investigate compatibility relations that are also defined by the relation properties they have. A **compatibility relation** is a relation that is reflexive and symmetric. Every equivalence relation is a compatibility relation, but here you will focus on compatibility relations that are not equivalence relations.

Part I.

1. Verify that the relation R on A is a compatibility relation.
   (a) A is the set of students at your college; $x \ R \ y$ if and only if x and y have taken the same course.
   (b) A is the set of all triangles; $x \ R \ y$ if and only if x and y have an angle with the same measure.
   (c) $A = \{1, 2, 3, 4, 5\}$; $R = \{(1, 1), (2, 2), (3, 3), (4, 4), (5, 5), (2, 3), (3, 2), (4, 1), (1, 4), (2, 4), (4, 2), (1, 2), (2, 1), (4, 5), (5, 4), (1, 3), (3, 1)\}$.

2. In Part I.1(c), the relation is given as a set of ordered pairs. A relation can also be represented by a matrix or a digraph. Describe how to determine if R is a compatibility relation using its
   (a) Matrix.
   (b) Digraph.

3. Give another example of a compatibility relation that is not an equivalence relation.

Part II.   Every relation has several associated relations that may or may not share its properties.

1. If R is a compatibility relation, is R^{-1}, the inverse of R, also a compatibility relation? If so, prove this. If not, give a counterexample.
2. If R is a compatibility relation, is $\overline{R}$, the complement of R, also a compatibility relation? If so, prove this. If not, give a counterexample.
3. If R and S are compatibility relations, is $R \circ S$ also a compatibility relation? If so, prove this. If not, give a counterexample.

Part III.   In Section 4.5, we showed that each equivalence relation R on a set A gives a partition of A. A compatibility relation R on a set A gives instead a covering of A. A **covering** of A is a set of subsets of A, $\{A_1, A_2, \ldots, A_k\}$, such that $\bigcup_{i=1}^{k} A_i = A$. We define a **maximal compatibility block** to be a subset B of A with each element of B related by R to every other element of B, and no element of $A - B$ is R-related to every element of B. For example, in Part I.1(c), the sets $\{1, 2, 3\}$ and $\{1, 2, 4\}$ are maximal compatibility blocks. The set of all maximal compatibility blocks relative to a compatibility relation R forms a covering of A.

1. Give all maximal compatibility blocks for the relation in Part I.1(c). Verify that they form a covering of A.

2. Describe the maximal compatibility blocks for the relation in Part I.1(b). The set of all maximal compatibility blocks form a covering of A. Is this covering also a partition for this example? Explain.

3. The digraph of a compatibility relation R can be simplified by omitting the loop at each vertex and using a single edge with no arrow between related vertices.

   (a) Draw the simplified graph for the relation in Part I.1(c).

   (b) Describe how to find the maximal compatibility blocks of a compatibility relation, given its simplified graph.

4. Find the covering of A associated with the relation whose simplified graph is given in

   (a) Figure B.1.

   (b) Figure B.2.

   (c) Figure B.3.

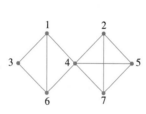 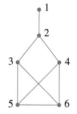 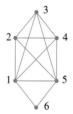

Figure B.1   Figure B.2   Figure B.3

5. Given the following covering of A, produce an associated compatibility relation R; that is, one whose maximal compatibility blocks are the elements of the covering.

   $\{\{1, 2\}, \{1, 3, 6, 7\}, \{4, 5, 11\}, \{5, 10\}, \{8, 5\}, \{2, 8, 9\}, \{3, 9\}, \{9, 10\}\}$

   Is there another compatibility relation that would produce the same covering of A?

EXPERIMENT 5

The θ-class of a function that describes the number of steps performed by an algorithm is referred to as the **running time** of the algorithm. In this experiment you will analyze several algorithms, presented in pseudocode, to determine their running times.

Part I.   The first algorithm is one method for computing the product of two $n \times n$ matrices. Assume that the matrices are each stored in an array of dimension 2 and that $A[i, j]$ holds the element of A in row i and column j.

◆ **ALGORITHM** MATMUL(A, B; C)

1.   **FOR** $I = 1$ **THRU** N
    a.   **FOR** $J = 1$ **THRU** N
        1. $C[I, J] \leftarrow 0$
        2.   **FOR** $K = 1$ **THRU** N
            a. $C[I, J] \leftarrow C[I, J] + A[I, K] \times B[K, J]$
END OF MATMUL

Assume that each assignment of a value, each addition, and each element multiplication take the same fixed amount of time.

    1. How many assignments will be done in the second **FOR** loop?
    2. How many element multiplications are done in the third **FOR** loop?
    3. What is the running time of MATMUL? Justify your answer.

Part II.   The following recursive algorithm will compute $n!$ for any positive integer n.

◆ **ALGORITHM** FAC(N)

1.   **IF** ($N = 1$) **THEN**
    a. $A \leftarrow 1$
2.   **ELSE**
    a. $A \leftarrow N \times$ FAC($N - 1$)
3.   **RETURN** (A)
END OF FAC

    1. Let S_n be the number of steps needed to calculate $n!$ using FAC. Write a recurrence relation for S_n in terms of S_{n-1}.
    2. Solve the recurrence relation in part 1 and use the result to determine the running time of FAC.

Part III.   The function SEEK will give the cell in which a specified value is stored in cells i through $i + n - 1$ (inclusive) of an array A. Assume that $i \geq 1$.

FUNCTION SEEK (*ITEM*, I, $I + N - 1$)
1.   CELL $\leftarrow 0$
2.   **FOR** $J = I$ **THRU** $I + N - 1$
    a.   **IF** ($A[J] = ITEM$) **THEN**

b.   CELL ← J
3.   **RETURN** (CELL)
END OF FUNCTION SEEK

1. How many cells are there from $A[i]$ to $A[i + n - 1]$ (inclusive)?
2. Give a verbal description of how SEEK operates.
3. What is the running time of SEEK? Justify your answer.

Part IV.   The algorithm HUNT will give the cell in which a specified value is stored in cells i through $i + n - 1$ (inclusive) of an array A. Assume that $i \geq 1$. To simplify the analysis of this algorithm, assume that n, the number of cells to be inspected, is a power of 2.

◆ **ALGORITHM**   HUNT (*ITEM, I, I + N − 1*)

1.   CELL ← 0
2.   **IF** ($N = 1$ **AND** $A[I] = ITEM$) **THEN**
     a.   CELL ← I
3.   **ELSE**
     a.   CELL1 ← HUNT(*ITEM, I, I + N/2 − 1*)
     b.   CELL2 ← HUNT(*ITEM, I + N/2, I + N − 1*)
4.   **IF** (CELL1 $\neq$ 0) **THEN**
     a.   CELL ← CELL1
5.   **ELSE**
     a.   CELL ← CELL2
6.   **RETURN** (CELL)
END OF HUNT

1. Give a verbal description of how HUNT operates.
2. What is the running time of HUNT? Justify your answer.
3. Under what circumstances would it be better to use SEEK (Part III) rather than HUNT? When would it be better to use HUNT rather than SEEK?

EXPERIMENT 6

The purpose of this experiment is to introduce the concept of a Markov chain. The investigations will use your knowledge of probability and matrices.

Suppose that the weather in Acia is either rainy or dry. We say that the weather has two possible **states**. As a result of extensive record keeping, it has been determined that the probability of a rainy day following a dry day is $\frac{1}{3}$, and the probability of a rainy day following a rainy day is $\frac{1}{2}$. If we know the weather today, then we can predict the probability that it is rainy tomorrow. In fact, if we know the state in which the weather is today, then we can predict the probability for each possible state tomorrow. A **Markov chain** is a process in which the probability of a system's being in a particular state at a given observation period depends only on its state at the immediate preceding observation period. Let t_{ij} be the probability that if the system is in state j at a certain observation period it will be in state i at the next period; t_{ij} is called a **transition probability**. It is convenient to arrange the transition probabilities for a system with n possible states as an $n \times n$ **transition matrix**. A transition matrix for Acia's weather is

$$
\begin{array}{cc}
\phantom{\mathbf{T} =} \text{D} & \text{R} \\
\mathbf{T} = \begin{bmatrix} \frac{2}{3} & \frac{1}{2} \\ \frac{1}{3} & \frac{1}{2} \end{bmatrix} \begin{array}{l} \text{D} \\ \text{R} \end{array}
\end{array}
$$

1. What is the sum of the entries in each column of $\mathbf{T}$? Explain why this must be the same for each column of any transition matrix.

The transition matrix of a Markov chain can be used to determine the probability of the system being in any of its n possible states at future times. Let

$$
\mathbf{P}^{(k)} = \begin{bmatrix} p_1^{(k)} \\ p_2^{(k)} \\ \vdots \\ p_n^{(k)} \end{bmatrix}
$$

denote the **state vector** of the Markov chain at the observation period k, where $p_j^{(k)}$ is the probability that the system is in state j at the observation period k. The state vector $\mathbf{P}^{(0)}$ is called the initial state vector.

2. Suppose today, a Wednesday, is dry in Acia and this is observation period 0.
   (a) Give the initial state vector for the system.
   (b) What is the probability that it will be dry tomorrow? What is the probability that it will be rainy tomorrow? Give $\mathbf{P}^{(1)}$.
   (c) Compute $\mathbf{TP}^{(0)}$. What is the relationship between $\mathbf{TP}^{(0)}$ and $\mathbf{P}^{(1)}$?

It can be shown that, in general, $\mathbf{P}^{(k)} = \mathbf{T}^k \mathbf{P}^{(0)}$. Thus the transition matrix and the initial state vector completely determine every other state vector.

3. Using the initial state vector from part 2, what is the state vector for
   (a) Friday?
   (b) Sunday?

(c) Monday?

(d) What appears to be the long-term behavior of this system?

In some cases the Markov chain reaches an equilibrium state, because the state vectors converge to a fixed vector. This vector is called the **steady-state vector**. The most common use of Markov chains is to determine long-term behavior, so it is important to know if a particular Markov chain has a steady-state vector.

4. Let

$$\mathbf{T} = \begin{bmatrix} 0 & 1 \\ 1 & 0 \end{bmatrix} \quad \text{and} \quad \mathbf{P}^{(0)} = \begin{bmatrix} \frac{1}{3} \\ \frac{2}{3} \end{bmatrix}.$$

Compute enough state vectors to determine the long-term behavior of this Markov chain.

A transition matrix **T** of a Markov chain is called **regular** if all the entries in some power of **T** are positive. If a Markov chain has a regular transition matrix, then the process has a steady-state vector. One way to find the steady-state vector, if it exists, is to proceed as in part 3; that is, calculate enough successive state vectors to identify the vector to which they are converging. Another method requires the solution of a system of linear equations. The steady-state vector **U** must be a solution of the matrix equation $\mathbf{TU} = \mathbf{U}$, and the entries of **U** have sum equal to 1.

5. Verify that the transition matrix for the weather in Acia is regular and that the transition matrix in part 4 is not regular.

6. Solve

$$\begin{bmatrix} \frac{2}{3} & \frac{1}{2} \\ \frac{1}{3} & \frac{1}{2} \end{bmatrix} \begin{bmatrix} x \\ y \end{bmatrix} = \begin{bmatrix} x \\ y \end{bmatrix}$$

with the condition that $x + y = 1$. Compare your solution with the results of part 3.

7. Consider a plant that can have red (R), pink (P), or white (W) flowers depending on the genotypes RR, RW, and WW. When we cross each of these genotypes with genotype RW, we have the following transition matrix.

		Flowers of parent plant		
		R	P	W
Flowers of offspring plant	R	0.5	0.25	0.0
	P	0.5	0.50	0.5
	W	0.0	0.25	0.5

Suppose that each successive generation is produced by crossing only with plants of RW genotype.

(a) Will the process reach an equilibrium state? Why or why not?

(b) If there is a steady-state vector for this Markov chain, what are the long-term percentages of plants with red, pink, and white flowers?

8. In Acia there are two companies that produce widgets, Widgets, Inc., and Acia Widgets. Each year Widgets, Inc., keeps one-fourth of its customers while three-fourths switch to Acia Widgets. Each year Acia Widgets keeps

two-thirds of its customers and one-third switch to Widgets, Inc. Both companies began business the same year and in that first year Widgets, Inc. had three-fifths of the market and Acia Widgets had the other two-fifths of the market. Under these conditions, will Acia Widgets ever run Widgets, Inc., out of business? Justify your answer.

EXPERIMENT 7

Ways to store and retrieve information in binary trees are presented in Sections 7.2 and 7.3. In this experiment you will investigate another type of tree that is frequently used for data storage.

A **B-tree of degree** k is a tree with the following properties:

1. All leaves are on the same level.
2. If it is not a leaf, the root has at least two children and at most k children.
3. Any vertex that is not a leaf or the root has at least $k/2$ children and at most k children.

Figure B.4

The tree in Figure B.4 is a B-tree of degree 3.

Part I.   Recall that the height of a tree is the length of the longest path from the root to a leaf.

1. Draw three different B-trees of degree 3 with height 2. Your examples should not also be of degree 2 or 1.
2. Draw three different B-trees of degree 4 (but not less) with height 3.
3. Give an example of a B-tree of degree 5 (but not less) with height 4.
4. Discuss the features of your examples in parts 1 through 3 that suggest that a B-tree would be a good storage structure.

Part II.   The properties that define a B-tree of degree k not only restrict how the tree can look, but also limit the number of leaves for a given height and the height for a given number of leaves.

1. If T is a B-tree of degree k and T has height h, what is the maximum number of leaves that T can have? Explain your reasoning.
2. If T is a B-tree of degree k and T has height h, what is the minimum number of leaves that T can have? Explain your reasoning.
3. If T is a B-tree of degree k and T has n leaves, what is the maximum height that T can have? Explain your reasoning.
4. If T is a B-tree of degree k and T has n leaves, what is the minimum height that T can have? Explain your reasoning.
5. Explain how your results in Part II, questions 3 and 4, support your conclusions in Part I.4.

EXPERIMENT 8

The purpose of this experiment is to investigate relationships among groups, sub-groups, and elements. Five groups are given as examples to use in the investigation. You may decide to look at other groups as well to test your conjectures.

S_3 is the group of permutations of $\{1, 2, 3\}$ with the operation of compo-sition. It is also the group of symmetries of a triangle. (See Section 9.4.)

D_4 is the group of symmetries of a square. (This group is presented in Ex-ercise 17, Section 9.4.)

S_4 is the group of permutations of $\{1, 2, 3, 4\}$ with the operation of compo-sition.

G_1 is the group whose multiplication table is given in Table B.1.

G_2 is the group whose multiplication table is given in Table B.2.

You may find it helpful to write out the multiplication tables for S_3, D_4, and S_4.

Table B.1

	1	2	3	4	5	6	7	8
1	1	2	3	4	5	6	7	8
2	2	5	4	7	6	1	8	3
3	3	8	5	2	7	4	1	6
4	4	3	6	5	8	7	2	1
5	5	6	7	8	1	2	3	4
6	6	1	8	3	2	5	4	7
7	7	4	1	6	3	8	5	2
8	8	7	2	1	4	3	6	5

Table B.2

	1	2	3	4	5
1	1	2	3	4	5
2	2	3	4	5	1
3	3	4	5	1	2
4	4	5	1	2	3
5	5	1	2	3	4

1. Identify the identity element e for each of the five groups.
2. For each of the five groups, do the following. For each element g in the group, find the smallest k for which $g^k = e$, the identity. This number k is called the **order of g**.
3. What is the relationship between the order of an element of a group and the order of the group? (The order of a group is the number of elements.)
4. For each of the five groups, find all subgroups of the group.
5. A group is called **cyclic** if its elements are the powers of one of the ele-ments. Identify any cyclic groups among the subgroups of each group.
6. What is the relationship between the order of a subgroup and the order of the group?
7. The groups G_1 and D_4 are both of order 8. Are they isomorphic? Explain your reasoning.

EXPERIMENT 9

Moore machines (Section 10.3) are examples of finite-state automata that recognize regular languages. Many computer languages, however, are not regular (type 3), but are context free (type 2). For example, a computer language may include expressions using balanced parentheses (a right parenthesis for every left parenthesis). A Moore machine has no way to keep track of how many left parentheses have been read to determine if the same number of right parentheses have also been read. A finite-state machine that includes a feature to do this is called a pushdown automaton.

A **pushdown automaton** is a sequence (S, I, F, s_0, T) in which S is a set of states, T is a subset of S and is the set of final states, $s_0 \in S$ is the start state, I is the input set, and F is a function from $S \times I \times I^*$ to $S \times I^*$. Roughly speaking, the finite-state machine can create a string of elements from the input set to serve as its memory. This string may be the empty string Λ. The transition function F uses the current state, the input, and the string to determine the next state and the next string. For example, $F(s_3, a, w) = (s_2, w')$ means that if the machine is in state s_3 with current memory string w and a is read, the machine will move to state s_2 with new string w'. In actual practice there are only two ways to change the memory string:

(1) From w to bw for some b in I; this is called pushing b on the stack.

(2) From bw to w for some b in I; this is called popping b off the stack.

A pushdown automaton accepts a string v if this input causes the machine to move from s_0 with memory string Λ to a final state s_j with memory string Λ.

1. Construct a Moore machine that will accept strings of the form $0^m 1^n$, $m \geq 0, n \geq 0$, and no others.

2. Explain why the Moore machine in part 1 cannot be modified to accept strings of the form $0^n 1^n$, $n \geq 0$, and no others.

3. Let $P = (S, I, F, s_0, T)$ with $S = \{s_0, s_1\}$, $I = \{0, 1\}$, $T = \{s_1\}$, and

$$F(s_0, 0, w) = (s_0, 0w), \quad F(s_0, 1, 0w) = (s_1, w),$$
$$F(s_1, 1, 0w) = (s_1, w)$$

where w is any string in I^*. Show that P accepts strings of the form $0^n 1^n$, $n \geq 0$, and no others.

4. Let $I = \{a, b, c\}$ amd $w \in \{a, b\}^*$. We define w^R to be the string formed by the elements of w in reverse order. For example, if w is $aabab$, then w^R is $babaa$. Design a pushdown automaton that will accept strings of the form wcw^R, and no others.

5. Let $G = (V, S, s_0, \mapsto)$ be a phrase structure grammar with $V = \{v_0, w, a, b, c\}$, $S = \{a, b, c\}$, and

$$\mapsto: \quad v_0 \mapsto av_0 b, \quad v_0 b \mapsto bw, \quad abw \mapsto c.$$

(a) Describe the language $L(G)$.

(b) Design a pushdown automaton whose language is $L(G)$. That is, it only accepts strings in $L(G)$.

EXPERIMENT 10

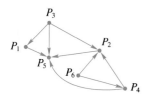

Figure B.5

Suppose that there are n individuals $P_1, P_2, \ldots, P_n$ some of whom can influence each other in making decisions. If P_3 influences P_5, it may or may not be true that P_5 influences P_3. In Figure B.5 we have drawn a digraph to describe the influence relations among the six members of a design team. Notice that the digraph has no loops; an individual does not influence herself or himself.

1. (a) Give the matrix for this relation.

   (b) Is there a leader in this design group? Justify your answer.

The relation described by the digraph in Figure B.5 is not transitive, but we can speak of two-stage influence. We say P_i has **two-stage influence** on P_j if there is a path of length 2 from P_i to P_j. Similarly, P_i has r-stage influence on P_j if there is a path of length r from P_i to P_j. In Section 4.3, a method for determining whether a path of length r exists from P_i to P_j is presented.

2. Use the matrix for the relation described by Figure B.5 to determine whether P_i has two-stage influence on P_j for each ordered pair of distinct members of the design team.

3. Consider a communication network among five sites with matrix

$$\begin{bmatrix} 0 & 1 & 0 & 0 & 0 \\ 0 & 0 & 1 & 0 & 1 \\ 0 & 0 & 0 & 1 & 0 \\ 1 & 1 & 0 & 0 & 0 \\ 0 & 0 & 1 & 1 & 0 \end{bmatrix}.$$

   (a) Can P_3 get a message to P_5 in at most two stages?

   (b) What is the minimum number of stages that will guarantee that every site can get a message to any other different site?

   (c) What is the minimum number of stages that will guarantee that every site can get a message to any site including itself?

A dictionary defines a clique as a small exclusive group of people. In studying organizational structures, we often find subsets of people in which any pair of individuals is related, and we borrow the word clique for such a subset. A **clique** in an influence digraph is a subset S of vertices such that

(1) $|S| \geq 3$.

(2) If P_i and P_j are in S, then P_i influences P_j and P_j influences P_i.

(3) S is the largest subset that satisfies (2).

4. Identify all cliques in the digraph in Figure B.6.

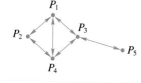

Figure B.6

If the digraph is small, cliques can be identified by inspection of the digraph. In general, it can be difficult to determine cliques using only the digraph. The algorithm CLIQUE identifies which vertices belong to cliques for an influence relation given by its matrix.

◆ALGORITHM CLIQUE

1　If $\mathbf{A} = \left[a_{ij} \right]$ is the matrix of the influence relation, construct the matrix $\mathbf{S} = \left[s_{ij} \right]$ as follows: $s_{ij} = s_{ji} = 1$ if and only if $a_{ij} = a_{ji} = 1$. Otherwise, $s_{ij} = 0$.

2　Compute $\mathbf{S} \odot \mathbf{S} \odot \mathbf{S} = \mathbf{C} = \left[c_{ij} \right]$.

3　P_i belongs to a clique if and only if c_{ii} is positive.

End of CLIQUE

5. Use CLIQUE and the matrix for the digraph in Figure B.6 to determine which vertices belong to a clique. Verify that this is consistent with your results for part 3. Explain why CLIQUE works.

6. Five people have been stationed on a remote island to operate a weather station. The following social interactions have been observed:

P_1 gets along with P_2, P_3, and P_4.
P_2 gets along with P_1, P_3, and P_5.
P_3 gets along with P_1, P_2, and P_4.
P_4 gets along with P_3 and P_5.
P_5 gets along with P_4.

Identify any cliques in this set of people.

7. Another application of cliques is in determining the chromatic number of a graph. (See Section 8.6.) Explain how knowing the cliques in a graph G can be used to find $\chi(G)$.

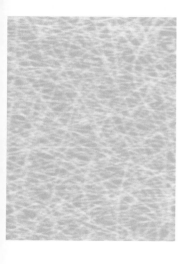

ANSWERS TO ODD-NUMBERED EXERCISES

NOTE: We have not included "solutions" for Coding Exercises for several reasons. We believe each department (or instructor) should have very specific programming standards and any code we presented would certainly violate someone's standards and set up unnecessary conflicts. More importantly, different programming languages support different constructions and what is a good choice in one can be a very bad one in another.

CHAPTER 1

Exercise Set 1.1, page 4

1. (a) True.   (b) False.   (c) False.
(d) False.   (e) True.   (f) False.

3. (a) {A, R, D, V, K}.   (b) {B, O, K}.
(c) {M, I, S, P}.

5. (a) False.   (b) True.   (c) False.
(d) True.   (e) False.   (f) False.

7. $\{x \mid x \text{ is a vowel}\}$

9. $\{x \mid x \in Z \text{ and } x^2 < 5\}$

11. (b), (c), (e).

13. { }, {BASIC}, {PASCAL}, {ADA}, {BASIC, PASCAL}, {BASIC, ADA}, {PASCAL, ADA}, {BASIC, PASCAL, ADA}.

15. (a) True.   (b) False.   (c) False.
(d) True.   (e) True.   (f) True.
(g) True.   (h) True.

17. (a) $\subseteq$ (b) $\subseteq$ (c) $\not\subseteq$
(d) $\subseteq$ (e) $\not\subseteq$ (f) $\subseteq$

19. {1, 2, 3}

21. Yes, Yes, the complement of a set would not be defined unambiguously.

23. (a) False.   (b) False.   (c) Insufficient information.
(d) False.   (e) True.   (f) True.

25. (a) $\{\{\ \}, \{3\}, \{7\}, \{2\}, \{3, 7\}, \{3, 2\}, \{7, 2\}, \{3, 7, 2\}\}$.
(b) 3.   (c) 8.

27. is one solution

29. $\varnothing \subseteq Z^+ \subseteq N \subseteq Z \subseteq \mathbb{Q} \subseteq \mathbb{R}$.

31. Let $b \in B$, $b \notin A$. Each subset of B either contains b or it does not. Let B_j be a subset of B. If it doesn't contain b, then B_j is one of the S_i. If it contains b, then the elements of B_j are all the elements of one of the S_i and b. In this case B_j is not one of the S_i, and B has exactly $2n$ subsets.

Exercise Set 1.2, page 11

1. (a) $\{a, b, c, d, e, f, g\}$.   (b) $\{a, c, d, e, f, g\}$.
(c) $\{a, c\}$.   (d) $\{f\}$.
(e) $\{b, g, d, e\}$.   (f) $\{a, b, c\}$.
(g) $\{d, e, f, h, k\}$.   (h) $\{a, b, c, d, e, f\}$.
(i) $\{b, g, f\}$.   (j) $\{g\}$.

3. (a) $\{a, b, c, d, e, f, g\}$.   (b) $\{\ \}$.
(c) $\{a, c, g\}$.   (d) $\{a, c, f\}$.
(e) $\{h, k\}$.   (f) $\{a, b, c, d, e, f, h, k\}$.

5. (a) $\{1, 2, 4, 5, 6, 8, 9\}$.   (b) $\{1, 2, 3, 4, 6, 8\}$.

455

(c) $\{1, 2, 4, 6, 7, 8\}$.   (d) $\{1, 2, 3, 4, 5, 9\}$.

(e) $\{1, 2, 4\}$.   (f) $\{8\}$.

(g) $\{2, 4\}$.   (h) $\{\ \}$.

7. (a) $\{1, 2, 3, 4, 5, 6, 8, 9\}$.   (b) $\{2, 4\}$.

(c) $\{1, 2, 4\}$.   (d) $\{8\}$.

(e) $\{3, 7\}$.   (f) $\{1, 3, 5, 6, 7, 8, 9\}$.

9. (a) $\{b, d, e, h\}$.   (b) $\{b, c, d, f, g, h\}$.

(c) $\{b, d, h\}$.   (d) $\{b, c, d, e, f, g, h\}$.

(e) $\{\ \}$.   (f) $\{c, f, g\}$.

11. (a) All real numbers except -1 and 1.

(b) All real numbers except -1 and 4.

(c) All real numbers except -1, 1, and 4.

(d) All real numbers except -1.

13. (a) True.   (b) True.   (c) False.   (d) False.

15. 1.

17. (a) $|A \cup B| = 10$, $|A| = 6$, $|B| = 7$, $|A \cap B| = 3$. Hence
$|A \cup B| = |A| + |B| - |A \cap B|$.

(b) $|A \cup B| = 11$, $|A| = 5$, $|B| = 6$, $|A \cap B| = 0$. Hence
$|A \cup B| = |A| + |B| - |A \cap B|$.

19. B must be the empty set.

21. $|A| = 6$, $|B| = 5$, $|C| = 6$, $|A \cap B| = 2$, $|A \cap C| = 3$,
$|B \cap C| = 3$, $|A \cap B \cap C| = 2$, $|A \cup B \cup C| = 11$. Hence
$|A \cup B \cup C| = |A| + |B| + |C| - |A \cap B| - |A \cap C|$
$- |B \cap C| + |A \cap B \cap C|$.

23. (a) 106.   (b) 60.

25. (a) 162.   (b) 118.   (c) 236.

(d) 290.   (e) 264.

27. A and to B.

29. (a)   (b) $A \cup B$.

(c) Let $x \in A \cup B$. Then $x \in A$ or $x \in B$. Since $A \subseteq C$
and $B \subseteq C$, $x \in A$ or $x \in B$ means that $x \in C$. Hence
$A \cup B \subseteq C$.

31. Yes. Suppose $x \in B$. Either $x \in A$ or $x \notin A$. If $x \in A$,
then $x \notin A \oplus B = A \oplus C$. But then x must be in C. If
$x \notin A$, then $x \in A \oplus B = A \oplus C$, and again x must be in
C. So $B \subseteq C$. A similar argument shows that $C \subseteq B$, so
$B = C$.

33. No. Let $A = \{1, 2, 3\}$, $B = \{4\}$, and $C = \{3, 4\}$. Then
$A \cup B = A \cup C$ and $B \neq C$.

35. (a) Let $x \in A \cup C$. Then $x \in A$ or $x \in C$, so $x \in B$ or
$x \in D$ and $x \in B \cup D$. Hence $A \cup C \subseteq B \cup D$.

(b) Let $x \in A \cap C$. Then $x \in A$ and $x \in C$, so $x \in B$ and
$x \in D$. Hence $x \in B \cap D$. Thus $A \cap C \subseteq B \cap D$.

37. We must subtract $|B \cap C|$, because each element in $B \cap C$
has been counted twice in $|A| + |B| + |C|$. But when we
subtract both $|B \cap C|$ and $|A \cap C|$, we have "uncounted"
all the elements of C that also belong to B and A. These
(and the similar elements of A and B) are counted again
by adding $|A \cap B \cap C|$.

Exercise Set 1.3, page 20

1. $\{1, 2\}$.

3. $\{a, b, c, \ldots, z\}$.

5. Possible answers include $xyzxyz\ldots$, $xxyyzzxxyyzz$, and
$yzxyzx\ldots$.

7. $5, 25, 125, 625$.

9. $2.5, 4, 5.5, 7$.

11. $a_n = a_{n-1} + 2$, $a_1 = 1$, recursive; $a_n = 2n - 1$, explicit.

13. $c_n = (-1)^{n+1}$, explicit.

15. $e_n = e_{n-1} + 3$, $e_1 = 1$, recursive.

17. $a_n = 2 + 3(n - 1)$.

19. A, uncountable; B, finite; C, countable; D, finite; E,
finite.

21. (a) 01101110.   (b) 01100000.

(c) 00100000.   (d) 11111011.

(e) 00010100.

23. $f_A(x) = 1$; $f_B(x) = 0$; $f_A(x)f_B(x) = 0$;
$f_A(x) + f_B(x) - 2f_A(x)f_B(x) = 1$.
$f_A(x) = 0$; $f_B(x) = 1$; $f_A(x)f_B(x) = 0$;
$f_A(x) + f_B(x) - 2f_A(x)f_B(x) = 1$.
$x \in A \cap B$ or $x \notin A \cup B$;
$f_A(x) + f_B(x) - 2f_A(x)f_B(x) = 0$.

25. (a) a and b are regular by RE2. ab is then regular by RE3
and so by RE5, $(ab)^*$ is regular. Applying RE2 and
RE3, we have $a + b(ab)^*$ and $a \times b$ are regular. By
RE4, $(a \times b \vee a)$ is regular. Using RE3,
$a + b(ab)^*(a \times b \vee a)$ is regular.

(b) By RE2, $a, b, +, \times$ are regular. By RE5, a^* is regular;
$(a^* \vee b)$ is regular by RE4. Using RE3, we have
$a + b \times (a^* \vee b)$ is regular.

(c) By RE2, a, b, $\vee$, $+$ are regular. Thus, by RE5, a^* and b^* are regular. Using RE3, we have a^*b and $\times ab^*$ are regular. By RE4, $(a^*b \vee +)$ is regular. By RE5, $(a^*b \vee +)^*$ is regular. And finally, by RE4, $((a^*b \vee +)^* \vee \times ab^*)$ is regular.

27. (a) No.   (e) No.   (c) Yes.

29. (a) $\{pr, prq, prqq, \dots ; qr, qrq, qrqq, \dots\}$.

 (b) $\{pr, pqqr, pqqqqr, \dots\}$.

31. T-numbers form the sequence $0, 3, 6, 9, \dots$. The T-numbers are the nonnegative multiples of 3.

33. $0, 1, 1, 3, 5, 11$.

Exercise Set 1.4, page 29

1. $20 = 6 \cdot 3 + 2$.

3. $3 = 0 \cdot 22 + 3$.

5. (a) $828 = 2^2 \cdot 3^2 \cdot 23$.   (b) $1666 = 2 \cdot 7^2 \cdot 17$.
 (c) $1781 = 13 \cdot 137$.   (d) $1125 = 3^2 \cdot 5^3$.
 (e) 107.

7. $d = 3; 3 = 3 \cdot 45 - 4 \cdot 33$.

9. $d = 1; 1 = 5 \cdot 77 - 3 \cdot 128$.

11. 1050.

13. 864.

15. (a) 6.   (b) 1.   (c) 0.
 (d) 1.   (e) 20.   (f) 14.

17. (a) $\{3, 9, 15, 21, \dots\}$.   (b) $\{1, 7, 13, 19, \dots\}$.

19. If $\mathrm{GCD}(a, c) = 1$, there are integers s, t such that $1 = sa + tc$. Thus $b = sab + tcb$. Since $c \mid ab$ and $c \mid c$, we have $c \mid (sab + tcb)$. That is, $c \mid b$.

21. Since $a \mid b$ and $c \mid b$, $ac \mid ab$ and $ac \mid bb$. We can write d as $sa + tb$ and $bd = sab + tbb$. Because ac divides both sab and tbb, ac also divides bd.

23. Clearly, $a \mid ab$ and $ab \mid ab$, so ab is a common multiple of a and ab. No smaller multiple of ab exists, so $ab = \mathrm{LCM}(a, ab)$.

25. Let $a = p_1^{a_1} p_2^{a_2} \cdots p_n^{a_n}$, $b = p_1^{b_1} p_2^{b_2} \cdots p_n^{b_n}$, and $k = p_1^{k_1} p_2^{k_2} \cdots p_n^{k_n}$, where some of the a_i, b_i, and k_i may be zero. Since $a \mid k$, $k_i \geq a_i$, $i = 1, 2, \dots, n$ and since $b \mid k$, $k_i \geq b_i$, $i = 1, 2, \dots, n$. If $c = \mathrm{LCM}(a, b)$, then $c = p_1^{c_1} p_2^{c_2} \cdots p_n^{c_n}$, where $c_i = \max(a_i, b_i)$. Then $k_i \geq c_i$, $i = 1, 2, \dots, n$ and $c \mid k$.

27. $p \mid a$ and $p \mid p$, so p is a common divisor of a and p. Hence $p \mid \mathrm{GCD}(a, p)$. But $\mathrm{GCD}(a, p) \mid p$, so $\mathrm{GCD}(a, p) = p$. (See Exercise 26.)

29. Consider $m = p_1^{a_1} p_2^{a_2} \cdots p_k^{a_k}$ and $n = q_1^{b_1} q_2^{b_2} \cdots q_l^{b_l}$, the prime factorizations of m and n. Since $\mathrm{GCD}(m, n) = 1$, $\{p_1, p_2, \dots, p_k\} \cap \{q_1, q_2, \dots, q_l\} = \{\ \}$. Thus for mn to be a perfect square, each a_i and b_j must be even. But this means m and n are each perfect squares.

31. (a) 563.   (b) 337.   (c) 153.

33. (a) 788.   (b) 870.

Exercise Set 1.5, page 37

1. (a) $-2, 1, 2$.   (b) $3, 4$.   (c) $4, -1, 8$.
 (d) $2, 6, 8$.

3. a is 3, b is 1, c is 8, and d is -2.

5. (a) $\begin{bmatrix} 4 & 0 & 2 \\ 9 & 6 & 2 \\ 3 & 2 & 4 \end{bmatrix}$.

 (b) $\mathbf{AB} = \begin{bmatrix} 7 & 13 \\ -3 & 0 \end{bmatrix}$.

 (c) Not possible.   (d) $\begin{bmatrix} 21 & 14 \\ -7 & 17 \end{bmatrix}$.

7. (a) $\mathbf{EB}$ is 3×2 and $\mathbf{FA}$ is 2×3; the sum is undefined.
 (b) $\mathbf{B} + \mathbf{D}$ does not exist.
 (c) $\begin{bmatrix} 10 & 0 & -25 \\ 40 & 14 & 12 \end{bmatrix}$ (d) $\mathbf{DE}$ does not exist.

9. (a) $\begin{bmatrix} 22 & 34 \\ 3 & 11 \\ -31 & 3 \end{bmatrix}$.   (b) $\mathbf{BC}$ is not defined.

 (c) $\begin{bmatrix} 25 & 5 & 26 \\ 20 & -3 & 32 \end{bmatrix}$.

 (d) $\mathbf{D}^T + \mathbf{E}$ is not defined.

11. Let $\mathbf{B} = \begin{bmatrix} b_{jk} \end{bmatrix} = \mathbf{I}_m \mathbf{A}$. Then $b_{jk} = \sum_{l=1}^{m} i_{jl} a_{lk}$, for $1 \leq j \leq m$ and $1 \leq k \leq n$. But $i_{jj} = 1$ and $i_{jl} = 0$ if $j \neq l$. Hence $b_{jk} = i_{jj} a_{jk}$, $1 \leq j \leq m$, $1 \leq k \leq n$. This means $\mathbf{B} = \mathbf{I}_m \mathbf{A} = \mathbf{A}$. Similarly, if $\mathbf{C} = \mathbf{AI}_n = \begin{bmatrix} c_{jk} \end{bmatrix}$, $c_{jk} = \sum_{l=1}^{n} a_{jl} i_{lk} = a_{jk} i_{kk} = a_{jk}$ for $1 \leq j \leq m$, $1 \leq k \leq n$.

13. $\mathbf{A}^3 = \begin{bmatrix} 27 & 0 & 0 \\ 0 & -8 & 0 \\ 0 & 0 & 64 \end{bmatrix}$ or $\begin{bmatrix} 3^3 & 0 & 0 \\ 0 & (-2)^3 & 0 \\ 0 & 0 & 4^3 \end{bmatrix}$,

 $A^k = \begin{bmatrix} 3^k & 0 & 0 \\ 0 & (-2)^k & 0 \\ 0 & 0 & 4^k \end{bmatrix}$.

15. The entries of $\mathbf{I}_n^T$ satisfy $i'_{kj} = i_{jk}$. But $i_{jk} = 1$ if $j = k$ and is 0 otherwise. Thus $i'_{kj} = 1$ if $k = j$ and is 0 if $k \neq j$ for $1 \leq j \leq n$, $1 \leq k \leq n$.

17. The jth column of $\mathbf{AB}$ has entries $c_{ij} = \sum_{k=1}^{n} a_{ik}b_{kj}$. Let $\mathbf{D} = [d_{ij}] = \mathbf{AB}_j$, where $\mathbf{B}_j$ is the jth column of $\mathbf{B}$. Then $d_{ij} = \sum_{m=1}^{n} a_{im}b_{mj} = c_{ij}$.

19. $\mathbf{A}^2 = \begin{bmatrix} 1 & 0 \\ 0 & 1 \end{bmatrix} = \mathbf{I}_2$.

21. (a) $(\mathbf{A} + \mathbf{B})^T = \mathbf{A}^T + \mathbf{B}^T$ by Theorem 3. Since $\mathbf{A}$ and $\mathbf{B}$ are symmetric, $\mathbf{A}^T + \mathbf{B}^T = \mathbf{A} + \mathbf{B}$ and $\mathbf{A} + \mathbf{B}$ is also symmetric.

    (b) $(\mathbf{AB})^T = \mathbf{B}^T\mathbf{A}^T = \mathbf{BA}$, but this may not be $\mathbf{AB}$, so $\mathbf{AB}$ may not be symmetric. Let $\mathbf{A} = \begin{bmatrix} 1 & 2 \\ 2 & 4 \end{bmatrix}$ and $\mathbf{B} = \begin{bmatrix} 3 & -1 \\ -1 & 2 \end{bmatrix}$. Then $\mathbf{AB}$ is not symmetric.

23. (a) The i, jth element of $(\mathbf{A}^T)^T$ is the j, ith element of $\mathbf{A}^T$. But the j, ith element of $\mathbf{A}^T$ is the i, jth element of $\mathbf{A}$. Thus $(\mathbf{A}^T)^T = \mathbf{A}$.

    (b) The i, jth element of $(\mathbf{A} + \mathbf{B})^T$ is the j, ith element of $\mathbf{A} + \mathbf{B}$, $a_{ji} + b_{ji}$. But this is the sum of the j, ith entry of $\mathbf{A}$ and the j, ith entry of $\mathbf{B}$. It is also the sum of the i, jth entry of $\mathbf{A}^T$ and the i, jth entry of $\mathbf{B}^T$. Thus $(\mathbf{A} + \mathbf{B})^T = \mathbf{A}^T + \mathbf{B}^T$.

    (c) Let $\mathbf{C} = [c_{ij}] = (\mathbf{AB})^T$. Then $c_{ij} = \sum_{k=1}^{n} a_{jk}b_{ki}$, the j, ith entry of $\mathbf{AB}$. Let $\mathbf{D} = [d_{ij}] = \mathbf{B}^T\mathbf{A}^T$, then

    $$d_{ij} = \sum_{k=1}^{n} b'_{ik}a'_{kj} = \sum_{k=1}^{n} b_{ki}a_{jk} = \sum_{k=1}^{n} a_{jk}b_{ki} = c_{ij}.$$

    Hence $(\mathbf{AB})^T = \mathbf{B}^T\mathbf{A}^T$.

25. (a) $\mathbf{A} \vee \mathbf{B} = \begin{bmatrix} 1 & 1 & 1 \\ 0 & 1 & 1 \\ 1 & 0 & 1 \end{bmatrix}$; $\mathbf{A} \wedge \mathbf{B} = \begin{bmatrix} 1 & 0 & 0 \\ 0 & 0 & 1 \\ 1 & 0 & 0 \end{bmatrix}$;

    $\mathbf{A} \odot \mathbf{B} = \begin{bmatrix} 1 & 1 & 1 \\ 0 & 0 & 1 \\ 1 & 1 & 1 \end{bmatrix}$.

    (b) $\mathbf{A} \vee \mathbf{B} = \begin{bmatrix} 0 & 1 & 1 \\ 1 & 1 & 0 \\ 1 & 0 & 1 \end{bmatrix}$; $\mathbf{A} \wedge \mathbf{B} = \begin{bmatrix} 0 & 0 & 1 \\ 1 & 1 & 0 \\ 1 & 0 & 0 \end{bmatrix}$;

    $\mathbf{A} \odot \mathbf{B} = \begin{bmatrix} 1 & 0 & 1 \\ 1 & 1 & 1 \\ 0 & 1 & 1 \end{bmatrix}$.

    (c) $\mathbf{A} \vee \mathbf{B} = \begin{bmatrix} 1 & 1 & 1 \\ 1 & 1 & 1 \\ 1 & 0 & 1 \end{bmatrix}$; $\mathbf{A} \wedge \mathbf{B} = \begin{bmatrix} 1 & 0 & 0 \\ 0 & 0 & 1 \\ 1 & 0 & 0 \end{bmatrix}$;

    $\mathbf{A} \odot \mathbf{B} = \begin{bmatrix} 1 & 1 & 1 \\ 1 & 0 & 0 \\ 1 & 1 & 1 \end{bmatrix}$.

27. Let $\mathbf{C} = [c_{ij}] = \mathbf{A} \vee \mathbf{B}$ and $\mathbf{D} = [d_{ij}] = \mathbf{B} \vee \mathbf{A}$.

    $$c_{ij} = \begin{cases} 1 & a_{ij} = 1 \text{ or } b_{ij} = 1 \\ 0 & a_{ij} = 0 = b_{ij} \end{cases} = d_{ij}$$

    Hence $\mathbf{C} = \mathbf{D}$.

29. Let $[d_{ij}] = \mathbf{B} \vee \mathbf{C}$, $[e_{ij}] = \mathbf{A} \vee (\mathbf{B} \vee \mathbf{C})$, $[f_{ij}] = \mathbf{A} \vee \mathbf{B}$, and $[g_{ij}] = (\mathbf{A} \vee \mathbf{B}) \vee \mathbf{C}$. Then

    $$d_{ij} = \begin{cases} 1 & b_{ij} = 1 \text{ or } c_{ij} = 1 \\ 0 & \text{otherwise} \end{cases}$$

    $$e_{ij} = \begin{cases} 1 & a_{ij} = 1 \text{ or } d_{ij} = 1 \\ 0 & \text{otherwise} \end{cases}$$

    But this means $d_{ij} = 1$ if $a_{ij} = 1$ or $b_{ij} = 1$ or $c_{ij} = 1$ and $d_{ij} = 0$ otherwise.

    $$f_{ij} = \begin{cases} 1 & a_{ij} = 1 \text{ or } b_{ij} = 1 \\ 0 & \text{otherwise} \end{cases}$$

    $$g_{ij} = \begin{cases} 1 & f_{ij} = 1 \text{ or } c_{ij} = 1 \\ 0 & \text{otherwise} \end{cases}$$

    But this means $g_{ij} = 1$ if $a_{ij} = 1$ or $b_{ij} = 1$ or $c_{ij} = 1$ and $g_{ij} = 0$ otherwise. Hence $\mathbf{A} \vee (\mathbf{B} \vee \mathbf{C}) = (\mathbf{A} \vee \mathbf{B}) \vee \mathbf{C}$.

31. Let $[d_{ij}] = \mathbf{B} \odot \mathbf{C}$, $[e_{ij}] = \mathbf{A} \odot (\mathbf{B} \odot \mathbf{C})$, $[f_{ij}] = \mathbf{A} \odot \mathbf{B}$, and $[g_{ij}] = (\mathbf{A} \odot \mathbf{B}) \odot \mathbf{C}$. Then

    $$d_{ij} = \begin{cases} 1 & b_{ik} = c_{kj} \text{ for some } k \\ 0 & \text{otherwise} \end{cases}$$

    and

    $$e_{ij} = \begin{cases} 1 & a_{il} = 1 = d_{lj} \text{ for some } l \\ 0 & \text{otherwise} \end{cases}$$

    But this means

    $$e_{ij} = \begin{cases} 1 & a_{il} = 1 = b_{lk} = c_{kj} \text{ for some } k, l \\ 0 & \text{otherwise.} \end{cases}$$

    $$f_{ij} = \begin{cases} 1 & a_{ik} = 1 = b_{kj} \text{ for some } k \\ 0 & \text{otherwise} \end{cases}$$

    and

    $$g_{ij} = \begin{cases} 1 & f_{il} = 1 = c_{lj} \text{ for some } l \\ 0 & \text{otherwise} \end{cases}$$

    But then

    $$g_{ij} = \begin{cases} 1 & a_{ik} = 1 = b_{kl} = c_{lj} \text{ for some } k, l \\ 0 & \text{otherwise} \end{cases}$$

    and $\mathbf{A} \odot (\mathbf{B} \odot \mathbf{C}) = (\mathbf{A} \odot \mathbf{B}) \odot \mathbf{C}$.

33. An argument similar to that in Exercise 32 shows that
$\mathbf{A} \vee (\mathbf{B} \wedge \mathbf{C}) = (\mathbf{A} \vee \mathbf{B}) \wedge (\mathbf{A} \vee \mathbf{C})$.

35. Since $c_{ij} = \sum_{t=1}^{n} a_{it}b_{tj}$ and $k \mid a_{it}$ for any i and t, k divides each term in c_{ij}, and thus $k \mid c_{ij}$ for all i and j.

Exercise Set 1.6, page 42

1. (a) Yes.   (b) Yes.

3. (a) No.   (b) Yes.

5. $A \oplus B = \{x \mid (x \in A \cup B) \text{ and } (x \notin A \cap B)\} = \{x \mid (x \in B \cup A) \text{ and } (x \notin B \cap A)\} = B \oplus A$.

7.

$x\ y\ z$	$y \square z$	$x \triangledown (y \square z)$	$x \triangledown y$	$x \triangledown z$	$(x \triangledown y) \square (x \triangledown z)$
0 0 0	0	0	0	0	0
0 0 1	1	0	0	0	0
0 1 0	1	0	0	0	0
0 1 1	0	0	0	0	0
1 0 0	0	0	0	0	0
1 0 1	1	1	0	1	1
1 1 0	1	1	1	0	1
1 1 1	0	0	1	1	0
		(A)			(B)

Since columns (A) and (B) are identical, the distributive property $x \triangledown (y \square z) = (x \triangledown y) \square (x \triangledown z)$ holds.

9. 5×5 zero matrix for $\vee$; 5×5 matrix of 1's for $\wedge$; $\mathbf{I}_5$ for $\odot$.

11. Let $\mathbf{A}, \mathbf{B}$ be $n \times n$ diagonal matrices. Let $\left[c_{ij} \right] = \mathbf{AB}$. Then $c_{ij} = \sum_{k=1}^{n} a_{ik}b_{kj}$, but $a_{ik} = 0$ if $i \neq k$. Hence $c_{ij} = a_{ii}b_{ij}$. But $b_{ij} = 0$ if $i \neq j$. Thus $c_{ij} = 0$ if $i \neq j$ and $\mathbf{AB}$ is an $n \times n$ diagonal matrix.

13. Yes, the $n \times n$ zero matrix, which is a diagonal matrix.

15. $-\mathbf{A}$ is the diagonal matrix with i, ith entry $-a_{ii}$.

17. $\begin{bmatrix} a & 0 \\ 0 & 0 \end{bmatrix} + \begin{bmatrix} b & 0 \\ 0 & 0 \end{bmatrix}$ or $\begin{bmatrix} a+b & 0 \\ 0 & 0 \end{bmatrix}$ belongs to M.

19. $\begin{bmatrix} a & 0 \\ 0 & 0 \end{bmatrix}^{T}$ or $\begin{bmatrix} a & 0 \\ 0 & 0 \end{bmatrix}$ belongs to M.

21. Yes, $\begin{bmatrix} 1 & 0 \\ 0 & 0 \end{bmatrix}$. By Exercise 16, we see this is the identity and $1 \in \mathbb{R}$.

23. If $a \neq 0$, then $\mathbf{A}^{-1} = \begin{bmatrix} \frac{1}{a} & 0 \\ 0 & 0 \end{bmatrix}$.

25. Yes.

27. No.

29. (a) Yes.   (b) Yes.   (c) Yes.

(d) Yes, $\begin{bmatrix} 0 \\ -1 \end{bmatrix}$.   (e) Yes, $\begin{bmatrix} -x \\ -y - 2 \end{bmatrix}$.

31. A.

33. $\overline{A \cap B}$.

Chapter 2

Exercise Set 2.1, page 51

1. (b), (d), and (e) are statements.

3. (a) It will not rain tomorrow and it will not snow tomorrow.

(b) It is not the case that if you drive, I will walk.

5. (a) I will drive my car and I will be late. I will drive my car or I will be late.

(b) $10 < NUM \leq 15$. $NUM > 10$ or $NUM \leq 15$.

7. (a) True.   (b) True.   (c) True.   (d) False.

9. (a) False.   (b) True.   (c) True.   (d) False.

11. (d) is the negation.

13. (a) The dish did not run away with the spoon and the grass is wet.

(b) The grass is dry or the dish ran away with the spoon.

(c) It is not true that today is Monday or the grass is wet.

(d) Today is Monday or the dish did not run away with the spoon.

15. (a) For all x there exists a y such that $x + y$ is even.

(b) There exists an x such that, for all y, $x + y$ is even.

17. (a) It is not true that there is an x such that x is even.

(b) It is not true that, for all x, x is a prime number.

19. 14: (a) False.   (b) True.
15: (a) True.   (b) False.
16: (a) False.   (b) True.
17: (a) False.   (b) True.
18: (a) False.   (b) False.   (c) False.   (d) False.

21.

p	q	$(p \vee q)$	$\vee$	$\sim q$
T	T	T	T	F
T	F	T	T	T
F	T	T	T	F
F	F	F	T	T
		(1)	↑	(2)

23.

p	q	r	$(\sim p \vee q)$	$\wedge$	$\sim r$
T	T	T	T		F
T	T	F	T		T
T	F	T	F		F
T	F	F	F		F
F	T	T	T		F
F	T	F	T		T
F	F	T	T		F
F	F	F	T		T
			(1)		↑

25.

p	q	r	$(p \downarrow q)$	$\wedge$	$(p \downarrow r)$
T	T	T	F	F	F
T	T	F	F	F	F
T	F	T	F	F	F
T	F	F	F	F	F
F	T	T	F	F	F
F	T	F	F	F	T
F	F	T	T	F	F
F	F	F	T	T	T
			(1)	↑	(2)

27. (a)

p	q	$p \triangle q$
T	T	F
T	F	T
F	T	T
F	F	F

(b)

p	$p \triangle$	$\sim p$
T	T	F
F	T	T
	↑	

29.

p	q	r	$(p\triangle q)$	$\triangle$	$(q\triangle r)$
T	T	T	F	F	F
T	T	F	F	T	T
T	F	T	T	F	T
T	F	F	T	T	F
F	T	T	T	T	F
F	T	F	T	F	T
F	F	T	F	T	T
F	F	F	F	F	F
			(1)	↑	(2)

31. (key $\neq$ "open") and ($t \geq$ limit)

Exercise Set 2.2, page 56

1. (a) $p \Rightarrow q$.   (b) $r \Rightarrow p$.

    (c) $q \Rightarrow p$.   (d) $\sim r \Rightarrow p$.

3. (a) If I am not the Queen of England, then $2 + 2 = 4$.

    (b) If I walk to work, then I am not the President of the United States.

    (c) If I did not take the train to work, then I am late.

    (d) If I go to the store, then I have time and I am not too tired.

    (e) If I buy a car and I buy a house, then I have enough money.

5. (a) True.   (b) False.

    (c) True.   (d) True.

7. (a) If I do not study discrete structures and I go to a movie, then I am in a good mood.

    (b) If I am in a good mood, then I will study discrete structures or I will go to a movie.

    (c) If I am not in a good mood, then I will not go to a movie or I will study discrete structures.

    (d) I will go to a movie and I will not study discrete structures if and only if I am in a good mood.

9. (a) If $4 > 1$ and $2 > 2$, then $4 < 5$.

    (b) It is not true that $3 \leq 3$ and $4 < 5$.

    (c) If $3 > 3$, then $4 > 1$.

11. (a)

p	q	$p \Rightarrow$	$(q \Rightarrow p)$
T	T	T	T
T	F	T	T
F	T	T	F
F	F	T	T
		↑	

    tautology

(b)

p	q	$q \Rightarrow$	$(q \Rightarrow p)$
T	T	T	T
T	F	T	T
F	T	F	F
F	F	T	T
		↑	

    contingency

13. Yes. If $p \Rightarrow q$ is false, then p is true and q is false. Hence $p \wedge q$ is false, $\sim(p \wedge q)$ is true, and $(\sim(p \wedge q)) \Rightarrow q$ is false.

15. (a) False.   (b) True.   (c) False.   (d) True.

17.

p	q	$p \wedge q$	$(p \downarrow p)$	$\downarrow$	$(q \downarrow q)$
T	T	T	F	T	F
T	F	F	F	F	T
F	T	F	T	F	F
F	F	F	T	F	T
		(A)		(B)	

Since columns (A) and (B) are the same, the statements are equivalent.

19. (a) (i).   (b) (iv).

21.

p	q	$\sim$	$(p \wedge q)$	$\sim p$	$\vee$	$\sim q$
T	T	F	T	F	F	F
T	F	T	F	F	T	T
F	T	T	F	T	T	F
F	F	T	F	T	T	T
			(A)		(B)	

Because columns (A) and (B) are the same, the statements are equivalent.

23. The statement $\exists x (P(x) \vee Q(x))$ is true if and only if $\exists x$ such that either $P(x)$ or $Q(x)$ is true, but this means either $\exists x\, P(x)$ or $\exists x\, Q(x)$ is true. This holds if and only if $\exists x\, P(x) \vee \exists x\, Q(x)$ is true.

25.

p	q	$(p \wedge q)$	$\Rightarrow p$
T	T	T	T
T	F	F	T
F	T	F	T
F	F	F	T
			↑

27.

p	q	$(p \wedge$	$(p \Rightarrow q))$	$\Rightarrow q$
T	T	T	T	T
T	F	F	F	T
F	T	F	T	T
F	F	F	T	T
				↑

Exercise Set 2.3, page 62

1. Valid: $((d \Rightarrow t) \wedge \sim t) \Rightarrow \sim d$.

3. Invalid.

5. Valid: $((f \vee \sim w) \wedge w) \Rightarrow f$.

7. Valid: $[(ht \Rightarrow m) \wedge (m \Rightarrow hp)] \Rightarrow [\sim hp \Rightarrow \sim ht]$.

9. Invalid.

11. (a) Invalid.   (b) Valid: $(\sim(p \Rightarrow q) \wedge p) \Rightarrow \sim q$

13. Suppose m and n are odd numbers. Then there exist integers j and k such that $m = 2j + 1$ and $n = 2k + 1$. $m + n = 2j + 1 + 2k + 1 = 2j + 2k + 2 = 2(j + k + 1)$. Since $j + k + 1$ is an integer, $m + n$ is even.

15. Suppose that m and n are odd. Then there exit integers j and k such that $m = 2j + 1$ and $n = 2k + 1$. $m \cdot n = 2j \cdot 2k + 2j + 2k + 1 = 2(2jk + j + k) + 1$. Since $2jk + j + k$ is an integer, $m \cdot n$ is odd and the system is closed with respect to multiplication.

17. If $A = B$, then, clearly, $A \subseteq B$ and $B \subseteq A$. If $A \subseteq B$ and $B \subseteq A$, then $A \subseteq B \subseteq A$ and B must be the same as A.

19. (a) If $A \subseteq B$, then $A \cup B \subseteq B$. But $B \subseteq A \cup B$. Hence $A \cup B = B$. If $A \cup B = B$, then since $A \subseteq A \cup B$, we have $A \subseteq B$.

(b) If $A \subseteq B$, then $A \subseteq A \cap B$. But $A \cap B \subseteq A$. Hence $A \cap B = A$. If $A \cap B = A$, then since $A \cap B \subseteq B$, we have $A \subseteq B$.

21. Any five consecutive integers can be represented by n, $n + 1, n + 2, n + 3, n + 4$. Their sum is $5n + 10$ or $5(n + 2)$. This is clearly divisible by 5.

23. Invalid. Multiplying by $x - 1$ may or may not preserve the order of the inequality.

25. Valid.

27. Let x and y be prime numbers, each larger than 2. Then x and y are odd and their sum is even (Exercise 13). The only even prime is 2, so $x + y$ is not a prime.

29. Suppose $x + y$ is rational. Then there are integers a and b such that $x + y = \frac{a}{b}$. Since x is rational, we can write $x = \frac{c}{d}$ with integers c and d. But now $y = x + y - x = \frac{a}{b} - \frac{c}{d}$. $y = \frac{ad - bc}{bd}$. Both $ad - bc$ and bd are integers. This is a contradiction since y is an irrational number and cannot be expressed as the quotient of two integers.

Exercise Set 2.4, page 69

Note: Only the outlines of the induction proofs are given. These are not complete proofs.

1. Basis step: $n = 1$ $P(1)$: $2(1) = 1(1 + 1)$ is true.
Induction step: $P(k)$: $2 + 4 + \cdots + 2k = k(k + 1)$.
$P(k + 1)$: $2 + 4 + \cdots + 2(k + 1) = (k + 1)(k + 2)$
LHS of $P(k + 1)$: $2 + 4 + \cdots + 2k + 2(k + 1) =$
$k(k + 1) + 2(k + 1) = (k + 1)(k + 2)$
RHS of $P(k + 1)$.

3. Basis step: $n = 0$ $P(0)$: $2^0 = 2^{0+1} - 1$ is true.
Induction step: LHS of $P(k + 1)$:
$1 + 2^1 + 2^2 + \cdots + 2^k + 2^{k+1} = (2^{k+1} - 1) + 2^{k+1} =$
$2 \cdot 2^{k+1} - 1 = 2^{k+2} - 1$ RHS of $P(k + 1)$.

5. Basis step: $n = 1$ $P(1)$: $1^2 = \dfrac{1(1 + 1)(2 + 1)}{6}$ is true.
Induction step: LHS of $P(k + 1)$:

$$1^2 + 2^2 + \cdots + k^2 + (k + 1)^2$$
$$= \frac{k(k + 1)(2k + 1)}{6} + (k + 1)^2$$
$$= (k + 1)\left(\frac{k(2k + 1)}{6} + (k + 1)\right)$$
$$= \frac{k + 1}{6}(2k^2 + k + 6(k + 1))$$
$$= \frac{k + 1}{6}(2k^2 + 7k + 6)$$

$$= \frac{(k+1)(k+2)(2k+3)}{6}$$

$$= \frac{(k+1)((k+1)+1)(2(k+1)+1)}{6}$$

RHS of $P(k+1)$.

7. Basis step: $n=1$ $P(1)$: $a = \dfrac{a(1-r^1)}{1-r}$ is true.
Induction step: LHS of $P(k+1)$:

$$a + ar + \cdots + ar^{k-1} + ar^k = \frac{a(1-r^k)}{1-r} + ar^k =$$

$$\frac{a - ar^k + ar^k - ar^{k+1}}{1-r} = \frac{a(1-r^{k+1})}{1-r} \quad \text{RHS of}$$
$P(k+1)$.

9. (a) LHS of $P(k+1)$: $1 + 5 + 9 + \cdots + (4(k+1) - 3) =$
$(2k+1)(k-1) + 4(k+1) - 3 = 2k^2 + 3k =$
$(2k+3)(k) = (2(k+1)+1)((k+1)+1)$ RHS of
$P(k+1)$.

(b) No; $P(1)$: $1 = (2 \cdot 1 + 1)(1-1)$ is false.

11. Basis step: $n=2$ $P(2)$: $2 < 2^2$ is true.
Induction step: LHS of $P(k+1)$:
$k + 1 < 2^k + 1 < 2^k + 2^k = 2 \cdot 2^k = 2^{k+1}$ RHS of
$P(k+1)$.

13. Basis step: $n=5$ $P(5)$: $1 + 5^2 < 2^5$ is true.
Induction step: LHS of $P(k+1)$:
$1 + (k+1)^2 = k^2 + 1 + 2k + 1 < 2^k + 2k + 1 <$
$2^k + k^2 + 1 < 2^k + 2^k = 2 \cdot 2^k = 2^{k+1}$ RHS of $P(k+1)$.

15. Basis step: $n=0$ $A = \{\ \}$ and $P(A) = \{\{\ \}\}$, so
$|P(A)| = 2^0$ and $P(0)$ is true.
Induction step: Use $P(k)$: If $|A| = k$, then $|P(A)| = 2^k$ to
show $P(k+1)$: If $|A| = k+1$, then $|P(A)| = 2^{k+1}$.
Suppose that $|A| = k+1$. Set aside one element x of A.
Then $|A - \{x\}| = k$ and $A - \{x\}$ has 2^k subsets. These
subsets are also subsets of A. We can form another 2^k
subsets of A by forming the union of $\{x\}$ with each subset
of $A - \{x\}$. None of these subsets are duplicates. Now A
has $2^k + 2^k$, or 2^{k+1}, subsets.

17. Basis step: $n=1$ $P(1)$: $\overline{A_1} = \overline{A_1}$ is true.
Induction step: LHS of $P(k+1)$:

$$\overline{\bigcap_{i=1}^{k+1} A_i} = \overline{\left(\bigcap_{i=1}^{k} A_i\right) \cap A_{k+1}}$$

$$= \overline{\bigcap_{i=1}^{k} A_i} \cup \overline{A_{k+1}} \quad \text{(De Morgan's laws)}$$

$$= \left(\bigcup_{i=1}^{k} \overline{A_i}\right) \cup \overline{A_{k+1}}$$

$$= \bigcup_{i=1}^{k+1} \overline{A_i} \quad \text{RHS of } P(k+1).$$

19. Basis step: $n=1$ $P(1)$: $A_1 \cup B = A_1 \cup B$ is true.
Induction step: LHS of $P(k+1)$:

$$\left(\bigcap_{i=1}^{k+1} A_i\right) \cup B = \left(\left(\bigcap_{i=1}^{k} A_i\right) \cap A_{k+1}\right) \cup B$$

$$= \left(\left(\bigcap_{i=1}^{k} A_i\right) \cup B\right) \cap (A_{k+1} \cup B)$$

(distributive property)

$$= \left(\bigcap_{i=1}^{k} (A_i \cup B)\right) \cap (A_{k+1} \cup B)$$

$$= \bigcap_{i=1}^{k+1} (A_i \cup B) \quad \text{RHS of } P(k+1).$$

21. Basis step: $n=1$ $P(1)$: $\mathbf{A}_1^T = \mathbf{A}_1^T$ is true.
Induction step: LHS of $P(k+1)$:
$(\mathbf{A}_1 + \mathbf{A}_2 + \cdots + \mathbf{A}_k + \mathbf{A}_{k+1})^T =$
$(\mathbf{A}_1 + \mathbf{A}_2 + \cdots + \mathbf{A}_k)^T + \mathbf{A}_{k+1}^T = \mathbf{A}_1^T + \mathbf{A}_2^T + \cdots + \mathbf{A}_k^T + \mathbf{A}_{k+1}^T$
RHS of $P(k+1)$.

23. Basis step: $n=1$ $P(1)$: $(\mathbf{AB})^1 = \mathbf{A}^1 \cdot \mathbf{B}^1$ is true.
Induction step: LHS of $P(k+1)$:
$(\mathbf{AB})^{k+1} = (\mathbf{AB})(\mathbf{AB})^k = \mathbf{AB} \cdot \mathbf{A}^k \mathbf{B}^k = \mathbf{BA} \cdot \mathbf{A}^k \mathbf{B}^k =$
$\mathbf{A} \cdot \mathbf{A}^k \mathbf{B} \cdot \mathbf{B}^k = \mathbf{A}^{k+1} \cdot \mathbf{B}^{k+1}$ RHS of $P(k+1)$.

25. Basis step: $n=28$ $P(28)$: 28 can be written as
$5 \cdot 4 + 8 \cdot 1$.
Induction step: We use $P(j)$ for $j = 28, 29, \ldots, k$ to show
$P(k+1)$. Consider $k+1 = (k+1-5) + 5 = (k-4) + 5$.
$P(k-4)$ guarantees that $k - 4 = 5a' + 8b$, a', b in $\mathbf{Z}^+$.
Hence, $k + 1 = 5a + 8b$, a, b in $\mathbf{Z}^+$.

27. Basis step: $n=1$ $P(1)$: If $GCD(a, b) = 1$, then
$GCD(a^1, b^1) = 1$ is true.
Induction step: Suppose $GCD(a, b) = 1$. Let
$d = GCD(a^{k+1}, b^{k+1})$. If $d \neq 1$, then let p be a prime
factor of d. Then $p \mid a^{k+1}$ and $p \mid b^{k+1}$. By Exercise 24,
$p \mid a$ and $p \mid b$. But this is a contradiction. Hence d must
be 1.

29. Loop invariant check:
Basis step: $n=0$ $P(0)$: $Y \times W_0 + Z_0 = X + Y^2$ is true
because $W_0 = Y$ and $Z_0 = X$.
Induction step: LHS of $P(k+1)$: $Y \times W_{k+1} + Z_{k+1} =$
$Y \times (W_k - 1) + (Z_k + Y) = Y \times W_k + Z_k = X + Y^2$
RHS of $P(k+1)$.
Exit condition check: When $W = 0$,
$Y \times W + Z = X + Y^2$ yields $Z = X + Y^2$.

31. Loop invariant check:
Basis step: $n=0$ $P(0)$: $R_0 \times N^{K_0} = N^{2M}$ is true,
because $R_0 = 1$ and $K_0 = 2M$.
Induction step: LHS of $P(k+1)$:
$R_{k+1} \times N^{K_{k+1}} = (R_k \times N) \times N^{K_k - 1} = R_k \times N^{K_k} = N^{2M}$
RHS of $P(k+1)$.
Exit condition check: When $K = 0$, $R \times N^K = N^{2M}$
yields $R \times N^0 = N^{2M}$ or $R = N^{2M}$.

33. Loop invariant check:
Basis step: $n = 0$ P(0): $R_0 + K_0 \times Y = X$ is true,
because $R_0 = X$ and $K_0 = 0$.
Induction step: LHS of P($k + 1$): $R_{k+1} + K_{k+1} \times Y =$
$(R_k - Y) + (K_k + 1) \times Y = R_k + K_k \times Y = X$ RHS of
P($k + 1$)
Exit condition check: $R < Y$. $R + K \times Y = X$ and
$R = 0$ if and only if $Y \mid X$.

Chapter 3

Exercise Set 3.1, page 77

1. 67,600.

3. 16.

5. 1296.

7. (a) 0.   (b) 1.

9. (a) $n!$ (b) $\dfrac{n!}{2}$ (c) $\dfrac{(n+1)!}{2}$

11. 120.

13. 4! or 24.

15. 30.

17. (a) 479,001,600.   (b) 1,036,800.

19. 240.

21. 360.

23. 39,916,800

25. 67,200.

27. $n \cdot {}_{n-1}P_{n-1} = n \cdot (n-1)(n-2)\cdots 2 \cdot 1 = n! = {}_nP_n.$

29. 190.

31. 2; 6; 12.

Exercise Set 3.2, page 81

1. (a) 1.   (b) 35.   (c) 4368.

3. ${}_nC_r = \dfrac{n!}{r!(n-r)!} = \dfrac{n!}{(n-(n-r))!(n-r)!} = {}_nC_{n-r}.$

5. 20,358,520

7. (a) 1.   (b) 360.

9. (a) One of size 0, 4 of size 1, 6 of size 2, 4 of size 3, and
1 of size 4.

(b) For each r, $0 \le r \le n$, there are ${}_nC_r$ subsets of size r.

11. (a) 980.   (b) 1176.

13. 2702.

15. 177,100

17. ${}_nC_{r-1} + {}_nC_r = \dfrac{n!}{(r-1)!(n-(r-1))!} + \dfrac{n!}{r!(n-r)!}$
$= \dfrac{n!r + n!(n-r+1)}{r!(n-r+1)!} = \dfrac{n!(n+1)}{r!(n+1-r)!}$
$= \dfrac{(n+1)!}{r!(n+1-r)!} = {}_{n+1}C_r.$

19. (a) 32.   (b) 5.   (c) 10.

21. (a) 2^n.   (b) ${}_nC_3$.   (c) ${}_nC_k$.

23. 525.

25. (a)

| 1 | 5 | 10 | 10 | 5 | 1 |

| 1 | 6 | 15 | 20 | 15 | 6 | 1 |

| 1 | 7 | 21 | 35 | 35 | 21 | 7 | 1 |

(b) Begin the row with a 1; write the sum of each
consecutive pair of numbers in the previous row,
moving left to right; end the row with a 1.

27. Exercise 17 shows another way to express the results of
Exercise 25(b) and Exercise 26.

Exercise Set 3.3, page 85

1. Let the birth months play the role of the pigeons and the
calendar months, the pigeonholes. Then there are 13
pigeons and 12 pigeonholes. By the pigeonhole principle,
at least two people were born in the same month.

3.

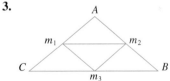

m_1, m_2, m_3 are the midpoints of sides AC, AB, and BC,
respectively. Let the four small triangles created be the
pigeonholes. For any five points in or on triangle ABC, at
least two must be in or on the same small triangle and thus
are no more than $\frac{1}{2}$ unit apart.

5. By the extended pigeonhole principle, at least
$\lfloor (50-1)/7 \rfloor + 1$ or 8 will be the same color.

7. Let 2161 cents be the pigeons and the six friends, the
pigeonholes. Then at least one friend has
$\lfloor (2161-1)/6 \rfloor + 1$ or 361 cents.

9. If repetitions are allowed, there are ${}_{16}C_5$ or 4368 choices.
At least $\lfloor 4367/175 \rfloor + 1$, or 25, choices have the same
cost.

11. You must have at least 49 friends.

13. Consider the first eight rows; one row must have at least 7 ones since there are 51 ones in all. Similarly, there is a column with at least 7 ones. The sum of the entries in this row and this column is at least 14.

15. Label the pigeonholes with $1, 3, 5, \ldots, 25$, the odd numbers between 1 and 25 inclusive. Assign each of the selected 14 numbers to the pigeonhole labeled with its odd part. There are only 13 pigeonholes, so two numbers must have the same odd part. One is a multiple of the other.

17. No. At least one pair of the 12 disks must add up to 21.

19. Consider the six sums $c_1, c_1 + c_2, c_1 + c_2 + c_3, \ldots,$ $c_1 + c_2 + c_3 + c_4 + c_5 + c_6$. If one of these has remainder 0 when divided by 6, then we are done. If none have remainder 0 when divided by 6, then two of them must give the same remainder. The positive difference of these two is a subsequence whose sum is divisible by 6.

Exercise Set 3.4, page 93

1. {HHH, HHT, HTH, THH, HTT, THT, TTH, TTT}.

3. {sb, sr, sg, cb, cr, cg}.

5. (a) $\{\ \}, \{1\}, \{2\}, \{3\}, \{2, 3\}, \{1, 2\}, \{1, 3\}, \{1, 2, 3\}$.

   (b) 2^n.

7. (a) The card is a red ace.

   (b) The card is black or a diamond or an ace.

9. (a) $\{(5, 1), (5, 2), (5, 3), (5, 4), (5, 5), (5, 6),$ $(1, 5), (2, 5), (3, 5), (4, 5), (6, 5)\}$.

   (b) $\{\ \}$.

11. (a) No, 3 satisfies both descriptions.

   (b) No, 2 satisfies both descriptions.

   (c) Yes, $E \cup F = \{3, 4, 5, 1, 2, 3\}$.

   (d) No, $E \cap F = \{3\}$.

13. (a) $\{dls, dln, dms, dmn, dus, dun, als, aln\}$.

   (b) $\{als, aln\}$.

   (c) $\{dls, dln, als\}$.

15. $\overline{E} \cup F = \{2, 6, 3\}, \overline{F} \cap G = \{4\}$.

17. {club}, {spade}, {diamond}, {heart}.

19. (a) $\frac{10}{18}$.   (b) $\frac{11}{18}$.   (c) $\frac{12}{18}$.   (d) $\frac{9}{18}$.

21. (a) 0.7.   (b) 0.   (c) 0.7.   (d) 1.

23. $p(A) = \frac{6}{11}, p(B) = \frac{3}{11}, p(C) = \frac{1}{11}, p(D) = \frac{1}{11}$.

25. $\frac{10}{32}$.

27. (a) $\frac{1}{6}$.   (b) 1.

29. (a) $\frac{11}{36}$.   (b) $\frac{27}{36}$.   (c) $\frac{6}{36}$.   (d) $\frac{21}{36}$.

31. (a) $\frac{35}{220}$.   (b) $\frac{80}{220}$.   (c) $\frac{210}{220}$.   (d) $\frac{210}{220}$.

33. $\dfrac{n + 1}{2}$.

35. $-\frac{11}{9}$ dollars.

Exercise Set 3.5, page 100

1. Yes, degree 1.

3. No.

5. No.

7. $s_1 = 2, s_2 = 3, s_n = s_{n-1} + 1$.

9. $A_1 = 100 \left(1 + \frac{0.06}{12}\right), A_n = \left(1 + \frac{0.06}{12}\right)(A_{n-1} + 100)$.

11. $a_n = 4(2.5)^{n-1}$.

13. $c_n = 3 + \dfrac{n(n + 1)}{2}$.

15. $e_n = -2(n - 1)$.

17. $a_n = \frac{8}{30} \cdot 5^n - \frac{20}{30}(-1)^n$.

19. $c_n = -\frac{19.7}{9}(-3)^n + \frac{12.2}{9}n(-3)^n$.

21. $e_n = 2(\sqrt{2})^n + (-\sqrt{2})^n$.

23. $a_n = r^{n-1}a_1 + s\left(\dfrac{r^{n-1} - 1}{r - 1}\right)$.

25. For $x^2 - r_1x - r_2 = 0$ to have a unique solution, s, we must have $r_1^2 = -4r_2$. Then $s = \dfrac{r_1}{2}$. We now show that $a_n = us^n + vns^n$ defines the same sequence as $a_n = r_1a_{n-1} + r_2a_{n-2}$ when u and v are chosen so that $a_1 = us + vs$ and $a_2 = us^2 + 2vs^2$. Consider

$$a_n = us^n + vns^n$$
$$= (us^{n-2} + v(n - 1)s^{n-2})s^2 + vs^n$$
$$= (us^{n-2} + v(n - 1)s^{n-2})(r_1s + r_2) + vs^n$$
$$= r_1(us^{n-1} + v(n - 1)s^{n-1})$$
$$\quad + r_2(us^{n-2} + v(n - 2)s^{n-2}) + r_2vs^{n-2} + vs^n$$
$$= r_1a_{n-1} + r_2a_{n-2} + vs^{n-2}(r_2 + s^2)$$
$$= r_1a_{n-1} + r_2a_{n-2} \quad \text{since } r_2 + s^2 = 0.$$

27. There are r places to insert an element in a sequence of length $r - 1$, but there are $n - (r - 1)$ different elements that could be used for the insertion.

29. $A_1 = 100(1.005),$
$A_n = 100(1.005)^n + 20,100(1.005^{n-1} - 1)$.

31. Basis step: $n = 1$ P(1): $b_1 < \frac{5}{2}$ is true, because b_1 is 1. Induction step: LHS of P(k + 1):
$b_{k+1} = b_k + 2b_{k-1} < \left(\frac{5}{2}\right)^k + 2\left(\frac{5}{2}\right)^{k-1} = \left(\frac{5}{2}\right)^{k-1}\left(\frac{5}{2} + 2\right) = $
$\left(\frac{5}{2}\right)^{k-1}\left(\frac{9}{2}\right) < \left(\frac{5}{2}\right)^{k-1}\left(\frac{25}{4}\right) = \left(\frac{5}{2}\right)^{k+1}$ RHS of P(k + 1).

33. $C_1 = 1, C_2 = 1, C_3 = 2, C_4 = 5, C_5 = 14$.

Chapter 4

Exercise Set 4.1, page 105

1. (a) x is 4.   (b) y is 3.

3. (a) x is 4; y is 6.   (b) x is 4; y is 2.

5. (a) $\{(a, 4), (a, 5), (a, 6), (b, 4), (b, 5), (b, 6)\}$.

(b) $\{(4, a), (5, a), (6, a), (4, b), (5, b), (6, b)\}$.

7. (a) $\{$(Fine, president), (Fine, vice-president), (Fine, secretary), (Fine, treasurer), (Yang, president), (Yang, vice-president), (Yang, secretary), (Yang, treasurer)$\}$.

(b) $\{$(president, Fine), (vice-president, Fine), (secretary, Fine), (treasurer, Fine), (president, Yang), (vice-president, Yang), (secretary, Yang), (treasurer, Yang)$\}$.

(c) $\{$(Fine, Fine), (Fine, Yang), (Yang, Fine), (Yang, Yang)$\}$.

9. gs, ds, gc, dc, gv, dv.

11. (outline) Basis step: $n = 1$ P(1): If $|A| = 3$ and $|B| = 1$, then $|A \times B| = 3$. $A \times B = \{(a_1, *), (a_2, *), (a_3, *)\}$. Clearly, $|A \times B| = 3$.
Induction step: Suppose $|B| = k > 1$. Let $x \in B$ and $C = B - \{x\}$. Then $|C| = k - 1 \geq 1$ and using P(k), we have $|A \times C| = 3(k - 1)$. $|A \times \{x\}| = 3$. Since $(A \times C) \cap (A \times \{x\}) = \{\ \}$ and $(A \times C) \cup (A \times \{x\}) = A \times B$, $|A \times B| = 3(k - 1) + 3$ or $3k$.

13.

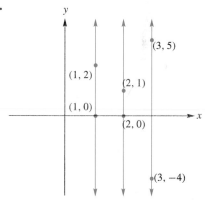

15.

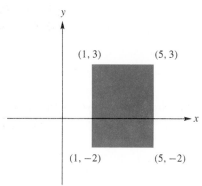

17. Let $(x, y) \in A \times B$, then $x \in A$ and $y \in B$. Since $A \subseteq C$ and $B \subseteq D$, $x \in C$ and $y \in D$. Hence $(x, y) \in C \times D$.

19. (a) Yes.   (b) No.

21. Answers will vary.

23. No. $|A| = 26$.

25. $\{\{1\}, \{2\}, \{3\}\}$, $\{\{1\}, \{2, 3\}\}$, $\{\{2\}, \{1, 3\}\}$, $\{\{3\}, \{1, 2\}\}$, $\{\{1, 2, 3\}\}$.

27. 3. There are 3 2-element partitions listed in the solution to Exercise 25.

29. Let $(x, y) \in A \times (B \cup C)$. Then $x \in A$, $y \in B \cup C$. Hence $(x, y) \in A \times B$ or $(x, y) \in A \times C$. Thus $A \times (B \cup C) \subseteq (A \times B) \cup (A \times C)$. Let $(x, y) \in (A \times B) \cup (A \times C)$. Then $x \in A$, $y \in B$ or $y \in C$. Hence $y \in B \cup C$ and $(x, y) \in A \times (B \cup C)$. So, $(A \times B) \cup (A \times C) \subseteq A \times (B \cup C)$.

Exercise Set 4.2, page 114

1. (a) No.   (b) No.   (c) Yes.
(d) Yes.   (e) Yes.   (f) Yes.

3. (a) No.   (b) No.   (c) Yes.
(d) Yes.   (e) No.   (f) Only if $n = 1$.

5. Domain: {IBM, Dell, COMPAQ, Gateway}, Range: {750C, 466V, 450SV, PS60};

$$\begin{bmatrix} 1 & 0 & 0 & 0 & 0 & 0 & 0 \\ 0 & 0 & 1 & 0 & 0 & 0 & 0 \\ 0 & 0 & 0 & 0 & 0 & 1 & 0 \\ 0 & 1 & 0 & 0 & 0 & 0 & 0 \\ 0 & 0 & 0 & 0 & 0 & 0 & 0 \end{bmatrix}.$$

7. Domain: $\{1, 2, 3, 4, 8\}$, Range: $\{1, 2, 3, 4, 8\}$;

$$\begin{bmatrix} 1 & 0 & 0 & 0 & 0 \\ 0 & 1 & 0 & 0 & 0 \\ 0 & 0 & 1 & 0 & 0 \\ 0 & 0 & 0 & 1 & 0 \\ 0 & 0 & 0 & 0 & 1 \end{bmatrix}.$$

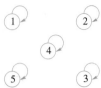

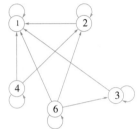

9. Domain: {1, 2, 3, 4, 6}, Range: {1, 2, 3, 4, 6};

$$\begin{bmatrix} 1 & 0 & 0 & 0 & 0 \\ 1 & 1 & 0 & 0 & 0 \\ 1 & 0 & 1 & 0 & 0 \\ 1 & 1 & 0 & 1 & 0 \\ 1 & 1 & 1 & 0 & 1 \end{bmatrix}.$$

11. Domain: {3, 5, 7, 9}, Range: {2, 4, 6, 8};

$$\begin{bmatrix} 0 & 0 & 0 & 0 \\ 1 & 0 & 0 & 0 \\ 1 & 1 & 0 & 0 \\ 1 & 1 & 1 & 0 \\ 1 & 1 & 1 & 1 \end{bmatrix}.$$

13. (a) No.   (b) No.   (c) Yes.

(d) Yes.   (e) No.   (f) No.

15. $\text{Dom}(R) = [-5, 5]$, $\text{Ran}(R) = [-5, 5]$.

17. (a) {1, 3}.   (b) {1, 2, 3, 6}.   (c) {1, 2, 4, 3, 6}.

19. $a \, R \, b$ if and only if $0 \le a \le 3$ and $0 \le b \le 2$.

21. $R =$
{(1, 1), (1, 2), (1, 4), (2, 2), (2, 3), (3, 3), (3, 4), (4, 1)}.

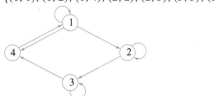

23. $R = \{(1, 2), (2, 2), (2, 3), (3, 4), (4, 4), (5, 1), (5, 4)\}$.

$$\begin{bmatrix} 1 & 0 & 0 & 0 & 0 \\ 0 & 1 & 1 & 0 & 0 \\ 0 & 0 & 0 & 1 & 0 \\ 0 & 0 & 0 & 1 & 0 \\ 1 & 0 & 0 & 1 & 0 \end{bmatrix}.$$

25.

Vertex	1	2	3	4	5
In-degree	1	2	1	3	0
Out-degree	1	2	1	1	2

27. The in-degree of a vertex is the number of ones in the column labeled by that vertex. The out-degree of a vertex is the number of ones in the row labeled by that vertex.

29. {(2, 3), (3, 6)}.

31. Delete any vertex labeled by an element of $A - B$. Then delete any edges that do not point to a vertex.

Exercise Set 4.3, page 120

1. 1, 2   1, 6   2, 3   3, 3   3, 4   4, 3   4, 5   4, 1   6, 4.

3. (a) 3, 3, 3, 3   3, 3, 4, 3   3, 3, 4, 5   3, 4, 1, 6
   3, 4, 1, 2   3, 4, 3, 3   3, 4, 3, 4   3, 3, 4, 1
   3, 3, 3, 4.

(b) In addition to those in part (a), 1, 2, 3, 3
   1, 2, 3, 4   1, 6, 4, 1   1, 6, 4, 5   2, 3, 3, 3
   2, 3, 3, 4   2, 3, 4, 3   2, 3, 4, 5   4, 1, 2, 3
   4, 1, 6, 4   6, 4, 3, 3   6, 4, 3, 4   6, 4, 1, 2
   6, 4, 1, 6   1, 6, 4, 3   2, 3, 4, 1   4, 3, 3, 3
   4, 3, 4, 3   4, 3, 4, 1   4, 3, 4, 5   4, 3, 3, 4.

5. One is 6, 4, 1, 6.

7. $\begin{bmatrix} 0 & 0 & 1 & 1 & 0 & 0 \\ 0 & 0 & 1 & 1 & 0 & 0 \\ 1 & 0 & 1 & 1 & 1 & 0 \\ 0 & 1 & 1 & 1 & 0 & 1 \\ 0 & 0 & 0 & 0 & 0 & 0 \\ 1 & 0 & 1 & 0 & 1 & 0 \end{bmatrix}.$

9. a, c a, b b, b b, f c, d c, e d, c d, b
   e, f f, d.

11. (a) a, c, d, c a, c, d, b a, c, e, f a, b, b, b
   a, b, b, f a, b, f, d.

(b) In addition to those in part (a), b, b, b, b
   b, b, b, f b, b, f, d b, f, d, b b, f, d, c
   c, d, c, d c, d, c, e c, d, b, b c, d, b, f
   c, e, f, d d, c, d, c d, c, e, f d, b, b, b
   d, b, b, f d, b, f, d d, c, d, b e, f, d, b
   e, f, d, c f, d, c, d f, d, c, e f, d, b, b
   f, d, b, f.

13. One is d, b, f, d.

15.

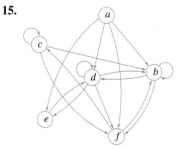

17. (a)
$$\begin{bmatrix} 0 & 1 & 1 & 1 & 1 & 1 \\ 0 & 1 & 1 & 1 & 1 & 1 \\ 0 & 1 & 1 & 1 & 1 & 1 \\ 0 & 1 & 1 & 1 & 1 & 1 \\ 0 & 1 & 1 & 1 & 1 & 1 \\ 0 & 1 & 1 & 1 & 1 & 1 \end{bmatrix}$$

(b) $\{(a, c), (a, d), (a, b), (a, e), (a, f), (b, b),$
$(b, c), (b, d), (b, e), (b, f), (c, b), (c, c),$
$(c, d), (c, e), (c, f), (d, b), (d, c), (d, d),$
$(d, e), (d, f), (e, b), (e, c), (e, d), (e, e),$
$(e, f), (f, b), (f, c), (f, d), (f, e), (f, f)\}.$

19. $x_i \; R^* \; x_j$ if and only if $x_i = x_j$ or $x_i \; R^n \; x_j$ for some n. The i, jth entry of $\mathbf{M}_{R^*}$ is 1 if and only if $i = j$ or the i, jth entry of $\mathbf{M}_{R^n}$ is 1 for some n. Since $R^\infty = \bigcup_{k=1}^{\infty} R^k$, the i, jth entry of $\mathbf{M}_{R^*}$ is 1 if and only if $i = j$ or the i, jth entry of $\mathbf{M}_{R^\infty}$ is 1. Hence $\mathbf{M}_{R^*} = \mathbf{I}_n \vee \mathbf{M}_{R^\infty}$.

21. 1, 7, 5, 6, 7, 4, 3.

23. 2, 3, 5, 6, 7, 5, 6, 4

25. The ij-entry of $\mathbf{M}_R \cdot \mathbf{M}_R$ is the number of paths from i to j of length two, because it is also the number of k's such that $a_{ik} = b_{kj} = 1$.

27. Direct; Boolean multiplication.

Exercise Set 4.4, page 127

1. Reflexive, symmetric, transitive.

3. None.

5. Irreflexive, symmetric, asymmetric, antisymmetric, transitive.

7. Transitive.

9. Antisymmetric, transitive.

11. Irreflexive, symmetric.

13. Reflexive.

15. Reflexive, antisymmetric, transitive.

17. Irreflexive, symmetric.

19. Symmetric.

21. Reflexive, symmetric, transitive.

23.

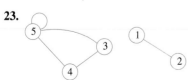

25. $\{(1, 5), (5, 1), (1, 6), (6, 1), (5, 6), (6, 5), (1, 2),$
$(2, 1), (2, 7), (7, 2), (2, 3), (3, 2)\}.$

27. Let $a_1, a_2, \ldots, a_n$ be the elements of the base set. The graph of R is connected if for each a_i, there is a 1 in the ith column of $(\mathbf{M}_R)_{\odot}^k$ for some k.

29. Let R be transitive and irreflexive. Suppose $a \; R \; b$ and $b \; R \; a$. Then $a \; R \; a$ since R is transitive. But this contradicts the fact that R is irreflexive. Hence R is asymmetric.

31. Let $R \neq \{ \; \}$ be symmetric and transitive. There exists $(x, y) \in R$ and $(y, x) \in R$. Since R is transitive, we have $(x, x) \in R$, and R is not irreflexive.

33. (Outline) Basis step: $n = 1$ P(1): If R is symmetric, then R^1 is symmetric is true.
Induction step: Use P(k): If R is symmetric, then R^k is symmetric to show P(k + 1). Suppose that $a \; R^{k+1} \; b$. Then there is a $c \in A$ such that $a \; R^k \; c$ and $c \; R \; b$. We have $b \; R \; c$ and $c \; R^k \; a$. Hence $b \; R^{k+1} \; a$.

35. First, suppose R is transitive. Let $a \; R^n \; b, n \geq 1$. Then there is a path $a, x_1, x_2, \ldots, x_{n-1}, b$ in R. Since R is transitive, we have $a \; R \; x_1 \wedge x_1 \; R \; x_2 \Longrightarrow a \; R \; x_2$; $a \; R \; x_2 \wedge x_2 \; R \; x_3 \Longrightarrow a \; R \; x_3, \ldots;$ $a \; R \; x_{n-1} \wedge x_{n-1} \; R \; b \Longrightarrow a \; R \; b$. Hence $R^n \subseteq R, n \geq 1$. Conversely, if $R^n \subseteq R, n \geq 1$, then $R^2 \subseteq R$ and R is transitive.

Exercise Set 4.5, page 132

1. Yes.

3. Yes.

5. No.

7. No.

9. Yes.

11. Yes.

13. $\{(a, a), (a, c), (a, e), (c, a), (c, c), (c, e), (e, a),$
$(e, c), (e, e), (b, b), (b, f), (b, d), (d, b), (d, d),$
$(d, f), (f, b), (f, d), (f, f)\}.$

15. $\{\{\ldots, -3, -1, 1, 3, 5, \ldots\}, \{\ldots, -4, -2, 0, 2, 4, \ldots\}\}.$

17. (a) $(a, b) \; R \; (a, b)$ because $ab = ba$. Hence R is reflexive. If $(a, b) \; R \; (a', b')$, then $ab' = ba'$. Then $a'b = b'a$ and $(a', b') \; R \; (a, b)$. Hence R is symmetric. Now suppose that $(a, b) \; R \; (a', b')$ and $(a', b') \; R \; (a'', b'')$. Then $ab' = ba'$ and $a'b'' = b'a''$.

$$ab'' = a\frac{b'a''}{a'} = ab'\frac{a''}{a'} = ba'\frac{a''}{a'} = ba''.$$

Hence $(a, b) \; R \; (a'', b'')$ and R is transitive.

(b) $\{\{(1, 1), (2, 2), (3, 3), (4, 4), (5, 5)\}, \{(1, 2), (2, 4)\},$
$\{(1, 3)\}, \{(1, 4)\}, \{(1, 5)\}, \{(2, 1), (4, 2)\},$
$\{(2, 3)\}, \{(2, 5)\}, \{(3, 1)\}, \{(3, 2)\}, \{(3, 4)\},$
$\{(3, 5)\}, \{(4, 1)\}, \{(4, 3)\}, \{(4, 5)\}, \{(5, 1)\},$
$\{(5, 2)\}, \{(5, 3)\}, \{(5, 4)\}\}.$

19. Let R be reflexive and circular. If $a\ R\ b$, then $a\ R\ b$ and $b\ R\ b$, so $b\ R\ a$. Hence R is symmetric. If $a\ R\ b$ and $b\ R\ c$, then $c\ R\ a$. But R is symmetric, so $a\ R\ c$, and R is transitive.

Let R be an equivalence relation. Then R is reflexive. If $a\ R\ b$ and $b\ R\ c$, then $a\ R\ c$ (transitivity) and $c\ R\ a$ (symmetry), so R is also circular.

21. $a\ R\ b$ if and only if $ab > 0$.

23. If z is even (or odd), then $R(z)$ is the set of even (or odd) integers. Thus, if a and b are both even (or odd), then $R(a) + R(b) = \{x \mid x = s + t, s \in R(a), t \in R(b)\} = \{x \mid x \text{ is even}\} = R(a + b)$. If a and b have opposite parity, then $R(a) + R(b) = \{x \mid x = s + t, s \in R(a), t \in R(b)\} = \{x \mid x \text{ is odd}\} = R(a + b)$.

Exercise Set 4.6, page 139

1.

VERT[1] = 9	(1, 6)	NEXT[9] = 10	(1, 3)
NEXT[10] = 1	(1, 2)	NEXT[1] = 0
VERT[2] = 3	(2, 1)	NEXT[3] = 2	(2, 3)
NEXT[2] = 0
VERT[3] = 6	(3, 4)	NEXT[6] = 4	(3, 5)
NEXT[4] = 7	(3, 6)	NEXT[7] = 0
VERT[4] = 0
VERT[5] = 5	(5, 4)	NEXT[5] = 0
VERT[6] = 8	(6, 1)	NEXT[8] = 0

3. On average, EDGE must look at the average number of edges from any vertex. If R has P edges and N vertices, then EDGE examines $\dfrac{\sum P_{ij}}{N} = \dfrac{P}{N}$ edges on average.

5.

$$\begin{bmatrix} 1 & 1 & 1 & 0 \\ 0 & 0 & 1 & 1 \\ 1 & 0 & 0 & 1 \\ 0 & 1 & 0 & 0 \end{bmatrix}$$

VERT	TAIL	HEAD	NEXT
1	1	1	2
4	1	2	3
6	1	3	0
8	2	3	5
	2	4	0
	3	1	7
	3	4	0
	4	2	0

7.

$$\begin{bmatrix} 0 & 1 & 1 & 0 \\ 0 & 1 & 1 & 0 \\ 0 & 0 & 0 & 1 \\ 1 & 0 & 1 & 1 \end{bmatrix}$$

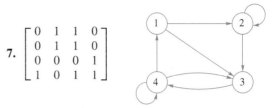

9.

VERT	TAIL	HEAD	NEXT
1	1	1	2
3	1	4	0
5	2	2	4
6	2	3	0
9	3	4	0
	4	1	7
	4	3	8
	4	5	0
	5	2	10
	5	5	0

11.

VERT	TAIL	HEAD	NEXT
1	a	a	2
4	a	b	3
6	a	d	0
9	b	b	5
	b	c	0
	c	b	7
	c	c	8
	c	d	0
	d	a	10
	d	b	11
	d	c	12
	d	d	0

Exercise Set 4.7, page 148

1. (a) $\{(1, 3), (2, 1), (2, 2), (3, 2), (3, 3)\}.$

(b) $\{(3, 1)\}.$

(c) $\{(1, 1), (1, 2), (2, 1), (2, 3), (3, 1), (3, 2), (3, 3)\}.$

(d) $\{(1, 2), (1, 3), (2, 3), (3, 3)\}.$

3. $\{(a, b) \mid a, b \text{ are sisters or } a, b \text{ are brothers}\}.$

5. $a\ (R \cup S)\ b$ if and only if a is a parent of b.

7. (a) $\{(2, 1), (3, 1), (3, 2), (3, 3), (4, 2), (4, 3),$
$(4, 4), (1, 4)\}.$

(b) $\{(1, 1), (1, 2), (2, 2), (2, 3), (2, 4), (4, 1), (3, 4)\}.$

(c) $\{(1, 1), (1, 2), (1, 3), (1, 4), (2, 2), (2, 3), (2, 4),$
$(3, 1), (3, 2), (3, 4), (4, 1), (4, 4)\}.$

(d) $\{(1, 1), (2, 1), (2, 2), (1, 4), (4, 1), (2, 3), (3, 2),$
$(1, 3), (4, 2), (3, 4), (4, 4)\}.$

9. (a) $\{(1, 1), (1, 4), (2, 2), (2, 3), (3, 3), (3, 4)\}.$

(b) $\{(1, 2), (2, 4), (3, 1), (3, 2)\}$.

(c) $\{(1, 1), (1, 2), (1, 3), (1, 4), (2, 1), (2, 4), (3, 1),$
    $(3, 2), (3, 3)\}$.

(d) $\{(1, 1), (2, 1), (4, 1), (4, 2), (1, 3), (2, 3), (3, 3)\}$.

11. (a) $\begin{bmatrix} 0 & 0 & 0 \\ 0 & 0 & 1 \\ 0 & 0 & 0 \\ 1 & 0 & 1 \end{bmatrix}$.   (b) $\begin{bmatrix} 1 & 1 & 1 \\ 1 & 1 & 1 \\ 1 & 1 & 1 \\ 1 & 1 & 1 \end{bmatrix}$.

(c) $\begin{bmatrix} 1 & 0 & 0 & 1 \\ 0 & 1 & 1 & 0 \\ 1 & 1 & 0 & 1 \end{bmatrix}$.   (d) $\begin{bmatrix} 1 & 0 & 1 \\ 0 & 1 & 0 \\ 0 & 1 & 0 \\ 0 & 0 & 0 \end{bmatrix}$.

13. (a) $\begin{bmatrix} 0 & 0 & 0 & 0 \\ 0 & 0 & 1 & 0 \\ 1 & 0 & 0 & 0 \\ 0 & 0 & 0 & 0 \end{bmatrix}$.   (b) $\begin{bmatrix} 1 & 1 & 1 & 0 \\ 0 & 0 & 1 & 0 \\ 1 & 1 & 0 & 0 \\ 0 & 1 & 1 & 1 \end{bmatrix}$.

(c) $\begin{bmatrix} 1 & 0 & 1 & 0 \\ 0 & 0 & 0 & 1 \\ 1 & 1 & 0 & 0 \\ 0 & 0 & 0 & 1 \end{bmatrix}$.   (d) $\begin{bmatrix} 1 & 0 & 1 & 1 \\ 1 & 1 & 0 & 1 \\ 0 & 0 & 1 & 1 \\ 1 & 1 & 0 & 1 \end{bmatrix}$.

15. $R \cap S = \{(a, a), (b, b), (b, c), (c, b), (c, c), (d, d), (e, e)\}$
    $\{\{a\}, \{b, c\}, \{d\}, \{e\}\}$.

17. (a) $\{(a, a), (a, d), (a, e), (b, b), (b, c), (b, e), (c, a),$
    $(c, b), (c, c), (d, b), (d, c), (d, d), (e, c), (e, e)\}$

(b) $\{(a, a), (a, d), (d, a), (a, e), (e, a), (b, c), (c, b),$
    $(b, e), (e, b), (c, a), (a, c), (c, c), (d, b), (b, d), (d, c),$
    $(c, d), (e, c), (c, e), (e, e)\}$

19. (a) Yes.   (b) Yes.

(c) $x\ (S \circ R)\ y$ if and only if $x \le 6y$

21. (a) Reflexive $a\ R\ a \wedge a\ S\ a \Rightarrow a\ S \circ R\ a$.
    Irreflexive   No $1\ R\ 2 \wedge 2\ S\ 1 \Rightarrow 1\ S \circ R\ 1$.
    Symmetric   No $1\ R\ 3, 3\ R\ 1, 3\ S\ 2, 2\ S\ 3$
    $\Rightarrow 1\ S \circ R\ 2$, but $2\ S \circ R\ 1$.
    Asymmetric   No $R = \{(1, 2), (3, 4)\}$ and
    $S = \{(2, 3), (4, 1)\}$ provide a counterexample.
    Antisymmetric   No $R = \{(a, b), (c, d)\}$ and
    $S = \{(b, c), (d, a)\}$ provide a counterexample.
    Transitive   No $R = \{(a, d), (b, e)\}$ and
    $S = \{(d, b), (e, c)\}$ provide a counterexample.

(b) No, symmetric and transitive properties are not
    preserved.

23. (a) $\begin{bmatrix} 1 & 1 & 0 & 1 & 1 \\ 0 & 1 & 0 & 1 & 1 \\ 1 & 1 & 0 & 1 & 1 \\ 1 & 1 & 0 & 1 & 1 \\ 1 & 1 & 0 & 0 & 1 \end{bmatrix}$.

(b) $\begin{bmatrix} 1 & 0 & 1 & 1 & 1 \\ 1 & 1 & 0 & 1 & 1 \\ 1 & 0 & 1 & 1 & 1 \\ 1 & 1 & 1 & 1 & 1 \\ 0 & 0 & 0 & 1 & 1 \end{bmatrix}$.

(c) $\begin{bmatrix} 1 & 1 & 0 & 1 & 1 \\ 1 & 1 & 0 & 0 & 1 \\ 0 & 1 & 0 & 1 & 1 \\ 1 & 1 & 0 & 1 & 1 \\ 1 & 1 & 0 & 0 & 1 \end{bmatrix}$.

(d) $\begin{bmatrix} 1 & 1 & 1 & 1 & 1 \\ 1 & 0 & 1 & 1 & 1 \\ 1 & 1 & 0 & 1 & 1 \\ 1 & 1 & 1 & 1 & 1 \\ 0 & 1 & 0 & 1 & 1 \end{bmatrix}$.

25. (a) Let $(x, y) \in (S \cup T) \circ R$. Then $x\ R\ z, z\ S \cup T\ y$ for
    some $z \in B$. Either $z\ S\ y$ or $z\ T\ y$ and $x\ S \circ R\ y$ or
    $x\ T \circ R\ y$. Hence $(x, y) \in S \circ R \cup T \circ R$. Now let
    $(x, y) \in S \circ R \cup T \circ R$. Say $(x, y) \in T \circ R$. Then
    $x\ R\ z, z\ T\ y$ for some $z \in B$. Thus $z\ S \cup T\ y$ and
    $(x, y) \in (S \cup T) \circ R$.

(b) Let $R = \{(x, z), (x, m)\}$, $S = \{(z, y)\}$, and
    $T = \{(m, y)\}$. Then $(x, y) \in (S \circ R) \cap (T \circ R)$, but
    $(S \cap T) \circ R = \{\ \}$.

27. (a) Let $\mathbf{M}_{R \cap S} = \begin{bmatrix} m_{ij} \end{bmatrix}, \mathbf{M}_R = \begin{bmatrix} r_{ij} \end{bmatrix}, \mathbf{M}_S = \begin{bmatrix} s_{ij} \end{bmatrix}$.
    $m_{ij} = 1$ if and only if $(i, j) \in R \cap S$. $(i, j) \in R$ if and
    only if $r_{ij} = 1$ and $(i, j) \in S$ if and only if $s_{ij} = 1$.
    But this happens if and only if the i, jth entry of
    $\mathbf{M}_R \wedge \mathbf{M}_S$ is 1.

(b) Let $\mathbf{M}_{R \cup S} = \begin{bmatrix} m_{ij} \end{bmatrix}, \mathbf{M}_R = \begin{bmatrix} r_{ij} \end{bmatrix}, \mathbf{M}_S = \begin{bmatrix} s_{ij} \end{bmatrix}$.
    $m_{ij} = 1$ if and only if $(i, j) \in R \cup S$. $(i, j) \in R$ if and
    only if $r_{ij} = 1$ or $(i, j) \in S$ if and only if $s_{ij} = 1$. But
    this happens if and only if the i, jth entry of
    $\mathbf{M}_R \vee \mathbf{M}_S$ is 1.

(c) The i, jth entry of $\mathbf{M}_{R^{-1}}$ is 1 if and only if
    $(i, j) \in R^{-1}$ if and only if $(j, i) \in R$ if and only if the
    j, ith entry of $\mathbf{M}_R$ is 1 if and only if the i, jth entry of
    $\mathbf{M}_R^T$ is 1.

(d) The i, jth entry of $\mathbf{M}_{\overline{R}}$ is 1 if and only if $(i, j) \in \overline{R}$ if
    and only if $(i, j) \notin R$ if and only if the i, jth entry of
    $\mathbf{M}_R$ is 0 if and only if the i, jth entry of $\overline{\mathbf{M}_R}$ is 1.

29. To form the digraph of R^{-1}, reverse the arrows in the
    digraph of R.

31. The edges of the digraph of $R \cup S$ are the edges that
    appear on either the digraph of R or the digraph of S.

Exercise Set 4.8, page 156

1. (a) $\begin{bmatrix} 1 & 1 & 1 \\ 1 & 1 & 1 \\ 1 & 1 & 1 \end{bmatrix}$.

(b) $\{(1, 1), (1, 2), (1, 3), (2, 1), (2, 2), (2, 3),$
   $(3, 1), (3, 2), (3, 3)\}$.

3. $W_1 = \begin{bmatrix} 1 & 0 & 0 & 1 & 0 \\ 0 & 1 & 0 & 0 & 0 \\ 0 & 0 & 0 & 1 & 1 \\ 1 & 0 & 0 & 1 & 0 \\ 0 & 1 & 0 & 0 & 1 \end{bmatrix}$, $W_2 = W_1 = W_3$.

5. Let R be reflexive and transitive. Suppose that $x \, R^n \, y$. Then $x, a_1, a_2, \ldots, a_{n-1}, y$ is a path of length n from x to y. $x \, R \, a_1 \wedge a_1 \, R \, a_2 \Rightarrow x \, R \, a_2$. Similarly, we have $x \, R \, a_k \wedge a_k \, R \, a_{k+1} \Rightarrow x \, R \, a_{k+1}$ and finally $x \, R \, a_{n-1} \wedge a_{n-1} \, R \, y \Rightarrow x \, R \, y$. Hence $R^n \subseteq R$. If $x \, R \, y$, then since R is reflexive we can build a path of length n, $x, x, x, \ldots, x, y$ from x to y and $x \, R^n \, y$.

7. $\begin{bmatrix} 1 & 0 & 0 & 1 \\ 1 & 1 & 0 & 1 \\ 0 & 0 & 1 & 0 \\ 0 & 0 & 0 & 1 \end{bmatrix}$.

9. $\begin{bmatrix} 1 & 0 & 0 & 1 \\ 0 & 1 & 1 & 0 \\ 0 & 1 & 1 & 0 \\ 1 & 0 & 0 & 1 \end{bmatrix}$.

11. $A \times A$.

13. $A/R = \{\{1, 2, 3\}, \{4, 5\}\}$, $A/S = \{\{1\}, \{2, 3, 4\}, \{5\}\}$, $A/(R \cup S)^{\infty} = A$.

15. The collection of elements in A/R and A/S can be separated into sub-collections of non-disjoint sets. Each element of $A/(R \cup S)^{\infty}$ is the union of the sets in one of these sub-collections.

17. $\{(1, 1), (1, 2), (1, 4), (2, 2), (3, 2), (3, 3), (4, 2),$
   $(4, 3), (4, 4), (1, 3)\}$.

19. We first show R^{∞} is transitive. Then we show it is the smallest relation that contains R. It is a direct proof.

Chapter 5

Exercise Set 5.1, page 168

1. (a) Yes. $\text{Ran}(R) = \{1, 2\}$.   (b) No.

3. Yes.

5. Each integer has a unique square that is also an integer.

7. Each $r \in \mathbb{R}$ is either an integer or it is not.

9. (a) 3   (b) 1   (c) $(x - 1)^2$

(d) $x^2 - 1$ (e) $y - 2$ (f) y^4

11. (a) Both.   (b) Neither.

13. (a) Both.   (b) Onto.

15. (a) Both.   (b) Onto.

17. (a) $(g \circ f)(a) = g\left(\dfrac{a + 1}{2}\right) = 2\left(\dfrac{a + 1}{2}\right) - 1 = a$.

(b) $(g \circ f)(a) = g(a^2 - 1) = \sqrt{a^2 + 1 - 1} = |a| = a$, since $a \geq 0$.

19. (a) $f^{-1}(b) = b^2 - 1$ (b) $f^{-1}(b) = \sqrt[3]{b - 1}$

21. No. $(a, 1), (a, 2) \in f^{-1}$.

23. $(g \circ f)(a) = \dfrac{2a + 1}{3}$; $(g \circ f)^{-1}(c) = \dfrac{3c - 1}{2}$;
   $f^{-1}(b) = \dfrac{b - 1}{2}$; $g^{-1}(c) = 3c$;
   $(f^{-1} \circ g^{-1})(c) = f^{-1}(3c) = \dfrac{3c - 1}{2}$.

25. $n!$.

27. g is one to one; f is one to one.

29. Let $g \circ f$ be one to one. Suppose $f(a) = f(b)$. Then $(g \circ f)(a) = (g \circ f)(b)$ and $a = b$. Hence f is one to one.

31. Suppose $O(a_1, f) \cap O(a_2, f) \neq \{\ \}$. Then $f^{k_1}(a_1) = f^{k_2}(a_2)$ for some k_1 and k_2. $(f^{-k_1} \circ f^{k_2})(a_1) = a_1 = f^{k_2 - k_1}(a_2)$. Hence $a_1 \in O(a_2, f)$ and $f^n(a_1) \in O(a_2, f)$ for all n. Similarly, $f^n(a_2) \in O(a_1, f)$ for all n. Thus $O(a_1, f) = O(a_2, f)$.

33. Since f is everywhere defined, $\text{Dom}(f) = A$. Suppose f is one to one. Then by Exercise 32, $|\text{Ran}(f)| = |\text{Dom}(f)| = n$. Since $|B| = n$, $\text{Ran}(f) = B$ and f is onto. Next, suppose f is onto. Then $\text{Ran}(f) = B$, $|\text{Ran}(f)| = n$ and $|\text{Dom}(f)| = n$. By Exercise 32, f must be one to one. Since (a) and (b) are equivalent, (a) and (b) are each equivalent to (c).

35. One possible one-to-one correspondence is $g: \mathbb{Z}^+ \to B$: $z \to 2z + 1$.

Exercise Set 5.2, page 173

1. (a) 7.   (b) 8.   (c) 3.

3. (a) 1.   (b) 0.   (c) 1.

5. (a) 2.   (b) −3.   (c) 14.
   (d) −18.   (e) 21.

7. (a) 26.   (b) 866.   (c) 74.   (d) 431.

9. (a) 2.   (b) 8.   (c) 32.   (d) 1024.

11. (a) 4.   (b) 7.   (c) 9.   (d) 10.

13. (a) 5; 6   (b) 6; 7

15. For any 5×5 matrix $\mathbf{M}$, $\mathbf{M}^T$ exists so t is everywhere defined. If $\mathbf{M}$ is a 5×5 matrix, then $t(\mathbf{M}^T) = \mathbf{M}$, so t is onto. Suppose $\mathbf{M}^T = \mathbf{N}^T$. Then $(\mathbf{M}^T)^T = (\mathbf{N}^T)^T$; that is, $\mathbf{M} = \mathbf{N}$ and t is one to one.

17. Every relation R on A defines a unique matrix $\mathbf{M}_R$ so f is everywhere defined and one to one. Any $n \times n$ Boolean matrix $\mathbf{M}$ defines a relation on A so f is onto.

19.

x	y	z	$f(x, y, z)$
T	T	T	T
T	T	F	F
T	F	T	T
T	F	F	F
F	T	T	T
F	T	F	T
F	F	T	T
F	F	F	F

21. (a) True.   (b) False.   (c) False.   (d) True.

23. (a) 31.   (b) 0.   (c) 36.

25. $\dfrac{\left\lfloor \frac{m}{k} + 2 \right\rfloor}{2}$.

27. $_{2n}C_n$.

29. There can be only one fewer, since p_i is the first such p_j; one more; $n - 1$ left and $n + 1$ right

31. $_{2n}C_{n-1}$

33. $\dfrac{_{2n}C_n}{n + 1}, \dfrac{(2n)!}{n!\,(n + 1)!}$.

Exercise Set 5.3, page 179

1. (a) The number of steps remains 1001.
   (b) The number of steps doubles.
   (c) The number of steps quadruples.
   (d) The number of steps increases eightfold.

3. $|n!| = |n(n - 1)(n - 2) \cdots 2 \cdot 1| \le 1 \cdot |n \cdot n \cdots n|, n \ge 1.$

5. $|8n + lg(n)| \le |8n + n| = 9|n|, n \ge 1.$

7. $|n\, lg(n)| \le |n \cdot n| = n^2, n \ge 1.$ Suppose there exist c and k such that $n^2 \le c \cdot n\, lg(n), n \ge k$. Choose $N > k$ with $N > c \cdot lg(N)$. Then $N^2 \le c \cdot N \cdot lg(N) < N^2$, a contradiction.

9. $|5n^2 + 4n + 3| \le |5n^2 + 500n|, n \ge 1$; $|5n^2 + 500n| \le 5|n^2 + 100n|$. We have $|n^2 + 100n| = |n^2 + 4 \cdot 25n| \le |n^2 + 4n^2|, n \ge 25$. But $|5n^2| \le |5n^2 + 4n + 3|.$

11. $\{f_5\}, \{f_6, f_{10}, f_{11}\}, \{f_7\}, \{f_4\}, \{f_8\}, \{f_1\}, \{f_2\}, \{f_3\}, \{f_9\}, \{f_{12}\}.$

13. $f(n) = 2 + 18 \cdot 6 + 1$ or $111 \quad \theta(1)$

15. $f(n) = 2 + n \cdot 5 + 1$ or $f(n) = 3 + 5n \quad \theta(n)$

17. $f(n) = 2 + 4lg(n) \quad \theta(lg\, n)$

19. $\theta(2^n)$.

21. Suppose $a < b$. Then $n^a \le n^b, n \ge 1$. Suppose $n^b \le c \cdot n^a$, for $n \ge k$. Choose $N > k$ with $N^{b-a} > c$. Then $N^b \le c \cdot N^a < N^{b-a} \cdot N^a = N^b$, a contradiction.

23. There exist c_1, k_1, c_2, k_2 such that $|f(n) + g(n)| \le |f(n)| + |g(n)| \le c_1|h(n)| + c_2|h(n)|$, $n \ge \max\{k_1, k_2\}$. Thus $f + g$ is $O(h)$.

Exercise Set 5.4, page 187

1. (a) Yes.   (b) No.

3. (a) Yes.   (b) No.

5. (a) $\begin{pmatrix} 1 & 2 & 3 & 4 & 5 & 6 \\ 3 & 4 & 1 & 2 & 6 & 5 \end{pmatrix}$.

   (b) $\begin{pmatrix} 1 & 2 & 3 & 4 & 5 & 6 \\ 2 & 5 & 6 & 3 & 1 & 4 \end{pmatrix}$.

7. (a) $\begin{pmatrix} 1 & 2 & 3 & 4 & 5 & 6 \\ 6 & 3 & 2 & 5 & 4 & 1 \end{pmatrix}$.

   (b) $\begin{pmatrix} 1 & 2 & 3 & 4 & 5 & 6 \\ 1 & 3 & 4 & 6 & 2 & 5 \end{pmatrix}$.

9. (a) $(1, 5, 7, 8, 3, 2)$.   (b) $(2, 7, 8, 3, 4, 6)$.

11. (a) $(a, f, g) \circ (b, c, d, e)$.
   (b) $(a, c) \circ (b, g, f)$.

13. (a) $(1, 6, 3, 7, 2, 5, 4, 8)$.
   (b) $(5, 6, 7, 8) \circ (1, 2, 3)$.

15. (a) $(2, 6) \circ (2, 8) \circ (2, 5) \circ (2, 4) \circ (2, 1)$.
   (b) $(3, 6) \circ (3, 1) \circ (4, 5) \circ (4, 2) \circ (4, 8)$.

17. (a) Even.   (b) Odd.

19. Suppose p_1 is the product of $2k_1 + 1$ transpositions and p_2 is the product of $2k_2 + 1$ transpositions. Then $p_2 \circ p_1$ can be written as the product of $2(k_1 + k_2) + 2$ transpositions. By Theorem 3, $p_2 \circ p_1$ is even.

21. (a) $(1, 5, 2, 3, 4)$.

(b) $(1, 4, 2, 5, 3)$.

23. (a) $(1, 2, 4)$.

(b) $\begin{pmatrix} 1 & 2 & 3 & 4 & 5 & 6 \\ 4 & 1 & 3 & 2 & 5 & 6 \end{pmatrix}$.

(c) $\begin{pmatrix} 1 & 2 & 3 & 4 & 5 & 6 \\ 4 & 1 & 3 & 2 & 5 & 6 \end{pmatrix}$.

(d) 3.

25. (a) Basis step: $n = 1$. If p is a permutation of a finite set A, then p^1 is a permutation of A is true.
Induction step: The proof in Exercise 16 shows that if p^{n-1} is a permutation of A, then $p^{n-1} \circ p$ is a permutation of A. Hence p^n is a permutation of A.

(b) If $|A| = n$, then there are $n!$ permutations of A. Hence, the sequence $1_A, p, p^2, p^3, \ldots$ is finite and $p^i = p^j$ for some $i \neq j$. Suppose $i < j$. Then $p^{-i} \circ p^i = 1_A = p^{-i} \circ p^j$. So $p^{j-i} = 1_A, j - i \in Z$.

27.

	1_A	p_1	p_2	p_3	p_4	p_5
1_A	1_A	p_1	p_2	p_3	p_4	p_5
p_1	p_1	1_A	p_4	p_5	p_2	p_3
p_2	p_2	p_3	1_A	p_1	p_5	p_4
p_3	p_3	p_2	p_5	p_4	1_A	p_1
p_4	p_4	p_5	p_1	1_A	p_3	p_2
p_5	p_5	p_4	p_3	p_2	p_1	1_A

29. $\{1_A\}, \{1_A, p_1\}, \{1_A, p_2\}, \{1_A, p_5\}, \{1_A, p_3, p_4\}, \{1_A, p_1, p_2, p_3, p_4, p_5\}$.

Chapter 6

Exercise Set 6.1, page 200

1. (a) No.   (b) No.

3. (a) Yes.   (b) Yes.

5. $\{(a, a), (b, b), (c, c), (a, b)\}$,
$\{(a, a), (b, b), (c, c), (a, b), (a, c)\}$,
$\{(a, a), (b, b), (c, c), (a, b), (c, b)\}$,
$\{(a, a), (b, b), (c, c), (a, b), (b, c), (a, c)\}$,
$\{(a, a), (b, b), (c, c), (a, b), (c, b), (c, a)\}$,
$\{(a, a), (b, b), (c, c), (a, c), (c, b), (a, b)\}$.

7. The structure of the proof is to check directly each of the three properties required for a partial order.

9.

11. $\{(1, 1), (2, 2), (3, 3), (4, 4), (1, 3), (1, 4), (2, 3), (2, 4), (3, 4)\}$.

13.

15.

17. $\begin{bmatrix} 1 & 1 & 1 & 1 & 1 \\ 0 & 1 & 0 & 0 & 0 \\ 0 & 0 & 1 & 0 & 0 \\ 0 & 0 & 0 & 1 & 0 \\ 0 & 0 & 0 & 0 & 1 \end{bmatrix}$.

19. ACE, BASE, CAP, CAPE, MACE, MAP, MOP, MOPE

21.

23. linear

25. If the main diagonal of $\mathbf{M}_R$ is all 1's, then R is reflexive. If $\mathbf{M}_R \odot \mathbf{M}_R = \mathbf{M}_R$, then R is transitive. If $a_{ij} = 1, i \neq j$ in $\mathbf{M}_R$, then a_{ji} must be 0 in order for R to be antisymmetric.

27.

9
8
7
6
5
3
2
1
4

29. For every $a, b \in A$, either $a\ R\ b$ or $b\ R\ a$. Hence either $b\ R^{-1}\ a$ or $a\ R^{-1}\ b$, and R^{-1} is a linear order.

31. Suppose $x < y$, $y < z$, $x, y, z \in A$. Then $x < z$ so $<$ is transitive. Clearly, $x \not< x$, $x \in A$; $<$ is irreflexive. Thus $<$ is a quasiorder.

33. $a\ R\ b$ if and only if $a \mid b$ and $a \neq b$.

35.

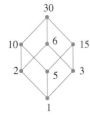

30
10   6   15
2   5   3
1

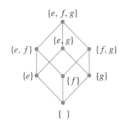

$\{e, f, g\}$
$\{e, f\}$ $\{e, g\}$ $\{f, g\}$
$\{e\}$ $\{f\}$ $\{g\}$
$\{\ \}$

Define F as follows: $F(1) = \{\ \}$; $F(2) = \{e\}$; $F(5) = \{f\}$; $F(3) = \{g\}$; $F(10) = \{e, f\}$; $F(6) = \{e, g\}$; $F(15) = \{f, g\}$; $F(30) = \{e, f, g\}$.

Exercise Set 6.2, page 206

1. Maximal: 3, 5;   Minimal: 1, 6.

3. Maximal: e, f;   Minimal: a.

5. Maximal: none;   Minimal: none.

7. Maximal: 1;   Minimal: none.

9. Greatest: f;   Least: a.

11. No greatest or least.

13. Greatest: none;   Least: none.

15. Greatest: 72;   Least: 2.

17. No, a may be maximal and there be an element of A, b, such that a and b are incomparable.

19. (a) True. There cannot be $a_1 < a_2 < \cdots$ since A is finite.
(b) False. Not all elements have to be comparable.
(c) True. There cannot be $\cdots, a_2, a_1$ since A is finite.

(d) False. Not all elements have to be comparable.

21. Suppose a and b are least elements of $(A, \leq)$. Then $a \leq b$ and $b \leq a$. Since $\leq$ is antisymmetric, $a = b$. Note: This is a restatement of Theorem 2.

23. (a) f, g, h.   (b) a, b, c.   (c) f.   (d) c.

25. (a) d, e, f.   (b) b, a (c) d.   (d) b.

27. (a) None.   (b) b.   (c) None.   (d) b.

29. (a) $x \in [2, \infty)$.   (b) $x \in (-\infty, 1]$.
(c) 2.   (d) 1.

31.

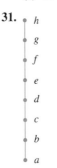

h
g
f
e
d
c
b
a

33. The least element of A is the label on the row that is all ones. The greatest element of A is the label on the column that is all ones.

Exercise Set 6.3, page 216

1. Yes, all the properties are satisfied.

3. No, GLB($\{e, b\}$) does not exist.

5. Yes, all the properties are satisfied.

7. No.

9.

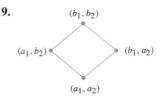

(b_1, b_2)
(a_1, b_2) (b_1, a_2)
(a_1, a_2)

11. For each $T_1, T_2 \subseteq T$, $T_1 \cap T_2$ and $T_1 \cup T_2$ are subsets of T so $P(T)$ is a sublattice of $P(S)$.

13. For any elements x, y of a linearly ordered poset, $x \leq y$ or $y \leq x$. Say $x \leq y$. Then $x = x \wedge y$ and $y = x \vee y$. Hence any subset of linearly ordered poset is a sublattice.

15.

17. Suppose $a \wedge b = a$.
$a \leq a \vee b = (a \wedge b) \vee b = (a \vee b) \wedge b \leq b$. Thus $a \leq b$.
Suppose $a \leq b$. $a \wedge b \leq a$ and $a \leq a$, $a \leq b$ gives
$a \leq a \wedge b$. Hence $a \wedge b = a$.

19. (a) 12.   Figure 6.44(a): 3   Figure 6.44(b): 3.

21. $(a_1, a_2) \wedge ((b_1, b_2) \vee (c_1, c_2)) = (a_1, a_2) \wedge (b_1 \vee c_1, b_2 \vee c_2) =$
$(a_1 \wedge (b_1 \vee c_1), a_2 \wedge (b_2 \vee c_2)) =$
$((a_1 \wedge b_1) \vee (a_1 \wedge c_1), (a_2 \wedge b_2) \vee (a_2 \wedge c_2)) =$
$((a_1, a_2) \wedge (b_1, b_2)) \vee ((a_1, a_2) \wedge (c_1, c_2))$.
A similar argument establishes the other distributive
property.

23. Suppose $a \wedge x = a \wedge y$ and $a \vee x = a \vee y$. Then
$y \leq y \vee (y \wedge a) = (y \wedge y) \vee (y \wedge a) = y \wedge (y \vee a) =$
$y \wedge (a \vee x) = (y \wedge a) \vee (y \wedge x) = (a \wedge x) \vee (y \wedge x) =$
$x \wedge (a \vee y) \leq x$. Hence $y \leq x$. A similar argument shows
$x \leq y$. Thus $x = y$.

25. $1' = 42, 42' = 1, 2' = 21, 21' = 2, 3' = 14, 14' = 3,$
$7' = 6, 6' = 7$.

27. Neither.

29. Distributive, but not complemented.

31. If $x = x'$, then $x = x \vee x = I$ and $x = x \wedge x = 0$. But by
Exercise 11, $0 \neq I$. Hence, $x \neq x'$.

33. Suppose $\mathcal{P}_1 \leq \mathcal{P}_2$. Then $R_1 \subseteq R_2$. Let $x \in A_i$. Then
$A_i = \{y \mid a \, R_1 \, y\}$ and $A_i \subseteq \{y \mid x \, R_2 \, y\} = B_j$ where
$x \in B_j$. Suppose each $A_i \subseteq B_j$. Then $x \, R_1 \, y$ implies
$x \, R_2 \, y$ and $R_1 \subseteq R_2$. Thus $\mathcal{P}_1 \leq \mathcal{P}_2$.

Exercise Set 6.4, page 224

1. No, it has 6 elements, not 2^n elements.

3. No, it has 6 elements, not 2^n elements.

5. Yes, it is B_3.

7. Yes, it is B_1.

9. Yes; $385 = 5 \cdot 7 \cdot 11$.

11. No, each Boolean algebra must have 2^n elements.

13. Suppose $a = b$.
$(a \wedge b') \vee (a' \wedge b) = (b \wedge b') \vee (a' \wedge a) = 0 \vee 0 = 0$.
Suppose $(a \wedge b') \vee (a' \wedge b) = 0$. Then $a \wedge b' = 0$ and
$a' \wedge b = 0$. We have $I = 0' = (a \wedge b')' = a' \vee b$. So a' is
the complement of b; $b' = a'$.

15. Suppose $a \leq b$. Then $a \wedge c \leq a \leq b$ and $a \wedge c \leq c$ so
$a \wedge c \leq b \wedge c$.

17. $(a \wedge b) \vee (a \wedge b') = a \wedge (b \vee b') = a \wedge I = a$.

19. $(a \wedge b \wedge c) \vee (b \wedge c) = (a \vee I) \wedge (b \wedge c) = I \wedge (b \wedge c) = b \wedge c$.

21. Suppose $a \leq b$. Then
$a \vee (b \wedge c) = (a \vee b) \wedge (a \vee c) = b \wedge (a \vee c)$.

23. R is reflexive because $m_{ii} = 1, i = 1, 2, \ldots, 8$. R is
antisymmetric since if $m_{ij} = 1$ and $i \neq j$, then $m_{ji} = 0$.
R is transitive, because $\mathbf{M}_R \odot \mathbf{M}_R$ shows that $R^2 \subseteq R$.

25. Complement pairs are $a, h; b, g; c, f; d, e$. Since each
element has a unique complement, (A, R) is
complemented.

27. (A, R) is not a Boolean algebra; complements are not
unique.

Exercise Set 6.5, page 229

1.

x	y	z	$x \wedge$	$(y \vee z')$
0	0	0	0	1
0	0	1	0	0
0	1	0	0	1
0	1	1	0	1
1	0	0	1	1
1	0	1	0	0
1	1	0	1	1
1	1	1	1	1
			↑	

3.

x	y	z	$(x \vee y')$	$\vee$	$(y \wedge (x' \vee y)$
0	0	0	0	0	0 1
0	0	1	0	0	0 1
0	1	0	0	1	1 1
0	1	1	0	1	1 1
1	0	0	1	1	0 0
1	0	1	1	1	0 0
1	1	0	0	1	1 1
1	1	1	0	1	1 1
			(1)	(4)	(3) (2)

5. $(x \vee y) \wedge (x' \vee y) = (x \wedge x') \vee y = 0 \vee y = y$.

7. $(z' \vee x) \wedge ((x \wedge y) \vee z) \wedge (z' \vee y)) =$
$(z' \vee (x \wedge y)) \wedge ((x \wedge y) \vee z) = (x \wedge y) \vee (z' \wedge z) =$
$(x \wedge y) \vee 0 = x \wedge y$.

9. $x \wedge z$.

11. $y \vee x'$.

13. (a)

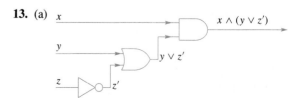

(b)
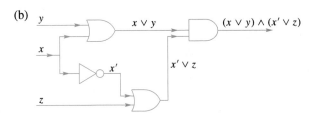

15. $(x \vee (y \wedge z))' \vee z'$

17. $((x \vee y) \wedge z)'$

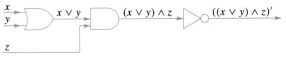

Exercise Set 6.6, page 239

1.

	y'	y
x'	1	0
x	0	1

3.

	y'		y	
x'	1	1	0	0
x	1	0	0	1

z / z'

5.

z'		z	
1	0	0	1
0	1	0	1
1	0	0	1
0	0	0	0

x' rows / x rows, y y', w / w'

7.

	y'		y	
x'	0	1	1	0
x	0	1	1	0

z / z'

9. $(x' \wedge y') \vee (x \wedge y)$

11. $z' \vee (x' \wedge z)$

13. $(z' \wedge y) \vee (x \wedge y') \vee (y' \wedge z)$

15. $(z \wedge x') \vee (w' \wedge x \wedge y) \vee (w \wedge x \wedge y')$

17. $(x' \wedge y') \vee (x \wedge y)$

19. $(x' \wedge y') \vee (x \wedge z')$

21. z

23. $(x' \wedge y' \wedge w') \vee (y \wedge z \wedge w') \vee (x' \wedge z' \wedge y \wedge w)$

Chapter 7

Exercise Set 7.1, page 248

1. Yes, the root is b.

3. Yes, the root is f.

5. No.

7. Yes, the root is t.

9. (a) $v_{12}, v_{10}, v_{11}, v_{13}, v_{14}$.

(b) $v_{10}, v_{11}, v_5, v_{12}, v_7, v_{15}, v_{14}, v_9$.

11. (a)

(b)

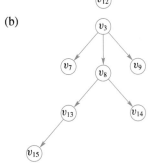

13. (T, v_0) may be an n-tree for $n \geq 3$. It is not a complete 3-tree.

15. (a) v_1, v_3.

(b) $v_6, v_7, v_8, v_{13}, v_{14}, v_{16}, v_{10}$.

17. (a) 4.   (b) 2.

19. The maximum number of vertices in a binary tree of height n is $1 + 2 + 2^2 + 2^3 + \cdots + 2^n = 2^{n+1} - 1$. (See Section 3.5, page 97.)

21. The total number of vertices is $1 + kn$, where k is the number of non-leaves and 1 counts the root, because every vertex except the root is an offspring. Since $l = m - k$, $l = 1 + kn - k$ or $1 + k(n - 1)$.

23. If both $v \, T \, u$ and $u \, T \, v$, then v, u, v is a cycle in T. Thus $v \, T \, u$ implies $u \, \tilde{T} \, v$. T is asymmetric.

25. Each vertex except the root has in-degree 1. Thus $s = r - 1$.

27. 4. The tree of maximum height has one vertex on each level.

29. Assume that the in-degree of $v_0 \neq 0$. Then there is a cycle that begins and ends at v_0. This is impossible. Hence the in-degree of v_0 must be 0.

Exercise Set 7.2, page 253

1.

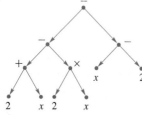

3.

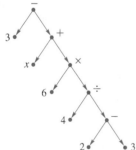

5.

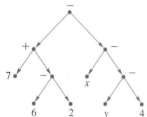

7.

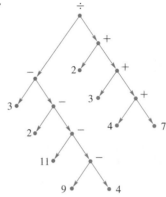

9.

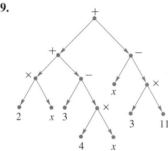

11.

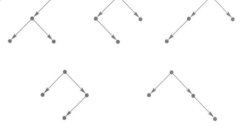

13. 721.

15.

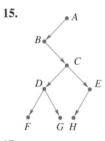

17.

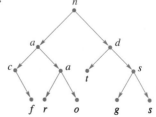

19.

LEFT	DATA	RIGHT
2	⊠	0
3	−	4
5	+	6
9	−	10
0	7	0
7	−	8
0	6	0
0	2	0
0	x	0
11	−	12
0	y	0
0	4	0

21.

LEFT	DATA	RIGHT
2	⊠	0
3	−	4
5	−	6
9	−	10
7	+	8
11	×	12
0	2	0
0	x	0
0	x	0
0	2	0
0	2	0
0	x	0

Exercise Set 7.3, page 261

1. $x\,y\,s\,z\,t\,u\,v$

3. $a\,b\,c\,g\,h\,i\,d\,k\,e\,j\,f$

5. TSAMZWEDQMLCKFNTRGJ

7. $2 + 3 - 1 \times 2$

9. $6\,4\,2\,1\,3\,5\,7$

11. $s\,y\,v\,u\,t\,z\,x$

13. $g\,h\,c\,i\,b\,k\,j\,f\,e\,d\,a$

15. ZWMADQESCNTFKLJGRMT

17. 4

19. $\frac{15}{16}$

21.

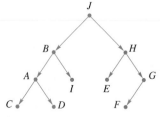

23.

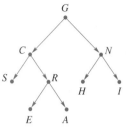

25. (a) The root must be labeled J; if J has a left child, it must be labeled B, otherwise the right child is labeled B.

(b) The root must be labeled G; if G has a left child, it must be labeled N, otherwise the right child is labeled N.

27.

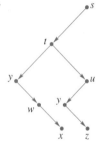

29.

$B(T)$ T

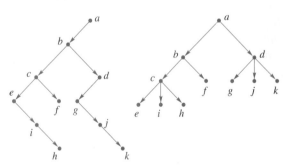

31.

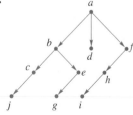

33. (a) •v_0 (b) •v_0 (c) •v_0

(d)

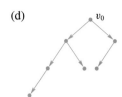

Exercise Set 7.4, page 270

1.

3.

5.

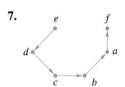

7.

9.

11.

13.

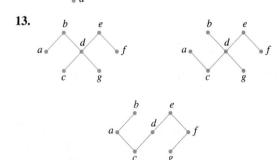

15. There are 5 spanning trees.

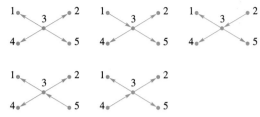

17.

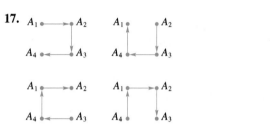

19. *n*

21. Prim's algorithm will produce a spanning tree; its symmetric closure is an undirected spanning tree for the relation.

Exercise Set 7.5, page 275

1.

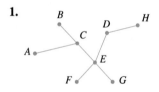

3.

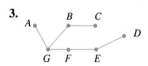

5.

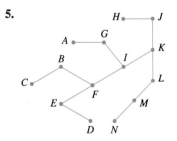

7.

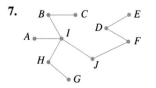

9.

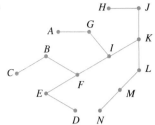

11. Suppose e_r is the required edge. Let $S = \{e_1, e_2, \ldots, e_k\} - \{e_r\}$ and $E = \{e_r\}$. Now apply Kruskal's algorithm to S and E.

13. Suppose (v_r, v_s) is the edge to be included. Step 1. Let $V = \{v_r\}$ and $E = \{(v_r, v_s)\}$. Steps 2 and 3 are unchanged.

15. Change Step 1 to read "Choose an edge e_1 in S of greatest weight."

17. Define v to be a farthest vertex of $V = \{v_1, v_2, \ldots, v_k\}$ if v is adjacent to some v_i in V and no other vertex is joined to a member of V by an edge of greater weight than (v, v_i). In Step 2 replace "nearest" with "furthest."

19. If each edge has a distinct weight, there will be a unique minimal spanning tree since only one choice can be made at each step.

21. (a) T is a tree with n vertices. (Theorem 3, Section 7.4)

(b) $T' \cup \{t_{k+1}\}$ would have n vertices and n distinct edges. (Theorem 2, Section 7.4)

(c) it is connected and acyclic (Theorem 1, Section 7.4)

(d) T is a spanning tree for R.

Chapter 8

Exercise Set 8.1, page 285

1. $V = \{a, b, c, d\}$, $E = \{\{a, b\}, \{b, c\}, \{b, d\}, \{c, c\}\}$

3. $V = \{a, b, c, d\}$, $E = \{\{a, b\}, \{b, c\}, \{d, a\}, \{d, c\}\}$. All edges are double edges.

5.

7. Degree of a is 2; degree of b is 3; degree of c is 3; degree of d is 1.

9. $a, c; a, b, c; a, c, d; a, c, e.$

11.

13. Only the graph given in Exercise 3 is regular.

15. One possible solution is

17.

$\{a, f\}$ ● ⟍
⟍
$\{e, b, d\}$ ●

$\{c\}$ ●

19. (a)

a ●
b ●
e ●
c ● — d ●

(b)

b ●
f ● e ●
c ● — d ●

21.

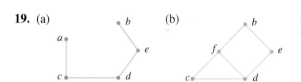

23. $n - 1$. The two "endpoints" have degree 1; the other $n - 2$ vertices each have degree 2. Hence the number of edges is

$$\frac{2(1) + 2(n - 2)}{2}$$

or $n - 1$, since each edge is counted twice in the sum of the degrees.

25. In a digraph there are no multiple edges between vertices. In a graph, the edges are not directed.

27. The sum of the degrees of all vertices with even degree is clearly even. Thus the sum of the degrees of all vertices of odd degree must also be even (using Exercise 26). But if there were an odd number of vertices of odd degree, the sum of their degrees would be odd, a contradiction.

Exercise Set 8.2, page 292

1. Neither. There are 4 vertices of odd degree.

3. Euler circuit. All vertices have even degree.

5. Euler path only, since exactly two vertices have odd degree.

7. Neither. The graph is disconnected.

9. Yes, all vertices have even degree.

11.

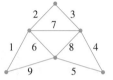

is one possible answer.

13. Yes. Note that if a circuit is required, it is not possible.

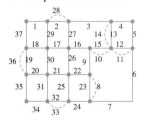

15. One more edge.

17. 7. One possible solution.

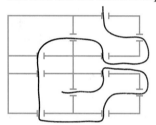

19. See the solution for Exercise 17. The consecutively numbered edges are one possible circuit.

Exercise Set 8.3, page 296

1. Neither.

3. Hamiltonian path, but no Hamiltonian circuit.

5. Hamiltonian circuit.

7. $A, B, D, F, G, H, E, C, A$

9. $A, B, C, E, D, F, J, G, H, I, A$

11. $C, A, B, D, F, G, H, E, C$

13. $I, H, G, J, F, D, E, C, B, A, I$

15. $D, B, A, C, E, H, G, F, D$

17. Choose any vertex, v_1, in $K_n, n \geq 3$. Choose any one of the $n - 1$ edges with v_1 as an endpoint. Follow this edge to v_2. Here we have $n - 2$ edges from which to choose. Continuing in this way we see there are $(n - 1)(n - 2) \cdots 3 \cdot 2 \cdot 1$ Hamiltonian circuits we can choose.

19. One example is

Exercise Set 8.4, page 305

1. One possible solution is

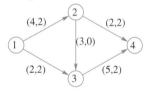

3. One possible solution is

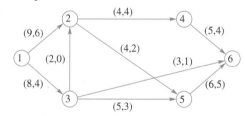

5. $value(F) = 6$

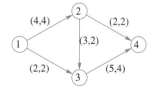

7. $value(F) = 13$

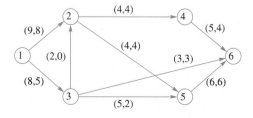

9. $value(F) = 16$

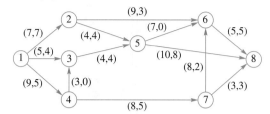

11. $\{(1, 2), (1, 3)\}$

13. $\{(2, 4), (3, 6), (5, 6)\}$

15. $\{(2, 5), (3, 5), (6, 8), (7, 8)\}$

Exercise Set 8.5, page 310

1. Yes. $M(s_2) = b_1$, $M(s_3) = b_3$, $M(s_4) = b_4$, $M(s_5) = b_2$.

3. $value(F) = 9$

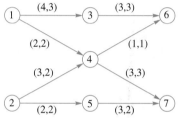

5. $value(F) = 17$

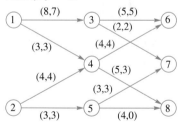

7. $\{(a, 1), (b, 2), (c, 4), (d, 3)\}$ is a maximal matching.

9. $\{(a, 5), (b, 2), (c, 3), (d, 1), (e, 4)\}$ is a maximal matching.

11. The matchings in Exercises 7, 9, and 10 are complete.

13. Let S be any subset of A and E the set of edges that begin in S. Then $k|S| \le |E|$. Each edge in E must terminate in a node of $R(S)$. There are at most $j|R(S)|$ such nodes. Since $j \le k$, $j|S| \le k|S| \le j|R(S)|$ and $|S| \le |R(S)|$. By Hall's marriage theorem, there is a complete matching for A, B, and R.

Exercise Set 8.6, page 315

1.

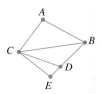

3. ME

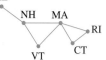

5. 2.

7. 2.

9. $P_G(x) = x(x - 1)(x^2 - 3x + 3)$; $P_G(0) = P_G(1) = 0$, $P_G(2) = 2$.

11. $P_G(x) = x(x - 1)(x^4 - 5x^3 + 10x^2 - 10x + 5)$; $P_G(0) = P_G(1) = 0$, $P_G(2) = 2$.

13. $P_G(x) = x(x - 1)(x - 2)^2$; $\chi(G) = 3$.

15. $P_G(x) = x(x - 1)(x - 2)(x - 3)$; $\chi(G) = 4$.

17. $P_G(x) = x(x - 1)(x - 2)^3$; $\chi(G) = 3$.

19. $P_G(x) = x(x - 1)^2(x - 2)(x^2 - 3x + 3)$; $\chi(G) = 3$.

21. (outline) Basis Step: $n = 1$ P(1): $P_{L_1}(x) = x$ is true, because L_1 consists of a single vertex.
Induction Step: We use P(k) to show P(k + 1). Let $G = L_{k+1}$ and e be an edge $\{u, v\}$ with $\deg(v) = 1$. Then G_e has two components, L_k and v. Using Theorem 1 and P(k), we have $P_{G_e}(x) = x \cdot x(x - 1)^{k-1}$. Merging v with u gives $G^e = L_k$. Thus $P_{G^e}(x) = x(x - 1)^{k-1}$. By Theorem 2, $P_{L_{k+1}}(x) = x^2(x - 1)^{k-1} - x(x - 1)^{k-1} = x(x - 1)^{k-1}(x - 1)$ or $x(x - 1)^k$.

Chapter 9

Exercise Set 9.1, page 323

1. Yes.

3. No.

5. No.

7. No.

9. Commutative, associative.

11. Not commutative, associative.

13. Commutative, associative.

15. Commutative, associative.

17. Commutative, associative.

19. One solution is

$*$	a	b	c
a	a	c	c
b	c	b	a
c	c	a	c

21. (a) a, a.   (b) c, b.   (c) c, a.   (d) Neither.

23.

$*$	a	b	c	d
a	a	b	c	d
b	b	a	c	d
c	c	d	c	d
d	d	c	c	d

25. $n^{\frac{n(n+1)}{2}}$ commutative operations.

27. (a) Associative: (1), (5), (8), (9), (10), (11), (15), (16).

(b) Idempotent: (5), (10), (11), (15).

29. A binary operation on a set S must be defined for every a, b in S. According to the earlier definition, $a * b$ may be undefined for some a, b in S. Any binary operation on a set S is a binary operation in the sense of Section 1.6.

Exercise Set 9.2, page 330

1. Semigroup: (b)   monoid: (b)

3. Semigroup: (a)   monoid: neither

5. Monoid: identity is 1; commutative.

7. Semigroup

9. Monoid: identity is S; commutative.

11. Monoid: identity is 12; commutative.

13. Monoid: identity is 0; commutative.

15. Neither.

17.

$*$	a	b	c
a	c	a	b
b	a	b	c
c	b	c	a

19. Let $f_1(a) = a$, $f_1(b) = a$; $f_2(a) = a$, $f_2(b) = b$; $f_3(a) = b$, $f_3(b) = a$; $f_4(a) = b$, $f_4(b) = b$. These are the only functions on S. It is not commutative.

$\circ$	f_1	f_2	f_3	f_4
f_1	f_1	f_1	f_4	f_4
f_2	f_1	f_2	f_3	f_4
f_3	f_1	f_3	f_2	f_1
f_4	f_1	f_4	f_4	f_4

21. (a) $abaccbababc$.

(b) $babcabacabac$.

(c) $babccbaabac$.

23. By Exercise 22, we need only check that $e \in S_1 \cap S_2$. But $e \in S_1$ and $e \in S_2$, because each is a submonoid of $(S, *)$.

25. Yes. Refer to Exercise 1.

27. Let $x, y \in S_1$.

$$(g \circ f)(x *_1 y) = g(f(x *_1 y))$$
$$= g(f(x) *_2 f(y))$$
$$= g(f(x)) *_3 g(f(y))$$
$$= (g \circ f)(x) *_3 (g \circ f)(y).$$

Hence $g \circ f$ is a homomorphism from $(S_1, *_1)$ to $(S_3, *_3)$.

29. Onto; homomorphism.

31. Let $x, y \in \mathbb{R}^+$. $\ln(x * y) = \ln(x) + \ln(y)$ so ln is a homomorphism. Suppose $x \in \mathbb{R}$. Then $e^x \in \mathbb{R}^+$ and $\ln(e^x) = x$ so ln is onto $\mathbb{R}^+$. Suppose $\ln(x) = \ln(y)$; then $e^{\ln(x)} = e^{\ln(y)}$ and $x = y$. Hence ln is one to one and an isomorphism between $(\mathbb{R}^+, \times)$ and $(\mathbb{R}, +)$.

Exercise Set 9.3, page 337

1. Let $(s_1, t_1), (s_2, t_2) \in S \times T$.
$(s_1, t_1) *'' (s_2, t_2) = (s_1 * s_2, t_1 *' t_2)$ so $*''$ is a binary operation. Consider $(s_1, t_1) *'' ((s_2, t_2) *'' (s_3, t_3)) = (s_1, t_1) *'' (s_2 * s_3, t_2 *' t_3) = (s_1 * (s_2 * s_3), t_1 *' (t_2 *' t_3)) = ((s_1 * s_2) * s_3, (t_1 *' t_2) *' t_3) = ((s_1, t_1) *'' (s_2, t_2)) *'' (s_3, t_3)$. Thus $(S \times T, *'')$ is a semigroup. $(s_1, t_1) *'' (s_2, t_2) = (s_1 * s_2, t_1 *' t_2) = (s_2 * s_1, t_2 *' t_1) = (s_2, t_2) *'' (s_1, t_1)$. Hence $*''$ is commutative.

3. Let $(s_1, t_1), (s_2, t_2) \in S \times T$. Then $f((s_1, t_1) *'' (s_2, t_2)) = f(s_1 * s_2, t_1 *' t_2) = s_1 * s_2 = f(s_1, t_1) * f(s_2, t_2)$. f is a homomorphism.

5. $*''$ is a binary operation, because both $*$ and $*'$ are. Consider $(s_1, t_1) *'' ((s_2, t_2) *'' (s_3, t_3))$.

$(s_1, t_1) *'' ((s_2, t_2) *'' (s_3, t_3)) = (s_1, t_1) *'' (s_2 * s_3, t_2 *' t_3)$
$$= (s_1 * (s_2 * s_3), t_1 *' (t_2 *' t_3))$$
$$= ((s_1 * s_2) * s_3, (t_1 *' t_2) *' t_3)$$
$$= ((s_1, t_1) *'' (s_2, t_2)) *'' (s_3, t_3)$$

Thus, $*''$ is associative.

7. Yes.

9. Yes.

11. Yes.

13. No.

15. Yes.

17. By Exercise 21, Section 4.7, we have that the composition of two equivalence relations need not be an equivalence relation.

19. $S/R = \{[0], [1]\}$, $[0] = \{0, \pm2, \pm4, \ldots\}$, $[1] = \{\pm1, \pm3, \pm5, \ldots\}$

$\oplus$	$[0]$	$[1]$
$[0]$	$[0]$	$[1]$
$[1]$	$[1]$	$[0]$

21. $S/R = \{[0], [1], [2], [3], [4]\}$,
$[a] = \{z \mid z = 5k + a, k \in Z\}, a = 0, 1, 2, 3, 4.$

$\oplus$	[0]	[1]	[2]	[3]	[4]
[0]	[0]	[1]	[2]	[3]	[4]
[1]	[1]	[2]	[3]	[4]	[0]
[2]	[2]	[3]	[4]	[0]	[1]
[3]	[3]	[4]	[0]	[1]	[2]
[4]	[4]	[0]	[1]	[2]	[3]

23. (a)

$\circledast$	$[a]$	$[b]$
$[a]$	$[a]$	$[b]$
$[b]$	$[b]$	$[b]$

(b) $f_R(e) = [a] = f_R(a)$, $f_R(b) = [b] = f_R(c)$.

25. This is a direct proof. For part (a), we check the three properties for an equivalence relation and then the property for a congruence relation. In part (b), we first check that $\overline{f}$ is a function and then the properties of an isomorphism are confirmed.

Exercise Set 9.4, page 348

1. No.

3. Yes; abelian; identity is 0; a^{-1} is $-a$.

5. No.

7. No.

9. No.

11. Yes; abelian; identity is { }; a^{-1} is a.

13. Since g_1, g_2, g_3 in S_3 each have order 2, they must be paired somehow with f_2, f_3, f_4 of Example 12 if the groups are isomorphic. But no rearrangement of the columns and rows labeled f_2, f_3, f_4 in Example 12 will give the "block" pattern shown by g_1, g_2, g_3 in the table for S_3. Hence the groups are not isomorphic.

15. (a) $\frac{8}{3}$.   (b) $\frac{-4}{5}$.

17.

$\circ$	f_1	f_2	f_3	f_4	f_5	f_6	f_7	f_8
f_1	f_1	f_2	f_3	f_4	f_5	f_6	f_7	f_8
f_2	f_2	f_3	f_4	f_1	f_8	f_7	f_5	f_6
f_3	f_3	f_4	f_1	f_2	f_6	f_5	f_8	f_7
f_4	f_4	f_1	f_2	f_3	f_7	f_8	f_6	f_5
f_5	f_5	f_7	f_6	f_8	f_1	f_3	f_2	f_4
f_6	f_6	f_8	f_5	f_7	f_3	f_1	f_4	f_2
f_7	f_7	f_6	f_8	f_5	f_4	f_2	f_1	f_3
f_8	f_8	f_5	f_7	f_6	f_2	f_4	f_3	f_1

19. Consider the sequence $e, a, a^2, a^3, \ldots$. Since G is finite, not all terms of this sequence can be distinct; that is, for some $i \leq j$, $a^i = a^j$. Then $(a^{-1})^i a^i = (a^{-1})^i a^j$ and $e = a^{j-i}$. Note that $j - i \geq 0$.

21. Yes.

23. Clearly, $e \in H$. Let $a, b \in H$. Consider
$(ab)y = a(by) = a(yb) = (ay)b = (ya)b = y(ab)$
$\forall y \in G$. Hence H is closed under multiplication and is a subgroup of G.

25. The identity permutation is an even permutation. If p_1 and p_2 are even permutations, then each can be written as the product of an even number of transpositions. Then $p_1 \circ p_2$ can be written as the product of these representations of p_1 and p_2. But this gives $p_1 \circ p_2$ as the product of an even number of transpositions. Thus $p_1 \circ p_2 \in A_n$ and A_n is a subgroup of S_n.

27. $\{f_1\}, \{f_1, f_2, f_3, f_4\}, \{f_1, f_3, f_5, f_6\}, \{f_1, f_3, f_7, f_8\}, \{f_1, f_5\}, \{f_1, f_6\}, \{f_1, f_3\}, \{f_1, f_7\}, \{f_1, f_8\}, D_4$.

29. $|xy| = |x| \cdot |y|$. Thus $f(xy) = f(x)f(y)$.

31. Suppose $f : G \rightarrow G : a \rightarrow a^2$ is a homomorphism. Then $f(ab) = f(a)f(b)$ or $(ab)^2 = a^2b^2$. Hence $a^{-1}(abab)b^{-1} = a^{-1}(a^2b^2)b^{-1}$ and $ba = ab$. Suppose G is abelian. By Exercise 18, $f(ab) = f(a)f(b)$.

33. Let $x, y \in G$.
$f_a(xy) = axya^{-1} = axa^{-1}aya^{-1} = f_a(x)f_a(y)$. f_a is a homomorphism. Suppose $x \in G$. Then $f_a(a^{-1}xa) = aa^{-1}xaa^{-1} = x$ so f_a is onto. Suppose $f_a(x) = f_a(y)$, then $axa^{-1} = aya^{-1}$. Now $a^{-1}(axa^{-1})a = a^{-1}(aya^{-1})a$ and $x = y$. Thus f_a is one to one and an isomorphism.

Exercise Set 9.5, page 353

1.

	$(\overline{0}, \overline{0})$	$(\overline{0}, \overline{1})$	$(\overline{0}, \overline{2})$	$(\overline{1}, \overline{0})$	$(\overline{1}, \overline{1})$	$(\overline{1}, \overline{2})$
$(\overline{0}, \overline{0})$	$(\overline{0}, \overline{0})$	$(\overline{0}, \overline{1})$	$(\overline{0}, \overline{2})$	$(\overline{1}, \overline{0})$	$(\overline{1}, \overline{1})$	$(\overline{1}, \overline{2})$
$(\overline{0}, \overline{1})$	$(\overline{0}, \overline{1})$	$(\overline{0}, \overline{2})$	$(\overline{0}, \overline{0})$	$(\overline{1}, \overline{1})$	$(\overline{1}, \overline{2})$	$(\overline{1}, \overline{0})$
$(\overline{0}, \overline{2})$	$(\overline{0}, \overline{2})$	$(\overline{0}, \overline{0})$	$(\overline{0}, \overline{1})$	$(\overline{1}, \overline{2})$	$(\overline{1}, \overline{0})$	$(\overline{1}, \overline{1})$
$(\overline{1}, \overline{0})$	$(\overline{1}, \overline{0})$	$(\overline{1}, \overline{1})$	$(\overline{1}, \overline{2})$	$(\overline{0}, \overline{0})$	$(\overline{0}, \overline{1})$	$(\overline{0}, \overline{2})$
$(\overline{1}, \overline{1})$	$(\overline{1}, \overline{1})$	$(\overline{1}, \overline{2})$	$(\overline{1}, \overline{0})$	$(\overline{0}, \overline{1})$	$(\overline{0}, \overline{2})$	$(\overline{0}, \overline{0})$
$(\overline{1}, \overline{2})$	$(\overline{1}, \overline{2})$	$(\overline{1}, \overline{0})$	$(\overline{1}, \overline{1})$	$(\overline{0}, \overline{2})$	$(\overline{0}, \overline{0})$	$(\overline{0}, \overline{1})$

3. Define $f : G_1 \rightarrow G_2 : (g_1, g_2) \rightarrow (g_2, g_1)$. By Exercise 4, Section 9.3, f is an isomorphism.

5.

	[0]	[1]	[2]
[0]	[0]	[1]	[2]
[1]	[1]	[2]	[0]
[2]	[2]	[0]	[1]

7. $\{[0]\}, \{[1]\}, \{[2]\}, \{[3]\}$.

9. $\{[0], [1], [2], [3]\}$.

11. $\{f_1, g_3\}, \{f_2, g_2\}, \{f_3, g_1\}$.

13. $\{f_1\}, \{f_2\}, \{f_3\}, \{g_1\}, \{g_2\}, \{g_3\}$.

15. $\{[0], [4]\}, \{[1], [5]\}, \{[2], [6]\}, \{[3], [7]\}$.

17. $\{(m + x, n + x) \mid x \in Z\}$ for $(m, n) \in Z \times Z$.

19. If N is a normal subgroup of G, Exercise 12 shows that
$a^{-1}Na \subseteq N$ for all $a \in G$.
Suppose $a^{-1}Na \subseteq N$ for all $a \in G$. Again the proof in
Exercise 12 shows that N is a normal subgroup of G.

21. $\{f_1\}, \{f_1, f_3\}, \{f_1, f_3, f_5, f_6\}, \{f_1, f_2, f_3, f_4\},$
$\{f_1, f_3, f_7, f_8\}, D_4$.

23. Suppose $f_a(h_1) = f_a(h_2)$. Then $ah_1 = ah_2$ and
$a^{-1}(ah_1) = a^{-1}(ah_2)$. Hence $h_1 = h_2$ and f_a is one to
one. Let $x \in aH$. Then $x = ah, h \in H$ and $f_a(h) = x$.
Thus f_a is onto and since it is everywhere defined as well,
f_a is a one-to-one correspondence between H and aH.
Hence $|H| = |aH|$.

25. Suppose $f(aH) = f(bH)$. Then $Ha^{-1} = Hb^{-1}$ and
$a^{-1} = hb^{-1}, h \in H$. Hence $a = bh^{-1} \in bH$ so
$aH \subseteq bH$. Similarly, $bH \subseteq aH$ so $aH = bH$. This
means f is one to one. If Hc is a right coset of H, then
$f(c^{-1}H) = Hc$ so f is also onto.

27. Consider $f(aba^{-1}b^{-1}) = f(a)f(b)f(a^{-1})f(b^{-1}) =$
$f(a)f(a^{-1})f(b)f(b^{-1}) = f(a)(f(a))^{-1}f(b)(f(b))^{-1}$
(by Theorem 5, Section 9.4) $= ee = e$. Hence
$\{aba^{-1}b^{-1} \mid a, b \text{ in } G_1\} \subseteq \ker(f)$.

29. Let $a \notin H$. The left cosets of H are H and aH. The right
cosets are H and Ha. $H \cap aH = H \cap Ha = \{\ \}$ and
$H \cup aH = H \cup Ha$. Thus $aH = Ha$. Since
$a \in H \Rightarrow aH = H$, we have $xH = Hx \ \forall x \in G$. H is a
normal subgroup of G.

31. Suppose $f: G \to G'$ is one to one. Let $x \in \ker(f)$. Then
$f(x) = e' = f(e)$. Thus $x = e$ and $\ker(f) = \{e\}$.
Conversely suppose $\ker(f) = \{e\}$. If $f(g_1) = f(g_2)$ then
$f(g_1 g_2^{-1}) = f(g_1)f(g_2^{-1}) = f(g_1)(f(g_2))^{-1} =$
$f(g_1)(f(g_1))^{-1} = e$. Hence $g_1 g_2^{-1} \in \ker(f)$. Thus
$g_1 g_2^{-1} = e$ and $g_1 = g_2$. Hence f is one to one.

Chapter 10

Exercise Set 10.1, page 364

1. $\{x^m y^n z, m \geq 0, n \geq 1\}$.

3. $\{a^{2n+1}, n \geq 0\} \cup \{a^{2n}b, n \geq 0\}$.

5. $\{\underbrace{((\ldots (}_{k} \underbrace{a + a + \cdots + a}_{n\ a\text{'s}} \underbrace{)\ldots)}_{k}, k \geq 0, n \geq 3\}$.

7. $\{x^m yz^n, m \geq 1, n \geq 0\}$.

9. (a), (c), (e), (h), (i).

11. $L(G)$ is the set of strings from $\{a, b, c, 1, 2, \ldots, 9, 0\}^*$
that begin with $a, b,$ or c.

13. $L(G) = \{(aa)^n bc^k (bb)^j b^k, n \geq 0, k \geq 1, j \geq 0\}$.

15.

v_0	v_0
$v_0 v_1$	$v_0 v_1$
$v_0 v_1 v_1$	$v_2 v_0 v_1$
$v_2 v_0 z$	xyv_1
xyz	xyz

17.

I	I
LW	LW
aW	LDW
aDW	$LDDW$
$a1W$	$LDDD$
$a1DW$	$aDDD$
$a1DD$	$a1DD$
$a10D$	$a10D$
$a100$	$a100$

19. $G = (V, S, v_0, \mapsto), V = \{v_0, v_1, 0, 1\}, S = \{0, 1\}$
$\mapsto: v_0 \mapsto 0v_1 1, v_0 \mapsto 1v_1 0, v_1 \mapsto 0v_1 1, v_1 \mapsto 1v_1 0,$
$v_1 \mapsto 01, v_1 \mapsto 10$

21. $G = (V, S, v_0, \mapsto), V = \{v_0, v_1, a, b\}, S = \{a, b\}$
$\mapsto: v_0 \mapsto aav_1 bb, v_1 \mapsto av_1 b, v_1 \mapsto ab$

23. $G = (V, S, v_0, \mapsto), V = \{v_0, x, y\}, S = \{x, y\}$
$\mapsto: v_0 \mapsto v_0 yy, v_0 \mapsto xv_0, v_0 \mapsto xx$

25. By using production rules 1, 2, and 4; production rule 4 to
$a^{3k-1}v_2; 3; a^{3k}av_3 \Rightarrow a^{3k+1}v_1 \Rightarrow a^{3k+2}v_2 \Rightarrow a^{3(k+1)}$.

27. Let $G_1 = (V_1, S_1, v_0, \mapsto_1)$ and $G_2 = (V_2, S_2, v_0', \mapsto_2)$.
Define $G = (V_1 \cup V_2, S_1 \cup S_2, v_0, \mapsto)$ as follows. If
$v_i \mapsto_1 wv_k$, then $v_i \mapsto wv_k$. If $v_i \mapsto_1 w$ where w consists
of terminal symbols, then $v_i \mapsto wv_0'$. All productions in
$\mapsto_2$ become productions in $\mapsto$.

Exercise Set 10.2, page 374

1. $\langle v_0 \rangle ::= x \langle v_0 \rangle \mid y \langle v_1 \rangle$
$\langle v_1 \rangle ::= y \langle v_1 \rangle \mid z$

3. $\langle v_0 \rangle ::= a \langle v_1 \rangle$
$\langle v_1 \rangle ::= b \langle v_0 \rangle \mid a$

5. $\langle v_0 \rangle ::= aa\langle v_0 \rangle \mid b\langle v_1 \rangle$
$\langle v_1 \rangle ::= c\langle v_2 \rangle b \mid cb$
$\langle v_2 \rangle ::= bb\langle v_2 \rangle \mid bb$

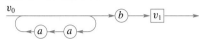

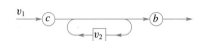

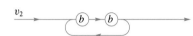

7. $\langle v_0 \rangle ::= x\langle v_0 \rangle \mid y\langle v_0 \rangle \mid z$

9. $\langle v_0 \rangle ::= a\langle v_1 \rangle$
$\langle v_1 \rangle ::= b\langle v_0 \rangle \mid a$

11. $\langle v_0 \rangle ::= ab\langle v_1 \rangle$
$\langle v_1 \rangle ::= c\langle v_1 \rangle \mid \langle v_2 \rangle$
$\langle v_2 \rangle ::= dd\langle v_2 \rangle \mid d$

13. $\langle v_0 \rangle ::= a\langle v_1 \rangle$
$\langle v_1 \rangle ::= a\langle v_2 \rangle$
$\langle v_2 \rangle ::= a\langle v_1 \rangle \mid a$

15. $(aa)^*aa$

17. $(()^*(a + a + (a +)^*a())^*$. Note: Right and left parentheses must be matched.

19. $(a \vee b \vee c)(a \vee b \vee c \vee 0 \vee 1 \vee \cdots \vee 9)^*$

21. $ab(d \vee (d(c \vee d)d))^*$

23. $ab(abc)^n b$, $n \geq 1$.

Exercise Set 10.3, page 380

1.

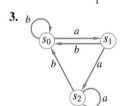

3.

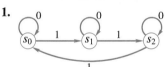

5.

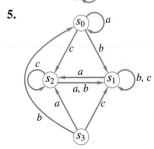

7.

	a	b
s_0	s_1	s_1
s_1	s_1	s_2
s_2	s_0	s_2

9.

	T	F
s_0	s_1	s_0
s_1	s_1	s_1
s_2	s_1	s_2

11.

	a	b	c
s_0	s_0	s_1	s_2
s_1	s_2	s_1	s_3
s_2	s_3	s_3	s_1
s_3	s_3	s_3	s_2

13. Let $x \in I$. Certainly $f_x(s) = f_x(s)$ for all $s \in S$. Thus $x \, R \, x$ and R is reflexive.
Suppose $x \, R \, y$. Then $f_x(s) = f_y(s) \, \forall s \in S$. But then $y \, R \, x$ and R is symmetric.
Suppose $x \, R \, y$, $y \, R \, z$. Then $f_x(s) = f_y(s) = f_z(s)$, $\forall s \in S$. Hence $x \, R \, z$ and R is transitive.

15. Using Exercise 14, we need only show that R is reflexive and symmetric. Let $s \in S$. $s = e * s$ so $f_e(s) = s$ and $s \, R \, s$. Suppose $x \, R \, y$. Then $f_z(x) = y$ for some $z \in S$. $y = z * x \Rightarrow z^{-1} * y = x$ and thus $f_{z^{-1}}(y) = x$. Hence $y \, R \, x$ and R is symmetric.

17. (a) Inspection of $\mathbf{M}_R$ shows that R is reflexive and symmetric. Since $\mathbf{M}_R \odot \mathbf{M}_R = \mathbf{M}_R$, R is transitive. Thus R is an equivalence relation. The table below shows that it is a machine congruence.

(b)

	0	1
[1]	[1]	[1]
[2]	[2]	[2]

19. Inspection of $\mathbf{M}_R$ shows that R is reflexive and symmetric. Since $\mathbf{M}_R \odot \mathbf{M}_R = \mathbf{M}_R$, R is transitive. Thus R is an equivalence relation. The digraph below shows that it is a machine congruence.

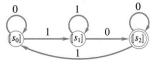

Exercise Set 10.4, page 385

1. $f_w(s_0) = s_2$, $f_w(s_1) = s_3$, $f_w(s_2) = s_0$, $f_w(s_3) = s_1$.

3. The number of 1's in w is divisible by 4.

5. The number of 1's in w is $2 + 4k$, $k \geq 0$.

7. $f_w(s_0) = s_0$, $f_w(s_1) = s_0$, $f_w(s_2) = s_0$.

9. All words ending in b.

11. Strings of 0's and 1's with $3 + 5k$ 1's, $k \geq 0$.

13. Strings of 0's and 1's that end in 0.

15. Strings of a's and b's that do not contain bb.

17. Strings of 0's and 1's that end in 01.

19. Strings xy and yz.

21. Let $w, u \in L(M)$. Then $f_w(s_0) = s \in T$, $f_u(s) \in T$. Hence $f_{w \cdot u}(s_0) \in T$, so $w \cdot u \in L(M)$.

Exercise Set 10.5, page 392

1. $G = (V, I, s_0, \mapsto)$, $V = \{s_0, s_1, s_2, s_3, 0, 1\}$, $I = \{0, 1\}$
$\mapsto : s_0 \mapsto 0s_0$, $s_0 \mapsto 1s_1$, $s_1 \mapsto 0s_1$, $s_1 \mapsto 1s_2$, $s_2 \mapsto 0s_2$,
$s_2 \mapsto 1s_3$, $s_2 \mapsto 1$, $s_3 \mapsto 0s_3$, $s_3 \mapsto 0$, $s_3 \mapsto 1s_0$.

3. $(0 \vee 1)^*1$

5. $G = (V, I, s_0, \mapsto)$, $V = \{s_0, s_1, s_2, a, b\}$, $I = \{a, b\}$
$\langle s_0 \rangle ::= a \langle s_0 \rangle \mid b \langle s_1 \rangle \mid a \mid b$
$\langle s_1 \rangle ::= a \langle s_0 \rangle \mid b \langle s_2 \rangle \mid a$
$\langle s_2 \rangle ::= a \langle s_2 \rangle \mid b \langle s_2 \rangle$

7.

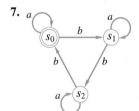

9.

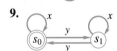

11.

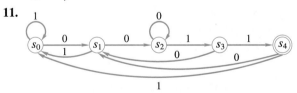

13.

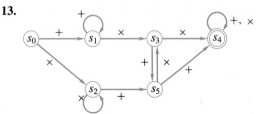

15.

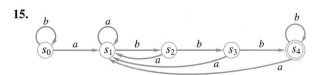

17.

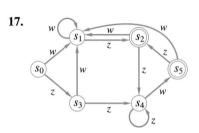

19.

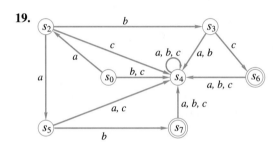

21.

	0	1
s_0	s_1	s_0
s_1	s_1	s_2
s_2	s_3	s_0
s_3	s_1	s_4
s_4	s_3	s_0

$T = \{s_4\}$

23.

	x	y
s_0	s_1	s_0
s_1	s_2	s_1
s_2	s_3	s_2
s_3	s_3	s_3

$T = \{s_2\}$

25. R is reflexive because $f_w(x) = f_w(x)$. R is symmetric because if $f_w(s_i)$, $f_w(s_j)$ are both (not) in T, then $f_w(s_j)$, $f_w(s_i)$ are both (not) in T. R is transitive because $s_i \ R \ s_j$, $s_j \ R \ s_k$ if and only if $f_w(s_i)$, $f_w(s_j)$, $f_w(s_k)$ are all in (not in) T.

Exercise Set 10.6, page 397

1. $R_0 = \{(s_0, s_0), (s_0, s_1), (s_1, s_0), (s_1, s_1), (s_2, s_2)\}$

3. $R_1 = \{(s_0, s_0), (s_1, s_1), (s_2, s_2), (s_3, s_3),$
$(s_4, s_4), (s_0, s_3), (s_3, s_0), (s_1, s_2), (s_2, s_1)\}$

5. $R_{127} = R_1$

7. $R_2 = R_1$

9. $R = \{(s_0, s_0), (s_1, s_1), (s_2, s_2)\}$

11. $R = R_1$ as given in Exercise 6.

13. $R = \{(s_0, s_0), (s_1, s_1), (s_2, s_2), (s_3, s_3), (s_4, s_4), (s_4, s_5),$
$(s_5, s_4), (s_5, s_5), (s_6, s_6), (s_3, s_6), (s_6, s_3)\}$

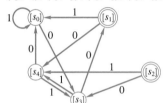

15.

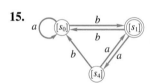

17.

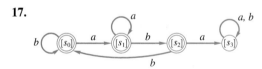

19. $P = \{\{a, g\}, \{f\}, \{b\}, \{c\}, \{d, e\}\}$

	0	1
$[a]$	$[a]$	$[c]$
$[b]$	$[a]$	$[d]$
$[c]$	$[f]$	$[d]$
$[d]$	$[a]$	$[d]$
$[f]$	$[a]$	$[f]$

21. M

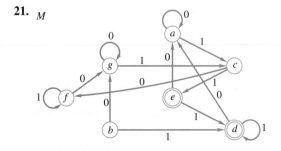

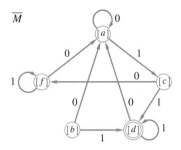

Chapter 11

Exercise Set 11.1, page 411

1. (a) 3.   (b) 2.   (c) 3.

3. (a) Yes.   (b) No.

5. (a) No.   (b) No.   (c) Yes.   (d) No.

7. (a) Yes.   (b) Yes.

9. By definition, $x \oplus y = (x_1 + y_1, x_2 + y_2, \ldots, x_n + y_n)$
and $x_i + y_i = 1$ if and only if $x_i \neq y_i$.

11. (a) 2.   (b) 6.

13. If $x = y$, they differ in 0 positions and $\delta(x, y) = 0$.
Conversely, if $\delta(x, y) = 0$, then x and y cannot differ in
any position and $x = y$.

15. 1.

17. (a) 3.   (b) 2 or fewer.

19. Let $a = 0000000$, $b = 0010110$, $c = 0101000$,
$d = 0111110$, $e = 1000101$, $f = 1010011$, $g = 1101101$,
$h = 1111011$.

$\oplus$	a	b	c	d	e	f	g	h
a	a	b	c	d	e	f	g	h
b	b	a	d	c	f	e	h	g
c	c	d	a	b	g	h	e	f
d	d	c	b	a	h	g	f	e
e	e	f	g	h	a	b	c	d
f	f	e	h	g	b	a	d	c
g	g	h	e	f	c	d	a	b
h	h	g	f	e	d	c	b	a

21. 2.

23. $\begin{bmatrix} 0 & 0 & 0 \\ 1 & 0 & 0 \\ 1 & 0 & 1 \\ 0 & 1 & 0 \end{bmatrix}.$

25. $\begin{bmatrix} 0 & 1 & 1 \\ 1 & 1 & 0 \\ 0 & 1 & 1 \end{bmatrix}.$

27. $e_H(000) = 000000$ $e_H(100) = 100100$
$e_H(001) = 001111$ $e_H(101) = 101011$
$e_H(010) = 010011$ $e_H(110) = 110111$
$e_H(011) = 011100$ $e_H(111) = 111000$

Exercise Set 11.2, page 420

1. (a) 011.   (b) 101.

3. (a) 11.   (b) 01.

5. 0.

7. 0.

9. 1.

11. (a) 01.   (b) 11.   (c) 10.

13. (a) 010.   (b) 110.   (c) 001 are possible answers.

15. (a) Suppose $x \in g_1 H \cap g_2 H$. Then $x = g_1 h_1 = g_2 h_2$, for some $h_1, h_2 \in H$. We have $g_1 = g_2 h_2 h_1^{-1} \in g_2 H$ since H is a subgroup. Hence $g_1 H \subseteq g_2 H$. Similarly, $g_2 = g_1 h_1 h_2^{-1}$ and $g_2 H \subseteq g_1 H$. Thus, $g_1 H = g_2 H$.

(b) Each element g of G belongs to gH. Part (a) guarantees that there is a set of disjoint left cosets whose union is G.

17. 00000, 00001, 00010, 00100, 01000, 10000, 01010 (or 10100), 00110 (or 11000).

19. 00, 01, 10, 11. Same order as in Exercise 16.

21. 000, 001, 010, 100, 011, 110, 011, 111. Same order as in Exercise 18.

23. (a) 00.   (b) 01.   (c) 10 are possible answers.

Appendix A, page 436

1. FUNCTION TAX (INCOME)
  1. **IF** (INCOME $\geq$ 30,000) **THEN**
    a.   TAXDUE $\leftarrow$ 6000
  2. **ELSE**
    a.   **IF** (INCOME $\geq$ 20,000) **THEN**
      1.   TAXDUE $\leftarrow$ 2500
    b.   **ELSE**
      1.   TAXDUE $\leftarrow$ INCOME $\times$ 0.1
  3. **RETURN** (TAXDUE)
  END OF FUNCTION TAX

3. 1.   SUM $\leftarrow$ 0
  2. **FOR** $I = 1$ **THRU** N
    a.   SUM $\leftarrow$ SUM $+ X[I]$
  3.   AVERAGE $\leftarrow$ SUM/N

5. 1.   DOTPROD $\leftarrow$ 0
  2. **FOR** $I = 1$ **THRU** 3
    a.   DOTPROD $\leftarrow$ DOTPROD $+ (X[I])(Y[I])$

7. 1.   RAD $\leftarrow (A[2])^2 - 4(A[1])(A[3])$
  2. **IF** (RAD < 0) **THEN**
    a.   PRINT ('ROOTS ARE IMAGINARY')
  3. **ELSE**
    a.   **IF** (RAD $= 0$) **THEN**
      1. $R1 \leftarrow -A[2]/(2A[1])$
      2.   PRINT ('ROOTS ARE REAL AND EQUAL')
    b.   **ELSE**
      1. $R1 \leftarrow (-A[2] + \text{SQ(RAD)})/(2A[1])$
      2. $R2 \leftarrow (-A[2] - \text{SQ(RAD)})/(2A[1])$

9. 1. **FOR** $I = 1$ **THRU** N
    a.   **IF** $(A[I] \neq B[I])$ **THEN**
      1. $C[I] \leftarrow 1$
    b.   **ELSE**
      1. $C[I] \leftarrow 0$

11. 1. **FOR** $I = 1$ **THRU** N
    a.   **IF** $(A[I] = 0$ **AND** $B[I] = 0)$ **THEN**
      1. $C[I] \leftarrow 1$
    b.   **ELSE**
      1. $C[I] \leftarrow 0$

13. 1.   SUM $\leftarrow$ 0
  2. **FOR** $I = 0$ **THRU** $2(N - 1)$ **BY** 2
    a.   SUM $\leftarrow$ SUM $+ I$

15. 1.   PROD $\leftarrow$ 1
  2. **FOR** $I = 2$ **THRU** $2N$ **BY** 2
    a.   PROD $\leftarrow$ (PROD) $\times I$

17. 1.   SUM $\leftarrow$ 0
  2. **FOR** $I = 1$ **THRU** 77
    a.   SUM $\leftarrow$ SUM $+ I^2$

19. 1.   SUM $\leftarrow$ 0
  2. **FOR** $I = 1$ **THRU** 10
    a.   SUM $\leftarrow$ SUM $+ (1/(3I + 1))$

21. MAX returns the larger of X and Y.

23. F returns $|X|$.

25. Assigns 1 to R if $N \mid M$ and assigns 0 otherwise.

27. $X = \sum_{I=1}^{N} I$; I is $N + 1$.

29. $X = 25$; $I = 49$.

ANSWERS TO
CHAPTER SELF-TESTS

CHAPTER 1 SELF-TEST, page 45

1. (a) (i) False. (ii) True. (iii) False.
(iv) True. (v) True.

(b) (i) False. (ii) True. (iii) False.
(iv) False. (v) True.

2. (a) $\{1, 2, 3, 4, \ldots, \}$. (b) $\{\ldots, -3, -2, -1, 0\} \cup B$.

(c) $\{2, 4\}$. (d) A.

(e) $\{2, 6, 10, 14, \ldots\}$.

3. (a) 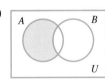 (b)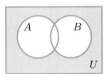

4. $A \cap B$ is always a subset of $A \cup B$. $A \cup B \subseteq A \cap B$ if and only if $A = B$.

5. (a) 6. (b) 18. (c) 41.

6. $0, 0, 1, -2, 9, -30$.

7. (a) | 1 | 1 | 1 | 1 | 0 | 1 | 0 | 1 | 0 | 1 |

(b) | 1 | 1 | 0 | 0 | 0 | 0 | 0 | 0 | 0 | 0 |

(c) | 1 | 1 | 0 | 0 | 0 | 1 | 0 | 0 | 0 | 0 |

(d) | 0 | 0 | 1 | 0 | 0 | 0 | 1 | 1 | 0 | 1 |

8. (a) Yes. (b) Yes. (c) No. (d) No.

9. $33 = 65(7293) - 108(4389)$.

10. (a) **AB** does not exist.

(b) $\mathbf{BA} = \begin{bmatrix} 4 & 12 & 8 \\ -7 & -15 & -10 \end{bmatrix}$.

(c) $\begin{bmatrix} 2 & -3 \\ 0 & 1 \end{bmatrix}$. (d) $\mathbf{A} + \mathbf{B}$ does not exist.

(e) $\begin{bmatrix} 7 & -1 \\ 3 & 3 \\ 2 & 2 \end{bmatrix}$.

11. (a) $\begin{bmatrix} 1 & 1 & 0 \\ 1 & 1 & 0 \\ 1 & 1 & 0 \end{bmatrix}$. (b) $\begin{bmatrix} 1 & 1 & 1 \\ 1 & 1 & 0 \\ 1 & 1 & 1 \end{bmatrix}$.

(c) $\begin{bmatrix} 1 & 0 & 0 \\ 0 & 1 & 0 \\ 0 & 1 & 0 \end{bmatrix}$.

12. **A** has a $\wedge$-inverse if and only if $\mathbf{A} = \begin{bmatrix} 1 & 1 \\ 1 & 1 \end{bmatrix}$, the $\wedge$-identity.

CHAPTER 2 SELF-TEST, page 72

1. (a) False. (b) True.

2. (a) False. (b) True.

3.

p	q	r	$(p \wedge \sim p)$	$\vee$	$(\sim$	$(q \wedge r))$
T	T	T	F	F	F	T
T	T	F	F	T	T	F
T	F	T	F	T	T	F
T	F	F	F	T	T	F
F	T	T	F	F	F	T
F	T	F	F	T	T	F
F	F	T	F	T	T	F
F	F	F	F	T	T	F
			(1)	↑ (3)		(2)

4. (a) $q \Rightarrow p$ $\sim q \Rightarrow \sim p$

(b) $q \Rightarrow (\sim r) \vee (\sim s)$ $\sim q \Rightarrow (r \wedge s)$

489

(c) $(p \vee s) \Rightarrow q$ $(\sim p \wedge \sim s) \Rightarrow \sim q$

5. (a) If $|2| = |-2|$, then $1 < -1$.
    If $|2| \neq |-2|$, then $1 \geq -1$.

    (b) If $|2| = |-2|$, then either $-3 \geq -1$ or $1 \geq 3$.
    If $|2| \neq |-2|$, then $-3 < -1$ and $1 < 3$.

    (c) If $1 < -1$ or $1 < 3$, then $|2| = |-2|$.
    If $1 \geq -1$ and $1 \geq 3$, then $|2| \neq |-2|$.

6. (a) False   True
    (b) False   True
    (c) True   True

7.
p	q	p xor q
T	T	F
T	F	T
F	T	T
F	F	F

8. (a) If an Internet business makes less money, then if I start an Internet business, then an Internet business is cheaper to start. An Internet business is not cheaper to start. Therefore, either an Internet business does not make less money or I do not start an Internet business. Valid.

    (b) If an Internet business is cheaper to start, then I will start an Internet business. If I start an Internet business, then an Internet business will make less money. An Internet business is cheaper to start. Therefore, an Internet business will make less money. Valid.

9. No, consider $6 \mid -6$ and $-6 \mid 6$. If both m and n are positive (or negative), then $n \mid m$ and $m \mid n$ guarantees that $n = m$.

10. Consider 7, 9, and 11. These are three consecutive odd integers whose sum is not divisible by 6.

11. Basis Step: $n = 0$. $P(0)$: $4^0 - 1$ is divisible by 3 is true since $3 \mid 0$.
    Induction Step: We use $P(k)$: 3 divides $4^k - 1$ to show $P(k + 1)$: 3 divides $4^{k+1} - 1$. Consider $4^{k+1} - 1 = 4(4^k - 1) + 3$. By $P(k)$, $3 \mid (4^k - 1)$ and we have $4^{k+1} - 1 = 3(a + 1)$ where $a = 4^k - 1$. So $3 \mid (4^{k+1} - 1)$.

12. Basis Step: $n = 1$. $P(1)$: $1 < \dfrac{(1+1)^2}{2}$ is true.
    Induction Step: We use $P(k)$:

    $$1 + 2 + 3 + \cdots + k < \frac{(k+1)^2}{2}$$

    to show $P(k + 1)$:

    $$1 + 2 + \cdots + (k+1) < \frac{(k+2)^2}{2}.$$

LHS of $P(k + 1)$:

$$1 + 2 + 3 + \cdots + k + (k+1) < \frac{(k+1)^2}{2} + (k+1)$$

$$= \frac{k^2 + 4k + 3}{2}$$

$$< \frac{k^2 + 4k + 4}{2}$$

$$= \frac{(k+2)^2}{2}$$

the RHS of $P(k + 1)$.

CHAPTER 3 SELF-TEST, page 102

1. (a) 32.   (b) 2,598,960.   (c) 20,160.

2. (a) 165,765,600.   (b) 118,404,002.

3. (a) 7776.   (b) 216.

4. 560.

5. 15,173,928.

6. $2 \cdot \dfrac{n!}{(n-2)! \, 2!} + n^2 = \dfrac{2 \cdot n! + 2! \, (n-2)! \, n^2}{(n-2)! \, 2!}$

    $= \dfrac{(n-2)! \, n(n-1+n)}{(n-2)!}$

    $= n(2n-1) \cdot \dfrac{(2n-2)!}{(2n-2)!} \cdot \dfrac{2}{2}$

    $= \dfrac{(2n)!}{2! \, (2n-2)!}$

    $= {}_{2n}C_2.$

7. $\lfloor \frac{50}{8} \rfloor + 1 = 7$ pieces, assuming the pepperoni slices are not cut.

8. At least two months must begin on the same day of the week. Let the seven days of the week be the pigeonholes and the twelve months of the year, the pigeons. Then by the pigeonhole principle, at least $\lfloor \frac{12}{7} \rfloor + 1$, or 2, months begin on the same day of the week.

9. $\frac{10}{32}$.

10. No, $p(A \cap B) = p(A) + p(B) - p(A \cup B) = 0.29 + 0.41 - 0.65 = 0.05$.

11. $b_n = -3^n + 4^n$.

12. $a_n = m^{n-1} a_1 - m^{n-2} - m^{n-3} - \cdots - m^2 - 1 = m^n - \dfrac{m^{n-1} - 1}{m - 1}.$

CHAPTER 4 SELF-TEST, page 159

1. (a) 12.

   (b) $\{(2, 1), (2, 2), (2, 3), (2, 4), (5, 1), (5, 2), (5, 3),$
       $(5, 4), (7, 1), (7, 2), (7, 3), (7, 4)\}.$

2. Let $U = \{1, 2, 3, 4, 5\}$, $A = \{1, 2, 3\}$, $B = \{2, 3\}$. Then
   $(2, 5) \in A \times B$, but $(2, 5) \notin \overline{A} \times \overline{B}$.

3. $\{\{a, b, c\}, \{d, e\}\}, \{\{a, b, d\}, \{c, e\}\}, \{\{a, b, e\}, \{c, d\}\},$
   $\{\{b, c, d\}, \{a, e\}\}, \{\{b, c, e\}, \{a, d\}\}, \{\{b, d, e\}, \{a, c\}\},$
   $\{\{c, d, e\}, \{a, b\}\}, \{\{a, c, d\}, \{b, e\}\}, \{\{a, c, e\}, \{b, d\}\},$
   $\{\{a, d, e\}, \{b, c\}\}, \{\{a, b, c, d\}, \{e\}\}, \{\{a, b, d, e\}, \{c\}\},$
   $\{\{b, c, d, e\}, \{a\}\}, \{\{a, b, c, e\}, \{d\}\}, \{\{a, c, d, e\}, \{b\}\}.$

4. (a)   (b) $\begin{bmatrix} 0 & 0 & 0 & 0 \\ 1 & 0 & 0 & 0 \\ 1 & 1 & 0 & 0 \\ 1 & 1 & 0 & 0 \end{bmatrix}.$

5. (a)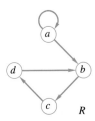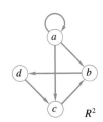

   (b) $\mathbf{M}_R = \begin{bmatrix} 1 & 1 & 0 & 0 \\ 0 & 0 & 1 & 0 \\ 0 & 0 & 0 & 1 \\ 0 & 1 & 0 & 0 \end{bmatrix}.$

   $\mathbf{M}_{R^2} = \begin{bmatrix} 1 & 1 & 1 & 0 \\ 0 & 0 & 0 & 1 \\ 0 & 1 & 0 & 0 \\ 0 & 0 & 1 & 0 \end{bmatrix}.$

   (c) $\mathbf{M}_{R^\infty} = \begin{bmatrix} 1 & 1 & 1 & 1 \\ 0 & 1 & 1 & 1 \\ 0 & 1 & 1 & 1 \\ 0 & 1 & 1 & 1 \end{bmatrix}.$

6. Reflexive, not irreflexive, not symmetric, not asymmetric, not antisymmetric, not transitive.

7. Not reflexive, not irreflexive, not symmetric, not asymmetric, antisymmetric, not transitive.

8. Since $a\ R\ b$ implies $b\ \not{R}\ a$, $a\ R\ b \wedge b\ R\ a$ is always false. Hence $a\ R\ b \wedge b\ R\ a \Rightarrow a = b$ is always true. R must be antisymmetric.

9. (a) $(u, v)\ R\ (u, v)$ since $u - v = u - v$ is true. Thus R is reflexive. If $(u, v)\ R\ (x, y)$, then $u - v = x - y$ and $(x, y)\ R\ (u, v)$. Thus R is symmetric. If $((u, v)\ R\ (x, y)) \wedge ((x, y)\ R\ (w, z))$, then $u - v = x - y = w - z$ and $(u, v)\ R\ (w, z)$. Hence, R is transitive.

   (b) $[(2, 3)] = \{(2, 3), (1, 2), (3, 4), (4, 5)\}.$

   (c) $A/R = \{[(2, 3)], [(2, 4)], [(2, 5)], [(2, 2)], [(2, 1)],$
       $[(3, 1)], [(4, 1)], [(5, 1)], [(1, 5)]\}.$

10. $\mathbf{M}_R = \begin{bmatrix} 0 & 1 & 1 & 0 \\ 1 & 0 & 1 & 0 \\ 1 & 0 & 0 & 1 \\ 0 & 1 & 0 & 1 \end{bmatrix}.$

11. (a) $R^{-1} = \{(a, a), (a, e), (b, a), (c, a), (c, b),$
        $(c, d), (e, c)\}.$

    (b) $R \circ S = \{(a, a), (a, b), (a, c), (a, d), (a, e), (b, c),$
        $(b, d), (c, d), (d, c), (d, d), (e, a), (e, b)\}.$

12. $\mathbf{M}_{R^\infty} = \begin{bmatrix} 1 & 1 & 1 & 1 & 1 \\ 1 & 1 & 1 & 1 & 1 \\ 1 & 1 & 1 & 1 & 1 \\ 1 & 1 & 1 & 1 & 1 \\ 1 & 1 & 1 & 1 & 1 \end{bmatrix}.$

CHAPTER 5 SELF-TEST, page 189

1. (a) Yes, $|R(x)| = 1$, $x \in A$.

   (b) No, $(1, b), (1, d) \in R^{-1}$.

2. Suppose $f(a) = f(b)$, then $-5a + 8 = -5b + 8$. But then $a = b$ so f is one to one. Let $r \in \mathbb{R}$. Then

$$f\left(\frac{r - 8}{-5}\right) = -5\left(\frac{r - 8}{-5}\right) + 8 = r - 8 + 8 = r.$$

   So f is onto.

3. (a) 16.   (b) -2.

4. (a) 17.   (b) -1.

5. (a) 0.   (b) 6.

6. True; false.

7. (a) 90.   (b) 106.   (c) 30; 4; 88.

8. $2n^2 + 9n + 5 \leq 2n^2 + n^2 + n^2$, $n \geq 9$. Choose 4 for c and 9 for k. Then $|2n^2 + 9n + 5| \leq 4|n^2|$, $n \geq 9$.

9. $\Theta(2^n)$.

10. $f(N) = 2 + 5\left(\dfrac{N + 1}{2}\right)$; $\Theta(n)$.

11. (a) $(1, 4, 5) \circ (2, 3, 6)$ $(1, 5) \circ (1, 4) \circ (2, 6) \circ (2, 3)$

(b) $(2, 6, 3) \circ (1, 5, 4)$.

12. (a) $\begin{pmatrix} 1 & 2 & 3 & 4 & 5 & 6 & 7 \\ 5 & 2 & 3 & 7 & 4 & 1 & 6 \end{pmatrix}$.

(b) $\begin{pmatrix} 1 & 2 & 3 & 4 & 5 & 6 & 7 \\ 4 & 3 & 2 & 5 & 6 & 7 & 1 \end{pmatrix}$.

(c) $p_1 = (1, 7, 6, 5, 4) \circ (2, 3) =$
$(1, 4) \circ (1, 5) \circ (1, 6) \circ (1, 7) \circ (2, 3)$ odd.

CHAPTER 6 SELF-TEST, page 243

1. (a) R is reflexive, antisymmetric, and transitive. Hence, R is a partial order on A.

(b) R is reflexive and transitive, but not antisymmetric. R is not a partial order.

2. (a)   (b)

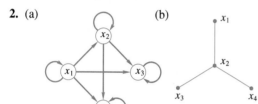

3. (a) Minimal: d, e maximal: a.

(b) Least: none   greatest: a.

4. (a) Upper bounds: $12, 24, 48$

(b) Lower bounds: 2.

(c) LUB$(B) = 12$.   (d) GLB$(B) = 2$.

5. (a) $a \wedge (b \vee c) = \begin{cases} a & \text{if } a \le b \vee c \text{ or } a \le b, c \quad (1) \\ b \vee c & \text{if } b \vee c \le a \text{ or } b, c \le a \quad (2) \end{cases}$

Thus, $(a \wedge b) \vee (a \wedge c) = \begin{cases} a \vee a \text{ or } a & (1) \\ b \vee c & (2) \end{cases}$

(b) $a \vee (b \wedge c) = \begin{cases} a & \text{if } (b \wedge c) \le a \text{ or } b, c \le a \quad (3) \\ b \wedge c & \text{if } a \le (b \wedge c) \text{ or } a \le b, c \quad (4) \end{cases}$

Thus, $(a \vee b) \wedge (a \vee c) = \begin{cases} a \wedge a \text{ or } a & (3) \\ b \wedge c & (4) \end{cases}$

6. $1' = 105; 3' = 35; 5' = 21; 7' = 15; 15' = 7; 21' = 5;$
$35' = 3; 105' = 1$.

7. (a) R is reflexive, because the main diagonal of $\mathbf{M}_R$ is all ones. R is antisymmetric, because if $m_{ij} = 1$, then $m_{ji} = 0$. $\mathbf{M}_{R^2} = \mathbf{M}_R$, so R is transitive.

(b)

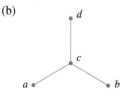

8. Since $a \le b \le b \vee d$ and $c \le d \le b \vee d$, we have $a \vee c \le b \vee d$. ($a \vee c$ is the LUB of a and c.) Also, $a \wedge c \le a \le b$ and $a \wedge c \le c \le d$ since $a \wedge c$ is the GLB of a and c. Thus, $a \wedge c \le b \wedge d$.

9. (a) (1) and (3) are not lattices; $b \vee c$ does not exist. (2) and (4) are lattices.

(b) (1), (2), and (3) are not Boolean algebras; the number of vertices is not a power of 2. (4) is B_3.

10. (a)

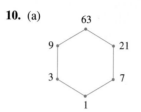

(b) D_{63} is not a Boolean algebra; there are 6 elements.

11. (a) $((x \wedge y) \vee (y \wedge z'))'$.   (b) $(y \wedge (x \vee z'))'$.

(c)

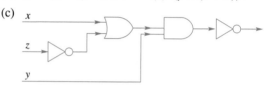

12. $(x \wedge y') \vee (y \vee z')$.

CHAPTER 7 SELF-TEST, page 278

1. It is not a tree; deleting either $(2, 3)$ or $(5, 3)$ will give a tree with root 4.

2. (a) 4.   (b) $v_4, v_{10}, v_6, v_9, v_8$.

(c) 4.   (d) v_6, v_8.

3. Every edge (v_i, v_j) in (T, v_0) belongs to a unique path from v_0 to v_j. Hence removing (v_i, v_j) would mean there is no path from v_0 to v_j.

4.

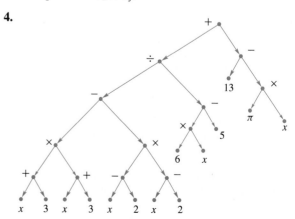

5.

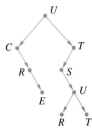

6. (a) UCRETSURT   (b) ERCRTUSTU

7.

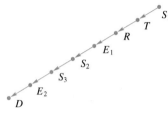

8.

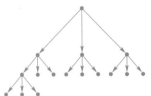

9.

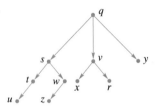

10. One solution is

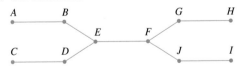

11. EA, AD, AG, GC, GB, BF

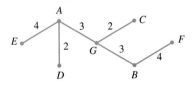

12. GC, AD, GB, GA, BF, AE

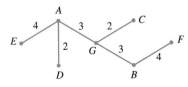

1.

2. One solution is

3.

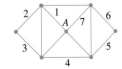

4. (a) Neither an Euler circuit nor path. There are more than 2 vertices of odd degree.

(b) An Euler circuit. All vertices have even degree.

5. (a) A Hamiltonian circuit.

(b) A Hamiltonian path, but not a circuit.

6. (a)

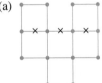

(b) One solution is

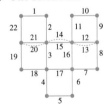

7. One solution is

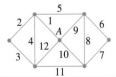

8. One solution is

9.

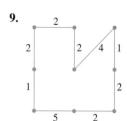

10. $value(F) = 6.$

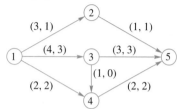

11. (a)

(b) 3.

12. $P_G(x) = x(x-1)(x-2)^3$
$P_G(0) = P_G(1) = P_G(2) = 0, P_G(3) = 6.$

CHAPTER 9 SELF-TEST, page 356

1. (a) Yes, $A * B$ is well-defined for all 2×2 Boolean matrices.

(b) Yes, this is ordinary addition for even numbers.

(c) Yes, 2^{ab} is defined uniquely for all a, b in Z^+.

2.

*	a	b	c	where $\square$ represents a, b, or c.
a	a	c	$\square$	
b	c	b	b	
c	$\square$	b	c	

3. (a) If $a, b \in Q$, then $a * b$ is also a rational number.

$$a * (b * c) = a * (b + c - bc)$$
$$= a + (b + c - bc) - a(b + c - bc)$$
$$= a + b + c - bc - ab - ac + abc.$$
$$(a * b) * c = (a + b - ab) * c$$
$$= a + b - ab + c - (a + b - ab)c$$
$$= a + b + c - ab - ac - bc + abc.$$

Hence, $*$ is associative. Zero is the identity for $(Q, *)$ which is a monoid.

(b) If $a \neq 1$, then

$$a * \frac{a}{a-1} = a + \frac{a}{a-1} - a\left(\frac{a}{a-1}\right)$$
$$= \frac{a^2 - a + a - a^2}{a-1}$$
$$= 0.$$

Thus all rational numbers except 1 have a $*$-inverse.

4. (a) and (b) are monoids. (c) is neither.

5. R has previously been shown to be an equivalence relation, because it is equality for the string lengths. Suppose $a \, R \, b$ and $\alpha \, R \, \beta$, then $length(a \cdot \alpha) = length(a) + length(\alpha) = length(b) + length(\beta) = length(b \cdot \beta)$. Thus $a \cdot \alpha \, R \, b \cdot \beta$ and R is a congruence relation.

6. No, $f(ab) = (ab)^{-1} = b^{-1}a^{-1} \neq f(a) \cdot f(b).$

7. $\{c, d, e\} = H = He = Hc = Hd$;
$Ha = \{a, b, f\} = Hb = Hf.$

8. Let $g \in G_2$ and $n \in f(N)$. Since f is onto, there is a $g' \in G_1$ such that $f(g') = g$. Since $n \in f(N)$, there is an $n' \in N$ such that $f(n') = n$. Then $gn = f(g')f(n') = f(g'n')$. N is normal in G_1 so $g'n' = n''g'$ for some $n'' \in N$. Then $f(g'n') = f(n''g') = f(n'')f(g') = f(n'')g \in f(N)g$. Thus $g \cdot f(N) \subseteq f(N) \cdot g$. Similarly, we can show $f(N) \cdot g \subseteq g \cdot f(N)$ and hence $g \cdot f(N) = f(N) \cdot g$ for all $g \in G_2$.

9. Suppose $x^2 = x$. Then $x^{-1}(xx) = x^{-1}x$ and $(x^{-1}x)x = e$. So $x = e$.

10. $f(a + b) = 2(a + b) = 2a + 2b = f(a) + f(b)$ so f is a homomorphism. For any even integer n, $n = 2k, k \in Z$, and $f(k) = n$ so f is onto. Suppose $f(a) = f(b)$. Then $2a = 2b$ and $a = b$. Hence f is one to one.

11. Since the identity e belongs to every subgroup, $e \in \bigcap_{i=1}^{k} H_i$.

Suppose h and h' belong to $\bigcap_{i=1}^{k} H_i$. Then h, h', and hh' belong to each H_i and $hh' \in \bigcap_{i=1}^{k} H_i$. Let $h \in \bigcap_{i=1}^{k} H_i$. Then h and h^{-1} belong to each H_i, since each H_i is a subgroup of G and so $h^{-1} \in \bigcap_{i=1}^{k} H_i$.

12. The left cosets of H are $rH = \{r \cdot 3^n, n \in Z\}$ for each $r \neq 3^k, k \in Z$, and H itself.

CHAPTER 10 SELF-TEST, page 399

1. (a) False.   (b) True.   (c) True.   (d) False.

2. $L(G) = \{a^n c^m db^n, n \geq 0, m \geq 0\}$.

3. (a)

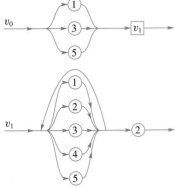

(b) $\langle v_0 \rangle ::= 1\langle v_1 \rangle \mid 3\langle v_1 \rangle \mid 5\langle v_1 \rangle$
   $\langle v_1 \rangle ::= 1\langle v_1 \rangle \mid 2\langle v_1 \rangle \mid 3\langle v_1 \rangle \mid 4\langle v_1 \rangle \mid 5\langle v_1 \rangle \mid 2$

4. $L(G) = \{(1 \vee 3 \vee 5)(1 \vee 2 \vee 3 \vee 4 \vee 5)^n 2, n \geq 0\}$.

5. $G = (V, S, v_0, \mapsto)$ with $V = \{v_0, v_1, 0, 1\}$, $S = \{0, 1\}$, and $\mapsto: v_0 \mapsto 0v_0, v_0 \mapsto 1v_1, v_1 \mapsto 0v_1, v_1 \mapsto 0$.

6.

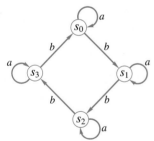

7. Strings of 0's and 1's with $3k + 1$ zeros, $k \geq 0$.

8. (a) R is easily seen to be an equivalence relation with equivalence classes $\{s_0\}$, $\{s_3\}$, $\{s_1, s_2, s_4\}$, $\{s_5, s_6\}$, and $\{s_7\}$.

	a	b	c
$[s_0]$	$[s_1]$	$[s_1]$	$[s_3]$
$[s_1]$	$[s_1]$	$[s_5]$	$[s_1]$
$[s_3]$	$[s_3]$	$[s_1]$	$[s_3]$
$[s_5]$	$[s_7]$	$[s_5]$	$[s_7]$
$[s_7]$	$[s_7]$	$[s_7]$	$[s_7]$

(b)

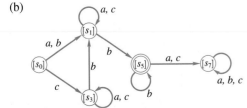

9. Strings with exactly $4k$, $k \geq 0$, b's.

10.

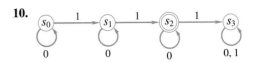

11. $\mathbf{M}_{R_0} = \begin{bmatrix} 1 & 1 & 0 & 0 \\ 1 & 1 & 0 & 0 \\ 0 & 0 & 1 & 1 \\ 0 & 0 & 1 & 1 \end{bmatrix}$.

$R_1 = \{(s_0, s_0), (s_1, s_1), (s_2, s_2), (s_3, s_3), (s_2, s_3), (s_3, s_2)\}$.

12. (a) $k = 1$.

(b) $R = \{(s_0, s_0), (s_1, s_1), (s_2, s_2), (s_3, s_3), (s_2, s_3), (s_3, s_2)\}$.

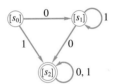

CHAPTER 11 SELF-TEST, page 423

1. (a) Yes.   (b) No.   (c) No.   (d) No.

2. (a) Yes.   (b) Yes.   (c) Yes.

3. One.

4. Let $c_1 = 00000$, $c_2 = 11110$, $c_3 = 01101$, $c_4 = 10011$, $c_5 = 01010$, $c_6 = 10100$, $c_7 = 00111$, $c_8 = 11001$.

$\oplus$	c_1	c_2	c_3	c_4	c_5	c_6	c_7	c_8
c_1	c_1	c_2	c_3	c_4	c_5	c_6	c_7	c_8
c_2	c_2	c_1	c_4	c_3	c_6	c_5	c_8	c_7
c_3	c_3	c_4	c_1	c_2	c_7	c_8	c_5	c_6
c_4	c_4	c_3	c_2	c_1	c_8	c_7	c_6	c_5
c_5	c_5	c_6	c_7	c_8	c_1	c_2	c_3	c_4
c_6	c_6	c_5	c_8	c_7	c_2	c_1	c_4	c_3
c_7	c_7	c_8	c_5	c_6	c_3	c_4	c_1	c_2
c_8	c_8	c_7	c_6	c_5	c_4	c_3	c_2	c_1

The table shows this subset is closed for $\oplus$, contains the identity for B^5, and contains the inverse of each element.

5. 0.

6. 11.

INDEX

Examples of Pseudocode Constructs

1. $X \leftarrow 0$
2. $Y \leftarrow 0$
3. **UNTIL** $(X \geq N)$
    a. $X \leftarrow X + 1$
    b. $Y \leftarrow Y + X$
4. $Y \leftarrow Y/2$

END OF ALGORITHM

FUNCTION SQR(N)
1. $X \leftarrow N$
2. $Y \leftarrow 1$
3. **WHILE** $(Y \neq N)$
    a. $X \leftarrow X + N$
    b. $Y \leftarrow Y + 1$
4. **RETURN** (X)

END OF FUNCTION SQR

1. **FOR** $I = 1$ **THRU** N
    a. **IF** $((A[I] = 1) \text{ OR } (B[I] = 1))$ **THEN**
        1. $C[I] \leftarrow 1$
    b. **ELSE**
        1. $C[I] \leftarrow 0$

FUNCTION $F(X)$
1. **IF** $(X < 1)$ **THEN**
    a. $R \leftarrow X^2 + 1$
2. **ELSE**
    a. **IF** $(X < 3)$ **THEN**
        1. $R \leftarrow 2X + 6$
    b. **ELSE**
        1. $R \leftarrow X + 7$
3. **RETURN** (R)

END OF FUNCTION F